# BUOYANCY-DRIVEN FLOWS

Buoyancy is one of the main forces driving flows on our planet, especially in the oceans and atmosphere. These flows range from buoyant coastal currents to dense overflows in the ocean, and from avalanches to volcanic pyroclastic flows on the Earth's surface. This book brings together contributions by leading world scientists to summarize our present theoretical, observational, experimental, and modeling understanding of buoyancy-driven flows.

This book strongly emphasizes the ocean, which displays an exceptionally wide range of buoyancy-driven flows. Buoyancy-driven currents play a key role in the global ocean circulation and in climate variability through their impact on deep-water formation. Correctly representing buoyancy-driven processes not currently resolved in the ocean components of climate models is a challenge. The limitations of current modeling techniques are examined, and recommendations are made for the proper physical parameterization of buoyancy-driven processes in order to accurately project long-term water mass evolution. Buoyancy-driven currents are also primarily responsible for the redistribution of fresh water throughout the world's oceans. In addition to fresh water, buoyancy-driven flows transport heat, nutrients, sediments, biogeochemicals, pollutants, and biological organisms along many continental shelves and thus have significant impacts on ecosystems, fisheries, and the coastal circulation.

This book is an invaluable resource for advanced students and researchers in oceanography, geophysical fluid dynamics, atmospheric science, and the wider Earth sciences who need a state-of-the-art reference on buoyancy-driven flows.

ERIC P. CHASSIGNET is professor in the Department of Earth, Ocean, and Atmospheric Sciences at Florida State University, Tallahassee; director of the Center for Ocean-Atmospheric Prediction Studies; and co-director of the Florida Climate Institute. He was awarded the 2008 National Oceanographic Partnership Program's Excellence in Partnering Award for his coordination of the U.S. Global Ocean Data Assimilation Experiment: Global Ocean Prediction with the hybrid Coordinate Ocean Model (U.S. GODAE HYCOM). Dr. Chassignet has published two previous books in collaboration with Dr. Verron: *Ocean Modeling and Parameterization* (1998) and *Ocean Weather Forecasting* (2004).

CLAUDIA CENEDESE is associate scientist with tenure in the Department of Physical Oceanography at the Woods Hole Oceanographic Institution (WHOI), faculty of the WHOI Geophysical Fluid Dynamics Summer Program, and co-director of the WHOI Geophysical Fluid Dynamics Laboratory.

JACQUES VERRON is director of research at the Centre National de Recherche Scientitifique (CNRS); chief scientist for the France-India SARAL/AltiKa altimetric satellite mission; and former director of the Laboratoire des Ecoulements Géophysiques et Industriels in Grenoble, France. Dr. Verron was awarded the silver medal from the CNRS in 1994 for his work on the development of operational oceanography.

DRS. CHASSIGNET, CENEDESE, AND VERRON served as co-directors of the 2010 Alpine Summer School on Buoyancy-Driven Flows held in Valsavarenche, Italy.

# BUOYANCY-DRIVEN FLOWS

*Edited by*

ERIC P. CHASSIGNET
*Florida State University*

CLAUDIA CENEDESE
*Woods Hole Oceanographic Institution*

JACQUES VERRON
*Centre National de la Recherche Scientifique*

**CAMBRIDGE**
**UNIVERSITY PRESS**

University Printing House, Cambridge CB2 8BS, United Kingdom

One Liberty Plaza, 20th Floor, New York, NY 10006, USA

477 Williamstown Road, Port Melbourne, VIC 3207, Australia

314-321, 3rd Floor, Plot 3, Splendor Forum, Jasola District Centre, New Delhi-110025, India

79 Anson Road, #06-04/06, Singapore 079906

Cambridge University Press is part of the University of Cambridge.

It furthers the University's mission by disseminating knowledge in the pursuit of education, learning and research at the highest international levels of excellence.

www.cambridge.org
Information on this title: www.cambridge.org/9781108446761



First published 2012
First paperback edition 2017

*A catalogue record for this publication is available from the British Library*

*Library of Congress Cataloging in Publication data*
Buoyancy-driven flows / [edited by] Eric Chassignet, Claudia Cenedese, Jacques Verron.
p. cm.
Includes bibliographical references and index.
ISBN 978-1-107-00887-8
1. Buoyant convection. 2. Ocean circulation. 3. Atmospheric circulation.
I. Chassignet, Eric P. II. Cenedese, Claudia, 1971– III. Verron, Jacques.
QC327.B86 2012
551.48–dc23 2011044342

ISBN 978-1-107-00887-8 Hardback
ISBN 978-1-108-44676-1 Paperback

# Contents

# Contributors

**Christophe Ancey** Ecole Polytechnique Fédérale de Lausanne, Lausanne, Switzerland
**Raphael Dussin** Laboratoire de Physique des Océans, CNRS-IRD-Ifremer-UBO, Brest, France
**Bruno Ferron** Laboratoire de Physique des Océans, CNRS-IRD-Ifremer-UBO, Brest, France
**William G. Large** National Center for Atmospheric Research, Boulder, CO, USA
**Sonya Legg** Atmosphere and Ocean Science Program, Princeton University, Princeton, USA
**Steve Lentz** Woods Hole Oceanographic Institution, Woods Hole, MA, USA
**Paul Linden** Department of Applied Mathematics and Theoretical Physics, University of Cambridge, Cambridge, UK
**Sylvie Malardel** Meteo-France, Toulouse, France, and ECMWF, Reading, UK
**Joseph Pedlosky** Woods Hole Oceanographic Institute, Woods Hole, MA, USA
**Michael A. Spall** Woods Hole Oceanographic Institute, Woods Hole, MA, USA
**Anne Marie Treguier** Laboratoire de Physique des Océans, CNRS-IRD-Ifremer-UBO, Brest, France
**Andy Woods** BP Institute, University of Cambridge, Cambridge, UK

# Introduction

Buoyancy is one of the main forces driving flows on our planet and buoyancy-driven flows encompass a wide spectrum of geophysical flows. In this book, contributions by leading world scientists summarize our present theoretical, observational, experimental, and modeling understanding of buoyancy-driven flows. These flows range from buoyant coastal currents to dense overflows in the ocean, and from avalanches to volcanic pyroclastic flows on the Earth's surface. By design, there is a strong emphasis on the ocean where a wide range of buoyancy-driven flows is observed. Buoyancy-driven currents play a key role in the global ocean circulation and in climate variability through deep-water formation. Formation of dense water usually occurs in marginal seas, which are either cooler (at high latitudes) or saltier (due to greater evaporation rates). These dense waters enter the ocean as a gravity-driven current, entrain surrounding waters as they descend along the continental slope, and modify the ocean's stratification as they become part of the global ocean circulation. Buoyancy-driven currents are also primarily responsible for the redistribution of fresh water throughout the world's oceans. In particular, buoyant coastal currents transport fresh water, heat, nutrients, sediments, biogeochemicals, pollutants, and biological organisms along many continental shelves and thus have significant impacts on ecosystems, fisheries, and coastal circulation.

In our examination of oceanic buoyancy-driven flows, we first provide a broad overview of our current understanding of these flows from observations, laboratory experiments, and idealized model configurations (Chapters 1–5). This is followed by an in-depth discussion on the importance of correctly representing processes in buoyancy-driven flows that are not currently resolved in the ocean component of climate models (Chapter 6) and on the difficulty of properly representing these flows in eddy-resolving ocean models (Chapter 7). Finally, oceanic buoyancy-driven flows are put in the context of a wider range of geophysical problems (atmospheric flows, volcanic flows, and avalanches; Chapters 8–10).

The dynamics regulating buoyancy-driven flows are almost equivalent regardless of whether the phenomena are coastal currents or avalanches. The driving forces are similarly represented in the equations of motion, but the latter may be characterized by terms specific to the kind of motion being investigated. For example, terms representing two-phase flows may be necessary when investigating volcanic flows, whereas terms representing mixing may be fundamental when investigating dense overflows.

The term "buoyancy-driven flow" (or alternatively, "density-driven flows") is used to identify those flows whose motion is forced by a horizontal density difference with the surrounding fluid. A flow is often referred to as *buoyant* when the moving fluid is lighter than the surrounding fluid and as *dense* when the fluid is heavier than the ambient fluid. Both currents, light and dense, may also be called "gravity currents" to indicate that the forcing term in the equations of motion is the gravitational term that arises from the density difference between the fluid in consideration and the surrounding fluid. When only two fluids are considered, the gravitational term contains the so-called reduced gravity, defined as

$$g' = g\frac{\Delta\rho}{\rho},$$

which substitutes the gravitational acceleration $g$ in the equations of motion to take into account that the fluid in consideration (lighter or denser) is embedded in another fluid with a slightly different density. For flows that cannot be simply represented by two fluids with different densities, the "reduced gravity" is replaced by an expression that considers the continuous change of density with depth, in contrast with the sharp change in density characteristic of two fluids. This expression can be written as

$$g' = N^2 H,$$

where $N^2 = \frac{g}{\rho}\frac{\partial\rho}{\partial z}$ is the Brunt-Väisälä frequency, or buoyancy frequency, which is the frequency at which a vertically displaced parcel will oscillate within a statically stable environment, and $H$ is the distance over which the vertical density gradient has been calculated.

Although the Earth's rotation influences many of the flows on our planet, our understanding of buoyancy-driven flows was first developed by ignoring the effects of planetary rotation on the fluid motion. An example of a flow that is density driven, but not strongly influenced by the Earth's rotation, is the sea breeze, a significant feature in coastal meteorology. During the day, the sun heats the land more than it heats the ocean; the air just above the land is warmer, and therefore lighter, than the air over the ocean. Hence, colder, denser air will flow from the ocean to the land. The dynamics of the sea breeze helped inspire Dr. John Simpson to devote his life to the understanding of gravity currents in the absence of rotation using mainly laboratory experiments (see Simpson 1997, for a review). One of the first theories to describe the motion of gravity currents was developed by Benjamin (1968) to predict the frontal velocity of a gravity current generated by the release of harmful gases in coal mines. Subsequent laboratory experiments (Gardner and Crow 1970; Lowe et al. 2005), using the so-called "lock release technique," verified the validity of the early theories by investigating an air cavity in a rectangular horizontal duct. Although these theories neglected the influence of the Earth's rotation, they have proven useful for understanding the behavior of a variety of gravity currents, including the dynamics regulating the natural ventilation in a building and the dynamics of ocean currents. Chapter 1 of this book discusses the fundamental aspects of nonrotating buoyancy-driven currents using theories and laboratory experiments. The limitations of earlier theories are discussed and new approaches are presented in order to examine some of the details in gravity currents that are still not fully understood, such as the generation of internal waves by a propagating gravity current in a stratified environment.

The introduction of the Earth's rotation does complicate the dynamics of buoyancy-driven flows (see Colin de Verdière 1988, for a discussion of large-scale flow patterns driven by a surface buoyancy flux). Chapter 2 illustrates how an ocean basin responds to buoyancy (i.e., cooling or heating) applied at its surface. The mechanism for propagation of information across the basin is a function of stratification and the variation of the Coriolis parameter with latitude (i.e., the $\beta$-effect). A full comprehension of buoyancy-driven flows in enclosed ocean basins requires a thorough understanding of the dynamics close to the boundaries, where dissipation effects are important. The buoyancy forcing applied at the free surface generates large vertical motions near the basin boundaries in regions that are not close to the forcing. Chapter 3 further investigates this response by adding more complexity to the problem. Numerical models are used to investigate the effect of buoyancy forcing under a more realistic scenario (i.e., one that includes unsteady flows and eddies). Building on the theoretical results of Chapter 2, Chapter 3 shows that baroclinic eddies can be an effective mechanism for moving the buoyancy response away from the forcing location. As in Chapter 2, the mean downwelling (i.e., vertical velocity) is found to be concentrated near the lateral boundary.

In a rotating environment, like the Earth, buoyancy forcing can be applied on the free surface of an ocean basin in the form of cooling and heating as discussed previously, but it can also be applied on the ocean basin boundary in the form of a buoyancy discharge, as it happens in the presence of a river estuary. The fresh water river outflow exiting at the ocean boundary will have a tendency to spread horizontally because of its positive buoyancy (i.e., the river outflow is lighter than the ocean water). However, in the presence of rotation, the buoyant waters cannot spread uniformly, and after a length scale of a few Rossby deformation radii, the Coriolis force turns the buoyant outflow to the right (left) in the Northern (Southern) Hemisphere. The presence of a coastline, then, is fundamental for the generation of buoyant "coastal" currents. Since there is no flow normal to the coast, there is no Coriolis force parallel to the coast, and the resulting motion is along the coast. The buoyant "coastal" current is forced to hug the coastline with the coast on its right (left) looking downstream in the Northern (Southern) Hemisphere. Buoyant coastal currents can be found in many parts of the world's oceans. Particularly striking examples are the Leeuwin Current (Griffith and Pearce 1985; Pearce and Griffith 1991), the Norwegian Coastal Current (Johannessen and Mork 1979), and the East Greenland Current (Wadhams et al. 1979). These buoyancy-driven currents are complex turbulent current systems, as can be seen in satellite images (see figure 5 of Legeckis 1978 or figure 12 of Wadhams et al. 1979). The fronts of these currents delineate boundaries between different water properties, and the stability of these fronts is of fundamental importance for the exchange of water properties (i.e., mixing) across the front (e.g., Cenedese and Linden 2002). Eddies (with scales of 10–100 km) detaching from an unstable current can transport a large volume of water across fronts and can be a very efficient mechanism for transferring water properties from one region to another, affecting, for example, the local fisheries and ecosystems.

The dynamics of buoyant coastal currents have been studied extensively (e.g., O'Donnell 1990; Yankovsky and Chapman 1997; Garvine 2001; Fong and Geyer 2002; Garcia Berdeal et al. 2002; Hetland 2005). Laboratory experiments have provided numerous insights into the dominant scales present in these flows as well as into their dynamics and stability (Griffiths and Linden 1981a; Lentz and Helfrich 2002). Chapter 4 shows that a buoyant coastal current can flow along a coastline in one of two forms: a surface-trapped current or a slope-controlled current. A surface-trapped current forms a shallow layer that intersects the sloping bottom topography close to shore; bottom topography has virtually no effect on its dynamics. Numerous studies have examined the behavior (formation, propagation, stability, etc.) of a surface-trapped current along a vertical wall (over a flat bottom) in various configurations (e.g., Griffiths and Linden 1981a; 1982; Griffiths and Hopfinger 1983; Chabert d'Hières et al. 1991; Garvine 1999; Fong and Geyer 2002; Geyer et al. 2004). A slope-controlled current is fundamentally different because bottom topography plays a leading dynamical role. The dynamics of a slope-controlled current were first described by Chapman and Lentz (1994) and then further investigated by Yankovsky and Chapman (1997) on the basis of numerical model calculations. Lentz and Helfrich (2002) confirmed the existence of the slope-controlled current using a laboratory model and developed a scaling theory that smoothly links the surface-trapped and slope-controlled currents.

The presence of bottom topography introduces a strong topographic $\beta$-effect, which influences the stability of buoyant coastal currents. For example, several studies (e.g., Cenedese and Linden 2002; Lentz and Helfrich 2002; Wolfe and Cenedese 2006) have shown that a sloping bottom tends to stabilize a current's front. Buoyant coastal currents over a flat bottom become unstable to nonaxisymmetric disturbances (Griffiths and Linden 1981a) that can be interpreted as a mix of baroclinic and barotropic instabilities. After the instability grows to large amplitude, dipoles form and propagate radially outwards. Griffiths and Linden (1981b) developed a simplified model for baroclinic instability, following Phillips (1954), which qualitatively agrees with their laboratory experiments. Griffiths and Linden (1981a) further developed the model to include frictional dissipation due to Ekman layers. Buoyant coastal currents often flow over a shelf break, where the flatter continental shelf ends and the steeper continental slope begins. The front in the Middle Atlantic Bight is an example of a buoyant coastal current's front flowing over the shelf break. This front, which originates in Nova Scotia and ends near Cape Hatteras, is the barrier between the warm and saline waters of the North Atlantic Ocean and the cooler, fresher waters of the continental shelf (Wright 1976; Linder and Gawarkiewicz 1998).

Buoyant coastal currents are also influenced dramatically by the wind, which can enhance or inhibit the offshore movement and dispersal of such currents. Upwelling, along shelf, wind forcing tends to flatten the current front, causing the buoyant current to thin and widen, whereas moderate downwelling winds steepen the front, causing the current to thicken and narrow. However, strong downwelling winds force vertical mixing that widens the current front, but leaves the current width almost unaltered. Several studies have investigated the response of a buoyant coastal current to wind events (Fong et al. 1997; Fong and Geyer 2001; Garcia Berdeal et al. 2002; Geyer et al. 2004; Lentz 2004; Hetland and Signell 2005) as discussed in detail in Chapter 4.

The densest waters in the oceans are generated at high latitudes where strong atmospheric cooling and the formation of ice with consequent brine rejection contribute to the formation of salty, cold water in the Nordic seas basins and along the continental shelf of Antarctica (Baines and Condie 1998; Ivanov et al. 2004). Dense warm and salty waters are also formed in marginal seas (e.g., the Mediterranean Sea and the Red Sea) where evaporation overcomes the input of fresh water from river runoff and precipitation. This dense water usually moves over a sill and/or through a constriction before generating a dense downslope current (called an overflow) over the continental slope. Major examples of these overflows can be found in the Denmark Strait (Dickson and Brown 1994; Girton and Sanford 2003; Käse et al. 2003), in the Faroe Bank Channel (Saunders 1990; Mauritzen et al. 2005; Fer et al. 2010), in the Baltic Sea (Arneborg et al. 2007; Umlauf and Arneborg 2009), in the Strait of Gibraltar (Baringer and Price 1997; Price et al. 1993), at the mouth of the Red Sea (Peters and Johns 2005; Peters et al. 2005), at various locations along the Arctic (Aagaard et al. 1981), and on the shelves of the Weddell Sea (Foster and Carmack 1976) and Ross Sea (Muench et al. 2009; Padman et al. 2009) in Antarctica. As the dense water descends over the continental slope, its basic dynamics can be described as a dense current driven by buoyancy and influenced by the Coriolis acceleration and bottom drag. These "stream tube models" have been successful at modeling the trajectory of these dense currents over the slope by balancing the above-mentioned forces (Smith 1975; Killworth 1977; Price and Baringer 1994). The dense current motion is deflected to the right when looking downslope in the Northern Hemisphere (to the left in the Southern Hemisphere) because of the Coriolis force. After a transition period, the dense current moves mainly along the slope, with a small downslope component due to the bottom drag. As the current moves over the slope, it entrains ambient water with a consequent decrease in density and increase in transport. The dense current eventually reaches either the bottom of the ocean or a level of neutral buoyancy where the ambient surrounding waters have the same density as the dense current. The dense current is then observed to spread horizontally to fill the bottom of the ocean, as in the case of the water formed on the Weddell Sea continental slope, which forms Antarctic Bottom Water (AABW). Alternatively, if the difference in density with the surrounding water is zero before the current reaches the bottom of the ocean, the current intrudes into the water column at the level of neutral buoyancy to form an intermediate layer, as observed with the Mediterranean overflow. Another dense current reaching a neutrally buoyant level is the product water of the Denmark Strait and Faroe Bank Channel overflows, which flows down the continental slope to the southern tip of Greenland to form North Atlantic Deep Water (NADW). This current is observed to hug the continental slope on the western side of the North Atlantic.

Chapter 5 reviews the fundamental characteristics of these overflows and the subsequent fate of the dense water as it moves over the slope. The final location and depth of these dense waters will depend on the amount of ambient water entrained by the dense current as it moves across a sill/constriction and descends over the continental shelf and slope. At the sill/constriction, the velocity of the currents is usually large compared to the velocity of the ambient waters surrounding them; hence, the currents can generate small-scale instabilities (i.e., turbulence) that lead to the entrainment of lighter ambient water, resulting in a decrease in density and an increase in transport. Strong entrainment is usually present near the location of the sill/constriction where the dense current maximum velocities are typically observed. For example, the Mediterranean overflow has been observed to entrain mainly within 50 km, or half a day, from the current exiting the Strait of Gibraltar, where the dense current velocity reaches its maximum (Price and Baringer 1994). This entrainment determines the main characteristics of temperature and salinity of the current as well as the depth of the neutrally buoyant layer into which the dense current will intrude (approximately 1000 m for the Mediterranean overflow water). However, entrainment has also been observed to occur over the slope where the current's velocity is much lower (Lauderdale et al. 2008; Cenedese et al. 2004; Cenedese and Adduce 2008; 2010). For example, the moderate entrainment that occurs along the Denmark Strait overflow's 1,000-km-long path on the continental slope between the Denmark Strait and Cape Farewell increases the volume transport of the current by the same amount as the intense entrainment that occurrs in the first 100 km near the actual Denmark Strait (Lauderdale et al. 2008). Hence, it is of fundamental importance to consider not only the supercritical entrainment occurring near the sill/constriction but also the subcritical entrainment occurring over the slope (Hughes and Griffiths 2006; Wåhlin and Cenedese 2006; Lauderdale et al. 2008). A correct prediction of the location, depth, and density of the NADW originating from the Denmark Strait overflow can be obtained only by considering the entrainment occurring immediately downstream of the sill as well as over the slope. The dynamics regulating the entrainment in a dense current were first investigated by Ellison and Turner (1959) using laboratory experiments and have recently been revisited in the context of overflows using numerical models (Ezer 2005; 2006; Legg et al. 2006; Özgökmen et al. 2006; 2009; Riemenschneider and Legg 2007; Chang et al. 2008) and laboratory experiments (Baines 2001; 2002; 2005; 2008; Cenedese et al. 2004; Wells and Wettlaufer 2005; Wells 2007; Cenedese and Adduce 2008). An accurate representation of the dynamics of buoyancy-driven flows in ocean numerical models is primarily dependent upon the horizontal and vertical resolution choices and the parameterization of unresolved subgrid scale processes (Chapters 6 and 7). In the case of overflows, the models' resolution at best marginally resolves the topography and is not fine enough to represent the small-scale processes related to the entrainment. The parameterization of the unresolved physics is model dependent and has a strong impact on the downstream evolution of water masses (Wu et al. 2007; Danabasoglu et al. 2010). This is an area of active research (Hallberg 2000; Xu et al. 2006; Jackson et al. 2008; Cenedese and Adduce 2010) that directly impacts our ability to represent the long-term evolution of oceanic water masses associated with climate variability and climate change (Chapter 6). Even when the model is eddy resolving (Chapter 7), the dynamics of overflows remain highly dependent on topographic details and eddy dynamics. Mesoscale eddies generated near the sill modulate the entrainment that takes place in the downslope flow of dense water. The simulation of these downslope flows of dense water is nontrivial and differs strongly among ocean models based on different vertical coordinate schemes (Griffies et al. 2000).

There are other areas in oceanography in which "buoyancy-driven" flows are manifested. In particular, much oceanographic and fluid dynamics research has focused on classical convection (e.g., Rayleigh–Bénard) and its oceanic manifestation. Although we have decided not to dedicate an entire chapter to this particular topic, Chapters 3 and 5 do provide an introduction to, as well as some examples of, convective flows.

Dynamics similar to those regulating oceanic buoyancy-driven flows also describe flows in the atmosphere, volcanoes, and avalanches (Chapters 8, 9, and 10, respectively). However, specific aspects of the medium may be fundamental in the dynamics of these different flows. For example, in the atmosphere (Chapter 8), the air is 1,000 times lighter than the water and is much more compressible; hence, some of the simplifications in the equations of motion made for the oceanic flows no longer hold. The introduction of moisture and phase transition adds complexity in the atmosphere. In volcanic flows (Chapter 9), the different phases of the fluids found in a magma chamber, and in particular the gas content in these phases, are important variables in the description of the flow. The erupting buoyant plumes are similar to oceanic buoyant plumes, and a correct representation necessitates including the effects of the particle content, also necessary for the description of turbidity currents in the ocean.

In avalanches, the fluid (air) is mixed with particles in the form of rocks and snow, and the flow can be dominated either by inertia or friction, as discussed in Chapter 10.

In summary, although we have made much progress in improving our understanding of oceanic buoyancy-driven flows, correctly representing buoyancy-driven processes not currently resolved in the ocean component of climate models remains a challenge. The credibility of these models is strongly limited by their ability to represent the climatologically important processes that occur on scales smaller than the climate models' grid scale (currently typically 100 km), and a correct physical parameterization of these buoyancy-driven processes will be essential for an accurate projection of the long-term water mass evolution in these climate models.

*This book was developed from the lectures of the 2010 Alpine Summer School on "Buoyancy-Driven Flows" (http://www.to.isac.cnr.it/aosta/), which was cosponsored by the National Science Foundation (NSF), le Centre National de la Recherche Scientifique (CNRS), Ifremer, the Institute of Atmospheric Sciences and Climate of the Italian National Research Council (ISAC-CNR), the Valsarenche municipality, and the National Park of Gran Paradiso. Contributions were peer-reviewed by the 2010 Alpine Summer School attendees.*

## References

Aagaard, K., L. K. Coachman, and E. C. Carmack, 1981. On the halocline of the Arctic Ocean. *Deep-Sea Res.* **28**, 529–45.

Arneborg, L., V. Fiekas, L.Umlauf, and H. Burchard, 2007. Gravity current dynamics and entrainment—A process study based on observations in the Arkona Basin. *J. Phys. Oceanogr.* **37**, 2094–113.

Baines, P. G., 2001. Mixing in flows down gentle slopes into stratified environments. *J. Fluid Mech.* **443**, 237–70.

Baines, P. G., 2002. Two-dimensional plumes in stratified environments. *J. Fluid Mech.* **471**, 315–37.

Baines, P. G., 2005. Mixing regimes for the flow of dense fluid down slopes into stratified environments. *J. Fluid Mech.* **538**, 245–67.

Baines, P. G., 2008. Mixing in downslope flows in the ocean—Plumes versus gravity currents. *Atmos. —Ocean* **46**, 405–19.

Baines P. G., and S. Condie, 1998. Observations and modeling of Antarctic downslope flows: A review. In: S. Jacobs and R. Weiss, (eds.), *Ocean, Ice and Atmosphere: Interactions at the Antarctic Continental Margin,* vol. **75**, pp. 29–49. American Geophysical Union, Washington, DC.

Baringer, M. O., and J. F. Price, 1997. Mixing and spreading of the Mediterranean outflow. *J. Phys. Oceanogr.* **27**, 1654–77.

Benjamin, T. B., 1968. Gravity currents and related phenomena. *J. Fluid Mech.* **31**, 209–48.

Cenedese, C., and C. Adduce, 2008. Mixing in a density-driven current flowing down a slope in a rotating fluid. *J. Fluid Mech.* **604**, 369–88.

Cenedese, C., and C. Adduce, 2010. A new parameterization for entrainment in overflows, *J. Phys. Oceanogr.* **40**, 1835–50.

Cenedese, C., and P. F. Linden, 2002. The stability of a buoyancy driven coastal current at the shelfbreak. *J. Fluid Mech.* **452,** 97–121.

Cenedese, C., J. A. Whitehead, T. A. Ascarelli, and M. Ohiwa, 2004. A dense current flowing down a sloping bottom in a rotating fluid. *J. Phys. Oceanogr.* **34**, 188–203.

Chabert d'Hières, G., H. Didelle, and D. Obaton, 1991. A laboratory study of surface boundary currents: Application to the Algerian Current. *J. Geophys. Res.* **96**, 12539–48.

Chang, Y. S., T. M. Özgökmen, H. Peters, and X. Xu, 2008. Numerical simulation of the Red Sea outflow using HYCOM and comparison with REDSOX observations. *J. Phys. Oceanogr.* **38**, 337–58.

Chapman, D. C., and S. J. Lentz, 1994. Trapping of a coastal density front by the bottom boundary layer. *J. Phys. Oceanogr*. **24**, 1464–79.

Colin de Verdière, A., 1988. Buoyancy driven planetary flows. *J. Marine Res.* **46**, 215–65.

Danabasoglu, G., W. G. Large, and B. P. Briegleb, 2010: Climate impacts of parameterized Nordic Sea overflows. *J. Geophys. Res. Oceans*, **115**, c11005, doi:10.1029/2010JC006243.

Dickson, R. R., and J. Brown, 1994. The production of North Atlantic deep water: Sources, rates and pathways. *J. Geophys. Res.* **12**, 319–41.

Ellison, T. H., and J. S. Turner, 1959. Turbulent entrainment in stratified flows. *J. Fluid Mech.* **6**, 423–48.

Ezer, T., 2005. Entrainment, diapycnal mixing and transport in three-dimensional bottom gravity current simulations using the Mellor–Yamada turbulence scheme. *Ocean Modelling* **9**, 151–68.

Ezer, T., 2006. Topographic influence on overflow dynamics: Idealized numerical simulations and the Faroe Bank Channel overflow. *J. Geophys. Res.* **111**, C02002, doi:10.1029/2005JC003195.

Fer, I., G. Voet, K. S. Seim, B. Rudels, and K. Latarius, 2010. Intense mixing of the Faroe Bank Channel overflow. *Geophys. Res. Lett.* **37**, L026042, doi:10.1029/2009GL041924.

Fong, D. A., W. R. Geyer, and R. P. Signell, 1997. The wind-forced response on a buoyant coastal current: Observations of the Western Gulf of Maine Plume. *J. Mar. Sys.* **12**, 69–81.

Fong, D. A., and W. R. Geyer, 2001. Response of a river plume during an upwelling favorable wind event. *J. Geophys. Res.* **106**, 1067–84.

Fong, D. A., and W. R. Geyer, 2002. The alongshore transport of freshwater in a surface-trapped river plume. *J. Phys. Oceangr.* **32**, 957–72.

Foster, T. D., and E. C. Carmack, 1976. Frontal zone mixing and Antarctic Bottom Water formation in the southern Weddell Sea. *Deep-Sea Res.* **23**, 301–17.

Garcia Berdeal, I., B. M. Hickey, and M. Kawase, 2002. Influence of wind stress and ambient flow on a high discharge river plume. *J. Geophys. Res.* **107**, 3130, doi:10.1029/2001JC000932.

Gardner, G. C., and I. G. Crow, 1970. The motion of large bubbles in horizontal channels. *J. Fluid Mech.* **43**, 247–255.

Garvine, R. W., 1999. Penetration of buoyant coastal discharge onto the continental shelf: A numerical model experiment. *J. Phys. Oceanogr.* **29**, 1892–909.

Garvine, R. W., 2001. The impact of model configuration in studies of buoyant coastal discharge. *J. Mar. Res.* **59**, 193–225.

Geyer, W. R., R. P. Signell, D. A. Fong, J. Wang, D. M. Anderson, and B. A. Keafer, 2004. The freshwater transport and dynamics of the Western Maine Coastal Current. *Cont. Shelf Res.* **24**, 1339–57.

Girton, J. B., and T. B. Sanford, 2003. Descent and modification of the overflow plume in Denmark Strait. *J. Phys. Oceanogr.* **33**, 1351–64.

Griffies, S. M., C. Böning, F. O. Bryan, E. P. Chassignet, R. Gerdes, H. Hasumi, A. Hirst, A.-M. Treguier, and D. Webb, 2000. Developments in ocean climate modelling. *Ocean Modelling* **2**, 123–92.

Griffiths, R. W., and E. J. Hopfinger, 1983. Gravity currents moving along a lateral boundary in a rotating frame. *J. Fluid Mech.* **134**, 357–99.
Griffiths R. W., and P. F. Linden, 1981a. The stability of buoyancy-driven coastal currents. *Dyn. Atmosph. Oceans* **5**, 281–306.
Griffiths R. W., and P. F. Linden, 1981b. The stability of vortices in rotating, stratified fluid. *J. Fluid Mech.* **105**, 283–316.
Griffiths, R. W., and P. F. Linden, 1982: Laboratory experiments on fronts. Part I. Density-driven boundary currents. *Geophys. Astrophys. Fluid. Dyn.* **19**, 159–187.
Griffiths, R. W., and A. F. Pearce, 1985. Satellite images of an unstable warm eddy derived from the Leeuwin Current. *Deep-Sea Res.* **32**, 1371–80.
Hallberg, R. W., 2000. Time integration of diapycnal diffusion and Richardson number-dependent mixing in isopycnal coordinate ocean models. *Mon. Wea. Rev.* **128**, 1402–19.
Hetland, R. D., 2005. Relating river plume structure to vertical Mixing. *J. Phys. Oceanogr.* **35**, 1667–88.
Hetland, R. D., and R. P. Signell, 2005. Modelling coastal current transport in the Gulf of Maine. *Deep-Sea Res. II* **52**, 2430–49.
Hughes, G. O., and R. W. Griffiths, 2006. A simple convective model of the global overturning circulation, including effects of entrainment into sinking regions. *Ocean Modelling*, **12**, 46–79.
Ivanov, V. V., G. I. Shapiro, J. M. Huthnance, D. L. Aleynik, and P. N. Golovin, 2004. Cascades of dense water around the world ocean. *Prog. Oceanogr.* **60**, 47–98.
Jackson, L., R. W. Hallberg, and S. Legg, 2008. A parameterization of shear-driven turbulence for ocean climate models. *J. Phys. Oceanogr.* **38**, 1033–53.
Johannessen, O. M., and M. Mork, 1979. Remote sensing experiment in the Norwegian Coastal Waters. Spring 1979. Rapport 3/79, Geophysisk Institutt, Universitetet i. Bergen.
Käse, R. H., J. B. Girton, and T. B. Sanford, 2003. Structure and variability of the Denmark Strait overflow: Model and observations. *J. Geophys. Res.* **108**, 3181, doi:10.1029/2002JC001548.
Killworth, P. D., 1977. Mixing on the Weddell Sea continental slope. *Deep-Sea Res.* **24**, 427–48.
Lauderdale, J. M., S. Bacon, A. C. Naveira Garabato, and N. P. Holliday, 2008. Intensified turbulent mixing in the boundary current system of Southern Greenland. *Geophys. Res. Lett.* **35**, L04611, doi:10.1029/2007GL032785.
Legeckis, R., 1978. A survey of worldwide sea surface temperature fronts detected by environmental satellites. *J. Geophys. Res.* **83**, 4501–22.
Legg, S., R. W. Hallberg, and J. B. Girton, 2006. Comparison of entrainment in overflows simulated by $z$-coordinate, isopycnal and nonhydrostatic models. *Ocean Modelling* **11**, 69–97.
Lentz, S., 2004. The response of buoyant coastal plumes to upwelling-favorable winds. *J. Phys. Oceanogr.* **34**, 2458–69.
Lentz, S. J., and K. R. Helfrich, 2002. Buoyant gravity currents along a sloping bottom in a rotating fluid. *J. Fluid Mech.* **464**, 251–78.
Linder, C. A., and G. Gawarkiewicz, 1998. A climatology of the shelfbreak front in the Middle Atlantic Bight. *J. Geophys. Res.* **103**, 18405–23.
Lowe, R. J., J. W. Rottman, and P. F. Linden, 2005. The non-boussiness lock-exchange problem. Part 1. Theory and experiments. *J. Fluid Mech.* **537**, 101–24.
Mauritzen, C., J. Price, T. Sanford, and D. Torres, 2005. Circulation and mixing in the Faroese Channels. *Deep-Sea Res. II* **52**, 883–913.
Muench, R., L. Padman, A. Gordon, and A. Orsi, 2009. A dense water outflow from the Ross Sea, Antarctica: Mixing and the contribution of tides. *J. Mar. Syst.* **77**, 369–87.

O'Donnell, J., 1990. The formation and fate of a river plume: A numerical model. *J. Phys. Oceanogr.* **20**, 551–69.

Özgökmen T. M., P. F. Fischer, and W. E. Johns, 2006. Product water mass formation by turbulent density currents from a high-order nonhydrostatic spectral element model. *Ocean Modelling* **12**, 237–67.

Özgökmen T. M., T. Iliescu, and P. F. Fischer, 2009. Reynolds number dependence of mixing in a lock-exchange system from direct numerical and large eddy simulations. *Ocean Modelling* **30**, 190–206.

Padman, L., S. L. Howard, A. Orsi, and R. Muench, 2009. Tides of the northwestern Ross Sea and their impact on dense outflows of Antarctic Bottom Water. *Deep-Sea Res. II* **56**, 818–34.

Pearce, A. F., and R. W. Griffiths, 1991. The mesoscale structure of the Leeuwin current: A comparison of laboratory models and satellite imagery. *J. Geophys. Res.* **96**, 16739–16757.

Peters, H., and W. E. Johns, 2005. Mixing and entrainment in the Red Sea outflow plume. Part II: Turbulence characteristics. *J. Phys. Oceanogr.* **35**, 584–600.

Peters, H., W. E. Johns, A. S. Bower, and D. M. Fratantoni, 2005. Mixing and entrainment in the Red Sea outflow plume. Part I: Plume structure. *J. Phys. Oceanogr.* **35**, 569–83.

Phillips, N. A., 1954. Energy transformation and meridional circulations associated with simple baroclinic waves in a two level quasi-geostrophic model. *Tellus* **3**, 61–5.

Price, J. F., and M. O. Barringer, 1994. Outflows and deep water production by marginal seas. *Prog. Oceanogr.* **33**, 161–200.

Price, J. F., M. O. Baringer, R. G. Lueck, G. C. Johnson, I. Ambar, G. Parrilla, A. Cantos, M. A. Kennelly, and T. B. Sanford, 1993. Mediterranean outflow mixing and dynamics. *Science* **259**, 1277–82.

Riemenschneider, U., and S. Legg, 2007. Regional simulations of the Faroe Bank Channel overflow in a level model. *Ocean Modelling*, **17**, 93–122, doi:10.1016/j.ocemod.2007.01.003.

Saunders, P. M., 1990. Cold outflow from the Faroe Bank Channel. *J. Phys. Oceanogr.* **20**, 29–43.

Simpson, J. E., 1997. *Gravity Currents in the Environment and the Laboratory,* 2nd ed., Cambridge University Press, Cambridge.

Smith, P. C., 1975. A streamtube model for bottom boundary currents in the ocean. *Deep-Sea Res.* **22,** 853–73.

Umlauf, L., and L. Arneborg, 2009. Dynamics of rotating shallow gravity currents passing through a channel. Part I: Observation of transverse structure. *J. Phys. Oceanogr.* **39**, 2385–401.

Wadhams, P., A. Gill, and P. F. Linden, 1979. Transects by submarine of the East Greenland Polar Front. *Deep-Sea Res.* **26**, 1311–327.

Wåhlin, A. K., and C. Cenedese, 2006. How entraining density currents influence the ocean stratification. *Deep-Sea Res. II* **53**, 172–93.

Wells, M. G., 2007. Influence of Coriolis forces on turbidity currents and sediment deposition. In: B. J. Geurts, H. Clercx, and W. Uijttewaal (eds.), *Particle-Laden Flow: From Geophysical to Kolmogorov Scales, ERCOFTAC Series*, 331–43. Springer-Verlag, Berlin.

Wells, M. G., and J. S. Wettlaufer, 2005. Two-dimensional density currents in a confined basin. *Geophys. Astrophys. Fluid Dyn.* **99**, 199–218.

Wolfe, C. L., and C. Cenedese, 2006. Laboratory experiments on eddy generation by a buoyant coastal current flowing over variable bathymetry. *J. Phys. Oceanogr.* **36**, 395–411.

Wright, W. R., 1976: The limits of shelf water south of Cape Cod, 1941 to 1972. *J. Mar. Res.* **34**, 1114.
Wu, W., G. Danabasoglu and W. G. Large, 2007. On the effects of parameterized Mediterranean overflow on North Atlantic ocean circulation and climate. *Ocean Modelling*, **19**, 31–52.
Xu, X., Y. S. Chang, H. Peters, T. M. Özgökmen, and E. P. Chassignet, 2006. Parameterization of gravity current entrainment for ocean circulation models using a high-order 3D nonhydrostatic spectral element model. *Ocean Modelling*, **14**, 19–44.
Yankovsky, A. E., and D. C. Chapman, 1997. A simple theory for the fate of buoyant coastal discharges. *J. Phys. Oceanogr.* **27**, 1386–401.

# 1

# Gravity Currents – Theory and Laboratory Experiments

PAUL LINDEN

## 1.1 Introduction

Gravity currents occur when there are horizontal variations in density in a fluid under the action of a gravitational field. A simple example that can be readily experienced is the gravity current that flows into a warm house through a doorway when it is opened on a cold, windless day. The larger density of the cold air produces a higher pressure on the outside of the doorway than on the inside, and this pressure difference drives the cold air in at the bottom and the warm air out at the top of the doorway. In addition to horizontal density variations, there must also be some feature to stop the fluid from either rising or falling indefinitely and to constrain the flow to be primarily horizontal. In many cases this is a solid boundary, such as the ground. In other situations it may be another feature of the density variations within the fluid, such as a density interface or vertical density stratification.

Gravity currents occur in gases when there are temperature differences, such as in the sea breeze, the flow of cool moist air from the sea to the land. On a warm day, the sun heats the land more than the sea, and, consequently, the air at low altitudes over the land is warmer than that over the sea. The sea breeze is a significant feature of coastal meteorology in many parts of the world. In some arid regions of the world, the effect of the sea breeze has been observed over 1,000 km from the coast.

Figure 1.1 shows dust suspended in the cold air advancing from underneath a thunderstorm in Leeton, New South Wales, in November 2002. The front shows the same convoluted structure – lobes and clefts – in the laboratory (Figure 1.2).

There is a fundamental reason for the study of gravity currents. In a gravitational field, spatial density variations in a fluid produce buoyancy forces. If the density varies in the horizontal direction, flow *always* results. As will be shown in Section 1.3, unless the horizontal density gradient is constant (i.e., the density varies linearly with horizontal distance), which, of course, is a very special case, the flow will develop into a gravity current with a sharp density front.

This chapter discusses laboratory experiments on gravity currents and the attempts to predict their speed of propagation. Despite the fact that a gravity current is a

Figure 1.1. A dust storm created by cold air flowing out from under a thunderstorm. For a color version of this figure please see the color plate section.

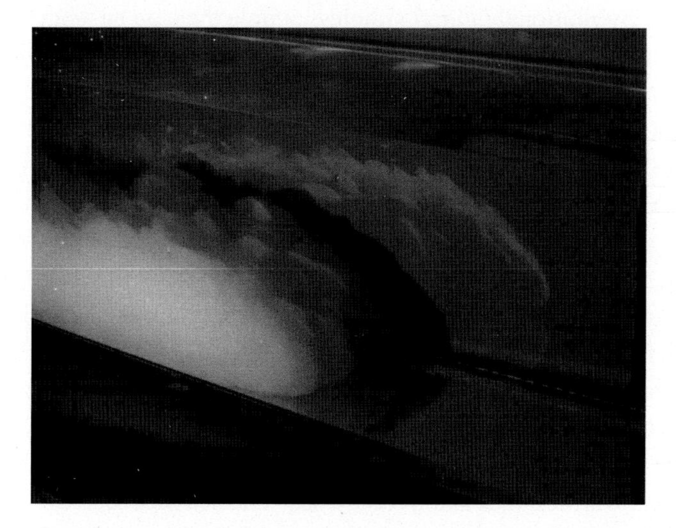

Figure 1.2. A saline laboratory gravity current flowing into fresh water. The current is made visible by milk added to the salt water. The lobes and clefts first reported by Simpson (1972) are clearly visible. The three dimensional structure persists behind the front and affects the structures at the top of the current.

turbulent, density-stratified, and time-dependent flow, remarkably, it has been possible to capture much of the physics in a single parameter, the *Froude number*, which is a nondimensional representation of the speed of the current. Here we provide a pedagogical account of this description and discuss its successes and limitations. The front of a gravity current, perhaps as a result of its particular dramatic form as seen in Figure 1.1, has been a source of fascination and challenge. This chapter is a story of the representation of this beautiful and complex flow by a single nondimensional number.

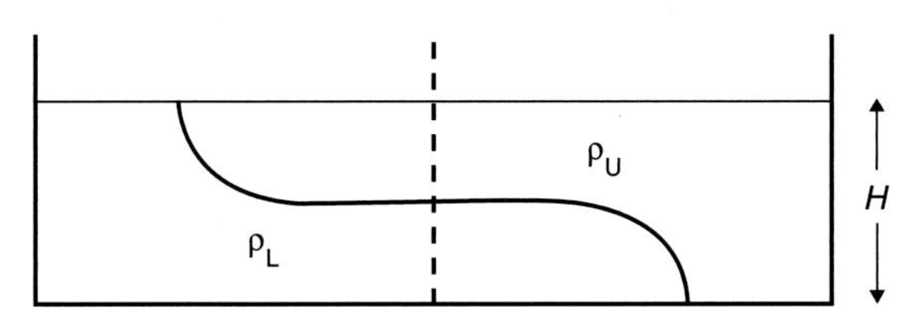


Figure 1.3. A schematic of a lock exchange experiment. In this case, fluid of density $\rho_U$ is separated by a vertical partition – the lock gate – from denser fluid with density $\rho_L$. Both fluids are initially at rest. When the gate is removed (vertically and carefully), a dense gravity current will flow along the bottom to the right and a buoyant current will flow along the top to the left.

## 1.2 Reduced Gravity

Many experiments start with the initial conditions shown in Figure 1.3 with fluid of depth $H$ with two different densities, $\rho_U < \rho_L$, separated by a vertical barrier. With the barrier in place, the fluid is at rest, and the difference in pressure on either side of the barrier at the bottom of the fluid is

$$\Delta p = g(\rho_L - \rho_U)H = g\Delta\rho H, \tag{1.1}$$

Once the barrier is removed – this is called a lock-exchange experiment – the dense fluid $\rho_L$ will be accelerated to the right along the bottom of the channel and the light fluid $\rho_U$ will be accelerated to the left along the top of the channel.

The acceleration is the *reduced gravity*

$$a = g\frac{\Delta\rho}{\rho} \equiv g', \tag{1.2}$$

where $\rho$ is a representative density. By dimensional analysis, the speed of a flow driven by a reduced gravity $g'$ in a fluid of depth $H$ is proportional to $\sqrt{g'H}$.

## 1.3 Frontogenesis

Consider a fluid with density that varies in the horizontal only $\rho = \rho(x)$. Restricting attention to two-dimensional motion $\mathbf{u} = (u, 0, w, t)$ in the $x$–$z$ plane in the absence of diffusion, for an incompressible fluid the conservation of mass equation is

$$\frac{\partial\rho}{\partial t} + u\frac{\partial\rho}{\partial x} = 0, \tag{1.3}$$

and the continuity equation is

$$\frac{\partial u}{\partial x} + \frac{\partial w}{\partial z} = 0. \tag{1.4}$$

The $x$-derivative of (1.3) and the substitution of $\frac{\partial u}{\partial x}$ from (1.4) give

$$\left(\frac{\partial}{\partial t}+u\frac{\partial}{\partial x}\right)\frac{\partial \rho}{\partial x}=\frac{\partial w}{\partial z}\frac{\partial \rho}{\partial x}. \tag{1.5}$$

If the initial horizontal density gradient $\frac{\partial \rho}{\partial x}|_0$ is constant in space, then (1.5) has a solution in which the horizontal density gradient does not change and $w=0$. In that case for a Boussinesq fluid, and in the absence of viscous effects, horizontal and vertical equations of motion reduce to

$$\frac{\partial u}{\partial t}=-\frac{1}{\rho_0}\frac{\partial p}{\partial x} \tag{1.6}$$

and

$$\frac{\partial p}{\partial z}=-g\rho. \tag{1.7}$$

Take the $z$-derivative of (1.6) and substitute for the pressure from (1.7) and obtain

$$\frac{\partial^2 u}{\partial t\partial z}=\frac{g}{\rho_0}\frac{\partial \rho}{\partial x}|_0. \tag{1.8}$$

Since continuity implies that $u=u(z,t)$, this equation may be integrated to give

$$u(z,t)=\frac{g}{\rho_0}\frac{\partial \rho}{\partial x}|_0 zt, \tag{1.9}$$

where the flow has been assumed to start from rest and that $u(0,t)=0$.

When the horizontal density gradient is constant in space, the motion causes the isopycnals to rotate at a constant speed toward the horizontal, creating an increasing vertical density gradient, and, from (1.3), the density field is

$$\rho(x,z,t)=-\tfrac{1}{2}\frac{g}{\rho_0}\left(\frac{\partial \rho}{\partial x}|_0\right)^2 zt^2. \tag{1.10}$$

The gradient Richardson number defined by

$$Ri\equiv-\frac{\frac{g}{\rho_0}\frac{\partial \rho}{\partial z}}{\left(\frac{\partial u}{\partial z}\right)^2} \tag{1.11}$$

in this case is exactly $\frac{1}{2}$ throughout the motion and the flow is stable. For *every* other form of horizontal stratification, (1.5) implies that the horizontal density gradient will increase if $\frac{\partial w}{\partial z}\frac{\partial \rho}{\partial x}>0$, and decrease if the sign of this term is negative.

Figure 1.4 shows the sharpening of a front in a laboratory experiment. The isopycnal surfaces are marked by dye, and as the flow proceeds along the tank (toward lower values of the vertical markers), the front catches up with the dye line immediately ahead of it and strengthens it, as can be seen qualitatively from the shadowgraph image.

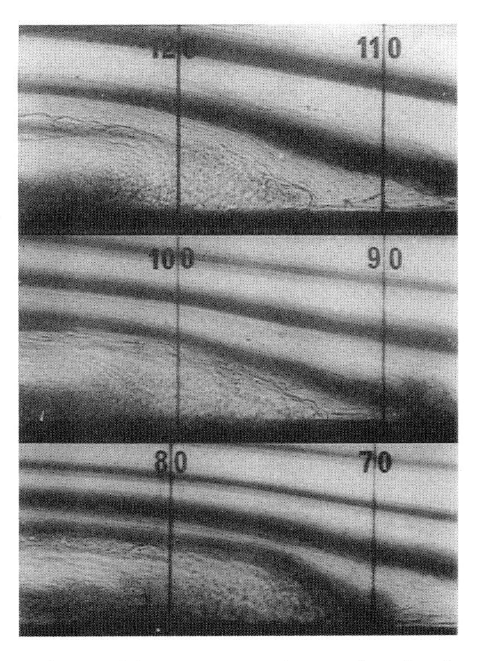


Figure 1.4. Sequences from a laboratory experiment showing frontogenesis. The dye lines mark surfaces of constant density and the vertical lines are 10 cm apart. The gravity-driven flow is from left to right and the flow is shown at three different times. At the earliest time (upper panel) the front has begun to form on the lower boundary and it is behind the dye line. At later times it catches up with the dye line and in the bottom panel the front has strengthened. Taken from Simpson and Linden (1989).

## 1.4 Nondimensional Parameters

As discussed previously, the most important nondimensional parameter for a Boussinesq gravity current is the Froude number $F_H$, defined as the ratio of the current speed $U$ to the long wave speed $\sqrt{g'H}$,

$$F_H = \frac{U}{\sqrt{g'H}}. \tag{1.12}$$

The second important parameter is the Reynolds number $Re$

$$Re \equiv \frac{UH}{\nu}, \tag{1.13}$$

where $\nu$ is the kinematic viscosity, which measures the importance of viscous dissipation on the current compared to its inertia.

The effects of diffusion of density are measured by the Peclet number $Pe$

$$Pe \equiv \frac{UH}{\kappa}, \tag{1.14}$$

where $\kappa$ is the molecular diffusivity. $Pe$ measures the relative magnitude of advection and diffusion of density.

For non-Boussinesq currents, the density ratio

$$\gamma \equiv \frac{\rho_U}{\rho_L} \tag{1.15}$$

is important.

## 1.5 Scaling Analysis

We begin the theoretical description with a scaling analysis of the gravity current that results from the release of a finite volume of dense fluid into a stationary unstratified ambient fluid. Consider a volume $V_0$ of dense fluid, density $\rho_L$, released at $t = 0$ from rest on a horizontal boundary in ambient fluid of density $\rho_U < \rho_L$. For simplicity, suppose the fluid is confined in a channel of unit width, and the properties of the flow are independent of the across-channel coordinate. The size of the release is $A_0 = DL_0$, the volume per unit channel width (Figure 1.5). As the current propagates, it may mix with the ambient fluid and change its density, and we denote the initial negative buoyancy of the dense fluid by $g'_0$.

Suppose that initially the flow rapidly accelerates to a speed large enough that viscous forces are unimportant, and so little fluid has left the lock that its initial volume is effectively infinite. Dimensional analysis shows that the velocity $U$ of the advancing current at time $t$ is given by

$$U = F(g'_0 D)^{1/2} f(t/T_a), \tag{1.16}$$

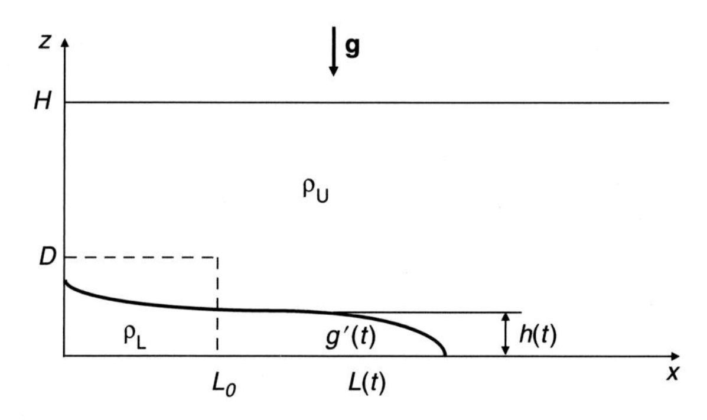


Figure 1.5. A sketch of the release of a finite volume of dense fluid into a less dense stationary environment of depth $H$. The dense fluid is initially held behind a lock gate at $x = L_0$, and the initial depth is $D$. The resulting flow is considered to be confined to channel of unit width perpendicular to the plane of the figure.

where $F$ is a dimensionless constant and

$$T_a = \sqrt{\frac{D}{g'_0}} \tag{1.17}$$

is the time scale associated with the acceleration from rest. After this acceleration the current is observed to travel with a constant speed

$$U = F_D(g'_0 D)^{\frac{1}{2}}, \tag{1.18}$$

where $F_D$ is a dimensionless constant. The length $L(t)$ of the current is given by

$$L(t) = L_0 + F_D(g'_0 D)^{\frac{1}{2}} t. \tag{1.19}$$

At later times, sufficient fluid has left the lock so the fact that the volume $A_0$, per unit channel width, is finite will influence the motion. This introduces a second time scale

$$T_V = \frac{L_0}{\sqrt{g'_0 D}}, \tag{1.20}$$

the time taken for a gravity wave with speed $\sqrt{g'_0 D}$ to travel the length $L_0$ of the lock. After this time, the finite volume of the lock now becomes a parameter. Then (1.16) becomes

$$U = F_D(g'_0 D)^{1/2} f(t/T_a, t/T_V). \tag{1.21}$$

Conservation of mass implies that $B_0 = g'_0 D L_0$, the initial negative buoyancy in the lock per unit width, remains constant. Since the dimensions of buoyancy per unit width $[B_0] = L^3\, T^{-2}$,

$$L = c B_0^{\frac{1}{3}} t^{\frac{2}{3}}, \tag{1.22}$$

where $c$ is a dimensionless constant (for high Reynolds numbers). The speed $U$ during this phase decreases as

$$U = \frac{2}{3} c B_0^{\frac{1}{3}} t^{-\frac{1}{3}}. \tag{1.23}$$

This time-dependent motion in (1.22) and (1.23) is called the *similarity phase* (Simpson 1997) as it also results from a similarity solution of the shallow water equations (see Section 1.7).

We can also derive (1.18) and (1.23) by assuming the front travels with a constant *local* Froude number, which is now based on the local depth $h(t)$ and the local buoyancy $g'(t)$ at the front (see Figure 1.5) rather than the initial values, so that $F = F_h = \frac{U}{\sqrt{g'h}}$.

We represent the current by a characteristic length $L(t)$ and depth $h(t)$ and suppose that it has a uniform buoyancy $g'(t)$ over its length and depth. Conservation of buoyancy per unit width is expressed as

$$g'(t)L(t)h(t) = c_B g'_0 A_0 = c_B B_0, \tag{1.24}$$

where $c_B$ is a shape constant, which would be unity if the current retained a rectangular shape. A constant local Froude number $F_h$ implies that

$$U = \frac{dL}{dt} = F_h(g'(t)h(t))^{1/2}. \tag{1.25}$$

Using (1.24) and integrating gives

$$L(t) = \left[\frac{3}{2}F_h(c_B B_0)^{1/2}t + (L_0)^{3/2}\right]^{2/3}. \tag{1.26}$$

In dimensionless form, (1.26) is

$$\frac{L(t)}{L_0} = \left[\frac{3}{2}F_h c_B{}^{\frac{1}{2}} t/T_V + 1\right]^{2/3}. \tag{1.27}$$

When $t \ll T_V$, equation (1.27) gives

$$\frac{L(t)}{L_0} \simeq 1 + F_h c_B{}^{\frac{1}{2}} t/T_V, \tag{1.28}$$

so the current travels at constant speed, and its length reduces to (1.19) if $F = F_h c_B{}^{\frac{1}{2}}$. When $t \gg T_V$,

$$\frac{L(t)}{L_0} \simeq \left(\frac{3F_h}{2}\right)^{\frac{2}{3}} c_B{}^{\frac{1}{3}}(t/T_V)^{2/3}, \tag{1.29}$$

which gives the same result as (1.22) if $c = \left(\frac{9}{4}c_B F_h{}^2\right)^{\frac{1}{3}}$. The agreement between this calculation and the dimensional analysis supports the assumption that the front of the current travels at a constant local Froude number and, therefore, that the front acts as a control on the flow.

As the current decelerates, the Reynolds number decreases and frictional effects eventually become important. The flow is then affected by the viscosity $\nu$ of the fluid, and a further time scale

$$T_\nu = \frac{\nu L_\nu{}^2}{g'_\nu h_\nu{}^3} \tag{1.30}$$

now enters the problem, where the subscript $\nu$ denotes the values of the depth, volume, and buoyancy of the current as it enters the viscous phase. Unless the fluid is very viscous, such as honey spreading on toast, these values will, in general, be different from the values of the initial release. Then the front speed may be written as

$$U = F(g'_\nu h_\nu)^{1/2} f(t/T_a, t/T_V, t/T_\nu). \tag{1.31}$$

In the viscous phase the horizontal pressure gradient driving the current is balanced by viscous stresses so that

$$\frac{\nu}{h(t)^2}\frac{dL}{dt} = \frac{c_\nu g'_\nu h(t)}{L(t)}, \tag{1.32}$$

where $c_\nu$ is a dimensionless shape constant. To proceed further, it is necessary to assume that volume is conserved

$$L(t)h(t) = c_A A_\nu, \tag{1.33}$$

where $c_A$ is a further shape constant. Substituting for $h(t)$ from (1.33) and solving the resulting differential equation gives

$$L(t) = \left[ 5 \frac{c_\nu c_A{}^3 g'_\nu A_\nu{}^3}{\nu} t + L_\nu{}^5 \right]^{1/5}, \tag{1.34}$$

where $L_\nu$ is the length of the current at the start of the viscous phase.

In dimensionless form, (1.34) is

$$\frac{L(t)}{L_\nu} = [5c_\nu t / T_\nu + 1]^{1/5}. \tag{1.35}$$

Thus, in the viscous phase, the current length will increase proportional to $t^{1/5}$.

The mechanism whereby the current changes from the constant velocity to the similarity phase was discovered by Rottman and Simpson (1983). Initially, the speed of the front is constant. Subsequently, the slope decreases so that $x$ increases like $t^{\frac{2}{3}}$, consistent with the results of the scaling analysis (1.29). For full-depth releases, Rottman and Simpson (1983) show that a bore propagates along the interface (see second and third images in Figure 1.6). The bore is observed to travel faster than the current head as can be seen in Figure 1.7. When the bore catches up with the front, the self-similar phase begins. If the depth of the lock $D < 0.6H$, the disturbance takes the form of a long expansion wave, rather than a bore.

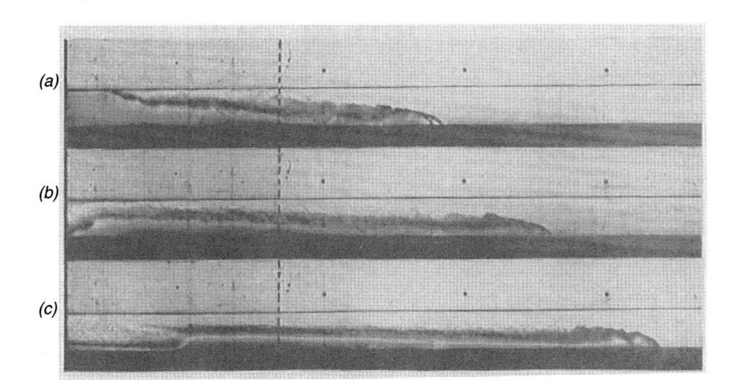


Figure 1.6. Shadowgraphs of a full–depth lock release. The location of the lock gate is shown by the vertical dotted line. In (a) a light surface current is propagating back into the lock. This reflects from the back wall of the lock and forms a bore, seen as the abrupt change in depth at the rear of the current in (b) and (c). While the bore is behind the front, the front travels at a constant speed, as indicated by its positions in (b) and (c), as the two images are taken at equal time intervals. Taken from Rottman and Simpson (1983).

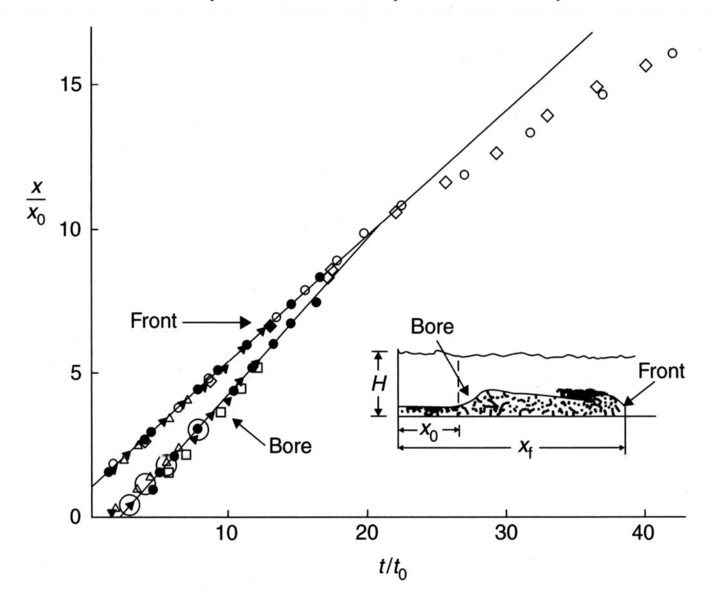


Figure 1.7. The front and bore positions as functions of time after release for a set of experiments with full-depth locks. The straight lines show that the front and bore travel at constant speeds until the bore catches the front, after which the front decelerates. Taken from Rottman and Simpson (1983).

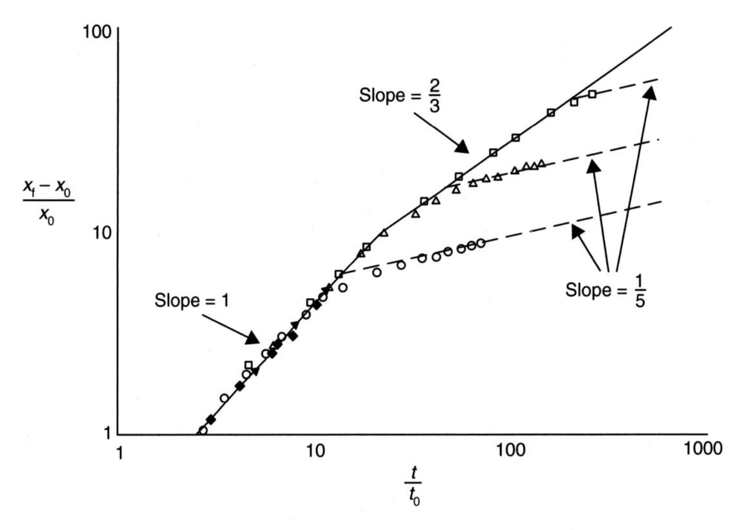


Figure 1.8. A logarithmic plot of dimensionless front positions against dimensionless time for three full-depth lock releases. The collapse of the data shows that they are well described by (1.16) during the constant velocity phase, when $x \sim t$ and by (1.26) during the similarity phase, when $x \sim t^{\frac{2}{3}}$. The different experiments enter the viscous phase at different times, and then obey (1.34), with $x \sim t^{\frac{1}{5}}$. Taken from Rottman and Simpson (1983).

As the current continues to decelerate, the Reynolds number decreases and viscous effects become increasingly important. In this final viscous phase, as shown in Figure 1.8 by the dashed lines, the front position grows as $t^{\frac{1}{5}}$, as predicted by (1.34).

## 1.6 Theories for the Froude Number

The scaling analysis given in Section 1.5 does not determine the quantitative speeds or scales. To obtain this quantitative information, additional theory is needed to determine the dimensionless constants, such as the Froude number, that arise. In the next subsections, we discuss the theories that have been developed to estimate the Froude number in the constant-velocity phase of the flow.

### *1.6.1 Yih's Theory*

The current derives its kinetic energy from the gravitational potential energy stored in the original density distribution. Yih (1965) calculated this energy exchange for a Boussinesq lock release (Figure 1.3). When $\rho_U \sim \rho_L$, symmetry implies that the current will initially occupy one-half the depth. In a time $\Delta t$, the fronts will have advanced a distance $U\Delta t$. The potential energy gained by the lighter fluid is $\frac{1}{4} g\rho_U H^2 U\Delta t$, and that lost by the denser fluid is $\frac{1}{4} g\rho_L H^2 U\Delta t$. The total kinetic energy gain is $\frac{1}{4}(\rho_U + \rho_L) H U^2 U\Delta t$. Equating these energies gives

$$U = \sqrt{\frac{g(\rho_L - \rho_U)H}{2(\rho_U + \rho_L)}}. \tag{1.36}$$

For the Boussinesq case ($\rho_U \approx \rho_L$), (1.36) implies that $F_H = \frac{1}{2}$.

### *1.6.2 Von Kármán's Theory*

The first attempt to obtain a relation for the Froude number of the front of the current was made by von Kármán in 1940. He considered the case of an infinitely deep ambient fluid of density $\rho_U$, and a current of density $\rho_L > \rho_U$. He assumed irrotational flow of a perfect fluid and that the velocities are horizontal and the pressure is hydrostatic away from the front. He also assumed that in the rest frame of the current the velocity everywhere inside the current is zero.

For an irrotational steady flow, Bernoulli's equation is

$$p + \frac{1}{2}\rho q^2 + g\rho z = \text{constant} \tag{1.37}$$

along streamlines, where $q$ is the speed of the flow, and $z$ is the height. First (1.37) is applied along the streamline along the bottom from the stagnation point $O$ to the

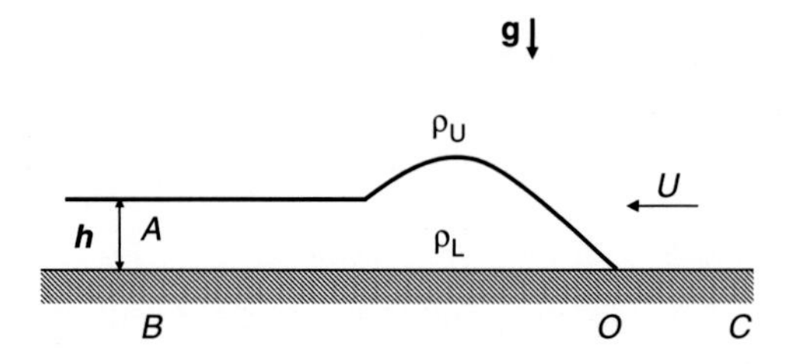


Figure 1.9. An idealized model of a perfect fluid gravity current in an infinitely deep fluid, viewed in the rest frame of the current, in which all the dense fluid inside the current is assumed to be at rest. Far downstream from the front the interface is assumed to be flat.

upstream point $C$ where $q = U$, the speed of the current in the laboratory frame (Figure 1.9). Hence,

$$p_O = p_C + \frac{1}{2}\rho_U U^2. \tag{1.38}$$

For a perfect fluid with no dissipation, (1.37) can also be applied along the interface from $O$ to $A$, giving

$$p_O = p_A + g\rho_L h, \tag{1.39}$$

where $h$ is the height of the interface far downstream from the front. Here we have taken the interface on the current side.

It is assumed that at some height well above the current, the pressure is constant with horizontal position. Otherwise, there would be a horizontal pressure gradient that would drive flow in the ambient fluid (see Section 1.2). As the pressure is hydrostatic, $p_C = p_A + g\rho_U h$. Eliminating $p_O$ from (1.38) and (1.39) and substituting for this pressure difference gives

$$U^2 = 2g\frac{\rho_L - \rho_U}{\rho_U}h. \tag{1.40}$$

Alternatively, Bernoulli's equation could be applied along the bottom inside the current from $O$ to $B$, giving

$$p_O = p_B, \tag{1.41}$$

since there is no flow inside the current. Also, $p_B = p_A + g\rho_L h$. Hence, $p_B = p_C + g(\rho_L - \rho_U)h$. From (1.38) and (1.41), we again obtain the same result (1.40) for the speed of the current, which may be expressed nondimensionally as

$$F_h = \frac{U}{\sqrt{g(1-\gamma)h}} = \sqrt{\frac{2}{\gamma}}, \tag{1.42}$$

where $\gamma = \frac{\rho_U}{\rho_L}$ is the density ratio, defined such that $0 < \gamma < 1$. For a Boussinesq flow, $\gamma \approx 1$, and $g(1-\gamma) = g'$, is the reduced gravity and (1.42) reduces to the much quoted result for the front Froude number of a Boussinesq current in an infinitely deep fluid

$$F_h = \frac{U}{\sqrt{g'h}} = \sqrt{2}. \tag{1.43}$$

This Froude number is based on the depth of the current away from the nonhydrostatic head region.

### 1.6.3 Benjamin's Theory

#### 1.6.3.1 Mass and Momentum Conservation

A current of density $\rho_L$ propagates into a fluid of density $\rho_U$, as shown in Figure 1.10. Denote the downstream depth of the current by $h$, and suppose that the velocity in the ambient fluid far behind the front where the interface is flat is $u_U$. It is further assumed that $u_U$ is independent of depth. Continuity implies that

$$UH = u_U(H-h). \tag{1.44}$$

Since there are no external horizontal forces acting on the flow, the net flux of momentum into a control volume including the front is zero. Consider the control volume consisting of the two vertical planes, one downstream at *BE* and one upstream of the front at *CD*, and the top and bottom boundaries of the channel.

The pressure distributions along the two lines *BE* and *CD* may be determined since the pressure is assumed to be hydrostatic. Along *BE*

$$p = \begin{cases} p_B - g\rho_L z, & 0 < z < h, \\ p_B - g\rho_L h - g\rho_U(z-h), & h < z < H, \end{cases} \tag{1.45}$$

where $p_B$ is the pressure at *B*. Similarly along *CD*,

$$p = p_C - g\rho_U z, \tag{1.46}$$

where $p_C$ is the pressure at *C*.

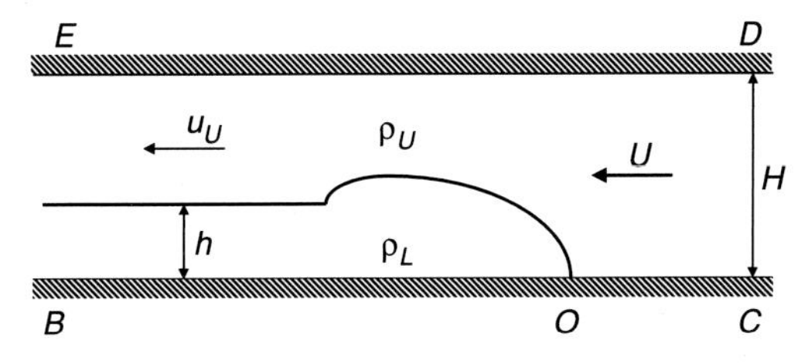


Figure 1.10. A schematic diagram of an idealized dense gravity current in a reference frame in which the current is at rest.

Conservation of the horizontal component of the momentum flux may be written as

$$\int_B^E p\,dz + \int_B^E \rho u^2 dz = \int_C^D p\,dz + \int_C^D \rho U^2 dz. \tag{1.47}$$

Substitution of (1.45) and (1.46) into the momentum balance (1.47) gives

$$p_B H + \frac{1}{2} g(\rho_L - \rho_U)h^2 - g(\rho_L - \rho_U)Hh + \rho_U u_U{}^2 (H - h) = p_C H + \rho_U U^2 H. \tag{1.48}$$

We define the pressure at the stagnation point $O$ to be $p_O$. Since the velocity within the current is zero, application of Bernoulli's equation along $BO$ gives $p_B = p_O$, and along $OC$ gives $p_C = p_O - \frac{1}{2}\rho_U U^2$. Substituting for $p_B$ and $p_C$ in (1.48) and using the continuity equation (1.44) gives

$$\frac{U^2}{g(1-\gamma)H} = \frac{1}{\gamma} f(h), \tag{1.49}$$

where

$$f(h) = \frac{h(2H-h)(H-h)}{H^2(H+h)}. \tag{1.50}$$

For Boussinesq currents, $\rho_U \approx \rho_L$, $\gamma \approx 1$, and $g(1-\gamma) = g'$, and the current speed may be written in terms of a Froude number $F_H$ based on the depth $H$ of the channel

$$F_H = \frac{U}{\sqrt{g'H}} = \sqrt{\frac{h(H-h)(2H-h)}{H^2(H+h)}}. \tag{1.51}$$

The right-hand side of (1.51) is plotted in Figure 1.11 as a function of $h/H$. The speed increases from zero as $h$ increases, reaches a maximum of $F_H = 0.527$ at $h_m = 0.347H$, and then decreases with further increase in $h$, until $h = \frac{1}{2}H$, when $F_H = \frac{1}{2}$. Values of $h > \frac{1}{2}H$ are not physically realizable as they require energy input into the control volume (Benjamin 1968).

The volume flux per unit width carried by the current is $Q = Uh$. From (1.51), the dimensionless volume flux $\frac{Q}{\sqrt{g'H^3}}$ is given by

$$\frac{Q}{\sqrt{g'H^3}} = \sqrt{\frac{h^3(H-h)(2H-h)}{H^4(H+h)}} \tag{1.52}$$

and is also plotted in Figure 1.11. The volume flux $Q$ increases monotonically with $h/H$ up to the limit $h/H = \frac{1}{2}$, at which value $\frac{Q}{\sqrt{g'H^3}} = \frac{1}{4}$.

#### *1.6.3.2 Energy Conservation*

The properties of the gravity current depend on its depth $h$. To specify, the depth a further condition is needed. One possibility explored by Benjamin (1968) is to

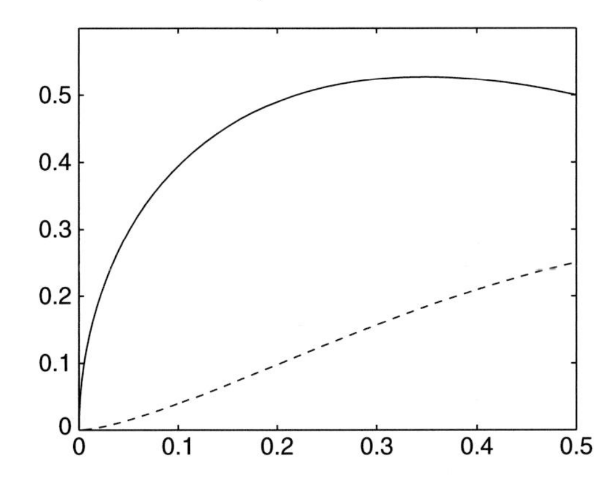


Figure 1.11. The Froude number $F_H$ (solid curve) and the dimensionless volume flux $\frac{Q}{\sqrt{g'H^3}}$ (dashed curve) plotted against the dimensionless current depth $\frac{h}{H}$.

apply energy conservation in the control volume. In this energy-conserving case, it is permissible to apply Bernoulli's equation (1.37) along streamlines in the flow.

Along the upper boundary *ED*,

$$p_E + \frac{1}{2}\rho_U {u_U}^2 = p_D + \frac{1}{2}\rho_U U^2. \tag{1.53}$$

From (1.45) and (1.46),

$$p_E - p_D = p_B - p_C - g(\rho_L - \rho_U)h. \tag{1.54}$$

Using the fact that $p_B - p_C = -\frac{1}{2}\rho_U U^2$ and substituting the pressure difference $p_E - p_D$ from (1.54) into (1.53) we get

$$\frac{1}{2}\rho_U {u_U}^2 = g(\rho_L - \rho_U). \tag{1.55}$$

From (1.44), the current speed $U$ is given by

$$U^2 = 2g\frac{(1-\gamma)}{\gamma}\frac{h(H-h)^2}{H^2}. \tag{1.56}$$

Substituting $p_A - p_O$ into (1.45) gives the same result as (1.55) for the upper layer speed $u_U$. Equating the two expressions (1.49) and (1.56) for the current speed $U$ gives two solutions for the current depth

$$\frac{h}{H} = 0 \quad \text{or} \quad \frac{h}{H} = \frac{1}{2}. \tag{1.57}$$

Hence an energy-conserving current occupies one-half the depth of the channel and travels with a nondimensional speed

$$\frac{U}{\sqrt{g'H}} = \tfrac{1}{2}. \tag{1.58}$$

### *1.6.3.3 Comparison with Experiment*

Figure 1.12 shows an air cavity that was the flow originally analyzed by Benjamin. It shows that the cavity occupies one-half the depth of the channel and has a smooth front, as expected from an energy-conserving flow. A Boussinesq current is shown in Figure 1.13 and compared with the depth of the energy-conserving current and the fastest moving current. Despite the mixing visible in the flow, the depth is close to the half-depth value.

Figure 1.14 shows a comparison of a Boussinesq full-depth lock exchange with Benjamin's approximate potential flow solution (Benjamin, 1968). The two front

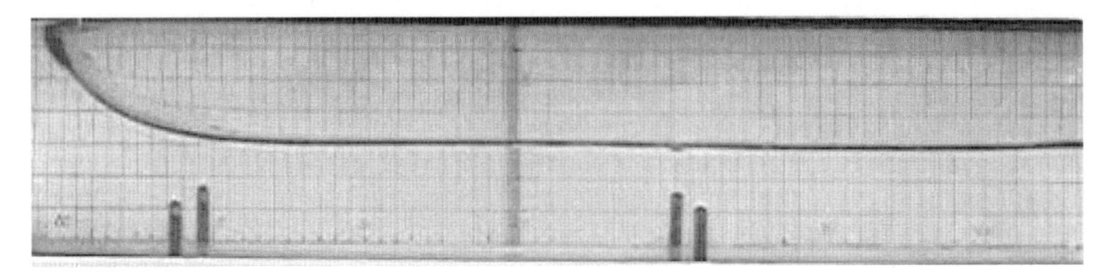

Figure 1.12. Air cavity in a rectangular horizontal duct (Gardner and Crow, 1970).

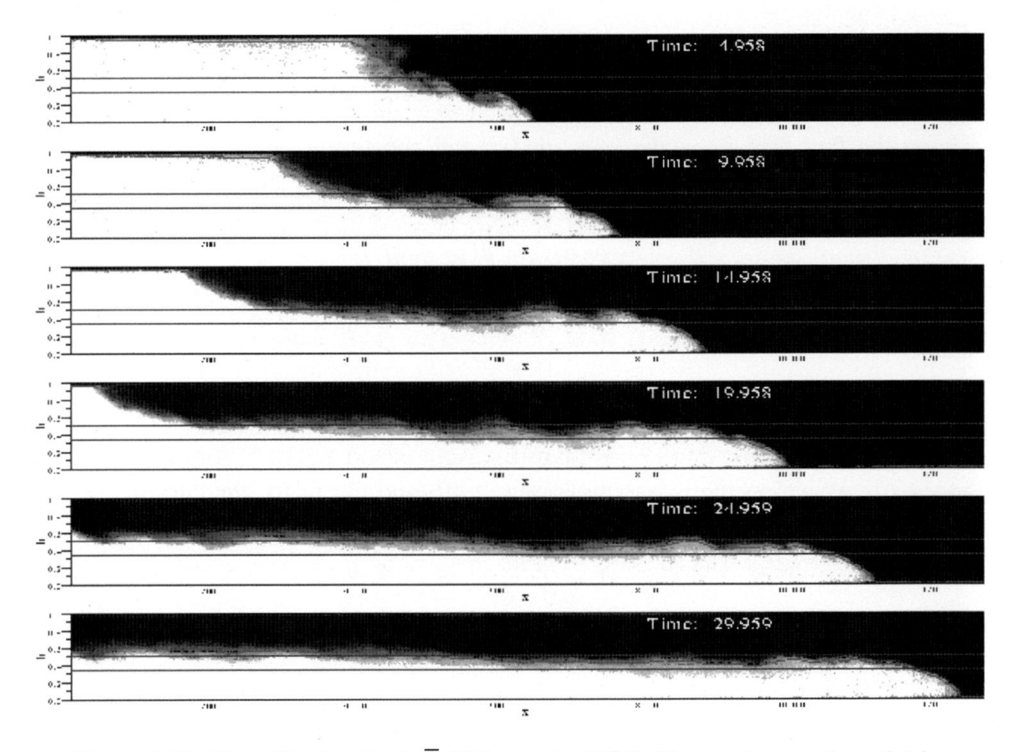


Figure 1.13. The effective depth $\overline{h}$ (Shin et al., 2004). For a color version of this figure please see the color plate section.

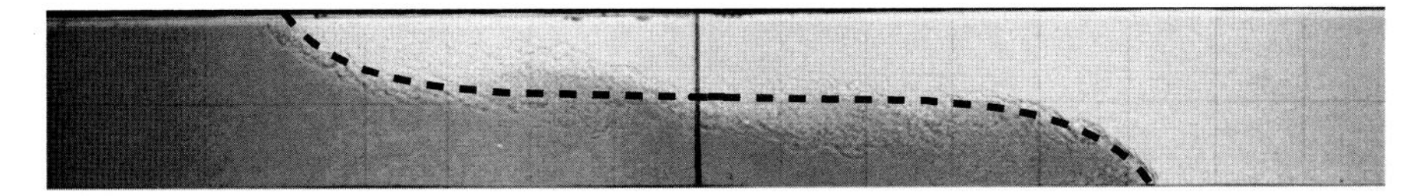

Figure 1.14. Comparison with Benjamin's potential flow solution – dashed line (Lowe et al., 2005).

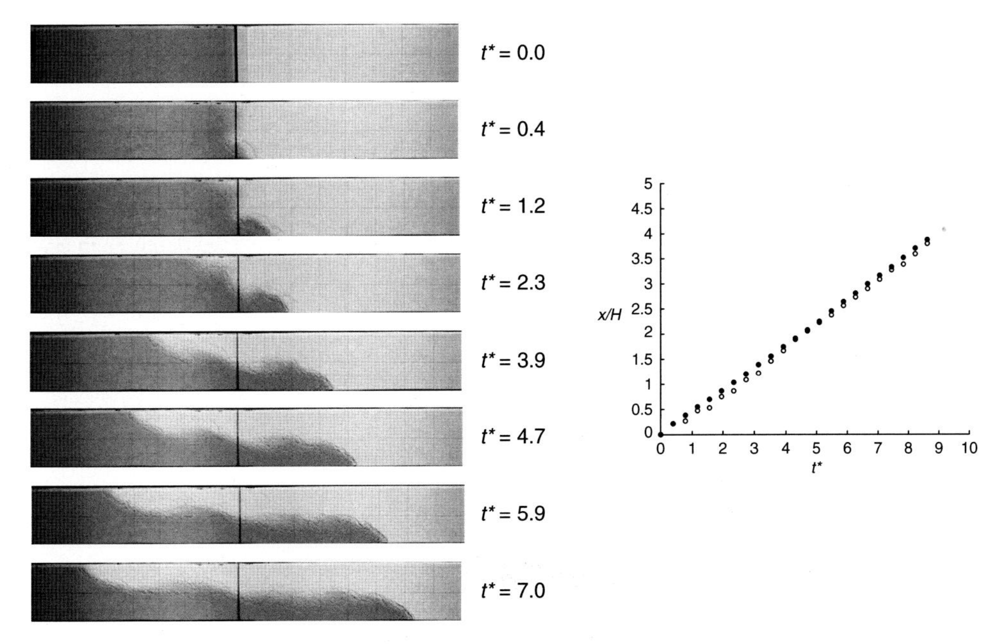


Figure 1.15. A sequence showing a Boussinesq full-depth lock exchange with $\gamma =$ 0.98 and the positions of the light and heavy fronts plotted against time. The value of $F_H = 0.48$ (Shin et al. 2004).

shapes have been joined by a straight line, and the composite fits the observed current shape well (Lowe et al., 2005).

Figure 1.15 shows a sequence of shadowgraph images of a full-depth lock exchange and the front positions. Both fronts move with the same (constant) speed and $F_H =$ 0.48. Thus, Benjamin's theory gives an excellent description of the constant-velocity phase of a full-depth lock exchange.

### *1.6.4 Energy-Conserving Theory*

Benjamin's theory described in Section 1.6.3 gives no indication of how the gravity current is established. Here we look at the formation from a prescribed set of initial conditions – the partial-depth lock exchange (Figure 1.16b) .

Figures 1.17 and 1.18 show the flow when the lock depth is $0.5H$ and $0.83H$, respectively. In both cases a dense current travels to the right as in the full-depth

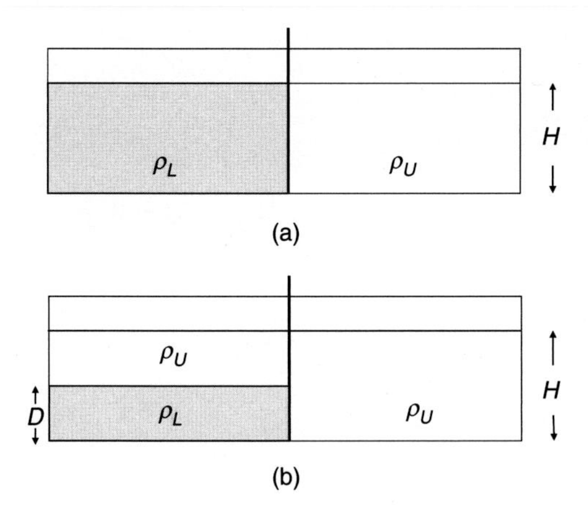


Figure 1.16. A schematic of the lock release initial conditions. The flow is started by removing the gate vertically. The dense fluid $\rho_L > \rho_U$ occupies the depth $H$ in a full-depth release (a) and a depth $D < H$ in a partial-depth release (b).

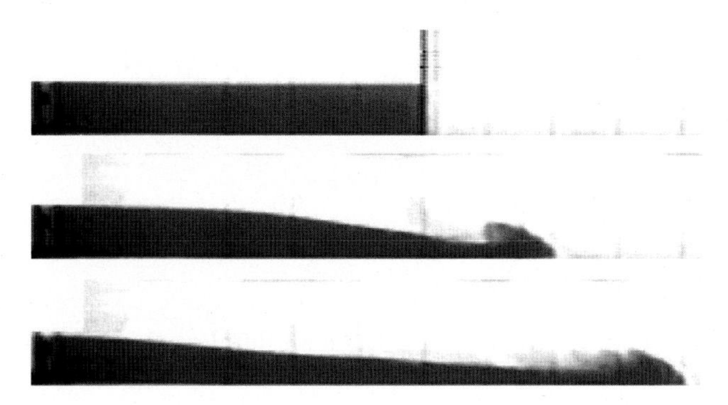

Figure 1.17. The flow from a partial-depth lock release. The initial depth $D = 0.5H$, and $\gamma = 0.989$.

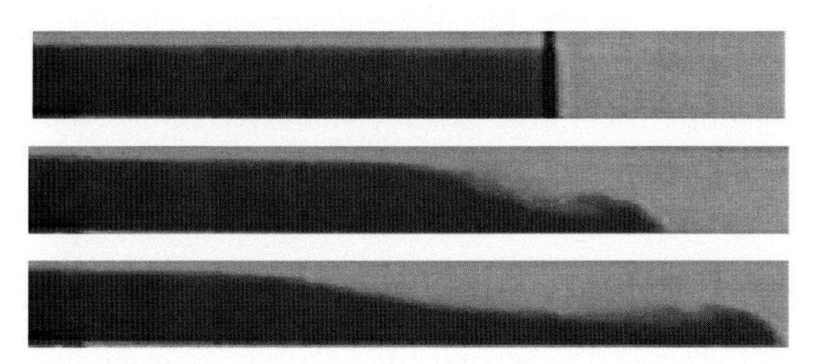

Figure 1.18. The flow from a partial-depth lock release. The initial depth $D = 0.83H$, and $\gamma = 0.990$.

release, but now the disturbance traveling to the left takes the form of a wave of rarefaction propagating on the interface. Apart from the fact that the dense current occupies significantly less than half the channel depth, qualitatively it looks much the same as the dense current shown in Figure 1.15, except there is a more pronounced raised head near the front of the shallower current (Figure 1.17).

#### 1.6.4.1 Partial-Depth Lock Releases

We now consider the case of partial-depth releases generated as shown in Figure 1.16b. In any hydraulic theory, it is necessary to include explicitly the flow features to be modeled. Following Shin et al. (2004), the model includes the gravity current propagating toward the right and a wave of depression moving toward the left on the interface between the two fluids, related to the light current in the full-depth lock exchange.

Dense fluid of height $D$ and density $\rho_L$ lies initially behind lock position. Light fluid of density $\rho_U$ lies on top of the dense fluid, as well as in front of the lock, so that the total height of fluid on both sides of the lock position is $H$. Fluid is initially at rest everywhere and lies between two smooth, rigid horizontal boundaries. When the dense fluid is released, it forms a gravity current that moves away from the lock at a constant speed $U$ from left to right. A disturbance, which travels in the opposite direction at constant speed $U_r$, is also formed, as shown in Figure 1.19b.

The fluid is assumed to be inviscid, irrotational, and immiscible. The shape of the interface is approximated by a horizontal middle section of height $h$ and two

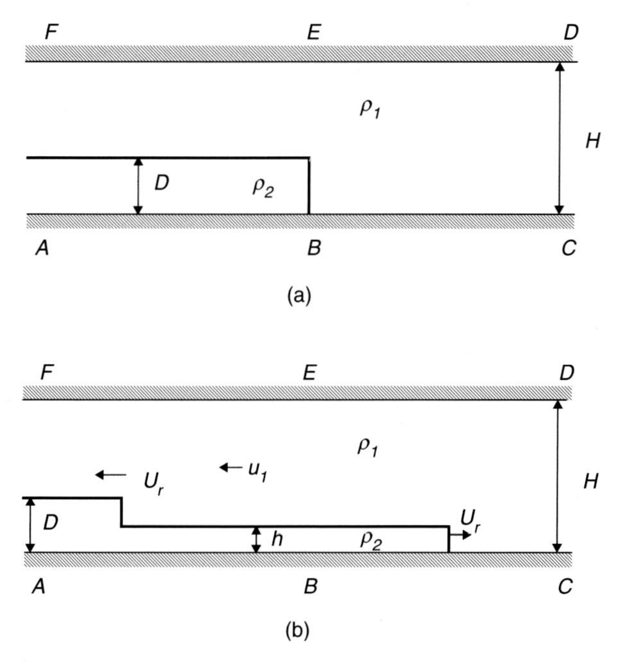


Figure 1.19. Schematic a partial-depth lock release in a channel (a) before release and (b) after release. The symbols in (b) are defined in the text.

advancing fronts. The two fronts are assumed to move at constant speeds and to have constant shapes in time. As in Benjamin's (1968) analysis, the exact shapes of the fronts do not matter as long as they remain steady. As discussed in Section 1.6.4.5, the backward disturbance is, in general, a rarefaction wave rather than a front of constant shape. Away from the advancing fronts, the flow is assumed to be horizontal, and, consequently, the pressure is hydrostatic.

#### *1.6.4.2 Mass and Momentum conservation*

We now apply conservation of mass across the current front and the disturbance front, respectively. This gives

$$UH = (U + u_U)(H - h) \tag{1.59}$$

and

$$U_r D = (U + U_r)h. \tag{1.60}$$

Thus,

$$u_U = \frac{Uh}{H - h} \tag{1.61}$$

and

$$U_r = \frac{Uh}{D - h}. \tag{1.62}$$

We now consider the momentum balance inside the fixed box $ACDF$ shown in Figure 1.19, which contains both the current and disturbance so that it includes all the fluid affected by the lock release at all times. The fluid outside the box is therefore always at rest. The rate of change of momentum $\dot{M}$ is

$$\dot{M} = \int_A^F p dz - \int_C^D p dz, \tag{1.63}$$

where the pressure distributions are given by hydrostatic formulae similar to (1.45) and (1.46). After integration, (1.63) yields

$$\dot{M} = \frac{1}{2} g(\rho_L - \rho_U) D^2 - (p_D - p_F) H, \tag{1.64}$$

where $p_D$ and $p_F$ are pressures at points $D$ and $F$, respectively.

The increase of momentum inside the control volume $ACDF$ is given by

$$\dot{M} = \rho_L (U + U_r) U h - \rho_U (U + U_r) u_U (H - h). \tag{1.65}$$

Using (1.61) and (1.62) and equating the two expressions (1.64) and (1.65), we find that the pressure difference along the upper boundary satisfies

$$(p_D - p_F) H = (\rho_L - \rho_U) \left[ U^2 \frac{Dh}{D - h} + \frac{1}{2} g D^2 \right]. \tag{1.66}$$

Assuming that energy is conserved in the top layer, the time-dependent Bernoulli equation can be applied there to find the pressure difference between $D$ and $F$. In this case Bernoulli's equation is

$$p_F + \rho_U \left.\frac{\partial \phi_U}{\partial t}\right|_F = p_D + \rho_U \left.\frac{\partial \phi_U}{\partial t}\right|_D, \tag{1.67}$$

where $\phi_U$ is a velocity potential for the upper layer flow.

In this model, the upper layer velocity is horizontal with $x$-component $u$ given by

$$u = \begin{cases} 0 & \text{for } x < x_r, \\ -u_U & \text{for } x_r < x < x_f, \\ 0 & \text{for } x > x_f, \end{cases} \tag{1.68}$$

where $x_r$ and $x_f$ are the positions along the $x$-axis of the disturbance front and the current front, respectively. A possible choice of $\phi_U$, which satisfies (1.68) and is continuous for all $x$, is given by

$$\phi_U = \begin{cases} 0 & \text{for } x < x_r, \\ -u_U(x - x_r) & \text{for } x_r < x < x_f, \\ -u_U(x_f - x_r) & \text{for } x > x_f. \end{cases} \tag{1.69}$$

Substituting this potential function into (1.67), one obtains

$$p_D - p_F = \rho_U u_U (\dot{x}_f - \dot{x}_r), \tag{1.70}$$

and so

$$p_D - p_F = \rho_U u_U (U + U_r). \tag{1.71}$$

Using the continuity equations (1.61) and (1.62) in (1.71), and substituting the result into (1.66), after some algebra one finds an expression for the speed of the current

$$\frac{U^2}{gH} = \frac{(\rho_L - \rho_U)D(D-h)(H-h)}{2hH(\rho_L(H-h) + \rho_U h)}. \tag{1.72}$$

### 1.6.4.3 Energy Conservation

As in Benjamin's analysis, we need another condition to determine the flow completely, and, as in Section 1.6.3, we assume that the *entire* flow is energy conserving. In this case, we conserve energy in the whole control volume $ACDF$ since there are clearly no energy fluxes in and out of its boundaries.

The rate of increase of energy $\dot{E}_G$ consists of the kinetic energy of the current and the disturbance and the increase of potential energy associated with the propagation of dense fluid to the right of $BE$, the initial lock location. This rate of energy gain is

$$\dot{E}_G = \frac{1}{2}\rho_L U^2 (U + U_r)h + \frac{1}{2}\rho_U {u_U}^2 (U + U_r)(H - h) + \frac{1}{2} g(\rho_L - \rho_U)Uh^2. \tag{1.73}$$

This energy is supplied by the loss of potential energy $\dot{E}_L$ as a result of the lowering of the interface to the left of the lock position *BE* at a rate

$$\dot{E}_L = \frac{1}{2} g(\rho_L - \rho_U) U_r (D^2 - h^2). \tag{1.74}$$

Assuming that energy is conserved inside *ACDF*, we equate (1.73) and (1.74) and obtain a further relation for the current speed

$$\frac{U^2}{gH} = \frac{(\rho_L - \rho_U)(D-h)(H-h)}{H(\rho_L(H-h) + \rho_U h)}. \tag{1.75}$$

The additional constraint of global energy conservation gives a unique value for the depth $h$ of the current. Comparing (1.72) and (1.75), we find that the only nontrivial case is

$$h = \frac{D}{2}. \tag{1.76}$$

Thus, an energy conserving gravity current produced by a partial-depth lock release has a depth that is half the initial lock depth before release. This result is consistent with Benjamin's result (1.57) for a full-depth release $D = H$. From (1.76), the speed of the current (nondimensionalized by the initial depth $H$) is given by

$$\frac{U^2}{gH} = \frac{(\rho_L - \rho_U) D (2H - D)}{2H(\rho_L(2H - D) + \rho_U D)}, \tag{1.77}$$

and the speed of the backward disturbance is equal to the current front speed

$$U_r = U. \tag{1.78}$$

Consequently, application of mass, momentum, and energy conservation yield a unique gravity current with properties determined by the initial densities and depths on the two sides of the lock. In general, the energy-conserving current does not occupy the half-depth of the channel, except in the case of a full-depth lock release.

Restricting attention to Boussinesq currents, $\gamma \approx 1$, (1.77) reduces to

$$\frac{U^2}{g'H} = \frac{D(2H - D)}{4H^2}. \tag{1.79}$$

In the limit of a full-depth lock $D = H$, (1.79) reduces to the Benjamin result $F_H = 1/2$ (1.58), so that both theories give the same result for that case. However, as has already been pointed out, for partial-depth locks, the results of the present theory differ significantly from Benjamin's result.

#### *1.6.4.4 Comparison with Experiments*

Shin et al. (2004) carried out experiments on high Reynolds numbers ($Re > 2{,}000$), Boussinesq partial–depth lock releases in a channel, and the measured speeds are given in terms of the Froude number based on the depth $D$ in the lock

$$F_D \equiv \frac{U}{\sqrt{g'D}}. \tag{1.80}$$

The depth of the current $h$ was measured at the position of the lock.

Figures 1.20 and 1.21 show the results of the current depths and Froude numbers as functions of the fractional depth of the lock $D/H$ compared with the theoretical values given by (1.76) and (1.77). The observed velocities and depths are within 10% of the theoretical values over the whole range of $D/H$.

### 1.6.4.5 Energy Transfers

As discussed in Section 1.6.4.1, Benjamin's theory agrees only with experiments for near half-depth, energy-conserving currents generated by full-depth lock releases, whereas the theory developed by Shin et al. (2004) agrees well with the observations over the full range of lock depths. This difference suggests that current and disturbance sides interact in a lock release, so Benjamin's analysis is not valid in general. The amount of interaction is difficult to quantify. Benjamin's (1968) model assumes that energy and momentum cannot be transferred between the two sides of the lock and does not consider the possibility of long waves entering the system. Waves can propagate along the interface between the denser and lighter fluid, and Shin et al. (2004) argue that these waves carry energy and momentum from the lock side to the current side. This transfer modifies the energy and momentum balance on the current side, so Benjamin's analysis is in general not appropriate.

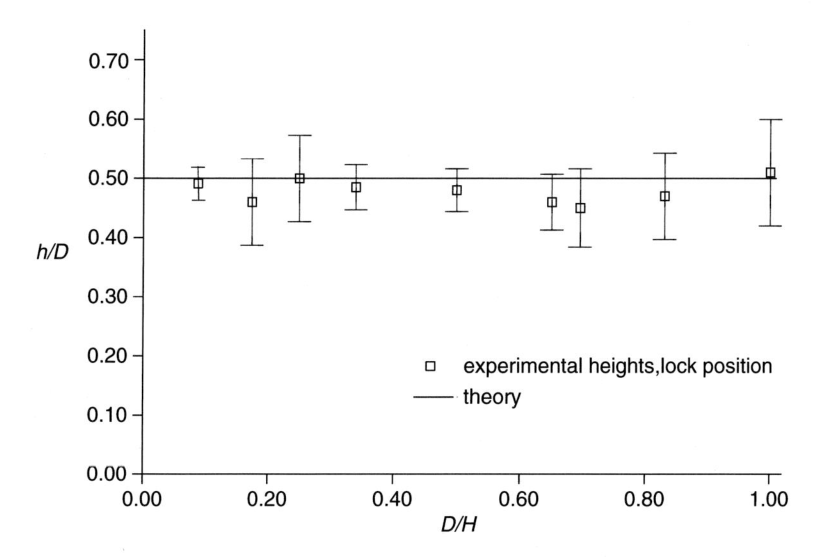


Figure 1.20. Comparison of measurements with the theoretical prediction (solid line) of the depth of the gravity current for partial-depth lock releases. Theory (1.76) predicts that $h/D = \frac{1}{2}$ for all lock depths. Taken from Shin et al. (2004).

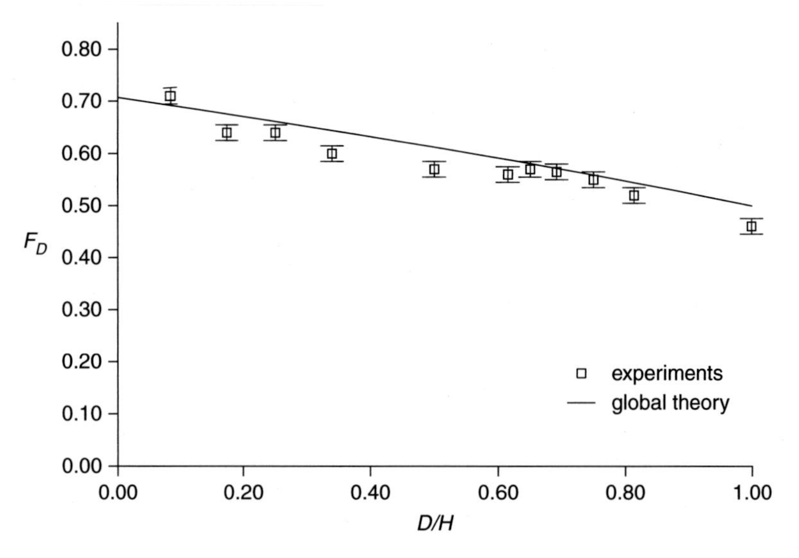


Figure 1.21. Gravity current speeds, expressed nondimensionally as Froude numbers based on the depth of the lock $D$, compared with the theoretical values (1.79) shown as the solid curve. Taken from Shin et al. (2004).

The speeds $c_\pm$ of long interfacial waves on an interface of depth $h$ between two layers traveling with speeds $u_U$ and $u_L$ is given by

$$c_\pm = \frac{u_U h + u_L(H-h)}{H} \pm \frac{1}{H}\sqrt{h(H-h)(g'H - (u_U - u_L)^2)}. \tag{1.81}$$

For the gravity current case, $u_L = U$ and $u_U = -U\frac{h}{H-h}$. Hence, in terms of the initial lock depth $D = 2h$, (1.81) is

$$\frac{c_\pm}{U} = \frac{2(H-D)}{2H-D} \pm \sqrt{\frac{2(H-D)}{2H-D}}, \tag{1.82}$$

and this relation is plotted in Figure 1.22.

The right traveling wave $c_+$ has speed $U$, when $\frac{2(H-D)}{2H-D} = \frac{3-\sqrt{5}}{2} = 0.382$, which occurs when $D = 0.76H$. For shallower locks, the current travels more slowly than long waves do, and energy and momentum may be transferred from the disturbance side to the current side. For locks with $D > 0.76H$, the current travels faster than all long waves. Thus, Benjamin's theory cannot apply for partial-depth releases with $D < 0.76H$, whereas for values of $D > 0.76H$ Benjamin's theory should be approximately correct. For example, taking a lock release of $D = 0.8H$, the two theories give values of the Froude number $F_H$ of the current as 0.24 and 0.27, respectively.

The physical reason for this dependence on the initial lock depth can be seen from (1.82), which shows that $\frac{c_+}{U}$ increases with decreasing $D/H$ and reaches a limit of 2 in the limit of very shallow locks $D \to 0$. This wave speed is the limit of the "dam break" flow derived using shallow water theory. This ability of the long interfacial

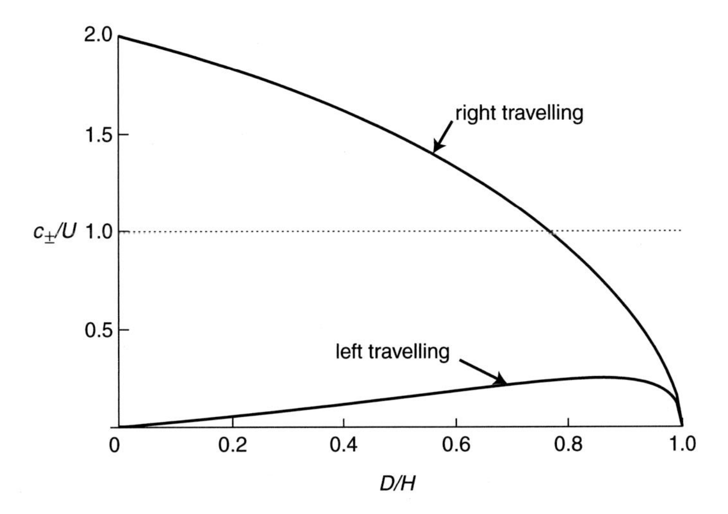


Figure 1.22. The speeds $\frac{c_\pm}{U}$ of waves on the top of the current plotted against the fractional depth $\frac{D}{H}$ of the release. Waves traveling to the left are always slower than the current, while waves traveling to the right are faster for $\frac{D}{H} < 0.76$.

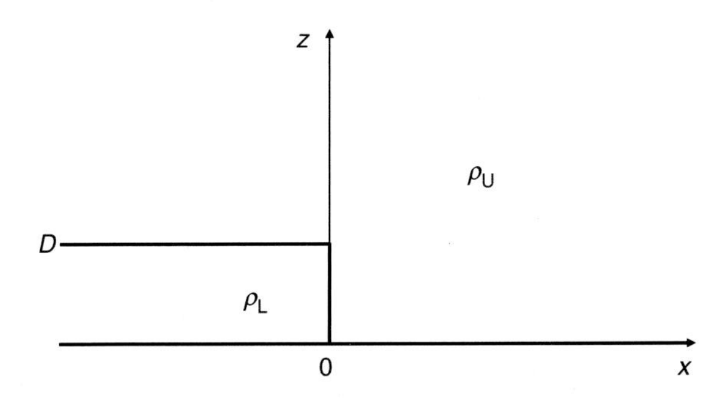


Figure 1.23. Definition sketch for one-layer initial conditions

waves to travel faster as the current becomes shallower results from two effects. First, the current itself travels more slowly as the depth decreases, and, second, the wave speed depends on the depths of both layers – long waves travel faster on deeper water – and so can propagate faster on the deeper upper layer in this limit.

Long waves travel in general much faster toward the right than toward the left: about 10 times faster for $D = 0.12H$ and about 35 times faster for $D = 0.05H$ (Figure 1.22). The asymmetry occurs because the velocity in the upper layer increases as the current becomes shallower in the laboratory frame. Interaction between disturbance and current sides is therefore expected to be strongest at lower fractional depths.

We now return to the case of a deep ambient fluid. In the limit $h/H \to 0$, the theory of Shin et al. (2004) predicts

$$F_h = 1, \tag{1.83}$$

(see (1.75) and (1.79)), which is well below the value ($\sqrt{2}$) found by von Kármán (1940) and Benjamin (1968).

## 1.7 Shallow Water Theory

In the limit of a one-layer flow momentum and mass conservation are

$$\frac{\partial u}{\partial t} + u\frac{\partial u}{\partial x} = -g'\frac{\partial h}{\partial x}, \tag{1.84}$$

$$\frac{\partial h}{\partial t} + \frac{\partial (uh)}{\partial x} = 0, \tag{1.85}$$

where $h(x,t)$ is the depth of the current and $u(x,t)$ is the horizontal velocity.

The initial conditions are

$$u(x,0) = 0 \quad -\infty < x < \infty, \tag{1.86}$$

and

$$h(x,t=0) = \begin{cases} D & \text{for } x < 0, \\ 0 & \text{for } x > 0. \end{cases} \tag{1.87}$$

Two further conditions are needed to specify the problem: the condition at the rear of the flow

$$u = 0, \text{when } h = D \tag{1.88}$$

and the front condition that is specified as a Froude number $F$

$$U = F\sqrt{g'h}. \tag{1.89}$$

The results of using the Benjamin front condition and the energy-conserving condition are shown in Figure 1.24. The experimental data fit much closer to the energy-conserving prediction, especially at low values of the fractional depth $D/H$ where the differences from the Benjamin solution are largest.

### *1.7.1 Similarity Solution*

The flow has five governing parameters $x$, $t$, $B$, $L_0$ and $F$. As there are two independent dimensions, $L$ and $T$, there are three dimensionless parameters

$$\eta = \frac{x}{B^{\frac{1}{3}}t^{\frac{2}{3}}}, \quad \tau = \frac{tB^{\frac{1}{2}}}{L_0^{\frac{3}{2}}}, \quad F.$$

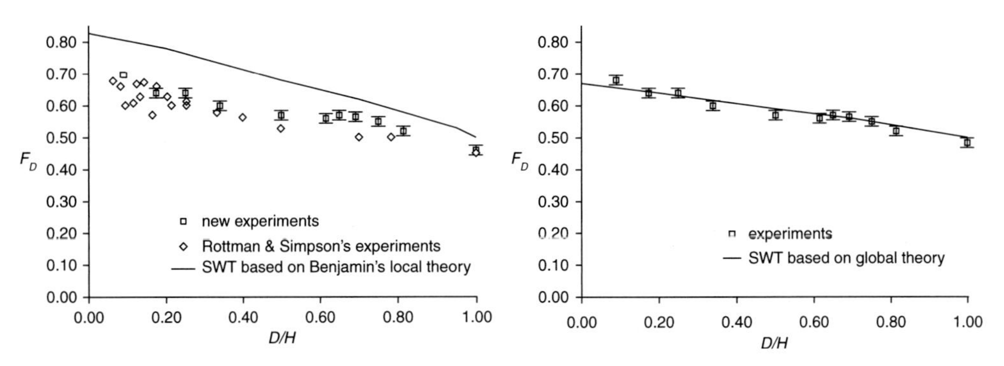


Figure 1.24. Froude numbers based on the depth of the lock $D$, compared with shallow water theory (solid curve) using a front condition based Benjamin's Froude number (left) and the energy-conserving theory (right).

Dimensional analysis implies

$$u = \frac{x}{t} U(\eta, F, \tau), \tag{1.90}$$

$$g'h = \frac{x^2}{t^2} H(\eta, F, \tau), \tag{1.91}$$

$$x_f = x X(\eta, F, \tau). \tag{1.92}$$

At later times, seek solution independent of $L_0$ (i.e, independent of $\tau$). Substituting into (1.85),

$$\frac{U'}{U - \frac{2}{3}} = -\frac{H'}{H} - \frac{3}{\eta}, \tag{1.93}$$

which integrates to

$$U - \frac{2}{3} = \frac{c}{H\eta^3}. \tag{1.94}$$

Since the buoyancy $g'h$ remains finite as $x, \eta \to 0$, $c = 0$ and $U = \frac{2}{3}$. Substituting into (1.84) gives

$$\eta H' + 2H = U - U^2 = \frac{2}{9}. \tag{1.95}$$

Hence,

$$H = \frac{1}{9} + \frac{K}{\eta^2}. \tag{1.96}$$

The Froude number front condition can be written as

$$\dot{x_f}^2 = F^2 \left( \frac{1}{9} + \frac{K}{\eta_f{}^2} \right), \tag{1.97}$$

and conservation of initial buoyancy

$$x_f{}^3 \left( \frac{1}{27} + \frac{K}{\eta_f{}^2} \right) = Bt^2. \tag{1.98}$$

Eliminating $K$ gives the governing equation

$$\dot{x_f}^2 x_f = 2F^2 \frac{1}{27}\frac{x_f^3}{t^2} + F^2 B. \tag{1.99}$$

By inspection this has the solution

$$x_f = \Gamma t^{\frac{2}{3}}, \tag{1.100}$$

where

$$\Gamma = \left(\frac{27F^2 B}{12-2F^2}\right)^{\frac{1}{3}}. \tag{1.101}$$

Dimensionally,

$$u = \frac{2}{3}\frac{x}{t}, \tag{1.102}$$

$$g'h = \frac{1}{t^2}\left[\frac{4}{9F^2}x_f^2 + \frac{1}{9}(x^2 - x_f^2)\right], \tag{1.103}$$

$$x_f = \left(\frac{27F_f^2}{12-2F^2}\right)^{1/3} B^{1/3} t^{2/3}, \tag{1.104}$$

$$h_f = \frac{12}{12-2F^2}\frac{V}{x_f}, \tag{1.105}$$

$$\frac{h}{h_f} = 1 - \frac{1}{4}F^2\left(1 - \left(\frac{x}{x_f}\right)^2\right). \tag{1.106}$$

*1.7.1.1 Comparison with Experiment*

Figures 1.25 and 1.27 show profiles of current height for full-depth and partial-depth lock releases, respectively. They show that initially the current front travels with constant speed, but decelerates at later times as the current enters the similarity phase. Profiles for the similarity phase are compared with the theoretical shallow water profiles (1.106) in Figures 1.26 and 1.28 for three values of the front Froude number 1.0, 1.2 and 1.4.

There is reasonable agreement between the theoretical curves and the laboratory measurements – certainly within the scatter of the latter.

From these data, three Froude numbers are calculated: $F_D$ based on the initial lock depth, $F_h$ based on the maximum height of the head, and $F_r$ based on the height of the current at the rear of the head as indicated in Figures 1.26 and 1.28. The results are shown in Figures 1.29 and 1.30 for the full-depth and the partial-depth releases, respectively.

The Froude number $F_D$ based on the initial lock depth is approximately constant during the constant-velocity phase ($x \preceq 10L_0$) as expected. The value $F_D = 0.45$ for the initial phase is consistent with previous measurements (see Shin et al. 2004). After

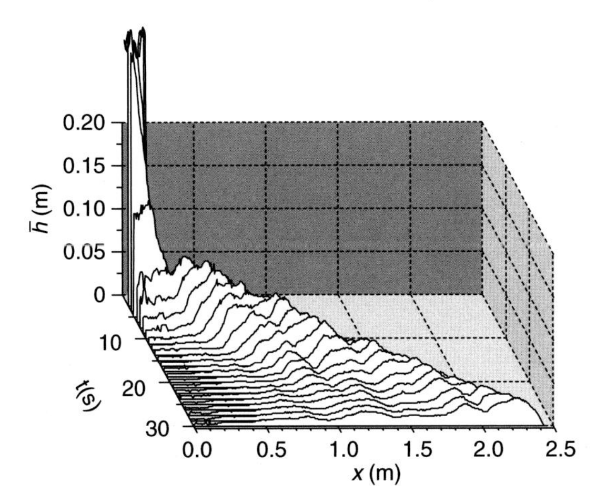


Figure 1.25. A time sequence of equivalent height profiles for a full-depth release $D/H = 1$ (Marino et al. 2005).

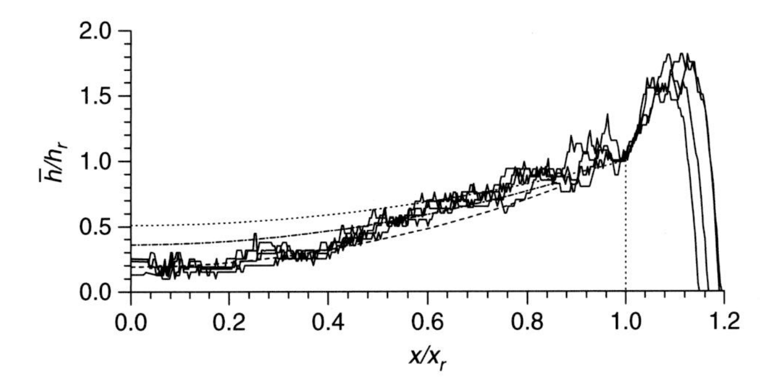


Figure 1.26. Profiles for a full-depth release in the similarity phase with $t/t_c = 21.35$, 23.60, 25.86, 28.11 and 31.11 (thick solid line). The profiles depicted by dotted, dash-dotted, and dashed lines correspond to similarity solutions to the shallow water equations with $F_f$ = 1.0, 1.2, and 1.4, respectively (Marino et al. 2005).

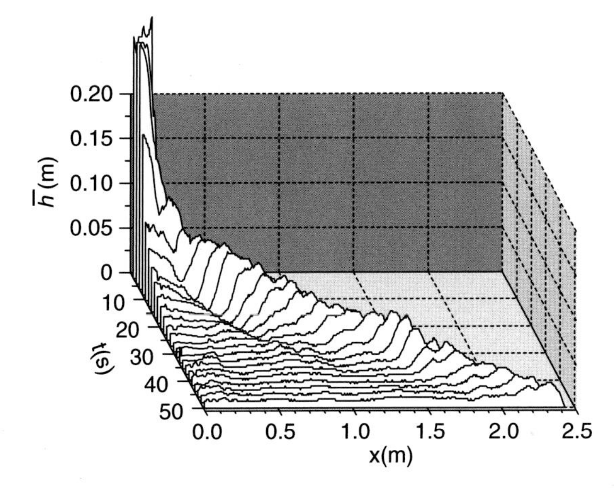


Figure 1.27. A time sequence of equivalent height profiles for a partial-depth release $D/H = 0.675$. (Marino et al. 2005).

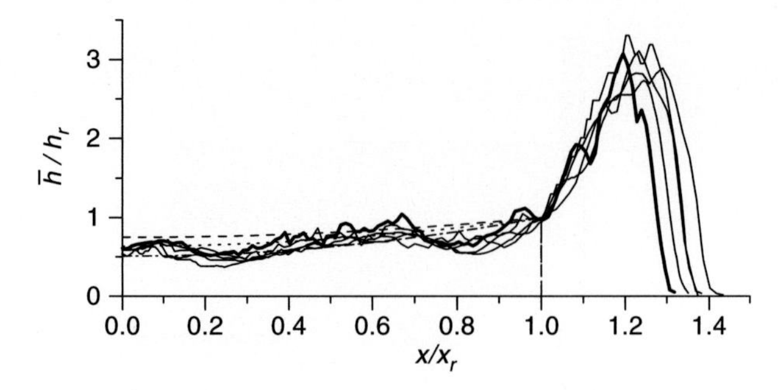


Figure 1.28. Profiles for a partial-depth release in the similarity phase with $t/t_c =$ 21.35, 23.60, 25.86, 28.11 and 31.11 (thick solid line). The profiles depicted by dotted, dash-dotted, and dashed lines correspond to similarity solutions to the shallow water equations with $F_f$ = 1.0, 1.2, and 1.4, respectively (Marino et al. 2005).

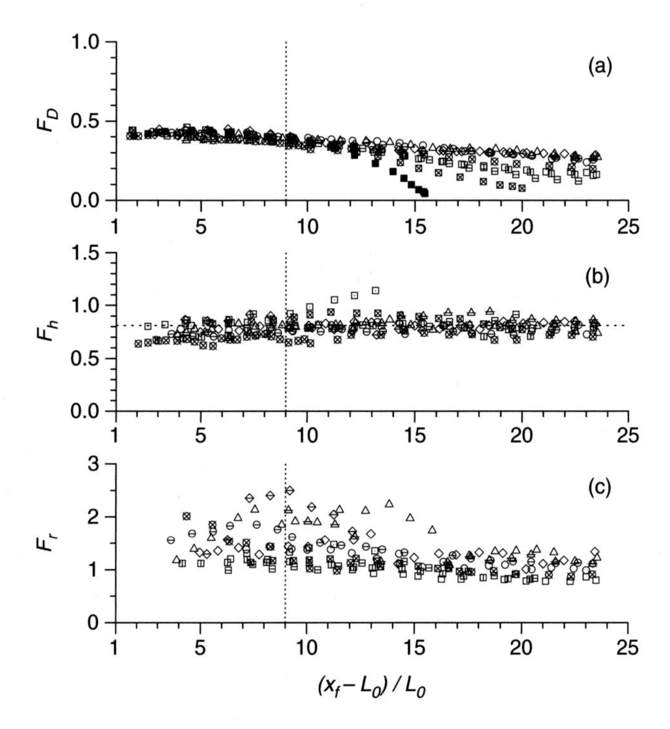


Figure 1.29. Evolution of the Froude number for the full-depth releases ($D/H = 1$). The horizontal line in (b) is the average value obtained in the similarity regime (Marino et al. 2005).

about 10 lock-lengths the similarity phase begins, and the value of $F_D$ decreases. During both the constant-velocity and the similarity phase, a reduced scatter in the values is observed. In particular, for $x_f \succeq 10L_0$, $F_h$ is approximately constant with a mean value $< F_h >= 0.81 \pm 0.10$. This measure of the Froude number collapses

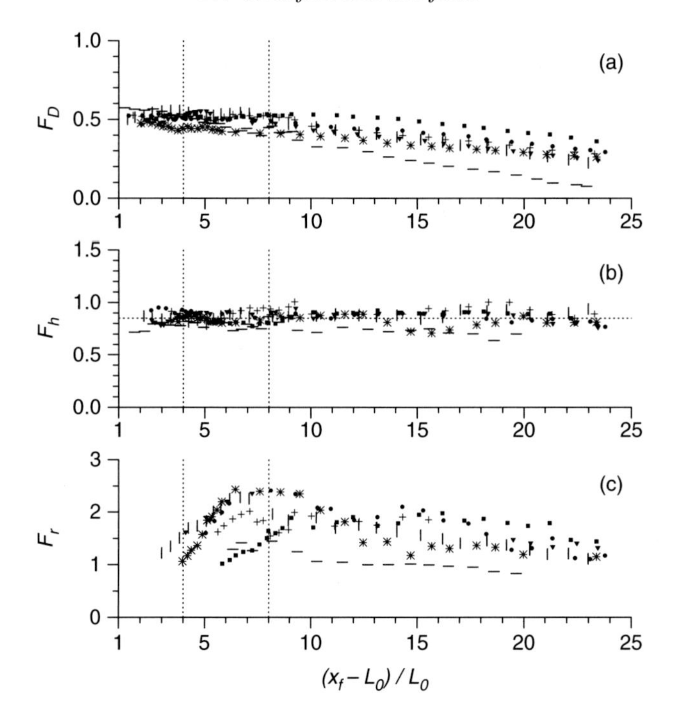


Figure 1.30. Evolution of the Froude number for the partial-depth releases ($D/H < 1$). The horizontal line in (b) is the average value obtained in the similarity regime (Marino et al. 2005).

the data at large $x_f$, in contrast with $F_D$, which applies only for the constant-velocity phase.

The Froude number $F_r$ based on the buoyancy at the rear of the head shows considerable variation among experiments and tends to decrease with the distance from the lock.

Thus, it seems that during the constant-velocity phase, when the initial conditions continue to influence the flow, the Froude number $F_D$ based on the initial lock depth provides the best estimate of the speed and that the value is consistent with the energy-conserving theory (1.79). Once the similarity phase is reached, the Froude number $F_h$ based on the maximum height of the head provides the better estimate of the speed, despite the fact that this region is in the nonhydrostatic part of the current, where shallow water theory is invalid.

## 1.8 Stratified Ambient Fluid

It seems that the use of a single parameter, the Froude number, to describe the complexities of these turbulent gravity currents provides a reasonable description of the current speed, even though, as discussed earlier, the identification of the Froude number with the appropriate theory is somewhat unsatisfactory. In concluding this chapter, it is worth exploring whether these ideas extend to a stratified ambient fluid.

Figure 1.31 shows two gravity currents traveling along the base of a stratified fluid with constant buoyancy frequency $N$. The supercritical current on the left travels faster than maximum speed linear internal waves $NH/\pi$, whereas the one on the right is subcritical to a range of internal wave speeds and so these waves propagate upstream ahead of the current.

These currents are initially observed to travel at constant speed after their release from a lock (Maxworthy et al. 2002). Here we explore the simple extension of the results in a uniform environment, *ignoring the effects of internal waves*.

Consider the case where the ambient fluid is stably stratified with constant buoyancy frequency $N$ defined by

$$N^2 = -\frac{g}{\rho_0}\frac{\mathrm{d}\rho}{\mathrm{d}z} = g\frac{\rho_B - \rho_T}{\rho_0 H}, \tag{1.107}$$

for a constant density gradient from a density $\rho_B$ at the bottom to $\rho_T$ at the top (Figure 1.32). In a Boussinesq fluid, we may take $\rho_0 = \rho_T$.

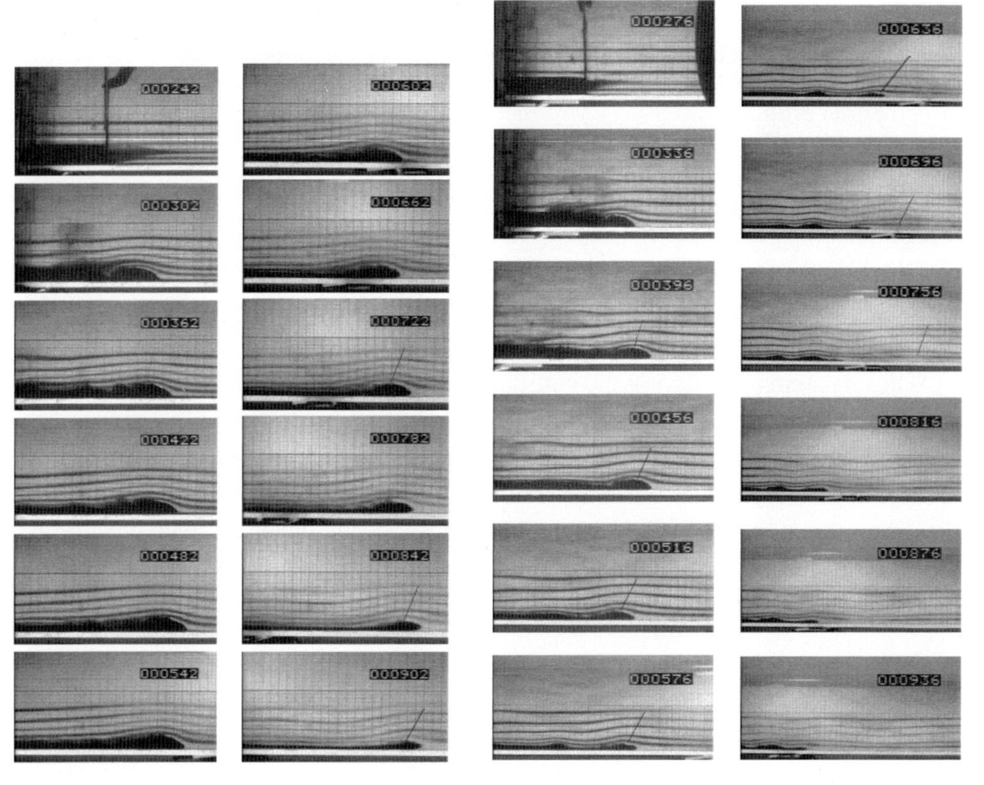


Figure 1.31. Gravity currents in a uniformly stratified fluid: left, a supercritical current and, right, a subcritical current (Maxworthy et al., 2002).

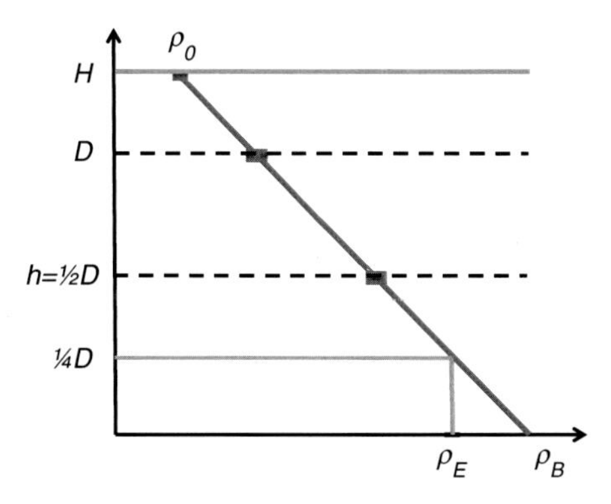


Figure 1.32. The equivalent density $\rho_E$ for a half-depth gravity current.

As pointed out by Ungarish and Huppert (2002), since the pressure is hydrostatic ahead of the current and the density gradient is constant, the pressure difference and the surrounding ambient fluid are determined by an "effective density" $\rho_E$ corresponding to an integral of the density over the depth of the current and given by

$$\rho_E = \rho(z = h/2). \tag{1.108}$$

For an energy-conserving current $h = D/2$ and so, from (1.107),

$$\rho_E = \rho(D/4) = \rho_B - \frac{\rho_0}{4g} N^2 D. \tag{1.109}$$

Consequently, denoting the density of the current by $\rho_C$, the effective reduced gravity $g'_E \equiv g\frac{\rho_C - \rho_E}{\rho_0}$ of the current is

$$g'_E = g'_N + \frac{1}{4} N^2 D, \tag{1.110}$$

where

$$g'_N = g\frac{\rho_C - \rho_B}{\rho_0} \tag{1.111}$$

is the reduced gravity of the current relative to the density at the base of the stratification.

We now wish to see whether the result for an unstratified environment (1.79) carries over to the case of a uniformly stratified ambient with buoyancy frequency $N$. The equivalent result is obtained by replacing $g'_C$ for a uniform environment by $g'_E$, defined by (1.110), in (1.79), in which case

$$F_D \equiv \frac{U}{\sqrt{g'_E D}} = \frac{1}{2}\sqrt{(2 - D/H)}. \tag{1.112}$$

Substituting (1.110) into (1.112) gives the Froude number in a stratified environment for all values of the lock height

$$F_N \equiv \frac{U}{\sqrt{(g'_N D + \frac{1}{4}N^2 D^2)(2 - D/H)}} = \frac{1}{2}. \tag{1.113}$$

In section 1.8.2, we will determine $F_N$ from laboratory experiments and numerical calculations and compare the results with the predicted value of 0.5.

Ungarish and Huppert (2002) define a stratification parameter

$$S \equiv \frac{\rho_B - \rho_0}{\rho_C - \rho_0}. \tag{1.114}$$

Since a gravity current, in contrast to an intrusion, requires $\rho_C \geq \rho_B$, $0 < S \leq 1$, with $S = 0$ corresponding to an unstratified ambient fluid and $S = 1$ corresponding to a current that has a density equal to the maximum environmental density. In terms of $S$,

$$g'_E = g'_C \left(1 - S + \frac{SD}{4H}\right), \tag{1.115}$$

where

$$g'_C = g\frac{\rho_C - \rho_0}{\rho_0}. \tag{1.116}$$

Hence,

$$F_N = \frac{U}{\sqrt{g'_C D(1 - S + SD/4H)(2 - D/H)}}. \tag{1.117}$$

### 1.8.1 Criticality

Gravity currents in a stratified ambient fluid can be either subcritical or supercritical. In the latter case, the speed of the current must exceed the speed of the fastest internal wave that can be supported by the stratification. For a constant $N$ fluid of depth $H$, the fastest linear wave is a vertical mode-1 long wave that has speed $c = NH/\pi$.

Using the theoretical prediction (1.113) for the speed of the current, the condition for a critical current is

$$\frac{D}{H}\left(2 - \frac{D}{H}\right)\left(\frac{4(1 - S_c)}{S_c} + \frac{D}{H}\right) = \frac{16}{\pi^2}. \tag{1.118}$$

Solution of (1.118) gives the critical stratification $S_c$ as a function of the lock aspect ratio $D/H$ (Figure 1.33). For a full-depth release, $S_c = 0.847$, and $S_c$ decreases with aspect ratio. Currents are predicted to be supercritical when $S < S_c$ and subcritical when $S > S_c$. For a fixed stratification $S < S_c$, deep locks produce supercritical currents, whereas all releases are subcritical for $S > 0.847$.

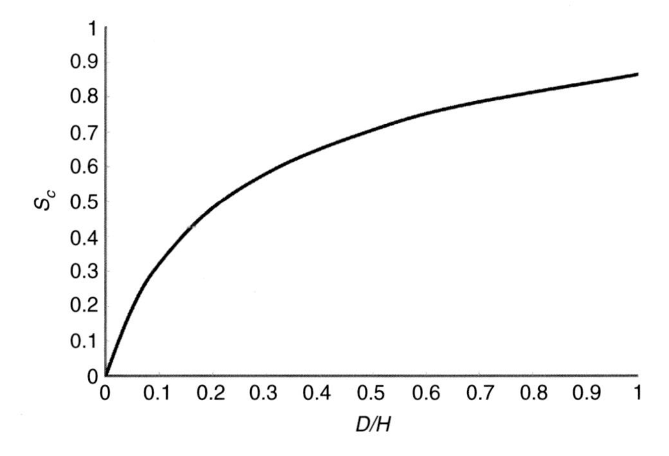


Figure 1.33. The critical stratification $S_c$ as a function of the lock-depth $D/H$ as given by (1.118).

### *1.8.2 Comparison with Data for Stratified Ambient Fluids*

#### *1.8.2.1 Current Speed*

Data on gravity currents in stratified ambient fluids are limited. Maxworthy et al. (2002) conducted a series of 36 experiments creating gravity currents in a stratified fluid with constant $N$ in a channel. The experiments covered lock depths $0.33 \le D/H \le 1$ and stratification strengths $1.35 \le N \le 2.1\ \mathrm{s}^{-1}$. In all cases, $\rho_C \ge \rho_B$ so the currents traveled along the lower boundary. They also carried out two-dimensional direct numerical simulations over the same ranges of lock depths and stratification parameters $S$. These experiments and simulations covered both subcritical and supercritical currents and both gave very similar results for the current speed. No information on the height of the currents was given in the paper. Birman et al. (2007) present four direct numerical simulations, again two-dimensional, for $S = 0.2$ and $D/H = 0.2$, 0.5 and 1.0, and for $S = 0.9$ and $D/H = 0.2$. White and Helfrich (2008) also present theoretical analysis and numerical simulations for stratified ambients, with the main emphasis on non-linear stratifications. There are four simulations for constant $N$, with $S = 0.91$ and $D/H = 0.5$, and $S = 1$ and $D/H = 0.4$, 0.7 and 0.9. Both the latter papers give speeds in the constant-velocity phase and some information on the current depths.

Figure 1.34 shows the Froude number $F_N$, defined by (1.113) as a function of the stratification parameter $S$ for the experiments and numerical simulations described previously. The data have been divided into subcritical (open symbols) and supercritical (closed symbols) currents, respectively, based on the ratio $\pi U/NH$ of the current speed to the mode-1 long-wave speed. The Froude numbers for the supercritical currents collapse over the full applicable range of $S$ for all lock depths $D/H$. They also fall on the theoretical value of 0.5 within the scatter of the data. The collapse is not as

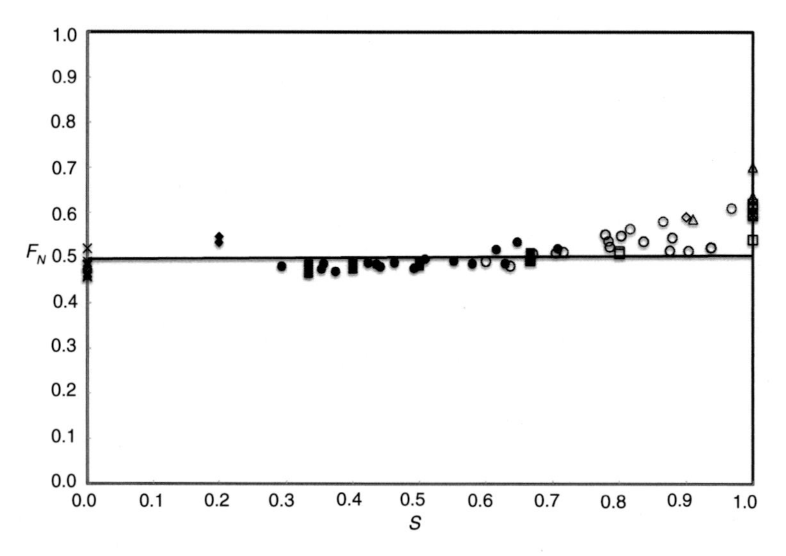


Figure 1.34. Froude numbers $F_N$ plotted against the stratification parameter $S$. ∘, Maxworthy et al. (2002); experiments: □, Maxworthy et al. (2002); simulations: ⋄, Birman et al. (2007): △, White and Helfrich (2008): ×, Shin et al. (2004). Subcritical currents, open symbols; supercritical currents, filled symbols.

tight for subcritical currents at the higher values of $S$, and there is more variation in the speeds and an increase of $F_N$ as $S \to 1$.

### *1.8.3 Current Depth*

As for the case of unstratified environments, measurements of the current depth are sparse. Figure 1.35 shows values of the height of the current $h/D$ extracted from Maxworthy et al. (2002), Birman et al. (2007) and White and Helfrich (2008). The data are divided into subcritical and supercritical currents and cover initial lock depths in the range $0.33 \leq D/H \leq 1$. Except for three cases in Birman et al. (2007) shown in their figure 4, I have estimated the height visually from published figures in these papers.

The data at $S = 0.2$ are from Birman et al. (2007). The two lower values are from a full-depth simulation. The smallest value ($h/D = 0.36$) is shown in their figure 4 and is derived from a density contour that is 'the average of the lock fluid and the ambient fluid at the bottom wall'. The value immediately above ($h/D = 0.47$) is my estimate from the top of the current shown in their figure 2. It is not clear why these values differ so much. The other two values are for partial-depth releases $D/H = 0.2$ and $0.5$. The other two supercritical cases are one each from an experiment and simulation by Maxworthy et al. (2002). Apart from the small value for $S = 0.2$, which differs from what I estimate from the visualization, the values are similar to, if somewhat above,

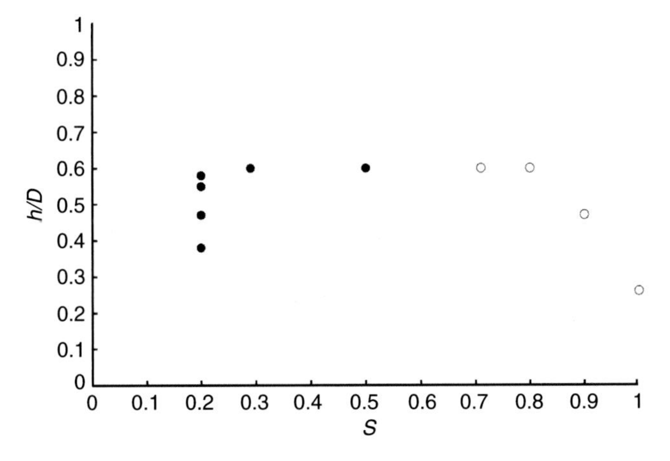


Figure 1.35. Dimensionless current heights $h/D$ plotted against the stratification parameter $S$. Subcritical currents, open symbols; supercritical currents, filled symbols.

the energy conserving value $h/D = 0.5$. This is true for the two subcritical currents for $S = 0.7$ and 0.8 from Maxworthy et al. (2002).

At larger values of $S$ the current height decreases, but the distortions of the top of the current by the internal waves make these estimates difficult, and in most of the cases in White and Helfrich (2008) impossible. A similar reduction in height with increasing $S$ was also found in the simulations in Birman et al. (2007) (their figure 12).

## 1.9 Summary and Conclusions

This chapter has reviewed theoretical and experimental studies of gravity currents. In an unstratified fluid, gravity current speeds and depths are quite well described by energy-conserving theories. Typically speeds are 5–10% lower, and depths, which are much less accurately measured, show similar deviations from the theoretical values. These discrepancies are probably a result of neglected dissipation and mixing and suggest that energy-conserving theories provide an order-one description of the flows. The attraction of this description is that it allows the currents to be described by a single dimensionless parameter.

It is remarkable that the description of the gravity current speed can be captured in a single dimensionless parameter, the Froude number. The value of the Froude number can be derived theoretically when the current travels at constant speed, using a steady hydraulic theory with a control volume that moves with the front of the current. This approach works well for the initial motion of currents generated by the release of a finite volume of fluid from a lock. At later times, the finite volume of the lock means that the currents decelerate, and this motion is captured by a similarity solution to the shallow water equations. This similarity solution requires the specification of a

Froude number at the front and experiments show that a Froude number based on the maximum height of the head gives the best agreement with observations.

In a stratified ambient, a whole new physical process is present – the ability of the current to generate internal gravity waves. Depending on the speed of the current, these waves can travel both upstream and downstream and can carry energy and momentum away from the current. Since the driving pressure difference between the current and the ambient fluid is a result of the difference in hydrostatic pressure at the base of the flow, we note that, for a linear vertical density gradient, the equivalent external density is the average of the ambient density over the height of the current. The results show that over a wide range of initial lock depths and stratification strengths, supercritical currents are well described by this modified theory. Thus, it appears that internal waves that are generated downstream (behind) the front of the current do not change the speed or height significantly.

Subcritical currents travel faster than predicted and tend to be shallower, although information on current heights is quite uncertain. It is not clear why the presence of internal waves ahead of the current cause an increase in speed, although this effect has been observed for intrusions in both a two-layer and a continuously stratified fluid (Cheong et al. 2006); (Flynn and Linden 2006); (Bolster et al. 2008). Whereas the currents travel more slowly than the speed of low-mode long waves, waves with shorter wavelengths travel more slowly and so can couple with the current as shown in White and Helfrich (2008). However, it seems that this interaction is quite complex and needs further study.

## Acknowledgments

This chapter was written to supplement lectures given at the AOSTA Alpine Summer School on "Buoyancy-driven Flows" held 21–30 June 2010. I am grateful to the organizers Claudia Cenedese, Eric Chassignet, and Jacques Verron for inviting me to lecture at this school and for their superb organization. I would like to thank the other lecturers and students for asking searching questions and for interesting discussions. I would like to thank Andrea Kuesters, Valentina Lombardi, Helena Nogueira, Charlotte Persson-Gulda, and Cristian Rendina for providing many helpful comments on an early draft of this chapter.

## References

Benjamin, T. B. 1968: Gravity currents and related phenomena. *J. Fluid Mech.* **31**, 209–248.

Birman, V. K., Meiburg, E., and Ungarish, M. 2007: On gravity currents in stratified ambients. *Phys. Fluids* **19** (086602).

Bolster, D. T., Hang, A. and Linden, P. F. 2008: The front speed of intrusions into a continuously stratified medium. *J. Fluid Mech.* **594**, 369–377.

Cheong, H. B., Kuenen, J. J. P., and Linden, P. F. 2006: The front speed of intrusive gravity currents. *J. Fluid Mech.* **552**, 1–11.

Flynn, M. R., and Linden, P. F. 2006: Intrusive gravity currents. *J. Fluid Mech.* **568**, 193–202.

Gardner, G. C., and Crow, I. G. 1970: The motion of large bubbles in horizontal channels. *J. Fluid Mech.* **43**, 247–255.

von Kármán, T. 1940: The engineer grapples with nonlinear problems. *Bull. Am. Math. Soc.,* **46**, 615–683.

Lowe, R. J., Rottman, J. W., and Linden, P. F. 2005: The non-Boussinesq lock-exchange problem. Part 1. Theory and experiments. *J. Fluid Mech.* **537**, 101–124.

Marino, B. M., Thomas, L. P., and Linden, P. F. 2005: The front condition for gravity currents. *J. Fluid Mech.* **536**, 49–78.

Maxworthy, T., Leilich, J., Simpson, J. E., & Meiburg, E. H. 2002: The propagation of a gravity current into a linearly stratified fluid. *J. Fluid Mech.* **453**, 371–394.

Rottman, J. W., and Simpson, J. E. 1983: Gravity currents produced by instantaneous release of a heavy fluid in a rectangular channel. *J. Fluid Mech.* **135**, 95–110.

Shin, J. O., Dalziel, S. B., and Linden, P. F. 2004: Gravity currents produced by lock exchange. *J. Fluid Mech.* **521**, 1–34.

Simpson, J. E. 1972: Effects of lower boundary on the head of a gravity current. *J. Fluid Mech.* **53**, 759–768.

Simpson, J. E. 1997: *Gravity Currents*, 2nd ed. Cambridge, Cambridge University Press.

Simpson, J. E., and Linden, P. F. 1989: Frontogenesis in a fluid with horizontal density gradients. *J. Fluid Mech. Digital Archive* **202**, 1–16.

Ungarish, M., and Huppert, H. E. 2002: On gravity currents propagating at the base of a stratified ambient. *J. Fluid Mech.* **458**, 283–301.

White, B. L., and Helfrich, K. R. 2008: Gravity currents and internal waves in a stratified fluid. *J. Fluid Mech.* **616**, 327–356.

Yih, C. S. 1965: *Dynamics of Nonhomogeneous Fluids*. New York, Macmillan.

# 2

# Theory of Oceanic Buoyancy-Driven Flows

JOSEPH PEDLOSKY

## 2.1 General Considerations and a Laboratory Example

### *2.1.1 Introduction*

Both the atmosphere and ocean are rotating and stratified, and for large-scale motions (an attribute that needs careful definition), both are important in shaping the dynamics. Both the mean circulation, whose scales usually reflect the forcing, and the inevitable eddy fields that result from the instability of those motions are determined by the effects of rotation and stratification. In this chapter, I am going to examine some particular aspects of the buoyancy-driven motion of a rotating stratified fluid with an eye to oceanic phenomena, in particular the important question of the relationship between the vertical motion and the surface buoyancy forcing. This question is of particular importance in the discussion of the oceans' role in climate since a key issue in delineating that role is how and *where* the sinking of surface-cooled water takes place. It should not be surprising that the presence of rotation and, in particular, the variation of the local normal component of that rotation (the *beta-effect*) renders the association of forcing and sinking sometimes nonintuitive. In sections 2.2 and 2.3, we will examine the nature of the circulation in simple models with an eye to uncovering in easily understandable circumstances the underlying physics that determines the structure of the vertical motion. Of course, the nature of the horizontal motion is coupled to the vertical motion and will be discussed as well.

In Section 2.1, we discuss the relatively straightforward motion of buoyancy-forced flows in "flat" geometries (i.e., ignoring the Earth's sphericity and the beta effect). This has the advantage, in addition to relative simplicity, of allowing laboratory examination of the theoretical results. In Section 2.2, we take up a more oceanographically relevant model including the beta-effect and show how that effect has strong consequences for the vertical regions of sinking. We also consider nonlinear effects.

Section 2.3 is reserved for a detailed study of certain boundary layer phenomena in a *weakly* stratified fluid as a very simple model of a polar mixed layer that shows the unexpectedly complex structure of the sinking motion near a coastal boundary.

The general philosophy of this chapter can be described as a belief, validated by extensive experience, that the patient and complete investigation of idealized model problems forms a sturdy foundation for the understanding of the more complex, natural phenomena that are often accessible to only nearly equally complex numerical models. Without the foundation built upon the understanding of simpler models, the more realistic problems would remain mysterious and beyond our comprehension.

Before we plunge into a detailed discussion of certain exemplary phenomena, it is useful to consider certain general issues related to both rotation and stratification. When we talk about buoyancy-forced motions, we are, of course, dealing with a stratified fluid. The degree of stratification may be independent of the amplitude of the buoyancy forcing, or it may be determined by the forcing. The first category lends itself to theories that are linearized about some preexisting stratification; the second is generally intrinsically nonlinear.

One way of approaching the subject is to recognize that both rotation and stratification impose certain *constraints* on the fluid motion. I am going to discuss these a bit now without detailed proof; the development of the lectures will certainly illustrate what I say now in case you have not seen these ideas before (which I think is unlikely). So, consider this discussion a reminder of what you probably already know.

For fluids that are nonstratified (i.e., of constant density), incompressible, inviscid, and rapidly rotating (we need to be more specific about what *rapidly rotating* actually means), the Taylor-Proudman theorem tells us that the velocity is independent of the direction of the rotation axis of the fluid, for example,

$$(\vec{\Omega} \cdot \nabla)\vec{u} = 0. \tag{2.1.1.1}$$

One way to interpret this constraint is in terms of the simple inertial waves in the system whose frequencies are

$$\omega_{inertial} = 2\vec{\Omega} \cdot \vec{K}/K, \tag{2.1.1.2}$$

so that the frequency depends on the projection of the wave vector ($K$ is its magnitude) along the rotation axis. For low-frequency motions, this must be small (or zero in the limit of steady motions) so the wave motion must be independent of that direction to make the dot product $\vec{\Omega} \cdot \vec{K} = 0$. We say that rotation adds a *stiffness* to the vertical direction along which information passes and which is the basis of the Taylor-Proudman theorem.

Now consider a *nonrotating* stratified fluid. If we can ignore for the moment the effects of diffusion, then the equation for conservation of density (essentially an energy conservation statement for an incompressible fluid) becomes

$$\frac{\partial \rho}{\partial t} + u\frac{\partial \rho}{\partial x} + v\frac{\partial \rho}{\partial y} + w\frac{\partial \rho}{\partial z} = 0. \tag{2.1.1.3}$$

If the density gradient antiparallel, in the resting state, to gravity is $-\frac{\partial \rho_s}{\partial z} > 0$ and is large enough (again, needs a precise definition), then the vertical velocity, $w$, will be suppressed, that is,

$$w\frac{\partial \rho}{\partial z} = 0, \quad \Rightarrow w = 0. \tag{2.1.1.4}$$

When coupled with the condition for incompressibility, this yields

$$\frac{\partial u}{\partial x} + \frac{\partial v}{\partial y} = 0. \tag{2.1.1.5}$$

If the motion were independent of one horizontal dimension, say the $y$-direction, we would immediately have the condition, from (2.1.1.5),

$$\frac{\partial u}{\partial x} = 0, \tag{2.1.1.6}$$

so that flow in the direction perpendicular to direction of gravity is independent of that direction, so-called *blocked flow*, in analogy with the Taylor-Proudman theorem. Again, thinking in terms of linear waves is helpful. The presence of gravity and the vertical density gradient supports waves that propagate information with frequency

$$\omega_{gravity} = N\frac{k}{\sqrt{k^2+m^2}}, \tag{2.1.1.7}$$

where $m$ is the wavenumber in the direction of gravity and $k$ is the horizontal wavenumber and where

$$N \equiv \left(-g\frac{1}{\rho_s}\frac{\partial \rho_s}{\partial z}\right)^{1/2}. \tag{2.1.1.8}$$

$N$ is called the buoyancy frequency. Note that for frequencies small with respect to $N$, the horizontal wavenumber $k$ must go to zero (i.e., the motion becomes a slug flow satisfying (2.1.1.6)). Loosely speaking, we can say that rotation tries to make the motion uniform in the vertical direction, while stratification, supporting waves that propagate horizontally, does the same in the perpendicular direction. The two constraints work at cross purposes, and a good deal of the dynamics of rotating, stratified fluids depends on the dominance or balance of these two fundamental constraints.

In geophysical systems like the ocean, the vertical scale of the motion, $D$, is usually small compared to a characteristic horizontal scale, $L$ so that $D/L \ll 1$. We shall find that the relative strength of rotation to stratification is measured by the parameter

$$S = \frac{N^2 D^2}{f^2 L^2}, \tag{2.1.1.9}$$

called the Burger number. In (2.1.1.9) $f = 2\Omega$. It is left for the student as an exercise to show that this is related to the ratio $\omega_{\text{gravity}}/\omega_{\text{inertial}}$.

In many situations, either the vertical scale or the horizontal scale may be given or known a priori. If $D$ is a given quantity, we can define a horizontal length scale on which the effects of rotation and stratification are of the same order. This scale is the *Rossby radius of deformation* (or more precisely the *internal* Rossby deformation radius – or simply, the deformation radius) and is defined as

$$L_d = \frac{ND}{f}. \tag{2.1.1.10}$$

On this scale the Burger number is unity, and rotation and stratification are of equal importance.

For many geophysical systems the central dynamical entity is the *potential vorticity*. For a rapidly rotating fluid, it can be shown that to a good approximation the potential vorticity for an incompressible fluid is

$$q = (f + \zeta)\frac{\partial \rho}{\partial z}, \tag{2.1.1.11}$$

where $\zeta$ is the vertical component of the vorticity due to the motion and $\rho$ is the density. If the density consists of the background density $\rho_S$ plus a small perturbation $\rho'$ and if $\zeta$ is much smaller than $f$ (these conditions are often met), then to the first approximation,

$$q = f\frac{\partial \rho_s}{\partial z} + \left(f\frac{\partial \rho'}{\partial z} + \zeta\frac{\partial \rho_s}{\partial z}\right) + \cdots. \tag{2.1.1.12}$$

The first term on the right-hand side is by far the largest, but it is, in the simplest case of constant $f$ and constant $N$, a simple constant and so its variation due to the motion is zero and is hence dynamically empty of significance. The dynamically interesting term is the second bracket. The second term in the bracket is of the order

$$\zeta\frac{\partial \rho_s}{\partial z} = O\left(-\rho_s \frac{U}{L}\frac{N^2}{g}\right). \tag{2.1.1.13a}$$

The first term on the right-hand side, which measures the change in the stratification (i.e., how the density surfaces have been stretched apart or compressed by the motion), can be estimated by

$$f\frac{\partial \rho'}{\partial z} = O\left(-\frac{f}{g}\frac{\partial^2 p'}{\partial z^2}\right) = O\left(\frac{f\,p'}{gD^2}\right), \tag{2.1.1.13b}$$

where $p'$ is the pressure perturbation. If the fluid is strongly rotating, the pressure perturbation is related to the velocity (we will see this presently) by the geostrophic balance, in which the horizontal pressure gradient is of the same order as the Coriolis acceleration, that is,

$$p' = O\left(\rho_s fUL\right). \tag{2.1.1.13c}$$

Thus, the ratio of the stretching term in the potential vorticity to the term due to the fluid's vorticity relative to the background rotation is, using (2.1.1.13a, b, c), just the

Burger number $S^{-1}$. Or, said another way, when the horizontal scale of the motion is of the order of the deformation radius, the relative vorticity and the stretching of density surfaces make equal contributions to the potential vorticity perturbation.

These are clearly imprecise measures, and in particular cases the balances can be modified, but it does give us a head start in understanding the nature of the competing mechanisms that we will encounter in buoyancy-forced flows in rotating systems.

### 2.1.2 A Laboratory Example: Formulation

Our first example is one that demonstrates the subtle connection between the influence of horizontal and vertical boundaries on the nature of a buoyancy-driven flow reflecting the constraints discussed in the previous section. To make the example concrete, we also will discuss a particular experiment that demonstrates these ideas (Pedlosky et al. 1997).

We consider flow in a right, circular cylinder of depth $L$ and radius $Lr_o$ as shown in Figure 2.1.

We will assume that the cylinder has been uniformly heated from above so as to establish a smooth and, in fact, linear temperature profile in the vertical coordinate parallel to $\Omega$. Aside from an irrelevant constant, the temperature in the resting state is

$$T = \Delta T_v z/L.$$

(Now, this neglects a small problem. The rotation will tend to bend the isotherms into parabolic bowls whose shape is inconsistent with a state of no motion. We will ignore this effect which will be possible if $\Omega^2 L/g \ll 1$.)

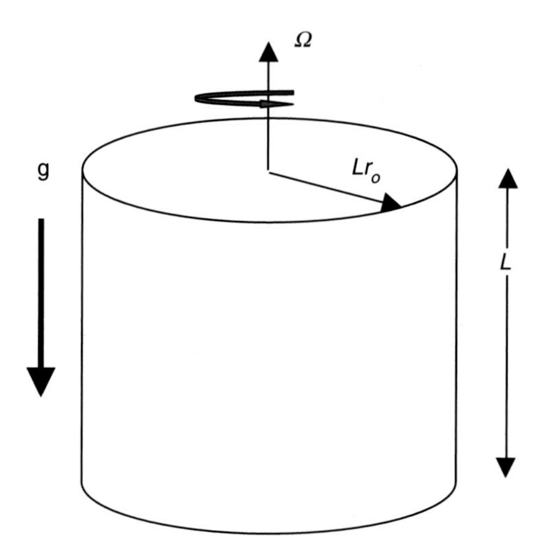


Figure 2.1. The cylinder in which fluid is driven by buoyancy forcing on the upper surface.

The following scales are introduced for nondimensionalization (asterisks denote dimensional quantities):

$$\begin{aligned}\vec{u}_* &= U\vec{u}, \;\; \vec{x}_* = L\vec{x}, \quad T_* = \Delta T_v (z_*/L) + \Delta T_h T(x, y, z), \\ \rho_* &= \rho_o [1 - \alpha(\Delta T_v z + \Delta T_h T)], \\ p_* &= \rho_o g \alpha \Delta T_v z_*^2 / 2L + \rho_o f U L \; p(x, y, z),\end{aligned} \tag{2.1.2.1}$$

where $\Delta T_h$ is the characteristic horizontal temperature variation and $\rho_0$ is the constant, undisturbed density before any temperature variation is imposed. Note that the first term in the last equation of (2.1.2.1) is the pressure due to the basic stratification in the rest state, while the second term is due to the motion.

I have chosen a simple, linear equation of state in which $\alpha$ is the coefficient of thermal expansion. The Coriolis parameter is $f = 2\Omega$. The characteristic velocity scale $U$ is related to $\Delta T_{\mathrm{h}}$ by the relation

$$U = \frac{\alpha g \Delta T_h}{f}, \tag{2.1.2.2}$$

which is an expression of our anticipation that the horizontal pressure gradient will be of the order of the Coriolis acceleration and that the vertical pressure gradient will be of the order of the gravitational buoyancy force.

For the dynamics, we will assume the fluid is moving slowly enough to be considered incompressible. We will make the Boussinesq approximation in which the density in the inertial accelerations can be replaced by $\rho_0$ (this assumes the density anomaly around the constant mean value is small) and that the heat equation is a balance between the advection of temperature and its diffusion (this ignores pressure effects in the energy equation, for example, our condition of incompressibility). In the problems we will discuss the motion will be steady so we will not be discussing the instability of our derived motion fields.

Under these approximations the governing, nondimensional equations are (I think it is important for each student to go through the equations and derive these from the incompressible Navier Stokes equations)

$$\begin{aligned}\varepsilon\vec{u} \cdot \nabla\vec{u} + \hat{k} \times \vec{u} &= -\nabla p + T\hat{k} + \frac{E}{2}\nabla^2\vec{u}, \\ \nabla \cdot \vec{u} &= 0, \\ \varepsilon\vec{u} \cdot \nabla T + wS &= \frac{E}{2\sigma}\nabla^2 T,\end{aligned} \tag{2.1.2.3a, b, c}$$

where

$$\varepsilon = \frac{U}{fL}, \quad E = \frac{2\nu}{fL^2}, \quad \sigma = \frac{\nu}{\kappa}, \quad S = \frac{\alpha g \Delta T_v}{f^2 L} \equiv \frac{N^2}{f^2}. \tag{2.1.2.4a, b, c, d}$$

The Prandtl number $\sigma$ is the ratio of the kinematic viscosity, $\nu$, to the thermal diffusivity, $\kappa$. The stratification parameter, $S$, is the Burger number we have already met.

The Ekman number, a measure of the friction, is $E$. The *Rossby number*, $\varepsilon$, measures the ratio of the vertical component of the relative vorticity with respect to the vorticity, $2\Omega$, due to the mean rotation. Note that when the relation between $U$ and the horizontal temperature gradient is used,

$$\frac{\varepsilon}{S} = \frac{\Delta T_h}{\Delta T_v}, \tag{2.1.2.5}$$

so if we want to keep the Rossby number small (as it is for many realistic oceanographic phenomena) for nontrivial stratification, we need to have the horizontal temperature gradients much less than the vertical gradients.

Generally, we will take no-slip conditions on solid surfaces and some combination of fixed temperature or fixed temperature gradient on the boundaries of the container. This may not be easy to do in an experimental situation, but we will come to that later. In what follows, we are going to specify a given temperature $T_u(r)$ on the upper boundary *in addition to the temperature that has established the stable stratification* $\Delta T_V$.

### 2.1.3 The Linear Problem

We will consider a problem in which the fluid is driven so gently that the Rossby number is small enough to allow the neglect of all nonlinear terms. When the flow occurs in a cylinder and the forcing and the motion are axially symmetric, the neglect of the nonlinearity is generally sensible because many of the nonlinear advective terms are identically zero. It is usually the advent of instabilities of the flow we are going to describe that more seriously limit the validity of the linearization. We consider flows in a right circular cylinder. Using an $r, \theta, z$ coordinate system with velocity components $u$ (radial), $v$ (azimuthal), and $w$ (vertical), the governing linear equations for axially symmetric motion are (Barcilon and Pedlosky 1967a, 1967b; Veronis 1967)

$$\begin{aligned}
-v &= -p_r + \frac{E}{2}\left[\nabla^2 u - u/r^2\right],\\
u &= \frac{E}{2}\left[\nabla^2 v - v/r^2\right],\\
0 &= -p_z + T + \frac{E}{2}\nabla^2 w,\\
\tfrac{1}{r}(ru)_r + w_z &= 0,\\
w\sigma S &= \frac{E}{2}\nabla^2 T,\\
\nabla^2 &= \frac{1}{r}\frac{\partial}{\partial r}r\frac{\partial}{\partial r} + \frac{\partial^2}{\partial z^2}.
\end{aligned} \tag{2.1.3.1a, b, c, d, e, f}$$

In equations (2.1.3.1), $T$ is the nondimensional temperature perturbation around the basic, stable stratification. The no-slip conditions on the top and bottom surfaces are satisfied by the addition of standard Ekman boundary layers whose structure is unchanged by the stratification as long as the $\sigma S \ll E^{-1/2}$, a condition easily satisfied. In that parameter range, the usual Ekman corrections to the interior flow occupy a region of height $E^{1/2} \ll 1$ in which the velocity in the interior of the fluid is brought to rest. The effect of friction in the boundary layer allows flow across the isobars from high to low pressure. This produces a convergence of the horizontal transport if there is a cyclonic region above the boundary layer in the interior leading to a vertical velocity pumped out of the bottom boundary layer

$$w_I(r,0) = \frac{E^{1/2}}{2r}(rv_I)_r, \tag{2.1.3.2}$$

where we have assumed axially symmetric motion in the interior. The right-hand side of (2.1.3.2) is proportional to the vorticity in the interior just above the boundary. So a low-pressure center (positive vorticity) leads to a convergence in the bottom Ekman layer and that convergence establishes a flux out of the boundary layer.

The same condition holds at the upper, solid surface. Naturally, a low-pressure center in the interior again yields a convergence in the boundary layer, but since the upper surface does not allow flow through the top, this convergence must produce a *downward* flow into the interior,

$$w_I(r,1) = -\frac{E^{1/2}}{2}\frac{1}{r}\frac{\partial}{\partial r} r\left(v_I(r,1) - v_T(r)\right), \tag{2.1.3.3}$$

where in (2.1.3.3) we have allowed the upper boundary to move with an azimuthal velocity $v_T(r)$. This would provide a mechanical driving in addition to the buoyancy driving we are interested in, but in some cases it is helpful in providing a simple solution useful for conceptual purposes. It will also allow us to emphasize the importance of the sidewall boundary layers.

We will get back to the question of the interaction with the sidewall shortly. Now we will go ahead and consider the interior region that is sandwiched between the two, thin, horizontal Ekman layers.

### 2.1.4 The Interior

Consider the interior region of the fluid (i.e., the region not adjacent to the boundaries where boundary layer scales can alter the balance of terms indicated by (2.1.3.1). Then for small values of the Ekman number and values of $S$ that are order one, the dominant terms in the equations yield

$$\begin{aligned} &v_I = p_{I_r}, \quad T_I = -p_{I_z}, \\ &u_I = \frac{E}{2}(\nabla^2 v_I - v_I/r^2), \\ &(ru_I)_r + w_{I_z} = 0, \\ &w_I = \frac{E}{2\sigma S}\nabla^2 T_I, \end{aligned} \tag{2.1.4.1a, b, c, d, e}$$

where the subscript $I$ reminds us that these variables represent adequately the variables in the fluid *external* to any boundary layers that are required to satisfy the boundary conditions. If our scaling for the temperature is sensible (i.e., if the temperature in the interior of the fluid is of the order of the applied temperature), then the vertical velocity in the interior should be $O(E/\sigma S)$. If the velocity adjacent to the horizontal surfaces were *O(1)*, then the vertical velocity pumped out of the boundary layer would be $O(E^{1/2})$, and since $S$ is an independent parameter, there is no reason to believe these two estimates have to be consistent. So, immediately the question of how the fluid adjusts to this possible discrepancy is raised.

To examine this question in detail, we start with the assumption that the Prandtl number $\sigma$ is order one and the Burger number $S$ is small. This means that the deformation radius is small compared to the radius of our circular basin, which is a fairly realistic parameter setting oceanographically. If $v_I$ is order one (it could be less but can't be greater because of thermal wind considerations), the scale for $u_I$ is order $E$; this is much less than $E/\sigma S$ so the continuity equation (2.1.4.1d) tells us that in that limit, to lowest order $w_I$ must be independent of $z$. Here is an example of the vertical coupling in $z$ due to the effects of strong rotation.

Using the boundary conditions (2.1.3.2) and (2.1.3.3) provided by the Ekman layers, we obtain

$$\begin{aligned} w_I &= \frac{1}{2}[w_I(r,1)+w_I(r,0)] \\ &= \frac{E^{1/2}}{4}\frac{1}{r}\frac{\partial}{\partial r}r\,[v_T - v_I(r,1)+v_I(r,0)]. \end{aligned} \tag{2.1.4.2}$$

At the same time the thermal wind equation is obtained from (2.1.4.1a) and (2.1.4.1b),

$$\frac{\partial v_I}{\partial z} = \frac{\partial T_I}{\partial r}, \tag{2.1.4.3}$$

which upon integration in $z$ yields

$$v_I(r,1) - v(r,0) = \frac{\partial}{\partial r}\int_0^1 T_I(r,z')dz', \tag{2.1.4.4}$$

and this allows us to write the vertical velocity entirely in terms of the interior horizontal temperature gradient, that is,

$$w_I = -\frac{E^{1/2}}{4}\frac{1}{r}\frac{\partial}{\partial r}\left(r\frac{\partial}{\partial r}\int_0^1 T_I(r,z')dz'\right) + \frac{E^{1/2}}{4}\frac{1}{r}\frac{\partial}{\partial r}(rv_T). \tag{2.1.4.5}$$

We use the following "trick," understood to mean a clever idea. We define the *pseudo-temperature:*

$$\theta = T_I + \lambda\int_0^1 T_I dz', \tag{2.1.4.6}$$

where

$$\lambda = \frac{\sigma S}{2E^{1/2}}, \tag{2.1.4.7}$$

which when inserted in (2.1.4.1e) yields the simple Poisson equation

$$\nabla^2 \theta = \lambda \frac{1}{r} \frac{\partial}{\partial r} (r v_T). \tag{2.1.4.8}$$

Once $\theta$ has been determined the actual temperature can be recovered by the easily proved relation

$$T_I = \theta - \frac{\lambda}{1+\lambda} \int_0^1 \theta dz. \tag{2.1.4.9}$$

The solution of (2.1.4.8) requires boundary conditions on $z = 0$ and 1 and a condition on the rim of the cylinder at $r = r_o$. Since the temperature perturbation in the Ekman boundary layer is small (it can be shown, using the temperature equation in the boundary layer and the scale for $w$ to be $O(E^{1/2}\sigma S)$), $T_I$ must satisfy the applied boundary conditions on the temperature on the upper and lower boundaries. On the other hand, we need to consider the boundary layers on the sidewalls, and these are not passive. The interaction of these boundary layers with the interior through the application of the sidewall boundary conditions will shape the nature of the interior flow and is another example of the lateral influence due to stratification.

It is also useful to exploit the axial symmetry of our problem to introduce a streamfunction for the circulation in the vertical plane such that

$$\begin{aligned} ru &= -\psi_z, \\ rw &= \psi_r \end{aligned} \tag{2.1.4.10}$$

in terms of which, and using (2.1.4.5), it follows that, since the interior vertical velocity is independent of $z$,

$$\psi_I(r) = \frac{E^{1/2}}{4} r v_T - \frac{E^{1/2}}{4(1+\lambda)} r \frac{\partial}{\partial r} \int_0^1 \theta dz, \tag{2.1.4.11}$$

so that once $\theta$ is determined the solution is complete.

### *2.1.5 Sidewall Boundary Layers $\sigma S \ll 1$*

The limit of small $\sigma S$ corresponds, generally, to the case where the deformation radius is small with respect to the basin scale and that, as we have mentioned, is a realistic limit. What is the nature of the boundary layer structure on the lateral boundaries then? First off, we need to find the horizontal scales of the layer. To do so, we first have to find the balances in the boundary layer among the terms in the equations of motion. Since the scale is shrunken in the radial direction, the terms in (2.1.3.1) will have different sizes when differentiated with respect to $r$. Assuming that in the boundary layers the dissipation terms (e.g., for temperature) are dominated by the second derivatives terms in $r$, we can considerably simplify the equations. We can also take advantage of the linearity of the problem by writing all variables in terms of the interior variables plus a *correction* to the variable that must go to zero as the interior is approached. So, in the following equations, the variables stand for corrections to the interior fields.

Over the range of stratification we will anticipate (an anticipation that can be checked a posteriori) that the azimuthal velocity will remain in geostrophic balance; hence, the set (2.1.3.1) can be approximated as

$$\begin{aligned} v &= p_r \\ u &= \frac{E}{2} v_{rr}, \\ 0 &= -p_z + T + \frac{E}{2} w_{rr}, \\ u_r + w_z &= 0, \\ \sigma S w &= \frac{E}{2} T_{rr}. \end{aligned} \qquad (2.1.5.1a, b, c, d, e)$$

All variables can be eliminated in favor of the pressure to obtain the sought-for master equation,

$$\overbrace{E^2 p_{6r}/4}^{(a)} + \overbrace{\sigma S p_{rr}}^{(b)} + \overbrace{p_{zz}}^{(c)} = 0. \qquad (2.1.5.2)$$

We will assume that derivatives with respect to $z$ are $O(1)$ in the sidewall layers and that each radial derivative is of order $1/\delta$, where $\delta$ is the boundary layer width, so that term (a), for example, is of order $E^2/\delta^6$. Now we can ask what balances are possible and we note that the full order of the equation (sixth order in $r$) must be preserved in any parameter setting. Let's suppose that terms (a) and (c) are of the same order. This clearly implies that

$$E^2/\delta^6 = O(1), \qquad \Rightarrow \delta = E^{1/3}. \qquad (2.1.5.3)$$

This is the scaled width of one of the Stewartson layers (see, Greenspan 1968, The *theory of rotating fluids*, for a full discussion). This layer exists for a homogeneous fluid as is obvious from (2.1.5.2) when $S$ is zero. To find the parameter limit of validity of that balance, we can evaluate the order of term (b) and compare it to terms (a) and (c). Term (b) is $\sigma S/E^{2/3}$ if $\delta = E^{1/3}$. Thus, for the balance (a)–(c) hold, we need

$$\sigma S < E^{2/3}. \qquad (2.1.5.4)$$

On the other hand, if the inequality is reversed, there are two possible balances. The balance of the first two terms yields a scale,

$$\delta = \frac{E^{1/2}}{(\sigma S)^{1/4}}. \qquad (2.1.5.5)$$

This is the thickness for the *buoyancy layer*. Note that in terms of dimensional quantities,

$$\delta_* = L\delta = \frac{(\nu\kappa)^{1/4}}{N^{1/2}} \tag{2.1.5.6}$$

and is *independent of rotation*. Remarkably, this scale can be found in Prandtl's *The Essentials of Fluid Mechanics* (1949). The ratio of term (c) to term (b) is

$$\frac{1}{\sigma S/\delta^2} = \frac{E}{(\sigma S)^{3/2}}, \tag{2.1.5.7}$$

so for the balance to be valid, $\sigma S > E^{2/3}$ (i.e., we must be in the complementary parameter regime compared to the situation where we had the balance (a)–(c). The remaining balance is (b)–(c), and it is clear that the scale must be

$$\delta = (\sigma S)^{1/2}, \tag{2.1.5.8}$$

and it is easy to show that this, too, requires, $\sigma S > E^{2/3}$. This layer is called the *hydrostatic layer.* Hence, in this range of stratification (clearly $\sigma S$ must be less than unity for this to represent a boundary layer) the original Stewartson layer splits into two parts: the buoyancy layer and the hydrostatic layer. The hydrostatic layer, which is the balance is described by a partial differential equation and its form is rather subtle.

Now that the scales are known, it is straightforward to find the balances in each of the dynamical equations. We will in this lecture consider only the case where $\sigma S > E^{2/3}$. Of course, for this to make sense we need $\sigma S \ll 1$. In the course of these lectures, we will need to consider the other two possibilities (i.e., $\sigma S \gg 1$ and also $\sigma S \ll E^{2/3}$). We will work through only the limit $\sigma S > E^{2/3}$ in detail. The balances in the other limits will be left to the student as an exercise. We take up the hydrostatic layer first.

### *2.1.6 The Hydrostatic Layer*

The nondimensional thickness of the hydrostatic layer is $(\sigma S)^{1/2}$; in dimensional units it is

$$\delta_h L = \left(\frac{\nu}{\kappa}\right)^{1/2} \frac{N}{f} L. \tag{2.1.6.1}$$

Aside from the Prandtl number factor, the dimensional scale is simply the Rossby deformation radius, and this scale pops up naturally in many oceanographic applications, especially in simple models of coastal upwelling where it is the scale over which strong vertical motions adjacent to a coastline can occur.

Introduce the boundary layer coordinate

$$\eta = \frac{r_o - r}{(\sigma S)^{1/2}} \tag{2.1.6.2}$$

and let the *correction* to the azimuthal velocity in the hydrostatic layer be

$$v_h = \tilde{v}(\eta, z) \tag{2.1.6.3a}$$

in terms of which (2.1.5.1b) implies that the proper scaling for the correction to the radial velocity is

$$u_h = \frac{E}{\sigma S}\tilde{u}(\eta, z), \tag{2.1.6.3b}$$

while, similarly, (2.1.5.1a) implies that

$$p_h = (\sigma S)^{1/2}\,\tilde{p}, \tag{2.1.6.3c}$$

and the temperature correction is

$$T_h = (\sigma S)^{1/2}\,\tilde{T}. \tag{2.1.6.3d}$$

The vertical velocity scale can be obtained from the continuity equation and (2.1.6.3b) so that

$$w_h = \frac{E}{(\sigma S)^{3/2}}\tilde{w}$$

and, obviously, $\psi_h = \frac{E}{\sigma S}\tilde{\psi}(\eta, z)$. It follows that the governing equations, ignoring terms of order $E^{2/3}/\sigma S$ or smaller, are

$$\begin{aligned} &\tilde{v} = -\tilde{p}_\eta, \quad \tilde{u} = \frac{1}{2}\tilde{v}_{\eta\eta}, \\ &\tilde{T} = \tilde{p}_z, \quad \tilde{w} = \frac{1}{2}\tilde{T}_{\eta\eta}, \\ &\tilde{u}_\eta = \tilde{w}_z. \end{aligned} \tag{2.1.6.4a, b, c, d, e}$$

Note that it is the hydrostatic balance, (2.1.6.4c) that gives this layer its name.

From the definition of the stream function, (2.1.6.4d) implies that

$$-r_o\tilde{w} = \tilde{\psi}_\eta = -\frac{r_o}{2}\tilde{T}_{\eta\eta}. \tag{2.1.6.5}$$

Since the tilde variables must vanish at infinity (in $\eta$), it follows that

$$\tilde{\psi} = -\frac{r_o}{2}\tilde{T}_\eta \tag{2.1.6.6}$$

so that the streamfunction in the boundary layer is tightly connected to the radial temperature gradient there.

Using (2.1.6.4b) and (2.1.6.4d) in the continuity equation yields

$$\tilde{v}_{\eta\eta} + \tilde{v}_{zz} = 0. \tag{2.1.6.7}$$

This is a partial differential equation and we need boundary conditions at $z = 0, 1$. The boundary layer width is $(\sigma S)^{1/2}$ and this will be much wider than the Ekman layer thickness $E^{1/2}$ as long as $\sigma S \gg E$. Since $\sigma S \gg E^{2/3}$, this obtains. This implies that, at the intersection of the hydrostatic layer and the Ekman layer, the $z$, derivatives in the viscous term dominate the radial derivatives so that as far as the Ekman layer is concerned, the hydrostatic layer just looks like an interior flow. We can therefore apply the Ekman boundary conditions for the hydrostatic layer, at $z = 0$, for example:

$$w_h = \frac{E}{(\sigma S)^{3/2}}\tilde{w} = \frac{E}{2(\sigma S)^{3/2}}\tilde{T}_{\eta\eta} = -\frac{E}{2(\sigma S)^{3/2}}\tilde{v}_{z\eta} \tag{2.1.6.8}$$

$$= \frac{E^{1/2}}{2}\frac{\partial}{\partial r}v_h = -\frac{E^{1/2}}{2(\sigma S)^{1/2}}\tilde{v}_\eta \tag{2.1.6.9}$$

integrating once in $\eta$ yields the boundary condition at $z = 0$,

$$\tilde{v}_z = 2\lambda\tilde{v} \tag{2.1.6.10a}$$

where $\lambda$ is given by (2.1.4.7). The same calculation at the top, $z = 1$, yields

$$\tilde{v}_z = -2\lambda\tilde{v}. \tag{2.1.6.10b}$$

Looking for solutions to (2.1.6.8) in the form $\tilde{v} = V(z)e^{-a\eta}$ yields an eigenvalue problem for $V$

$$\begin{aligned} &V_{zz} + a^2 V = 0 \\ &V_z = \mp 2\lambda V \quad at \quad z = \left\{ \begin{array}{c} 1 \\ 0 \end{array} \right\} \end{aligned} \tag{2.1.6.11}$$

I will not discuss the eigenvalue problem in detail except to note that for small values of $\lambda$, the lowest eigenvalue for $a$ is very small, in fact, $a_o \approx \left(\frac{2\sigma S}{E^{1/2}}\right)^{1/2}$, and this gives rise to a solution nearly independent of $z$ with a characteristic scale in the radial direction of $E^{1/4}$. This calculation is left to the student. This is the second boundary layer scale of Stewartson for homogeneous fluids. For that eigenvalue to obtain, we need to have $\sigma S \ll E^{1/2}$. Otherwise, and this is the case we will concentrate on, the full structure of the hydrostatic layer has the scale of $\sqrt{\sigma S}$, as anticipated. Note that as $\sigma S$ approaches unity, this layer will fill the interior and become part of the interior flow.

### 2.1.7 The Buoyancy Layer

Given the scale of the buoyancy layer, $\delta_b = \frac{E^{1/2}}{(\sigma S)^{1/4}}$, we can obtain the relative scaling of the correction functions in the buoyancy layer as

$$\begin{aligned} &T_b = \delta_b \hat{T}, \quad w_b = \frac{E}{\sigma S \delta_b}\hat{w}, \quad u_b = \frac{E}{\sigma S}\hat{u}, \\ &v_b = \frac{E}{(\sigma S)^{3/2}}\hat{v}, \quad p_b = \frac{E}{\delta_b(\sigma S)^{3/2}}\hat{p}. \end{aligned} \tag{2.1.7.1a, b, c, d, e}$$

We define the boundary layer coordinate in the buoyancy layer as

$$\xi = {}^{(r_o - r)}\!/_{\delta_b}, \tag{2.1.7.2}$$

and then, as long as $\sigma S \gg E^{2/3}$, the following balances result:

$$\hat{v} = -\hat{p}_\xi, \quad \hat{u} = \frac{1}{2}\hat{v}_{\xi\xi}, \quad \hat{w} = \frac{1}{2}\hat{T}_{\xi\xi}, \quad \hat{u}_\xi = -\hat{w}_z, \qquad \text{(2.1.7.3a, b, c, d, e)}$$
$$0 = \hat{T} + \frac{1}{2}\hat{w}_{\xi\xi}.$$

The last equation, in which the buoyancy force is balanced by the frictional force due to the relatively strong vertical velocity, distinguishes the buoyancy layer; it is *nonhydrostatic*. Note, too, that a correction to the streamfunction for the circulation in the vertical plane exists in the buoyancy layer, and it is

$$\psi_b = \frac{E}{\sigma S}\hat{\psi}, \qquad \hat{\psi} = -\frac{r_o}{2}\hat{T}_\xi, \qquad \text{(2.1.7.4a, b)}$$

so that once again the streamfunction is closely related to the temperature gradient in the boundary layer. Eliminating the vertical velocity between (2.1.7.3 c) and (2.1.7.3e) yields

$$\hat{T}_{4\xi} + 4\hat{T} = 0, \qquad \text{(2.1.7.5)}$$

which has the same form as the Ekman layer governing equation. The similar solutions here are

$$\hat{T} = Ae^{-\xi}\cos\xi + Be^{-\xi}\sin\xi,$$
$$\hat{w} = Ae^{-\xi}\sin\xi - Be^{-\xi}\cos\xi. \qquad \text{(2.1.7.6a, b)}$$

From the continuity equation, we obtain the velocity in the radial direction,

$$\hat{u} = -A_z e^{-\xi}(\cos\xi + \sin\xi)/2 + B_z e^{-\xi}(\cos\xi - \sin\xi)/2 \qquad \text{(2.1.7.7)}$$

so that this velocity depends on the variation of the coefficients $A, B$ in the vertical direction.

Equivalently,

$$\hat{\psi} = \frac{-r_o}{2}\left[-Ae^{-\xi}\{\cos\xi + \sin\xi\} + Be^{-\xi}\{\cos\xi - \sin\xi\}\right]. \qquad \text{(2.1.7.8)}$$

The coefficients $A$ and $B$ and the solution in the hydrostatic layer (in principle a sum of the normal modes that are solutions of (2.1.6.11)) are determined by matching the full solution to the boundary conditions on the sidewall. It is here that the boundary layer control (or lack of it) of the interior flow is manifested. We will be particularly interested in the size of the vertical mass flux that does (or does not) take place in the sidewall boundary layers with respect to the interior. Note that our scaling allows all three to be of the same order, that is, $O(\mathrm{E}/\sigma S)$.

### 2.1.8 Matching the boundary conditions at $r = r_o$

To complete the solution and determine the flow in the basin, we need to complete the interior solution, determine the flow in the boundary layer(s), and, in particular, determine how the surface heating determines the motion as a function of the background stratification.

The control of the interior is manifested by the boundary condition on the interior variable, the pseudotemperature, $\theta$, that must be satisfied on $r = r_o$. This is in turn obtained by the application of the physical boundary conditions at the outer wall to the total solution, the interior plus the correction functions for the buoyancy and hydrostatic layers. In this section, we will go through this matching in detail as an example of how it's done. In further discussion of other problems, the discussion will be abbreviated. The conditions on the outer wall are several; the first is the no-slip condition on the azimuthal velocity. At the rim, using the sum of the interior and boundary layer corrections, we obtain

$$v_I(r_o, z) + \tilde{v}(0, z) + \frac{E}{(\sigma S)^{3/2}} \hat{v}(0, z) = 0. \tag{2.1.8.1}$$

For the present problem, we will consider a cylinder whose sidewalls are insulated so that the radial heat flux is zero and so that the radial temperature gradient vanishes there, thus suppressing the explicit dependence on $z$ and remembering the condition is evaluated on the rim (which is the origin of coordinates for the boundary layer correction functions),

$$T_{I_r} - \tilde{T}_\eta - \hat{T}_\xi = 0, \tag{2.1.8.2}$$

The condition of no normal flow is equivalent to the condition that the total streamfunction for the circulation in the vertical plane is constant. Without loss of generality, the constant is taken to be zero so that at the outer rim,

$$\psi_I + \frac{E}{(\sigma S)} \tilde{\psi} + \frac{E}{(\sigma S)} \hat{\psi} = 0. \tag{2.1.8.3}$$

From the interior temperature equation, the interior streamfunction might be expected to be as small as $\frac{E}{(\sigma S)}$, but from the Ekman conditions leading to (2.1.3.2), it might be as large as $E^{1/2}$; however, this depends on the details of the solution.

The final condition is the vanishing of the vertical velocity on the rim (no-slip condition on $w$):

$$w_I + \frac{E}{(\sigma S)^{2/3}} \tilde{w} + \frac{E}{\sigma S \delta_b} \hat{w} = 0. \tag{2.1.8.4}$$

From (2.1.6.7) and (2.1.7.4) which relate the streamfunction to the temperature gradient, it follows that (2.1.8.3) is equivalent to

$$\psi_I - \frac{E}{(\sigma S)} \frac{r_o}{2} \tilde{T}_\eta - \frac{E}{(\sigma S)} \frac{r_o}{2} \hat{T}_\xi = 0. \tag{2.1.8.5}$$

In the parameter regime of interest (for which we have the hydrostatic layer) $\sigma S \gg E^{2/3}$, and so the last term in (2.1.8.1), the contribution of the buoyancy layer to the azimuthal velocity on the rim, is negligible, so on the rim,

$$v_I(r_o, z) + \tilde{v}(0, z) = 0, \tag{2.1.8.6a}$$

but since both the interior and the hydrostatic layer have the azimuthal velocity in thermal wind balance, the $z$-derivative of (2.1.8.6) implies that

$$T_{I_r} - \tilde{T}_\eta = 0 \tag{2.1.8.6b}$$

so that

$$\hat{T}_\xi = 0 \text{ on } r = r_O \tag{2.1.8.7}$$

while at the same time it follows from (2.1.8.4) that

$$\hat{w} = 0 \text{ on } r = r_O. \tag{2.1.8.8}$$

These two conditions applied to (2.1.7.6a, b) imply that the buoyancy layer does not make a nontrivial contribution to the solution and *at lowest order is absent.* At the same time, (2.1.8.5) and (2.1.8.7) yield

$$\psi_I(r_o) - \frac{r_o}{2}\frac{E}{\sigma S}\tilde{T}_\eta = 0,$$

$$\text{on } r = r_o$$

$$\boxed{\psi_I(r_o) = \frac{E}{\sigma S}\frac{r_o}{2}T_{I_r}} \tag{2.1.8.9a, b}$$

The matching conditions have thus produced a boundary condition on the *interior variables alone* on the outer rim. Note that since the interior streamfunction is independent of $z$ to lowest order, the interior temperature gradient on the outer rim must also be independent of $z$. From (2.1.4.11),

$$\frac{E^{1/2}}{4}r_o v_T - \frac{E^{1/2}}{4}r_o\int_0^1 T_{I_r}dz - \frac{r_o}{2}\frac{E}{\sigma S}\int_0^1 T_{I_r}dz = 0, \tag{2.1.8.10}$$

where I have exploited the fact that the temperature gradient in $r$ is independent of $z$ at the rim. With the definition of $\lambda$ and the independence of $T_{I_r}(r_o)$ on $z$ it follows that the interior temperature gradient on the rim must satisfy

$$\frac{\partial}{\partial r}T_I(r_o) = \frac{\lambda}{1+\lambda}v_T(r_o), \qquad \lambda = \frac{\sigma S}{2E^{1/2}}, \tag{2.1.8.11}$$

or in terms of the pseudotemperature $\theta$,

$$\frac{\partial}{\partial r}\theta(r) = \lambda v_T(r_o), \qquad r = r_o, \tag{2.1.8.12}$$

which serves as the boundary condition for (2.1.4.8).

The physical basis for this result is rather simple. The flow that radially flows outward in the upper and lower Ekman layers rises (or descends) in the hydrostatic layer. In so doing, the temperature anomalies generated must diffuse into the interior (the cylinder wall is insulated), and this determines the interior temperature gradient at the wall (really at the edge of the hydrostatic layer) to absorb the heat flux. Note that this depends only on the driving action of the differentially rotating upper lid at the outer rim since the Ekman layer flux is determined locally in $r$ by the discrepancy between the interior flow and the velocity difference of the lid and the interior flow. We will now discuss two examples, one a bit unrealistic but very revealing, the other more realistic but more complex while amenable to laboratory confirmation.

### *2.1.9 The Purely Mechanically Driven Flow*

Although we are interested in buoyancy-driven flows, the case of mechanical driving is of interest because (1) it's simple and (2) it affords us a clear example of the interactions between the boundary layer flow and the interior circulation.

To solve (2.2.4.8) we also need boundary conditions on the upper and lower boundaries for $\theta$. A very simple condition is obtained if we impose the condition that the vertical heat flux is constrained so that it is given by the background temperature gradient so that there is no additional anomalous temperature gradient, that is, that

$$T_{I_z} = 0, \quad z = 0, 1. \tag{2.1.9.1}$$

This is a rather difficult condition to satisfy in practice but leads to a very simple solution to our problem. The condition (2.1.9.1) is also equivalent to

$$\theta_z = 0, \quad z = 0, 1 \tag{2.1.9.2}$$

and a solution that satisfies (2.1.9.2), (2.1.4.8), and (2.1.8.12) is just

$$\begin{aligned} &\theta_r(r) = \lambda v_T(r) \\ &\Rightarrow \\ &T_{I_r} = \frac{\lambda}{1+\lambda} v_T(r) \end{aligned} \tag{2.1.9.3}$$

so that the interior temperature gradient vanishes when the fluid is homogeneous ($\lambda = 0$) and is just equal to $v_T(r)$ for large stratification (compared to $E^{1/2}$). From the thermal wind equation $v_{I_z} = T_{I_r}$, it follows that

$$v_I = \frac{\lambda}{1+\lambda} z v_T + V_o(r). \tag{2.1.9.4}$$

The "constant" of integration $V_O$ can be determined by using the boundary condition at $z = 0$ (2.1.3.2) and the temperature equation (recall that $T_I$ is independent of $z$ in the solution)

$$\begin{aligned} w(z=0) &= \frac{E^{1/2}}{2} \frac{1}{r} \frac{\partial r v(r,0)}{\partial r} = \frac{E}{2\sigma S} \frac{\partial^2}{\partial r^2} T_I = \frac{E}{2\sigma S} \frac{\lambda}{1+\lambda} \frac{1}{r} \frac{\partial r v_T}{\partial r} \\ \Rightarrow V_o &= \frac{1}{2(1+\lambda)} v_T \end{aligned} \tag{2.1.9.5}$$

so that

$$v_I = \frac{\lambda}{1+\lambda} v_T(z - 1/2) + v_T/2. \qquad (2.1.9.6)$$

When the stratification is zero (i.e., when $\lambda$ is zero), the azimuthal velocity in the interior is independent of $z$ (Taylor-Proudman theorem) and is equal to the average of the upper and lower boundary velocities so that the Ekman flux in the upper and lower layers is equal and opposite. As the stratification increases the temperature anomaly introduced by diffusion from the hydrostatic layer produces a thermal wind in the interior that tilts the velocity profile. For large $\lambda$ (i.e., when $E^{1/2} \gg \frac{E}{\sigma S}$), the shear of the velocity has reached an asymptotic value so that the azimuthal velocity at the lower boundary is zero and at the upper boundary is equal to the velocity of the upper lid. Thus, in the limit of strong stratification, *the interior velocity satisfies the no-slip conditions, and there is no need for an Ekman layer at lowest order, and the Ekman layer disappears.* The Ekman layer is choked off by the interior. One way to think about this is as follows: If there were an $O(1)$ discrepancy between the interior and the boundary velocities, the vertical velocity pumped out of the Ekman layers would be $O(E^{1/2})$, but the interior, because of the stratification, can only absorb a vertical velocity of $O(E/\sigma S)$. Hence, the interior adjusts, with the aid of the sidewall layer, to remove that discrepancy. Indeed, now that the interior azimuthal velocity is known, the interior vertical velocity is found from (2.1.4.1)

$$w_I = \frac{E^{1/2}}{4(1+\lambda)} \frac{1}{r} \frac{\partial r v_T}{\partial r}. \qquad (2.1.9.7)$$

For zero stratification the vertical velocity is $O(E^{1/2})$but falls to a value of $O(E/\sigma S)$ as the stratification increases and the Ekman layers become choked off. It is important to note that if the Ekman layers are expunged, the viscous dissipation occurring in those layers is also diminished.

Experimental confirmation of these ideas was presented by Linden (1977). Figure 2.2 shows the measured azimuthal flow for the unstratified case.

The azimuthal velocity is almost exactly the mean of the upper and lower boundary velocities (0.5 in nondimensional units).

When the fluid is strongly stratified the azimuthal velocity is shown in Figure 2.3 and varies linearly from the upper to the lower boundary velocity precisely as predicted.

### *2.1.10 The Buoyancy Driven Flow in the Cylinder*

In this section, we will consider the same cylindrical geometry but now the fluid will be driven only by an applied heating. In our simplest example, we suppose that we have been able to arrange the heating so that in addition to the basic stratification we apply an additional temperature gradient (in $z$) equivalent to a heat addition at the

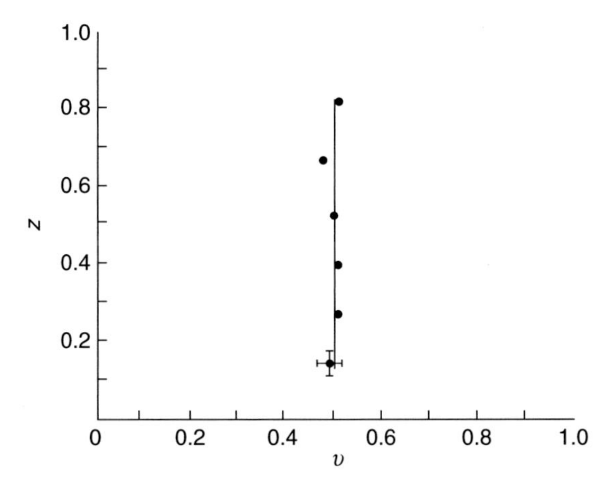


Figure 2.2. The measured $v(z)$ in Paul Linden's experiment for the unstratified case (Linden 1977).

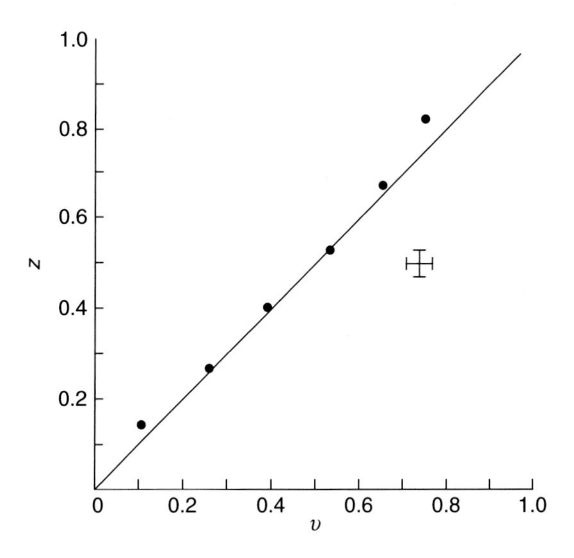


Figure 2.3. The azimuthal velocity for large values of stratification $\sigma S/E^{1/2}$~5.75 $10^4$ (from Linden 1977).

upper boundary of the form,

$$T_z(r, 1) = h(r) = h_o J_o(k_1 r/r_o), \quad z = 1, \tag{2.1.10.1}$$

where the amplitude of the heating is $h_o$ and the shape is a Bessel function of zero order and the parameter, $k_1$ is the first zero of $J_1(k_1)$ so that $k_1$ is also the first zero of the *derivative* of $J_o(k)$. It follows that the integral of the heating over the area of the upper lid is zero; that is, as much heat is taken out of the top as put in. On the lower boundary, the *anomaly* of $T_z$ is taken to be zero so all the heat is put in and extracted at the upper surface.

In the absence of mechanical forcing, the boundary condition on $r = r_o$ for the *interior solution* (see (2.1.8.11)) is just

$$\frac{\partial T}{\partial r} = 0, \quad r = r_o. \tag{2.1.10.2}$$

This implies that the boundary layer correction to the temperature gradient at the wall will be zero, which in turn implies that in this purely buoyancy-driven case (with insulating sidewall conditions) there will be no vertical mass flux in the sidewall layers, (i.e., the circulation in the $r - z$ plane must close within the interior).

When translated into a condition on the pseudotemperature, $\theta$, the solution of (2.1.4.8) is simply

$$\theta = h_o \frac{r_o}{k_1} J_o(k_1 r/r_o) \frac{\cosh(k_1 z/r_o)}{\sinh(k_1/r_o)}, \tag{2.1.10.3}$$

from which the other variables can be determined, that is,

$$T = h_o \frac{r_o}{k_1} J_o(k_1 r/r_o) \left[ \frac{\cosh(k_1 z/r_o)}{\sinh(k_1/r_o)} - \frac{\lambda}{\lambda+1} \frac{r_o}{k_1} \right],$$

$$v = -h_o J_1(k_1 r/r_o) \frac{r_o}{k_1} \left[ \frac{\sinh(k_1 z/r)_o}{\sinh(k_1/r_o)} - \frac{\lambda}{\lambda+1} z - \frac{1}{2(\lambda+1)} \right], \tag{2.1.10.4a, b, c}$$

$$w = h_o \frac{E^{1/2}}{4(\lambda+1)} J_o(k_1 r/r_o).$$

Where the final term in (2.1.10.4 b) is determined by the application of the matching to the Ekman pumping at $z = 0$ and gives rise to a term independent of $z$. Note that for small stratification ($\lambda \ll 1$), the vertical velocity is $O(E^{1/2})$ and is equal to the velocity pumped out of or into the Ekman layers. This requires that $v(r, 0) = -v(r, 1)$, and this can be checked in (2.1.10.4b) for all values of $\lambda$. As the stratification increases, the magnitude of the vertical velocity decreases until, for very large $\lambda$, $w$ becomes of order $E/\sigma S$, which is the magnitude of $w$ required by the temperature equation (2.1.4.1e) when the temperature is order one. How does this occur if the Ekman layers pump a vertical velocity of order $E^{1/2}$ if $v$ is order one at $z = 0$ and 1? As in the mechanically driven case, this happens because the vertical motion in the interior is a function of horizontal position, $r$, and so introduces horizontal temperature gradients and, hence, a shear of the azimuthal velocity such that for large $\lambda$ the value of $v$ on the upper and lower boundaries matches the no-slip condition, vitiating the need for an Ekman boundary layer and hence any Ekman layer pumping.

Figure 2.4a shows the solution for $v$ contoured in the $r$,$z$ plane for $\lambda = 100$. The contour interval is a tenth of its maximum absolute value of about 0.06.

Note that the velocity is everywhere positive. A profile of $v$ at the mid radius, $r = r_o/2$ is shown in Figure 2.4b.

Note that the velocity of the interior flow vanishes at $z = 0, 1$ for this large value of $\lambda$, just as in the case of the mechanically driven flow. The velocity is positive even

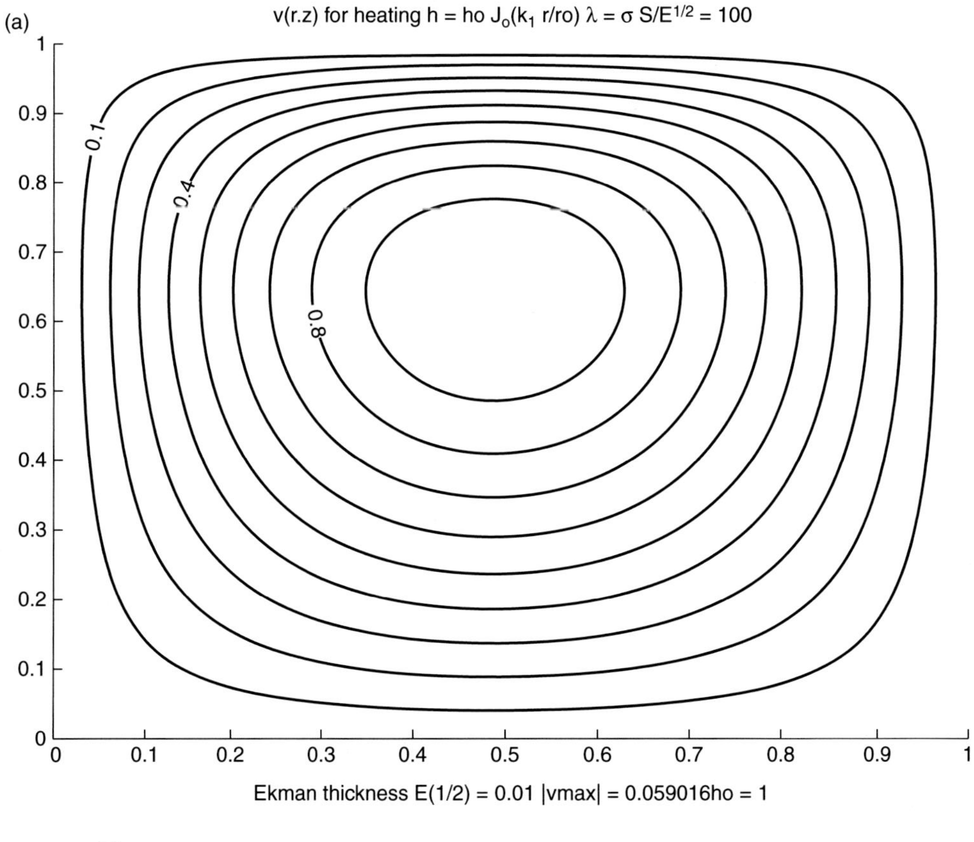


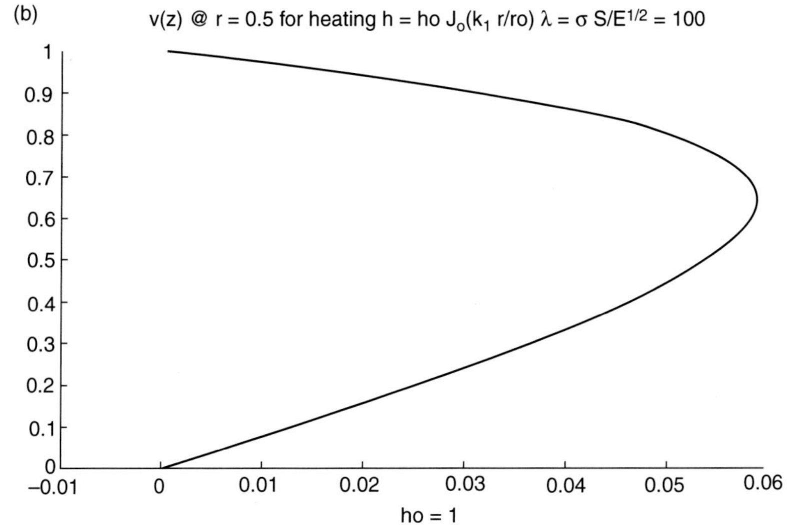


Figure 2.4 (a) Contours of $v(r, z)$ for $\lambda = 100$. (b) The velocity profile at $r/r_o = ½$. (c) The temperature (anomaly) provided by the heating and cooling at the upper surface. Negative anomalies (dashed lines) display the reversal of the radial temperature gradient near the lower boundary.

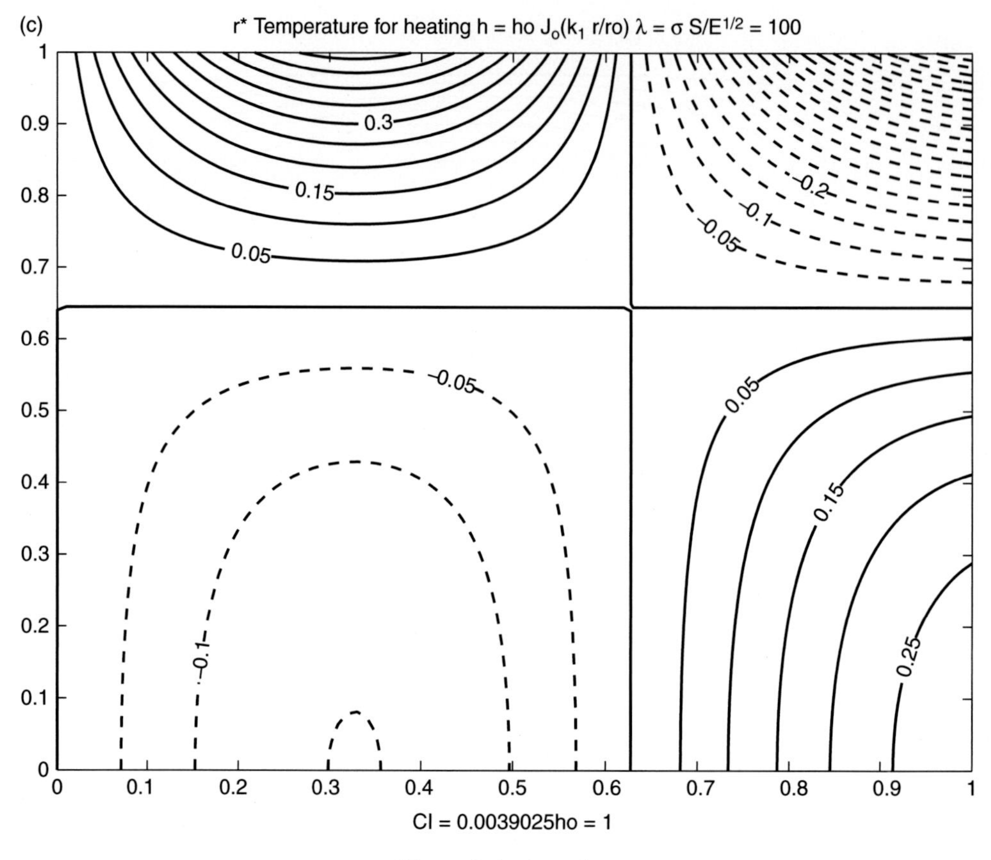


Figure 2.4. (*cont.*)

though the radial temperature gradient imposed by the heating at the upper surface is negative; that is, the fluid heated near the center is warmer than the fluid near the rim. All that does is make the *shear* negative near the upper surface. However, for large stratification, the interior azimuthal velocity must vanish at $z = 1$, so the velocity must be positive *below* $z = 1$. Since $v$ must also vanish at $z = 0$, it achieves a maximum at mid-depth before going to zero at the bottom. This further implies that the temperature gradient must be reversed at the bottom. The solution for $T(r, z)$ shown in Figure 2.4c confirms that this, in fact, occurs.

For smaller values of $\lambda$ the interior velocity does not vanish at the upper and lower boundaries and Ekman layers are required there to satisfy the boundary conditions. Figure 2.5a shows the profile of $v$ at $r = r_o/2$ for the same heating function but for $\lambda = 1.0$.

For this lower value of stratification the azimuthal velocity is negative near the upper, heated surface. It is still true that the largest velocities in the interval of $z$ are positive. For even smaller values of $\lambda$ (i.e., for a value 0.10), the profile of $v$ becomes

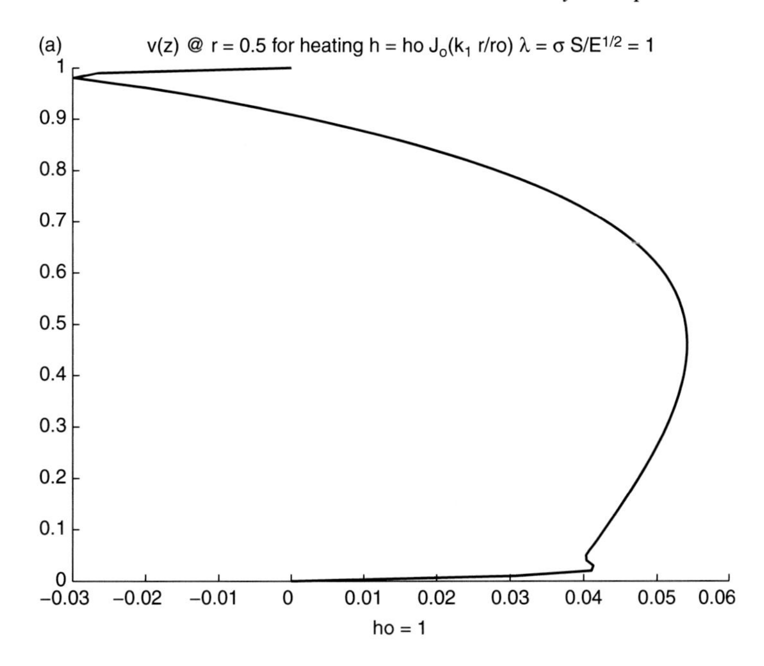


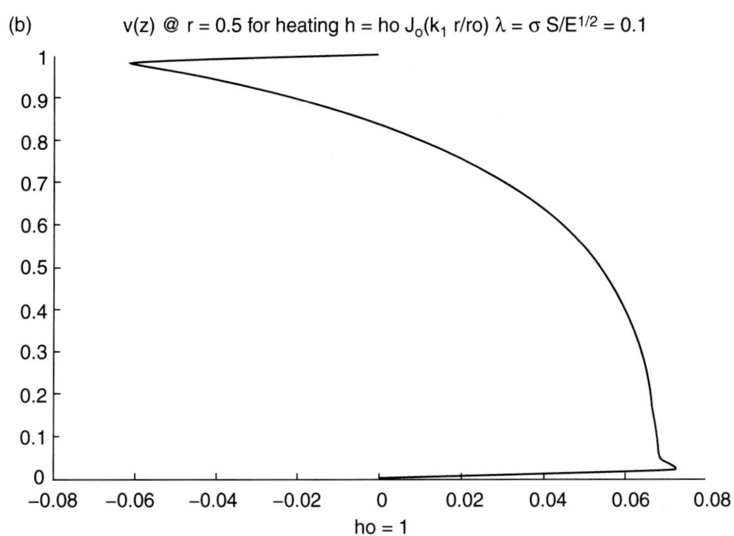


Figure 2.5 (a) The profile of $v(z)$ for $\lambda = 1.0$. Note the appearance of Ekman layers needed to correct the interior flow near $z = 0$ and 1. (b) As in Figure 2.5a but for $\lambda = 0.1$.

ever more negative in the upper region of the domain under the heating as shown in Figure 2.5b.

The *form* of the vertical velocity is not a function of $\lambda$ for this simple example. Figure 2.6 shows the relationship between $w$ and the heating function $h(r)$.

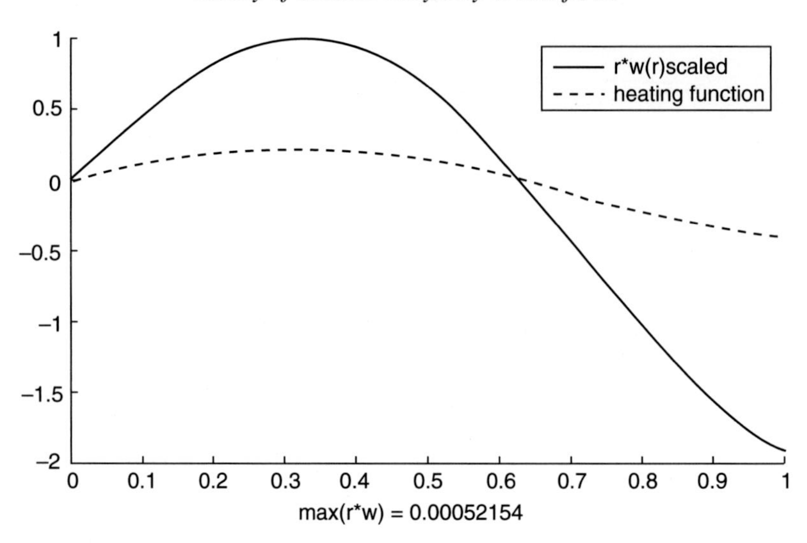


Figure 2.6. The vertical velocity $w$ (solid line) and the shape of the heating function, $h(r)$.

In Figure 2.6 the vertical velocity has been multiplied by $r$ in order to give a more accurate picture of the vertical volume flux as a function of $r$. As one might expect, the region of rising fluid corresponds exactly to the region of heating, while the region of cooling corresponds to sinking motion. Although this is the relationship one would intuitively expect, we shall see in our further discussion that this is *not* generally true. But, it is the case in this, our simplest of models. As (2.1.10.4c) shows, the magnitude of $w$ decreases as the stratification increases, falling from $O(E^{1/2})$ for small values of $\lambda$ to values of $O(E/\sigma S)$ as $\lambda$ becomes much greater than unity. Thus, the vertical velocity for $\lambda = 100$ will be about 100 times smaller than its value for $\lambda = 0.1$.

### *2.1.11 A Laboratory Example*

In this section, we consider the same cylindrical geometry, but now the fluid is forced by an applied *temperature anomaly* on the upper boundary and not a heating anomaly as in the example in Section 2.1.10 (this is experimentally easier) and again with no mechanical forcing. Thus, the boundary conditions we will consider are

$$\begin{aligned} T &= T_u(r), \qquad z = 1, \\ T &= 0, \qquad z = 0. \end{aligned} \tag{2.1.11.1a, b}$$

This system was examined experimentally (Pedlosky et al. 1997) and spanned a rather wide range of values of $S$. Indeed, it was rather natural in the laboratory situation to consider values of $\sigma S$ that were *large* compared to unity as well as small values. When $\sigma S$ is large, the hydrostatic layer has already filled the interior. Indeed, as we shall show, a metamorphosis of that layer occurs and a boundary layer of scale $(\sigma S)^{-1/2}$ now exists in the vicinity of the upper, heated boundary, and it plays a role similar to the $(\sigma S)^{1/2}$ layer on the sidewall for small values of $S$. I will not go through the derivation of the full boundary layer analysis. Instead, we restrict attention to the region outside the upper and lower Ekman layers and outside the sidewall buoyancy layers. The equations of motion are

$$\begin{aligned} v &= p_r, \qquad u = \frac{E}{2}\left[\nabla^2 v - {}^{v}/_{r^2}\right], \\ T &= p_z, \qquad w = \frac{E}{2\sigma S}\left[\nabla^2 T\right], \\ &\frac{1}{r}\frac{\partial}{\partial r}(ru) + \frac{\partial w}{\partial z} = 0 \end{aligned} \qquad \text{(2.1.11.2a, b, c, d, e)}$$

for linear, axisymmetric flow. In (2.1.11.2a), I have assumed that the azimuthal velocity remains in geostrophic balance in the interior (i.e., external to any Ekman layers).

All variables now can be eliminated in terms of the pressure that satisfies

$$\begin{aligned} &\nabla^2\left[\sigma S \nabla_h^2 + \frac{\partial^2}{\partial z^2}\right] p = 0, \\ &\nabla^2 = \frac{1}{r}\frac{\partial}{\partial r} r \frac{\partial}{\partial r} + \frac{\partial^2}{\partial z^2}, \quad \nabla_h^2 = \frac{1}{r}\frac{\partial}{\partial r} r \frac{\partial}{\partial r}. \end{aligned} \qquad \text{(2.1.11.3a, b, c)}$$

It is clear that for small $\sigma S$ there will be a boundary layer structure in $r$ with the scale of the hydrostatic layer, while for large values of the parameter a boundary layer with the scale $(\sigma S)^{-1/2}$ is possible in $z$ near the horizontal boundaries. In the latter case, following the same matching procedures as in the last section, we can show that for large $\sigma S$ this thermal boundary layer is the region in which the thermal forcing is trapped near the upper surface and in which the circulation in the vertical plane takes place. Beneath that region the radial velocity is negligible and the azimuthal velocity satisfies (2.1.11.2b) with $u = 0$ leading to a Couette-type velocity profile forced by the thermal flow in the upper region as a solution of the conduction equation. For large values of $\sigma S$, the circulation in the vertical plane is very weak in the interior, of order $E$, and is closed in a very weak Ekman layer on the upper surface. As long as $E^{1/2} \ll \sigma S \ll E^{-1}$, the governing equation is (2.1.11.3a), and the appropriate boundary conditions are

$$\begin{aligned} p_r &= 0, & z &= 0, 1 \text{ (no Ekman layers)}, \\ p_z &= 0, & z &= 0, \\ p_z &= T_u(r) & z &= 1, \\ p_r &= 0, & r &= r_o \end{aligned} \qquad \text{(2.1.11.4a, b, c, d)}$$

The first condition reflects the fact that in the parameter range $\sigma S \gg E^{1/2}$, there are no Ekman layers to *lowest order*. The second and third conditions are the imposed thermal conditions at the upper and lower surfaces. The last condition states that for the insulated sidewall of the cylinder in the absence of mechanical forcing, the hydrostatic layer is absent (i.e., when $v_T = 0$, see (2.1.8.11)), while for $\sigma S \gg 1$ there is no possibility of boundary layer aside from the buoyancy layer, which plays no role when the wall is insulated.

A solution to (2.1.11.3a) can be found for the full parameter range of interest in the form

$$p = \sum_{n=1} P_n(z) J_o(k_n r/r_o), \tag{2.1.11.5}$$

where $k_n$ is the $n$th zero of the first-order Bessel function and

$$P_n = A_n e^{-q_n(z-1)} + B_n e^{-q_n z} + C_n e^{-q_n d(z-1)} + D_n e^{-q_n dz}, \tag{2.1.11.6}$$

where

$$q_n = k_n/r_o, \quad d = (\sigma S)^{1/2}. \tag{2.1.11.7a, b}$$

The coefficients $A_n$, $B_n$, $C_n$, $D_n$ are easily obtained by matching the boundary conditions at $z = 0, 1$. Once the pressure is known the remaining fields, including the streamfunction for the circulation in the vertical plane, can be determined.

Jack Whitehead set up an experiment to examine these ideas and these boundary layer balances. The apparatus is shown in Figure 2.7 along with some sample profiles of the measured azimuthal velocity using a dye streak method.

The temperature distribution on the upper lid, forcing the flow, was somewhat difficult to control and instead of $T_u(r)$ being specified, the lid temperature was measured. An example showing the radial distribution is shown in Figure 2.8. Note that there are regions in $r$ in which the radial temperature gradient is positive and others (closer to the center) where it is negative instead of the single sign of our simple example.

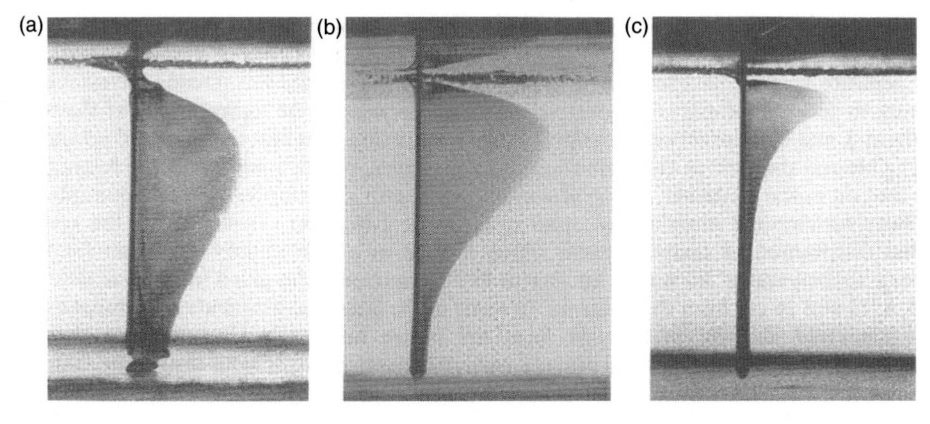


Figure 2.7. Dye streaks indicating the vertical structure of the azimuthal flow for three different stratifications, increasing from left to right (from Pedlosky et al. 1997).

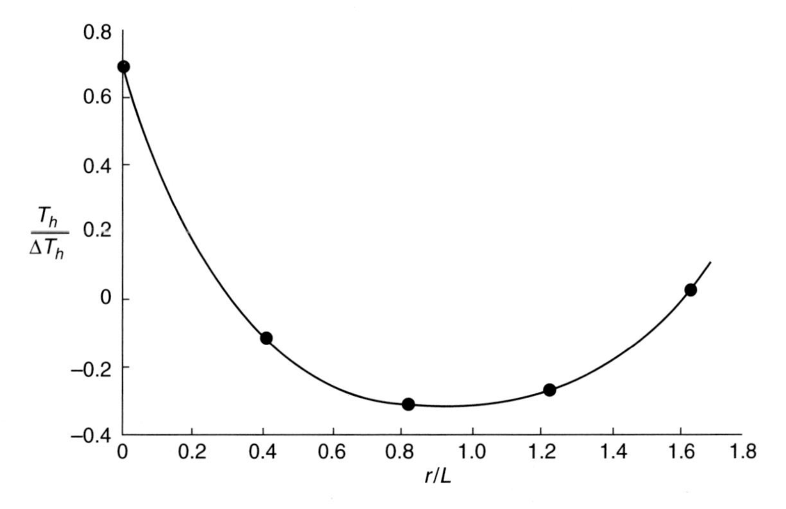


Figure 2.8. The measured temperature on the upper lid used as an upper boundary condition for the solution of (2.1.11.3) (From Pedlosky et al. 1997).

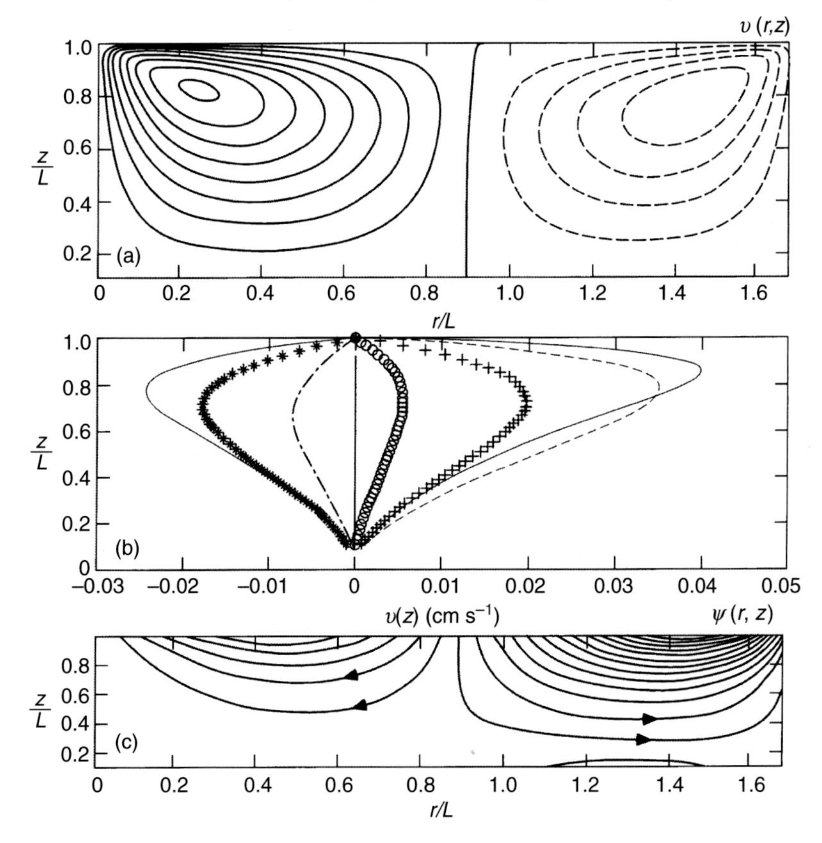


Figure 2.9. The azimuthal flow is contoured in the first panel, the profiles of $v$ at various $r$ are shown in the second panel and $\psi$ outside the upper Ekman layer is shown in the third panel (from Pedlosky et al. 1997).

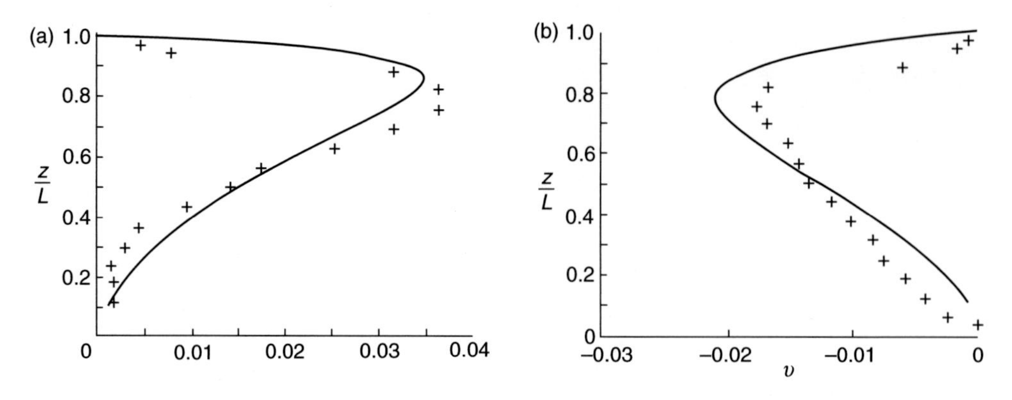


Figure 2.10. The theoretical predictions are shown by the solid lines and the measured profiles in the experiments are shown with + (from Pedlosky et al. 1997).

The stratification was established by setting the apparatus in a thermal bath to set the lower temperature, while the upper temperature was set by two cylindrical tubes at slightly different temperatures. This established the radial temperature gradient of the lid as well as the vertical stratification.

For the parameter value $\lambda = 42$, the predicted fields are shown in Figure 2.9. The azimuthal velocity, in the upper panel, has two cores reflecting the change in sign of the velocity due to the change in sign of the radial temperature gradient at the upper lid. The second panel shows the velocity profiles at selected radii and the third panel shows the meridional streamfunction.

The comparison between the predicted velocity profiles and the measured ones were generally very good. For the previous case, Figure 2.10 displays the comparison between the experimental and theoretical predictions for $v$.

In the next Section, we will consider a buoyancy-driven flow on the $\beta$ plane, and we will see that the vertical motion induced by heating and cooling will not, in general, be so closely linked geographically to the applied heating or cooling.

## 2.2 Buoyancy-Driven Flows in Beta-Plane Basins: The Relation Between Buoyancy Forcing and the Location of Vertical Motion

### *2.2.1 Introduction*

In (Section 2.1), the vertical motion driven by buoyancy forcing on the surface of the fluid was accompanied by vertical motion of a very intuitive structure. The region of rising motion occurred directly under the heating and the downwelling motion occurred under the cooling. Also, with insulating conditions on the sidewalls, very little vertical motion occurred in the boundary layers. We shall see that neither of these results is generally true and all that is needed to alter the relationships between vertical motion and the buoyancy forcing is a mechanism for the lateral propagation of information.

The oceanographic relevance of these considerations is almost obvious, and the issue has been raised in a paper by Pedlosky and Spall (2005).

We are going to use the model of that paper but we will be examining a somewhat different example to focus on the colocation question. The effect of $\beta$ on the horizontal structure of the wind-driven circulation is, as is well known, dramatic. It is the fundamental reason for the western intensification of the circulation and the appearance of strong currents such as the Gulf Stream and Kuroshio in their respective ocean basins. For a fuller discussion of the beta plane and its dynamical consequences, see Pedlosky (1987).

If we reexamine closely the dynamics of the sidewall boundary layers studied in Section 2.1, we will observe that the net vertical motion in those layers is proportional to the boundary layer's horizontal temperature gradient at the boundary and so to the tangential velocity near the boundary. If the $\beta$-effect leads to a similar intensification for buoyancy-driven motion we might expect a similar intensification of the currents near the western boundary. Those baroclinic currents, with their temperature gradients, would lead to an intensification of regions of rising or sinking motion.

In general, although attention is often focused on the upwelling component of the thermohaline circulation in the oceanic interior and its relation to mixing, I also want to emphasize the strong role of mixing in determining the location and structure of the sinking branch of the circulation. This is motivated by a range of applications in which buoyancy forcing and the beta-effect are important, including high-latitude polar seas, marginal seas, and basin-scale thermohaline circulations. The fundamental physical constraint is the connection between vertical motion and vorticity production by stretching. To get strong vertical motion, there must be an agent to balance the vorticity production by the stretching of planetary vorticity. That is difficult to do in the fluid interior, but in boundary layers where dissipation becomes important, the vorticity that is locally produced could be locally dissipated.

To examine this question in the simplest possible format I am going to use the model of Pedlosky and Spall (2005) shown in Figure 2.11.

### 2.2.2 The Model Formulation

The model is shown in Figure 2.11. Two layers of slightly different density are separated by an interface whose elevation above its rest level is $\eta\ (x, y)$ where $x$ is a coordinate to the east and $y$ is a coordinate to the north. The rest thicknesses of each layer, $H_1$ and $H_2$, are chosen to be equal ($H$). The flow is driven by buoyancy forcing alone, and in the absence of topography in our model, the resulting circulation can be shown to be purely baroclinic (i.e., the barotropic component of velocity $H_1\vec{u}_1 + H_2\vec{u}_2 = 0$). When (and if) fluid crosses the interface between the two layers

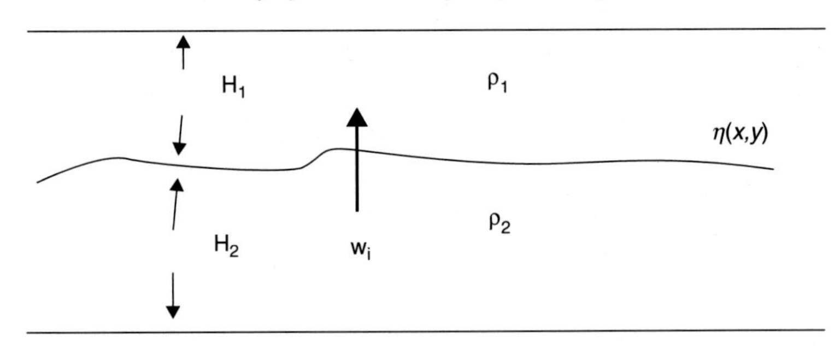


Figure 2.11. The two-layer model. The motion is driven by a flow of fluid *across* the interface.

from the lower to the upper layer, it changes density and becomes lighter; it therefore represents a nonadiabatic *transformation* of the fluid density (i.e., the fluid is heated). If the fluid penetrates into the lower layer from the upper layer, the fluid is being cooled. In our layer model, this will be our representation of diabatic heating and cooling.

We represent the remaining baroclinic horizontal velocity as

$$\vec{u} = \vec{u}_1 - \vec{u}_2. \tag{2.2.2.1}$$

Below we will show plots of this velocity as if it were the velocity in the upper layer. You must keep in mind that there is an equal and opposite velocity in the lower layer.

Again, to keep the model as simple as possible, we will linearize the dynamics, noting that the *baroclinic* pressure is related to the interface deformation by

$$p_2 - p_1 = \rho_o g' \eta + \text{const.}, \tag{2.2.2.2}$$

where $\rho_o$ is the mean density of the two layers. This leads to the linear equations of motion,

$$\begin{aligned} f u &= g' \eta_y + A \nabla^2 v, \\ -f v &= g' \eta_x + A \nabla^2 u, \\ u_x + v_y &= 2 \frac{w_i}{H}. \end{aligned} \tag{2.2.2.3a, b, c}$$

Note the signs of the pressure gradient terms. They are reversed from the normal signs because of (2.2.2.2). Integration of the continuity equation in each layer, using the fact that the horizontal velocity is independent of $z$ within each layer and then subtracting the resulting two equations, leads to (2.2.2.3c). Note the factor of 2. In this linear, steady model, the vertical velocity at the interface, $w_i$, is also the cross-isopycnal velocity, and it is that velocity that is the driver of the baroclinic circulation. Again, a positive (negative) value of that velocity represents heating (cooling). In (2.2.2.3 a,b) $f = f_o + \beta y$, $A$ is the coefficient of momentum mixing, our simplistic way of including turbulent eddy mixing of momentum.

The key feature of the model is the representation of the cross-isopycnal velocity. I choose the following parameterization:

$$w_i = \gamma(\eta - h) - \kappa \nabla^2 \eta. \tag{2.2.2.4}$$

The first term on the right-hand side is a crude representation of *vertical* mixing. The function $h$ is a target profile to which the interface relaxes on a time scale $\gamma^{-1}$. If $h$ is negative, as in Figure 2.12, then the target profile tends to force the interface to drop below the undisturbed level of the interface and so produce, locally, a region of enhanced upper layer fluid. That is, we can interpret a negative $h$ as a heating. Similarly, if $h$ is positive, it will represent a local cooling. According to (2.2.2.4), if the interface height is less than the target profile, then there will be negative cross-isopycnal motion. In Figure 2.12, the interface (dashed) is negative but algebraically greater than $h$, so a positive cross-isopycnal velocity will be forced representing the transformation of dense to light fluid (heating). In our model, $h$ will be our buoyancy forcing.

The second term in (2.2.2.4) represents the horizontal diffusion of layer thickness by unresolved mesoscale eddies. The coefficient $\kappa$ will be taken to be very small, and the effect of this diffusion term will be considered negligible everywhere *except* in sidewall boundary layers where we will interpret it as a nonadiabatic flux analogous to the temperature diffusion we noted in the laboratory model in the previous section.

I introduce the following nondimensional scaling indicating nondimensional variables with primes:

$$\begin{aligned}
&(x, y) = L(x', y'), \qquad (u, v) = U(u', v'), \quad w_i = U\frac{H}{L}w_i', \\
&h = h_o\theta(x, y), \qquad \eta = \frac{f_o U L}{g'}\eta', \\
&U = \frac{g' h_o}{f_o L}.
\end{aligned} \tag{2.2.2.5a, b, c, d, e, f}$$

The length scale $L$ is the scale of the basin, and $H$ is the undisturbed depth of each layer. The last relation fixes the scaling for the horizontal velocity in terms of the amplitude of the applied buoyancy forcing. The function $\theta$ describes the spatial shape

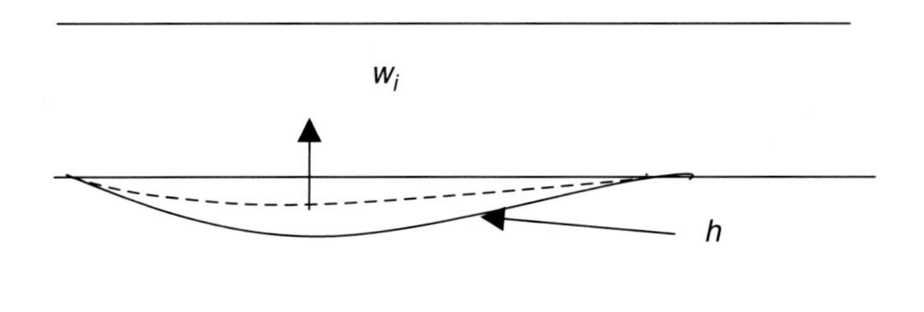


Figure 2.12. The *target* profile, $h$, for the interface (solid). The interface height is dashed. This represents a region of heating and so positive cross-isopycnal velocity.

of the buoyancy forcing; in this scaling, it has an order one amplitude. Using this scaling, the equations can be written in nondimensional form as (after dropping the primes on dimensionless variables),

$$\begin{aligned} f u &= \eta_y + b\delta_m^3 \nabla^2 v, \\ -f v &= \eta_x + b\delta_m^3 \nabla^2 u, \\ u_x + v_y &= 2w_i, \\ w_i &= b\left[\frac{\eta - \theta}{\delta_T} - \delta_K \nabla^2 \eta\right], \end{aligned} \qquad (2.2.2.6a, b, c, d)$$

where the key nondimensional parameters are

$$\begin{aligned} \delta_m &= (A/\beta)^{1/3}/L, \quad \delta_T = \frac{\beta L_d^2}{\gamma L}, \quad \delta_K = \frac{\kappa}{\beta L_d^2 L} \\ b &= \frac{\beta L}{f_o}, \\ L_d^2 &= \frac{g' H}{f_o^2}. \end{aligned} \qquad (2.2.2.7a, b, c, d, e)$$

The first three parameters are measures of length. The first is the nondimensional Munk boundary layer thickness, the ratio of the possible viscous boundary layer on the western boundary to the basin scale, clearly a small parameter. The second, $\delta_T$, is a thermal boundary scale and is a measure of how far a long, baroclinic Rossby wave can travel westward before it is damped by the relaxation process whose damping rate is $\gamma$. The third is a diffusion scale and measures the ratio of the time it takes a long Rossby wave to cross the basin, $L/\beta L_d^2$, compared to a diffusion time, $L^2/\kappa$, on the scale of the basin. We will consider a parameter regime in which the Munk scale and the diffusion scale are small. The thermal boundary scale, $\delta_T$, may be order one or less. The $\beta$-plane parameter, $b$, is considered small so that quasi-geostrophic beta-plane dynamics will be valid; that is, $f$ is locally considered a constant in the momentum equations to lowest order in $b$.

In most of the following development, we are interested in the limit of weak viscosity and so will order the parameters such that

$$\delta_m \ll \delta_K \ll \delta_T = O(1). \qquad (2.2.2.8)$$

For comparison with the $f$-plane model of the results of Section 2.1, it is useful to note the following equivalents:

$$\begin{aligned} E &= \frac{A}{f_o L^2} = b\delta_m^3, \\ S &= \frac{L_d^2}{L^2}, \quad \sigma = A/\kappa, \end{aligned}$$

$$E/\sigma S = b\delta_K,$$
$$\gamma / f_o S = b/\delta_T. \qquad (2.2.2.8a, b, c, d, e)$$

The first is the Ekman number, the second is the stratification parameter, and the third is the Prandtl number.

An additional boundary layer scale enters the problem and is familiar from the f-plane models of the first lecture. We define this scale as

$$\delta_h = (\sigma S)^{1/2} = \left(\delta_m^3 / \delta_K\right)^{1/2}. \qquad (2.2.2.9)$$

This is the scale for the hydrostatic layer. Its width is independent of $\beta$ and in the $f$-plane models is the scale over which the strong vertical motions near the boundary occur. We will see the reemergence of this boundary layer with the same function in the beta-plane model. It is important to note that in the $f$-plane models the characteristic vertical transport in the interior and in the boundary layers is $O\left(E/\sigma S\right) = b\delta_K$, and this gives us a useful point of comparison for the beta-plane case where we will find vertical transports that are larger by a factor of $\delta_K^{-1}$.

### 2.2.3 Interior Solution

We consider the motion to take place in a rectangular basin whose width (scaled by $L$) runs from $x = 0$ to $x = x_e$, while in the north–south direction the basin extends from $y = 0$ to $y = 1$. Outside of boundary layers on the basin's perimeter, we anticipate that the spatial scales of the dimensionless variables are order 1 so that the terms in (2.2.2.6) are appropriately measured by the parameters before each term. So, for small values of $\delta_m$ and $\delta_K$, we can approximate the equations in the interior by

$$\begin{aligned} -fv &= \eta_x, \\ fu &= \eta_y, \\ u_x + v_y &= 2w_i, \\ w_i &= \frac{b}{\delta_T}(\eta - \theta), \end{aligned} \qquad (2.2.3.1a, b, c, d)$$

leading to the *Sverdrup vorticity balance*,

$$bv = -2fw_i, \qquad (2.2.3.2)$$

where for small $b$ (beta plane) we can consider $f$ on the right-hand side to be essentially unity. With (2.2.3.1a) and (2.2.3.1c), this can be rewritten as

$$\eta_x - \frac{2f^2}{\delta_T}(\eta - \theta) = 0, \qquad (2.2.3.3)$$

where, again, within the quasi-geostrophic, beta-plane approximation, $f$ in (2.2.3.3) is equal to one.

For the buoyancy forcing, lets consider a forcing,

$$\theta = -\sin(\pi x/x_e)\sin(2\pi y). \tag{2.2.3.4}$$

The buoyancy forcing is a maximum (or minimum, depending on $y$) for values of longitude, $x$, in the very center of the basin (i.e., at $x = x_e/2$), and it will be interesting to check on where the maximum upwelling or sinking happens with respect to that position. The buoyancy forcing has a target profile that is negative in the range $0 \leq y \leq 1/2$ and positive in the range $1/2 \leq y \leq 1$. The former region corresponds to a region of heating while the more northerly half of our basin is being cooled. Note that there is no *net* heating or cooling.

The solution to (2.2.3.3) forced by (2.2.3.4) is simply

$$\eta = -\frac{\sin(2\pi y)}{\left[\frac{4}{\delta_T^2} + \frac{\pi^2}{x_e^2}\right]}\left\{\frac{4}{\delta_T^2}\sin(\pi x/x_e) + \frac{2\pi}{x_e\delta_T}\left(\cos(\pi x/x_e) + e^{-2(x_e-x)/\delta_T}\right)\right\}$$
$$+ N_e e^{-2(x_e-x)/\delta_T}, \tag{2.2.3.5}$$

where $N_e$ is at this stage an arbitrary constant and represents the interface height on the eastern boundary. We have anticipated the well-known fact that satisfaction of the no-normal flow condition by the interior solution must be done on the eastern boundary. Note that the fundamental equation in the interior in $\eta$ is only first order in $x$ so its solution in the interior can't satisfy boundary conditions on both $x = 0$ and $x_e$. Note also that we have used a homogenous solution to force the particular solution to go to zero on the eastern boundary. That homogeneous solution decays *westward* on the scale $\delta_T$ and can be interpreted as an arrested long, baroclinic Rossby wave excited at the eastern boundary by the boundary condition. There is an additional portion of the homogeneous solution with an amplitude $N_e$ that is undetermined at this stage.

The vertical velocity across the interface obtained from (2.2.3.5) and thus valid in the interior of the fluid follows from (2.2.3.1d),

$$w_i = \frac{b}{\delta_T}\frac{\sin(2\pi y)}{\left[\frac{4}{\delta_T^2} + \frac{\pi^2}{x_e^2}\right]}\left[\frac{\pi^2}{x_e^2}\sin(\pi x/x_e) - \frac{2\pi}{x_e\delta_T}\left[\cos(\pi x/x_e) + e^{-2(x_e-x)/\delta_T}\right]\right]$$
$$+ \frac{b}{\delta_T}N_e e^{-2(x_e-x)/\delta_T}. \tag{2.2.3.6}$$

We are going to show, for the forcing in this example, that $N_e$ is zero but that will have to wait until we discuss the boundary layer structure on the western boundary at

$x = 0$. Note that the *interior* meridional velocity is given directly from (2.2.3.6) using the Sverdrup relation (2.2.3.2).

### 2.2.4 Boundary Layer Structure

#### 2.2.4.1 The diffusion layer

In the parameter range $\delta_m \ll \delta_K$, the boundary layer structure near the western boundary consists, as in Section 2.1, of a set of two nested boundary layers. The wider one has a nondimensional thickness

$$\delta_K = \frac{\kappa}{\beta L_d^2 L}, \tag{2.2.4.1}$$

while within it there is a narrower layer whose width is

$$\delta_h = \left(\frac{\delta_m^3}{\delta_K}\right)^{1/2}, \tag{2.2.4.2}$$

which we noted earlier is the scale of the hydrostatic layer of the previous lecture and which is independent of $\beta$. The ratio of the two boundary layer widths is

$$\frac{\delta_h}{\delta_K} = \left(\frac{\delta_m}{\delta_K}\right)^{3/2} \ll 1 \tag{2.2.4.3}$$

and so is small under the restriction $\delta_m \ll \delta_K$.

Since our problem is linear, we can modify the interior solution in the boundary layers by adding *correction* functions to the interior solution.

Thus, in the diffusion layer, the wider of the two boundary layers, the correction functions, denoted by a subscript $K$, can be written

$$[u_K, v_K, w_{iK}, \eta_K] = \frac{1}{\delta_K}[\delta_K \tilde{u}, \tilde{v}, \tilde{w}_i, \delta_K \tilde{\eta}], \tag{2.2.4.4}$$

where the tilde variables are order one. The relative order of the variables is determined by the boundary layer scale and so is locally determined independent of the precise shape of the enclosing basin.

This implies that the zonal velocity is of the same order, order one, as the interior zonal velocity but that the meridional velocity is much larger, order $\delta_K^{-1}$. This is because the net meridional flux in the interior is returned in the western boundary layer, giving rise to large velocities inversely proportional to the width of the region of return flow. As we will see, it is this boundary layer that closes the horizontal circulation. Note, too, that the *vertical* velocity is much larger, by the same factor as the interior vertical velocity. The variables in this diffusion layer are functions of the boundary layer variable

$$\xi = x/\delta_K. \tag{2.2.4.5}$$

To order $(\delta_m/\delta_K)$ the equations for the correction functions are

$$\begin{aligned} f\tilde{u} &= \tilde{\eta}_y, \\ f\tilde{v} &= -\tilde{\eta}_\xi, \\ \tilde{u}_\xi + \tilde{v}_y &= 2\tilde{w}_i, \\ \tilde{w}_i &= -b\tilde{\eta}_{\xi\xi}, \end{aligned} \tag{2.2.4.6}$$

so that in this layer the dominant term producing the vertical velocity is the final term on the right-hand side of (2.2.2.6 d), our parameterization of horizontal mixing.

Eliminating all variables with respect to the interface height correction,

$$\tilde{\eta}_{\xi\xi} - \frac{1}{2f^2}\tilde{\eta}_\xi = 0. \tag{2.2.4.7}$$

The functions marked by the tilde are correction variables. They must therefore vanish for large values of the boundary layer coordinate where the solution must merge smoothly to the interior solution. That being the case, the only solution of (2.2.4.7) satisfying that requirement is

$$\tilde{\eta} = C_K e^{-\xi/2}, \tag{2.2.4.8}$$

where again, within the quasi-geostrophic beta-plane approximation, $f$ (nondimensional) can be approximated by unity. The form of (2.2.4.8) demonstrates that this boundary layer can only occur on the *western* boundary since its single decaying solution decays to the east. It is important to note that the *total* vertical mass flux correction in this boundary layer is simply

$$\delta_K \int_0^\infty \frac{1}{\delta_K}\tilde{w}(\xi, y)d\xi = -b\int_0^\infty \tilde{\eta}_{\xi\xi} d\xi = b\tilde{\eta}_\xi(0, y), \tag{2.2.4.9}$$

so that the mass flux in the boundary layer is determined by the horizontal gradient, at the wall, of the interface height. This is the analogue in the two-layer system of the horizontal temperature gradient at the wall and similar to the same result we found in Section 2.1.

To satisfy the full boundary conditions, we are going to need an additional correction in the thinner $\delta_h$ layer to satisfy the no-slip condition. However, we can satisfy the no-normal flow condition with just the $\delta_K$ layer. This is equivalent, because both the interior and the diffusion layer are in geostrophic balance for $u$, to making $\eta$ a constant on $x = 0$. Doing so leads to a solution for $\eta$ in the interior plus the diffusion layer.

$$\eta = -\frac{\sin(2\pi y)}{\left[\frac{4}{\delta_T^2} + \frac{\pi^2}{x_e^2}\right]} \left\{ \begin{array}{l} \frac{4}{\delta_T^2}\sin(\pi x/x_e) \\ +\frac{2\pi}{x_e\delta_T}\left(\cos(\pi x/x_e) + e^{-2(x_e - x)/\delta_T}\right)(1 - e^{-\xi/2}) \end{array} \right\} + N_e e^{-2(x_e - x)/\delta_T}. \tag{2.2.4.10}$$

The correction to the vertical velocity in the diffusion layer is

$$w_{iK} = -\frac{b}{\delta_K\left[\frac{4}{\delta_K^2} + \frac{\pi^2}{x_e^2}\right]}\left\{\frac{\pi}{2x_e\delta_T}\left[1 + e^{-2x_e/\delta_T}\right]e^{-\xi/2}\right\}\sin 2\pi y. \tag{2.2.4.11}$$

In this quasi-geostrophic system, the total horizontal southward transport in the interior is returned in the $\delta_K$ layer. However, since both the components of the velocity are geostrophic, it follows that the *sum* of the interior and diffusion layer variables satisfies the Sverdrup balance (2.2.3.2). It then follows directly that since the integral of $v$ across the basin must vanish for each $y$,

$$b\int_0^{x_e} v\,dx = 0 = -2\int_0^{x_e} w_i\,dx, \tag{2.2.4.12}$$

so that the integral of the vertical, cross-isopycnal flux in the interior must be equal and opposite to the flux in the diffusion layer, and this can be verified directly from the explicit solutions. Where then does the net flux occur? We have to examine that very carefully, but it is clear that it will not be occurring directly under the heating and cooling that occur in the middle of the basin.

#### 2.2.4.2 *The Hydrostatic Layer*

The thinnest layer has a thickness $\delta_h = \left(\frac{\delta_m^3}{\delta_K}\right)^{1/2}$, which is always less than the diffusion layer thickness as long as $\delta_m \ll \delta_K$, which is the region of parameter space we are concentrating on. Recall that this length is independent of $\beta$ and satisfies the same dynamics as the hydrostatic layer on the $f$ plane (no $\beta$). In this layer, the corrections to the solution consisting of the sum of the interior and diffusion layer variables are given by

$$(u_h, v_h, w_{ih}, \eta_h) = \left(\frac{\delta_h}{\delta_K}\hat{u}, \frac{1}{\delta_K}\hat{v}, \frac{b}{\delta_h}\hat{w}, \frac{\delta_h}{\delta_K}\hat{\eta}\right). \tag{2.2.4.13}$$

In this layer, the correction to the zonal velocity is very small, which is why the diffusion layer must, on its own, satisfy the no-normal flow condition. On the other hand, the meridional velocity is of the same order as the strong meridional flow in the $\delta_K$ layer and it will be used to balance that velocity to satisfy the no-slip condition on $v$. The meridional velocity in this sublayer is as large as the velocity in the diffusion layer, but, of course, the meridional transport in this layer is smaller than in the diffusion layer by the ratio (small) of the two layer thicknesses. Note that the magnitude of the vertical transport in the hydrostatic layer is the same order as in the diffusive layer, which, as we have seen, is the same as the total vertical transport in the interior. All the variables in (2.2.4.13) are functions of $s = x/\delta_h$ in terms of which the leading order balances are

$$
\begin{aligned}
f\hat{u} &= \hat{\eta}_y + b\left(\frac{\delta_m}{\delta_h}\right)^3 \hat{v}_{ss}, \\
f\hat{v} &= -\hat{\eta}_s, \\
\hat{u}_s + \hat{v}_y &= 2b\frac{\delta_K}{\delta_h}\hat{w}, \\
\hat{w} &= -\hat{\eta}_{ss},
\end{aligned}
\qquad (2.2.4.14\text{a, b, c, d})
$$

which, upon eliminating all variables in favor of $\hat{\eta}$, leads to the governing equation in the hydrostatic layer,

$$\hat{\eta}_{ss} - 2\hat{\eta} = 0, \qquad (2.2.4.15)$$

which is similar to the partial differential equation for the hydrostatic layer of Section 2.1 except that now the structure in $z$ is frozen to be a first baroclinic mode. The exponential structure in the horizontal variable is, as before, symmetric in that variable (i.e., the equation is unaltered if $s \to -s$), so this boundary layer could equally well exist on the eastern boundary. However, its amplitude is much larger on the western boundary because its function is to bring to zero the large meridional velocity in the diffusive $\delta_K$ layer, which exists only on the western boundary. The solution to (2.2.4.15) is

$$\hat{\eta} = D(y)e^{-s\sqrt{2}}. \qquad (2.2.4.16)$$

It is important to note that, again, the vertical mass flux carried in the hydrostatic layer is given by the integral of (2.2.4.14d) so that

$$\delta_h \int_0^\infty w_{ih} ds = b \int_0^\infty \hat{w}_i ds = b\hat{\eta}_s(0, y), \qquad (2.2.4.17)$$

so that the sum of the vertical transports of the diffusive and hydrostatic layers, $W_K + W_h$ is simply

$$W_K + W_h = b[\hat{\eta}_s(0,) + \tilde{\eta}_\xi(0,)]. \qquad (2.2.4.18)$$

It is interesting to note, in connection to the first lecture, that the term

$$b\left(\frac{\delta_m}{\delta_h}\right)^3 = \frac{E}{(\sigma S)^{3/2}} = \left(\frac{E^{2/3}}{\sigma S}\right)^{3/2} \ll 1 \qquad (2.2.4.19)$$

in the parameter regime in which the hydrostatic layer of Section 2.1 is valid. In deriving (2.2.4.19), the identification $S = \frac{L_d^2}{L^2}$ has been used.

### 2.2.5 Matching

To complete the solution and determine the boundary layer contribution to the solution, we need to use all the boundary layer corrections to the interior solution to satisfy the boundary condition on $x = 0$. So, for example, the condition on the zonal velocity is (here we temporarily use a subscript $I$ to indicate the interior solution)

$$u_I + \tilde{u} + \frac{\delta_h}{\delta_K}\hat{u} = 0, \quad x = 0, \tag{2.2.5.1}$$

so that to lowest order it is, as we have anticipated, the diffusion layer that satisfies the normal flow condition and the contribution from the hydrostatic layer is negligible at lowest order. The zonal velocity generated in the hydrostatic layer is too small to enter the condition to order ${}^{\delta_h}/_{\delta_K}$.

The no-slip condition for the meridional velocity is

$$v_I + \frac{1}{\delta_K}\left(\tilde{v} + \hat{v}\right) = 0, \quad x = 0, \tag{2.2.5.2}$$

so this implies that $D$ in (2.2.4.16) is determined by annulling the meridional velocity in the $\delta_K$ layer at $x = 0$. Thus,

$$D = -\frac{\pi}{2^{1/2} x_e \delta_T}\left(1 + e^{-2x_e/\delta_T}\right)\sin(2\pi y)\frac{1}{\left[\frac{4}{\delta_T^2} + \frac{\pi^2}{x_e^2}\right]}, \tag{2.2.5.3}$$

so that $v$ can be written

$$v = \frac{\sin 2\pi y}{\left[\frac{4}{\delta_T^2} + \frac{\pi^2}{x_e^2}\right]}\left[\begin{array}{l} \frac{4\pi}{\delta_T^2 x_e}\{\cos \pi x/x_e - e^{-2^{1/2}x/\delta_h}\} \\ + \frac{2\pi}{x_e \delta_T}\left\{-\frac{\pi}{x_e}\sin(\pi x/x_e) + \frac{2}{\delta_T}e^{-2(x_e - x)/\delta_T}\right\}\left\{1 - e^{-2^{1/2}x/\delta_T}\right\} \\ + \frac{\pi}{x_e \delta_T \delta_K}\left\{\cos(\pi x/x_e) + e^{-2(x_e - x)/\delta_T}\right\}\left(e^{-x/2\delta_K} - e^{-2^{1/2}x/\delta_h}\right) \end{array}\right], \tag{2.2.5.4}$$

where the dominant terms are clearly the final terms in the square bracket of order $\delta_K^{-1}$.

This means that to lowest order $[\hat{\eta}_s(0, y) + \tilde{\eta}_\xi(0, y)] = 0$ so that the *vertical mass flux in the diffusion layer is equal to and opposite the vertical flux in the hydrostatic layer*. That is, the diffusion layer vertical flux is equal and opposite to both the total interior vertical mass flux and the hydrostatic layer's vertical mass flux. In turn, this implies that the net vertical mass flux could be calculated from the interior solution.

Now that condition, for the interior, is simply, to conserve total mass,

$$\int_0^1 \int_0^{x_e} (\eta - h)\, dx\, dy = 0, \tag{2.2.5.5}$$

but given the form of $h$ (i.e., its $\sin 2\pi y$ structure), it integrates to zero. Since to lowest order mass is conserved in each layer, the constraint (2.2.5.5) follows. Examining the form of the interior interface $\eta$ each term will similarly integrate to zero *except for the term in* $N_e$. Therefore, to satisfy the global conservation of mass, we must choose that hitherto arbitrary constant to be zero, as mentioned earlier, and that completes the solutions. Please refer to Pedlosky and Spall (2005) for another example where the forcing requires a nonzero value of the interface on the eastern boundary.

### *2.2.6 An Example*

We will use the forcing (2.2.3.4) to illustrate the nature of the solution, and we will see that it contains some surprising, even nonintuitive features.

One of the most important features of the steady oceanic circulation on the beta-plane is its western intensification. Ever since Stommel's original work (Stommel, 1948), there has been a clear understanding of the connection between the western intensification and the $\beta$-effect. The western intensification due to the $\beta$-effect reflects the physics of long Rossby waves propagating energy to the west (see Pedlosky1965). The western intensification is independent of the nature of the forcing and Figure 2.13, which shows the response of the interface height to the applied buoyancy forcing, shows the phenomenon very clearly.

Figure 2.13 shows that the interface pattern is displaced westward with respect to the forcing, which has its center, of both heating and cooling, in the center of each gyre. The two gyres are antisymmetric in this linear model. The subtropical gyre in the interval $0 \leq y \leq 1/2$ has a baroclinic circulation that is anticyclonic (recall that this means the deeper layer is moving cyclonically) and corresponds to a depression of the interface, that is, an increase in the volume of warm water due to the heating. The northern gyre in $1/2 \leq y \leq 1$ has a cyclonic baroclinic circulation and a raised interface corresponding to more dense water resulting from the northern cooling. The intensification to the west leads to strong western boundary currents (recall that for quasi-geostrophic motion the interface height is a streamfunction for the baroclinic velocity).

The displacement of the interface perturbation to the west of the maximum of the forcing means that the cross-isopycnal velocity will not be a simple reflection of the target profile of the buoyancy forcing; see (2.2.3.6). Figure 2.14 shows the interior distribution of $w_i$.

The curious and unexpected feature of Figure 2.14 is the region to the west of $x$ of just over 0.2, in which the fluid is *being cooled in the subtropical gyre and heated in the subpolar gyre*. As we shall see shortly, this is a consequence of the shift to the west of the interface elevation and the parameterization of vertical mixing by the simple but sensible recipe (2.2.3.1 d). We can understand this better by examining the profiles of the interface as shown in Figure 2.15a.

Figure 2.15a shows the interior distribution of the interface across the basin in the middle of the subpolar gyre as well as the cooling profile there.

The profile of the forcing as a function of $x$ at $y = 0.75$ (i.e., at the midlatitude of the subpolar gyre) is shown as the black curve in Figure 2.15b. This is where the

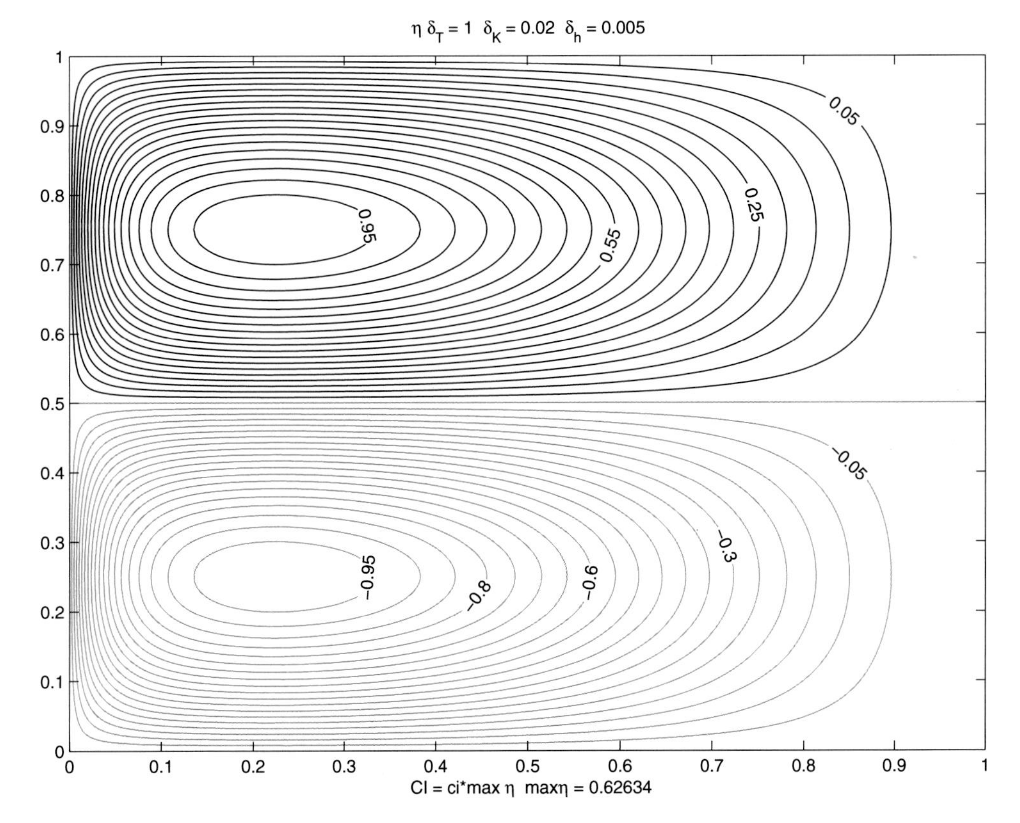


Figure 2.13. The interface height in response to the forcing (2.2.3.4). The contours shown in grey represent a negative interface anomaly (i.e., the lowering of the interface). Black contours denote a rise in the interface. Contours are given in units of $1/20^{\text{th}}$ of the maximum nondimensional value. 0.62634.

maximum forcing (i.e., the maximum of $h$) occurs and we would expect that to be where the maximum of cooling (i.e., the maximum of the transformation of upper layer water to lower layer water) to take place. This is not what we see!

Note that because of the beta-effect the maximum of the interface upward displacement is moved westward with respect to the maximum of the cooling. In particular, near the western side of the basin, but still external to the boundary layers there, the interface height actually exceeds the target height, $h$. This implies by (2.2.3.4) a *positive* cross-isopycnal velocity (i.e., a local heating). This is what we observe in Figure 2.15b, a region of interior heating in the western side of the subpolar gyre. The same structure occurs in the subtropical gyre, where cooling occurs to the west. We also observe in Figure 2.15a that the interface displacement has slid so much to the west that the difference between the interface height and $h$ is largest to the east of the cooling maximum, leading to an *eastward* shift of the cross-isopycnal velocity with respect to the forcing as shown in Figure 2.15a. This entire shift of the actual cooling

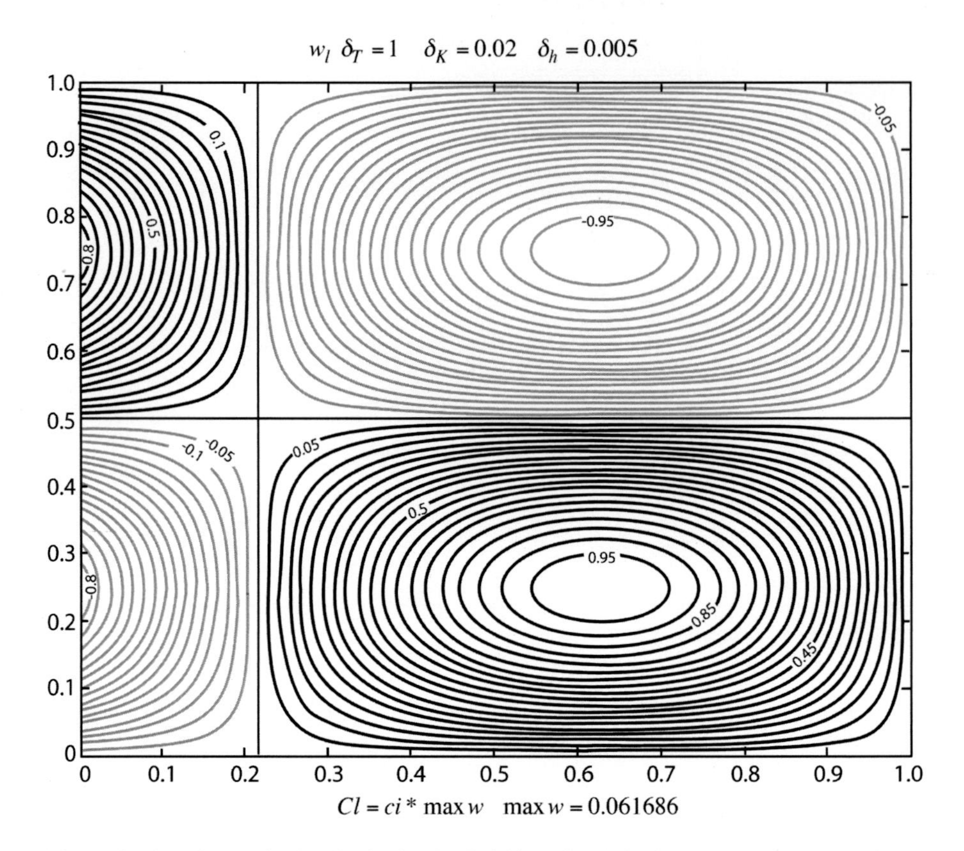


Figure 2.14. The vertical velocity in the fluid interior. Black contours correspond to positive values of $w_i$ (i.e., heating), and grey contours correspond to negative values. As in the previous figure, contour intervals are 1/20th of the maximum.

from the profile of the buoyancy forcing profile is due to the propagation properties of the beta-effect. The underlying wave nature of the system as a consequence has strong implications for the steady circulation.

In the quasi-geostrophic system the required closure of the horizontal circulation in the western $\delta_K$ layer implies, since (2.2.3.2) holds equally well in the diffusion layer as in the interior, that balancing the horizontal transports between the interior and the diffusion layer implies that the vertical transports of the two regions are always equal and opposite. I take this as a fundamental property of the system. A major implication is that the *net downwelling occurs in the narrowest region, the hydrostatic layer*. Since the downwelling in this region is, in fact, equal to the downwelling over the entire interior domain of the gyre, the vertical velocity in this region must be the largest of any part of the domain. This is clearly seen in Figure 2.15b, where the vertical velocity (in black) is many times larger than in either the interior or the diffusion layer.

A similar result was found for a quite different form of forcing in the work of Pedlosky and Spall (2005). Indeed, in that work, the Rossby wave propagation mechanism

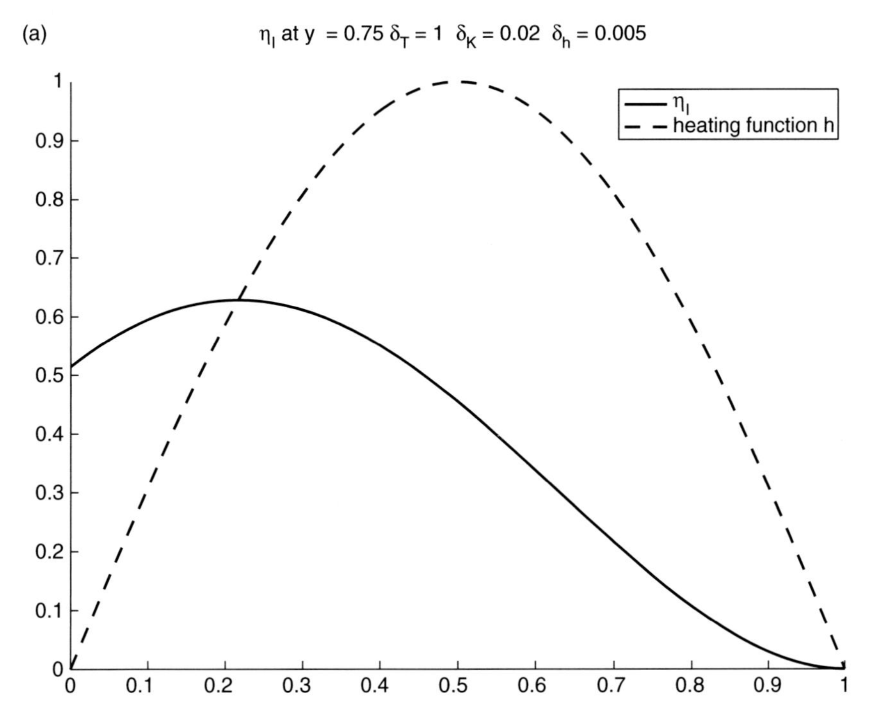


Figure 2.15 (a) The distribution of interface height (black line) and cooling target profile (dashed line) at $y = 0.75$. (b) The cross-isopycnal velocity as a function of $x$ at $y = \frac{3}{4}$ (long dashed line). The interface height is shown with the short dashed line and the profile of the forcing is in black.

in the nonlinear extension of the present analysis was replaced by a vigorous eddy field that, however, led to a similar displacement of the fields and again, the major vertical velocities occurred in the western boundary of the domain.

The western intensification, as seen in Figure 2.13 is also clearly evident in Figure 2.16, where the meridional velocity is shown as a function of $x$ at the same value of $y$ (i.e., 0.75).

The maximum value of $v$ occurs in the western boundary current. In the subpolar gyre, the current (in the upper layer) is southward for this purely buoyancy driven flow.

Thus, in contradistinction to the results in Section 2.1, where we examined flow on the $f$ plane, where $\beta = 0$, the region of strongest downwelling occurs not under the maximum of the negative buoyancy forcing but at the western boundary where friction allows the dissipation of the vorticity that would otherwise be produced by such strong vortex tube stretching.

The solution for the total $w_i$ is composed of the three components:

$$w_I = \frac{b}{\delta_T} \frac{\sin 2\pi y}{\left(\frac{4}{\delta_T^2} + \frac{\pi^2}{x_e^2}\right)} \left[ \frac{\pi^2}{x_e^2} \sin \pi \frac{x}{x_e} - \frac{2\pi}{x_e \delta_T} \left[ \cos \pi x / x_e + e^{-2(x_e - x)/\delta_T} \right] \right],$$

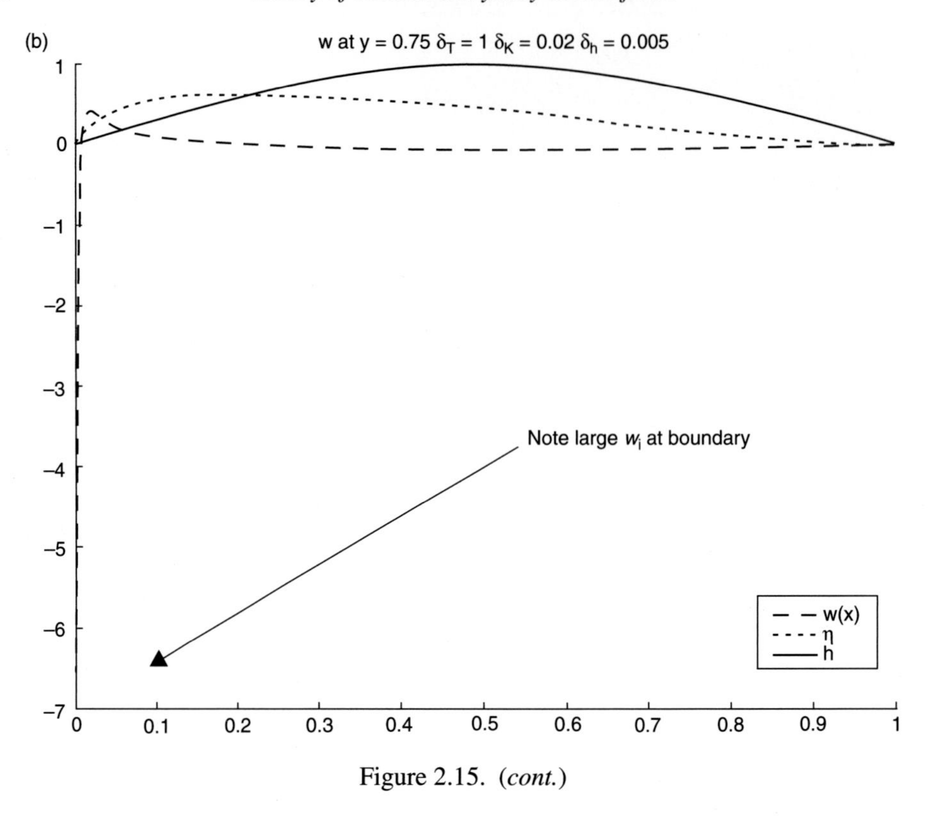


Figure 2.15. (*cont.*)

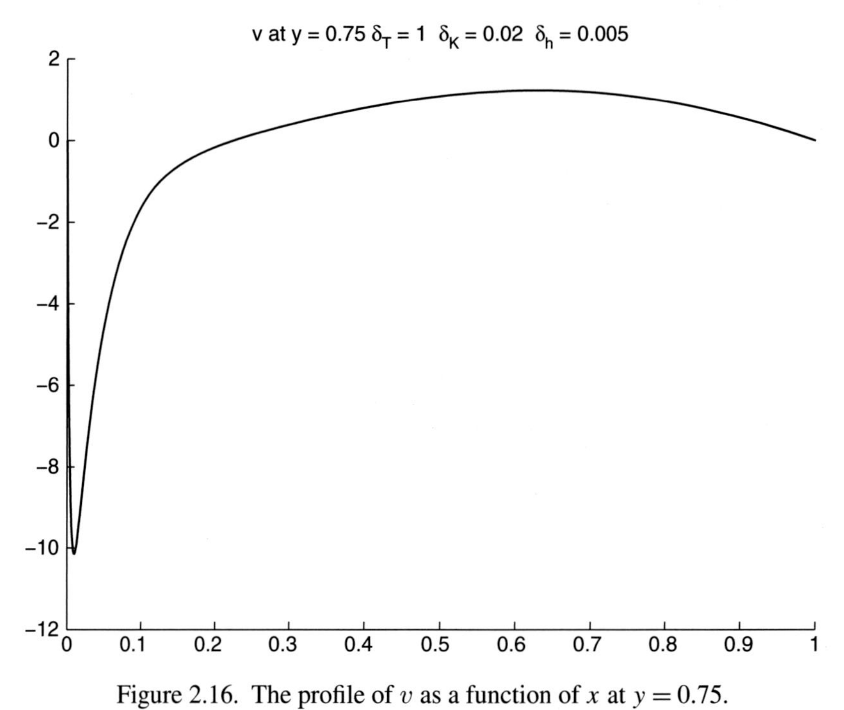


Figure 2.16. The profile of $v$ as a function of $x$ at $y = 0.75$.

$$w_K = \frac{b}{\delta_K} \frac{\sin 2\pi y}{\left(\frac{4}{\delta_T^2} + \frac{\pi^2}{x_e^2}\right)} \left[ -\frac{\pi}{2x_e\delta_T} \left(1 + e^{-2x_e/\delta_T}\right) e^{-x/2\delta_K} \right],$$

$$w_h = \frac{b}{\delta_h} \frac{\sin 2\pi y}{\left(\frac{4}{\delta_T^2} + \frac{\pi^2}{x_e^2}\right)} \left[ \frac{2^{1/2}\pi}{x_e\delta_T} \left(1 + e^{-2x_e/\delta_T}\right) e^{-2^{1/2}x/\delta_h} \right].$$

(2.2.6.1a, b, c)

Recent numerical work by Hristova (2009) includes the effects of eddies in reshaping the structure of the buoyancy-driven circulation. As in the problem of the wind-driven circulation, the eddies are responsible for both gross and subtle changes in the current structure, but the overall pattern is recognizable from the linear theory.

### *2.2.7 Nonlinear Theory*

We can make an attempt to extend our results of the previous section to include nonlinearity, although as we shall see, probably the most important aspect of the nonlinear problem is the development of a strong eddy field. Mike Spall touches on that aspect of the problem in Chapter 3 (this volume). In this section I am going to essentially reconsider the problem treated in this chapter *without* assuming the interface displacements are small, although I will assume that the velocities are small enough to keep the motion geostrophic. That is, I will be assuming *planetary geostrophy* only. To make matters as simple as possible, I will employ a 2½ layer model, that is, as shown in Figure 2.17, a model consisting of two homogeneous layers of different densities over a third, *resting* layer of greater density. The dynamics are assumed to be hydrostatic, geostrophic, and incompressible, much as in the previous section, except, as mentioned earlier, without assuming that the interface deformations are small compared to the mean thickness.

Use of the hydrostatic equations in each of the moving layers, along with the condition that the horizontal pressure gradient in layer 3 is zero (since it is assumed

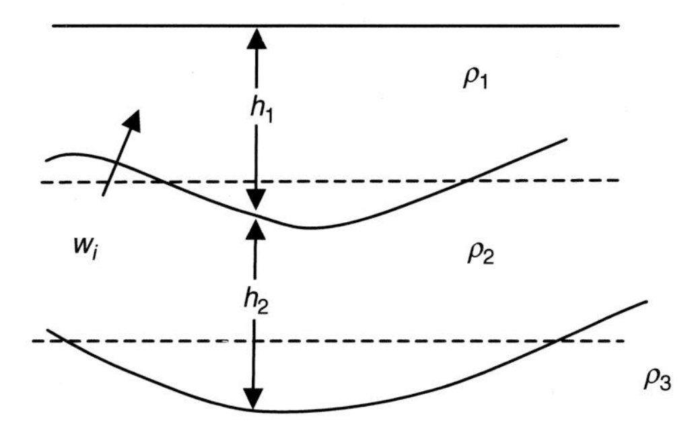


Figure 2.17. The 2½ layer model: two moving layers over a resting abyss with deep density $\rho_3$.

at rest), leads to the following geostrophic relations:

$$\begin{aligned} f v_2 &= \gamma_2 h_x, \quad f u_2 = -\gamma_2 h_y, \\ f v_1 &= \gamma_2 h_x + \gamma_1 h_{1x}, \quad f u_1 = -\gamma_2 h_y - \gamma_1 h_{1y}, \end{aligned} \tag{2.2.7.1a, b, c, d}$$

where

$$\gamma_n = \frac{\rho_{n+1} - \rho_n}{\rho_0}, \quad h = h_1 + h_2, \; n = 1, 2, \tag{2.2.7.2a, b}$$

and where $\rho_0$ is the mean density of the fluid in this Boussinesq model. As before, the buoyancy forcing will be represented by a cross-isopycnal velocity between layers 1 and 2, in terms of which the mass conservation statement for each layer in the steady state is

$$(h_n u_n)_x + (h_n v_n)_y = -(-1)^n w_i, \quad n = 1, 2. \tag{2.2.7.3}$$

The Sverdrup relation, when integrated vertically, implies that in the absence of wind stress and its accompanying Ekman pumping, the total barotropic meridional transport in the ocean interior will be zero. For our two-layer model, this yields

$$\begin{aligned} & h_1 v_1 + h_2 v_2 = 0, \\ & \rightarrow \frac{1}{2f} \frac{\partial}{\partial x} \{\gamma_2 h^2 + \gamma_1 h_1^2\} = 0, \end{aligned} \tag{2.2.7.4a, b}$$

or integrating in $x$,

$$h^2 + \frac{\gamma_1}{\gamma_2} h_1^2 = H^2 + \frac{\gamma_1}{\gamma_2} H_1^2. \tag{2.2.7.5}$$

In (2.2.7.5), the *constants* $H$ and $H_1$ are the values of the total depth $h$ of the moving fluid and the depth of the upper layer on the eastern boundary where the layer depths must be constant to avoid flow into the boundary. Note that (2.2.7.5) yields $h_1$ as a function of $h$.

The vorticity equation for layer 2 is derived by eliminating the pressure (i.e., layer thicknesses) between (2.2.7.1a,b,c,d) to obtain

$$\frac{df}{dy} v_2 + f(u_{2x} + v_{2y}) = 0, \tag{2.2.7.6}$$

and with the use of the mass conservation equation this becomes

$$u_2 \frac{\partial}{\partial x} \frac{f}{h_2} + v_2 \frac{\partial}{\partial y} \frac{f}{h_2} = \frac{f}{h_2^2} w_i. \tag{2.2.7.7}$$

In the absence of cross-isopycnal flow the *potential vorticity*, $f/h_2$, would be conserved on streamlines in the lower layer. Nonadiabatic effects of heating and cooling will alter the potential vorticity (pv) on fluid columns in each layer. The student is invited to derive the similar equation for the upper layer. Using (2.2.7.7), the $x$ derivative of (2.2.7.5) and the obvious relation $h_2 = h - h_1$, we can finally obtain, after a bit of algebra,

$$c_r h_x = \frac{\gamma_1}{\gamma_2}\frac{h_1}{h} w_i, \tag{2.2.7.8}$$

where

$$c_r = \beta \frac{\gamma_1 h_1 h_2}{f^2 h} \equiv \beta L_d^2, \tag{2.2.7.9}$$

which is the nonlinear generalization of the long, baroclinic Rossby wave speed to the west and where the deformation radius is a function of position through the variations of the layer depths. Now, (2.2.7.8) can be written as first-order differential equation for $h$

$$\frac{\beta}{f^2} h_2 \frac{\partial h}{\partial x} = \frac{1}{\gamma_2} w_i. \tag{2.2.7.10}$$

Since

$$\begin{aligned} h_1 &= \left[\frac{\gamma_2}{\gamma_1}\left(H^2 - h^2\right) + H_1^2\right]^{1/2}, \\ h_2 &= h - h_1. \end{aligned} \tag{2.2.7.11a, b}$$

It follows that (2.2.7.10) could be written in terms of $h$ alone if the cross-isopycnal velocity can be parameterized in terms of the layer thicknesses.

We can choose to use the same formulation as in the previous sections. That is, we write

$$w_i = \gamma(\eta - \eta_o), \tag{2.2.7.12}$$

where $\gamma$ (not to be confused with the reduced gravities $\gamma_n$) is the same decay time scale for the Rossby wave due to vertical mixing as in the linear model, and $\eta_o(x, y)$ is the target profile for the interface between layers one and two and $\eta$ is actual the interface perturbation. I have slightly changed notation because of the use of $h$ for layer thickness. Note that the interface perturbation is just $H_1 - h_1$.

If we scale layer depths with $H$, the total depth of the layers on the eastern boundary, and we scale $x$ with the meridional scale of the region, we can rewrite our system as

$$\begin{aligned} \frac{\partial h}{\partial x} &= \frac{1}{\delta_T}(1 + by)^2 \left[H_1 - h_1 - \eta_o\right]/h_2, \\ h_1 &= \left\{\Gamma_{12}(1 - h^2) + H_1^2\right\}, \\ h_2 &= h - h_1, \\ \delta_T &= \frac{\gamma_2 \beta H}{f_o^2 \gamma L}, \quad b = \frac{\beta L}{f_o}, \quad \Gamma_{12} = \frac{\gamma_1}{\gamma_2}, \end{aligned} \tag{2.2.7.13a, b, c, d, e, f}$$

and recall that all depths are scaled by $H$.

After choosing the target profile for the upper layer thickness, we integrate the above equation *westward* from the eastern boundary where the layer thicknesses are specified.

We will consider the region $0 \le y \le 1$ as the subpolar gyre of the two-gyre problem we have already discussed and choose a target profile, which, if it were extended to the region $-1 \le y \le 0$, would be the antisymmetric forcing of the subpolar gyre we consider here. That is, we choose the target profile for the interface to be

$$\eta_o = h_o \sin \pi y \sin \pi x/x_e \tag{2.2.7.13}$$

in terms of which the cross-isopycnal velocity is

$$w_i = \frac{1}{\delta_T}(H_1 - h_1 - \eta_o). \tag{2.2.7.14}$$

The parameter $h_o$ is the relative amplitude parameter of the forcing, and for this nonlinear problem, the nature of the results will depend on the forcing.

For small values of the forcing (e.g., for $h_o = 0.1$), and the same parameters as in the linear problem ($\delta_T = 1$), the circulation is much as shown in the northern portion of Figure 2.13. What is more revealing is examining the profiles of the layer depths (e.g., the interfaces) and the induced cross-isopycnal velocity. Figure 2.18 shows the result for this small value of forcing, and it is essentially the same as the linear result.

Note that, as before, the beta-effect has produced a displacement of the maximum interface perturbation from the midpoint of the gyre at $x/x_e = ½$ to a position to the west. The cross-isopycnal velocity similarly shows cooling to the east and heating (i.e.,

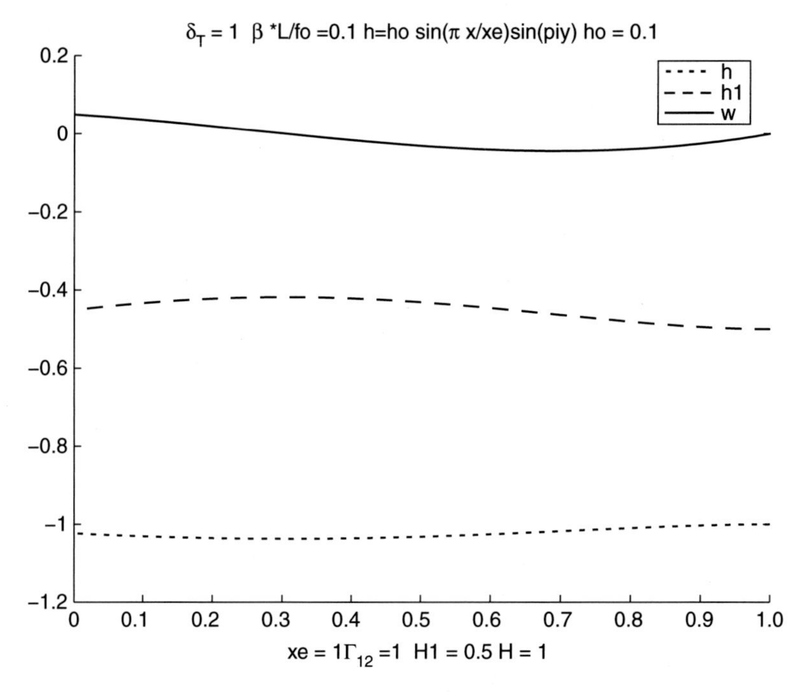


Figure 2.18. The layer thicknesses for the case $h_o = 0.1$. The solid line curve shows the (nondimensional) cross isopycnal velocity.

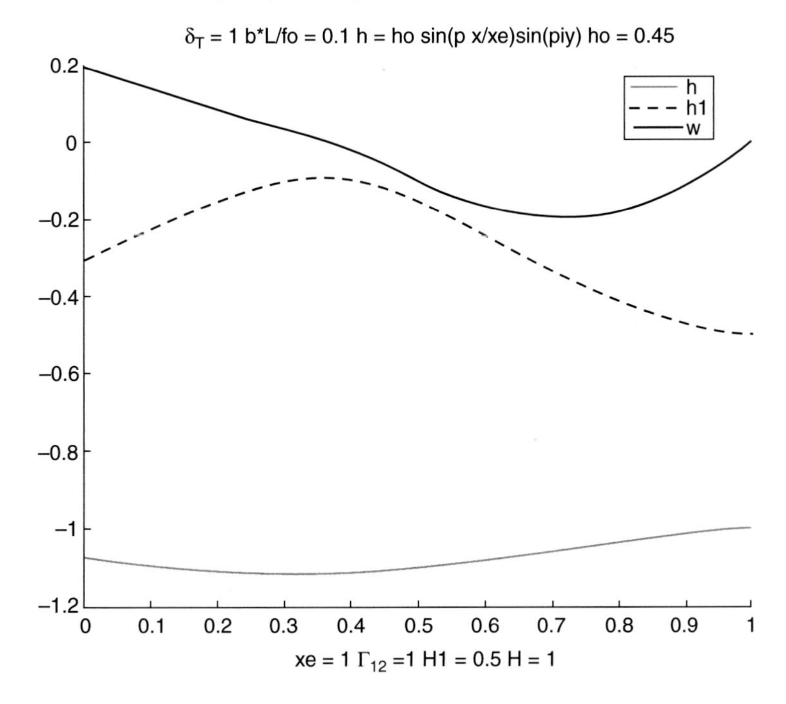


Figure 2.19. As in Figure 2.18 but now the cooling has increased, $h_o = 0.45$.

positive $w_i$ to the west due to the westward propagation of the interface perturbation). When the forcing, in this case cooling, is increased, there is increased flow from the upper layer into the lower layer as a result, and the upper layer becomes very thin as shown in Figure 2.19.

Now the upper layer is strongly distorted by the cooling and becomes very shallow in the western portion of the basin before the induced heating restores its thickness somewhat. At values of the cooling parameter $h_o$ slightly greater than this value, the thickness of layer 1 goes to zero and layer 2 outcrops within the basin. The simple theory here then breaks down as the representation of the forcing is no longer valid. A more complex theory is required. Also note that we have not included the western boundary current whose dynamics becomes quite difficult to unravel in the nonlinear limit. All these limits of validity point to directions of possible future work.

## 2.3 Buoyancy Forced Flows with Weak Stratification: Downstream Variation Effects

### *2.3.1 Introduction*

The location of the sinking of cooled water in polar regions is one of the fundamental issues that needs clarification for the theory of the ocean's overturning circulation. Recent work on that sinking (e.g., Pedlosky and Spall 2005) has emphasized the enhancement of the sinking in the vicinity of lateral boundaries of the basin where the vorticity produced by stretching can be dissipated by friction. There have been many other studies of the process, in which the sinking is supposed to occur in boundary regions with significant vertical stratification. We examined similar processes in the first two of these lectures. Recall that the primary parameter restriction on both of those lectures was that the parameters measuring friction and stratification satisfied

$$\sigma S \gg E^{2/3} \tag{2.3.1.1}$$

and this was a requirement for the existence of the sidewall boundary layer we named the hydrostatic layer, whose width was the Rossby deformation radius slightly modified by the Prandtl number $\sigma$. We are going to examine now a problem in which the inequality in (3.1.1) is reversed. (i.e., a case of thermal forcing for which the background stratification is very weak.)

The motivation is provided by a recent numerical study (Spall 2008) that considered the downwelling induced by buoyancy loss (cooling) in a boundary current in polar regions. The cooled boundary current is transformed by the cooling suffered by the fluid as it moves in the downstream direction and induces a boundary-intensified region of sinking. The oceanographic context is the cooling experienced by the boundary current in the Greenland Sea as it flows counterclockwise around the boundary of its basin. To study the process in a simpler geometry, Spall considered a current that was introduced at the entrance to a channel, and cooling, uniform in the down-channel direction, produced an evolution of the current in that direction such that an along-channel pressure gradient in geostrophic balance drives fluid to the right-hand boundary of the channel (looking downstream) where it underwent strong sinking. The principal physical novelty is the downwelling induced by the interior *geostrophic* onshore flow.

In contrast to earlier work the model develops a mixed layer of very weak but nonzero vertical stratification in which the sinking occurs but the lateral temperature gradients in the layer drive a geostrophic flow forcing the downwelling. This contributes to making the sinking region extremely narrow, and the narrowness of the sinking region is such that only a few grid points in the calculation represents the boundary layer, so its spatial resolution is marginal. Although it is not thought that this affects the overall strength of the downwelled fluid, it appears conceptually of interest to try to resolve the structure of the dynamics with a simple analytical model and that is the subject of the present lecture. The details can be found in Pedlosky (2009) from which much of the present lecture is taken. It will also give us an opportunity to examine the nature of the circulation in the limit $\sigma S \ll E^{2/3}$. (Although the parameter $D/L$, the ratio of the layer depth to its width, leads to a minor change in this condition.)

One of the curious features of Spall's numerical results is the nonmonotonic behavior of the along-channel flow near the boundary. The numerical model has a double boundary layer structure in which a broad Prandtl-type boundary layer appears to act to satisfy the no-slip condition on the along-channel flow, and yet as the boundary is approached within this layer, $u$, the along-channel flow, *increases* before finally being brought to zero in a very narrow region within the no-slip layer. In the discussion which follows a very simple linear model of a weakly stratified fluid, cooled at the upper surface, is employed to discuss, in particular, the inner region of the boundary layer where the overshoot of $u$ occurs and where the strong sinking is found. The use of this linear model is suggested by the relative insensitivity in the numerical model of Spall (2008) to the degree of nonlinearity. Indeed, Spall suggested the layer was a modified form of the nonhydrostatic Stewartson layer (Stewartson 1957) found in the theory of homogeneous rotating fluids and which we touched on in Lecture 1. There is much that is unrealistic in the analytic model, and yet its ability to reproduce salient features of the full numerical model implies that those features are robust and not dependent on the nonlinear nature of the original calculation.

### 2.3.2 The Model

We consider the flow in a channel of width $L$ and depth $D$ where $D \ll L$. The fluid in the channel is cooled at the surface at a rate $H$ such that at the upper surface, $z = D$,

$$\kappa_v \frac{\partial T}{\partial z} = H(x, y), \tag{2.3.2.1}$$

where $H$ is negative and $T$ is the temperature *anomaly* above a weak background vertical gradient, that is,

$$T_{total} = \Delta T_v z/D + T. \tag{2.3.2.2}$$

$\kappa_v$ is the thermal diffusivity in the vertical direction and $x$ and $y$ are the along-channel and cross-channel coordinates, respectively. The cooling anomaly will be small enough so that even though the basic stratification is small, the stratification including the anomaly will remain statically stable.

The independent variables are scaled

$$(x, y, z) = (Lx', Ly', Dz'). \tag{2.3.2.3}$$

The temperature anomaly is scaled, using (2.3.2.1)

$$T = \frac{H_o D}{\kappa_v} T', \tag{2.3.2.4}$$

where $H_o$ is the characteristic magnitude of the buoyancy forcing. The horizontal and vertical velocities and the pressure are scaled in expectation of a geostrophic and hydrostatic balance holding over most of the domain, that is,

$$(u, v, w) = \frac{g\alpha H_o D^2}{f\kappa_v L}\left(u', v', \frac{D}{L}w'\right),$$
$$p = \rho_o \frac{g\alpha H_o D^2}{\kappa_v} p', \qquad (2.3.2.5a, b)$$

where $\alpha$ is the coefficient of thermal expansion, $g$ is the acceleration due to gravity, $\rho_o$ is the constant reference density, and $f$ is the constant Coriolis parameter. The Rossby number of the flow

$$\varepsilon = \frac{g\alpha H D^2}{f^2 \kappa_v L^2} \qquad (2.3.2.6)$$

will be assumed small enough so that linear theory will be uniformly applicable. The parameter measuring the background stratification, $S$, the Burger number, is

$$S = \frac{g\alpha \Delta T_v D}{f^2 L^2}. \qquad (2.3.2.7)$$

The nondimensional linearized equations of motion for this incompressible fluid on the $f$ plane are

$$u = -p_y + \frac{E_H}{2}\nabla^2 v + \frac{E_v}{2} v_{zz},$$
$$-v = -p_x + \frac{E_H}{2}\nabla^2 u + \frac{E_v}{2} u_{zz},$$
$$0 = -p_z + T + \frac{D^2}{L^2}\left[\frac{E_H}{2}\nabla^2 w + \frac{E_v}{2} w_{zz}\right], \qquad (2.2.8a, b, c, d, e)$$
$$0 = u_x + v_y + w_z,$$
$$wS = \frac{E_H}{2\sigma_H}\nabla^2 T + \frac{E_V}{2\sigma_v} T_{zz}.$$

The primes on the dimensionless quantities have been dropped, and the Laplacian operator in (2.3.2.8) is the horizontal Laplacian, $\nabla^2 = \frac{\partial^2}{\partial x^2} + \frac{\partial^2}{\partial y^2}$. Subscripts denote differentiation with respect to the independent variable. The parameters

$$(E_H, E_v) = \frac{2}{f L^2}(\nu_H, \nu_v L^2/D^2) \qquad (2.3.2.9)$$

and the Prandtl numbers are the ratios

$$\sigma_H = {}^{\nu_H}/_{\kappa_H}, \quad \sigma_v = {}^{\nu_v}/_{\kappa_v}. \qquad (2.3.2.10a, b)$$

Different viscosity coefficients have been introduced for lateral and vertical momentum mixing, but given the simplicity of the model and the weak stratification that will be assumed, such detailed assumptions about the anisotropy of the mixing are problematic and not emphasized. The theory is qualitatively unchanged if the mixing coefficients are isotropic.

The boundary conditions are

$$\begin{aligned} T_z &= H, \ z = 1, \\ T &= 0, \quad z = 0, \\ T_y &= u = v = w = 0, y = 0, 1, \\ u_z &= v_z = w = 0, \ z = 0, 1. \end{aligned} \tag{2.2.11a, b, c, d}$$

The thermal conditions represent a nonuniform heating at the upper boundary and a fixed temperature at the lower boundary, which is our substitute in this simple model for a fairly passive fluid layer beneath. The sidewalls are thermally insulated.

The last condition in (2.3.2.11) expresses the condition of *no stress* at the top and bottom of the fluid layer, which seems appropriate as a model of a cooled mixed layer not in contact with rigid horizontal boundaries. That condition, through the use of Ekman layers at the top and the bottom of the layer, is translated into a condition on the vertical velocity at the edge of the very thin Ekman layers,

$$w = \mp \frac{E_v}{2} \frac{\partial}{\partial z} \left[ v_x - u_y \right], \ z = \begin{pmatrix} 1 \\ 0 \end{pmatrix}, \tag{2.3.2.12}$$

where $z = 0$ and 1 are understood to be at the edge of the Ekman layer. Note that the vertical velocity pumped out by an order one velocity is $O(E_v)$ and so much smaller than the $O(E_v^{1/2})$ velocity that one would have if the boundaries were solid and no-slip.

We will be particularly interested in the limit where $S$ is small, but will insist that the temperature anomaly is small enough to maintain a stable stratification consistent with our linearization. Our interest will be focused on the region between the Ekman layers and, in particular, on the boundary layers on the sides of the channel where we anticipate the major vertical motion will occur. We will consider the parameter limit of weak stratification expressed by (see Section 2.3.4.4)

$$\sigma_H S \ll E_H^{2/3} (D/L)^{2/3}, \tag{2.3.2.13}$$

where the disparity between $D$ and $L$ leads to the alteration of the laboratory parameter restriction described by (2.3.1.1).

### 2.3.3 The Interior

In the fluid *interior*, the scales for the variables introduced in the last section are presumed to give an accurate measure of the relative importance of the individual terms in the equations of motion. Of course, this need not be true in boundary layers where the length scale changes and terms depending on those derivative will change in magnitude. However, for the interior where the scaling leading to (2.3.2.8) is appropriate, the horizontal momentum equations reduce to geostrophic balance and the vertical momentum equation is simply the hydrostatic balance. Denoting interior-dependent variables with a subscript $I$,

$$u_I = -p_{Iy}, \quad v_I = p_{Ix}, \quad T_I = p_{Iz}. \qquad (2.3.3.1a, b, c)$$

It follows from the first two equations that the horizontal velocity is nondivergent, so that $w_I$ must be independent of $z$ to lowest order. Examining the vorticity balance in the interior shows that the divergence of the horizontal velocity will be $O(E_v)$ at the most. Hence, as long as $\sigma S \ll 1$, it follows that the vertical advection of temperature is negligible in the thermal equation (2.3.2.8e), which then becomes

$$\frac{E_H}{2\sigma_H}\nabla^2 T_I + \frac{E_V}{2\sigma_v} T_{I_{zz}} = 0. \qquad (2.3.3.2)$$

The interior temperature thus satisfies the conduction equation.

Once the solution for (2.3.3.2) is found, the horizontal velocities can be determined up to a barotropic ($z$-independent) constant of integration from the thermal wind relations. The vertical velocity in the interior is very weak, of order $E_H$, $E_v$ (which for simplicity we will assume are of the same order). However, it is not possible to determine the solution of (2.3.3.2) until boundary conditions are specified on the sidewalls.

At the horizontal boundaries, the interior temperature must satisfy the conditions (2.3.2.11a, b). To find the appropriate boundary conditions for (2.3.3.2) on the vertical boundaries and to determine the barotropic component of the interior flow, it is necessary to examine the boundary layers at $y = 0$ and 1.

We will also need to consider the possibility of a mode of motion in the interior that is purely barotropic (i.e., a mode independent of $z$ for which $T$ and $w$ are zero in order to supplement the baroclinic velocities that can be derived from the solution of (2.3.3.2). For such motion, which I will label with subscript $b$, the vorticity equation derived from (2.3.2.8a, b) is simply

$$E_H \nabla^2 \zeta_{Ib} = 0, \quad \zeta_{Ib} = \nabla^2 p_{Ib}. \qquad (2.3.3.3a, b)$$

### 2.3.4 The Sidewall Boundary Layer for $\sigma_H S \ll E_H^{2/3}(D/L)^{2/3}$

For a very weakly, nearly homogeneous fluid, the structure of the sidewall boundary layers in linear theory has been described by Stewartson (1957) (see also Greenspan 1968, pp. 327), although the theory needs some alteration to deal with the differing mixing coefficients in the vertical and horizontal directions and the smallness of $D/L$. Fundamentally, though, the basic ideas are not significantly altered. In the original theory, there are two possible boundary layers; an outer layer with thickness that depends on the quarter power of the friction and which acts to satisfy the *no-slip* condition and a second thinner layer of thickness $E^{1/3}$. The outer $E_H^{1/4}$ layer depends on vorticity dissipation in Ekman layers on solid horizontal surfaces bounding the fluid on at least one horizontal boundary. The layer arises when the horizontal diffusion of vorticity, proportional to $E_H \nabla^2 \zeta$, is balanced against the vortex tube stretching by Ekman layers on the horizontal surfaces bounding the layer. That effect is of order $E_v^{1/2} \zeta$ for a rigid horizontal surface. Equating those two effects requires variations on the scale $\delta = \left( \frac{E_H}{E_V^{1/2}} \right)^{1/2}$. For the case where the two Ekman numbers are of the same order, the scale is the quarter power for the Ekman number. However, when the bounding surfaces are *no stress*, the velocity pumped out of the Ekman layer is $O(E_V)$, and for Ekman numbers of the same order that gives rise to a length scale that is order one (i.e., the scale of the interior motion). That is why we have to expect solutions of (2.3.3.3) to be significant. Those solutions essentially represent the spread of the ¼power layer into the interior. The absence of this outer quarter power layer will force a strong constraint on the interior flow.

To find the surviving boundary layer scale, we anticipate that the long shore velocity will remain in geostrophic balance, so that the boundary layer *correction* to $u_I$ satisfies

$$u_b = -p_{by}, \tag{2.3.4.1}$$

while the cross-channel velocity, which we expect to have a small correction in the boundary layer, will satisfy

$$-v_b = -p_{bx} + \frac{E_H}{2} u_{byy}, \tag{2.3.4.2}$$

where we have assumed that the dominant friction term comes from the rapid derivatives in the cross-channel direction. This means $v$ has two portions: a geostrophic portion and an ageostrophic portion due to the frictional term. Eliminating the pressure between (2.3.4.1) and (2.3.4.2) yields, with the continuity equation,

$$\frac{E_H}{2} u_{byyy} = w_{bz}, \tag{2.3.4.3}$$

in which the vorticity diffusion (the dominant term in the vorticity is just $-u_{by}$) is balanced by vortex tube stretching. Taking a $z$ derivative of the vertical equation of motion,

$$\frac{E_H}{2} \frac{D^2}{L^2} w_{byyz} = p_{bzz} - T_z, \tag{2.3.4.4}$$

and eliminating $T$ with the aid of the thermal equation (2.3.2.8e) and the vorticity equation, we eventually obtain, with the geostrophic relation for $u_{Ib}$,

$$\overbrace{\frac{E_H^2}{4} \frac{D^2}{L^2} p_{byyyyyy}}^{a} + \overbrace{p_{bzz}}^{b} + \overbrace{2\sigma_H S \; p_{byy}}^{c} = 0 \tag{2.3.4.5}$$

If each derivative with $y$ yields a term of order $\delta^{-1}$, where $\delta$ is the nondimensional boundary layer thickness, then the ratio of term $c$ to term $b$ is of order

$$\frac{c}{b}=O\left(\frac{\sigma_H S}{\delta^2}\right). \tag{2.3.4.6}$$

To have term $b$ be as large as term $a$ requires a boundary layer thickness

$$\delta=E_H^{1/3}\left(\frac{D}{L}\right)^{1/3} \tag{2.3.4.7}$$

and this balance is valid if term $c$ is negligible, which from (2.3.4.6) requires the ordering relationship

$$\sigma_H S \ll E^{2/3}\left(\frac{D}{L}\right)^{2/3}, \tag{2.3.4.8}$$

which is the weak stratification limit. If the inequality is reversed, we obtain the boundary layer structure we found in the previous two Sections. Instead, we have here the classical $E^{1/3}$ of Stewartson slightly modified by the aspect ratio $D/L$. In dimensional units, this layer thickness is

$$\delta_{*b}=\left(\frac{2\nu_H}{f}\right)^{1/3} D^{1/3} \tag{2.3.4.9}$$

and note that it is independent of $L$. We can now define the *correction* fields in the boundary layer near $y=0$ as

$$\begin{aligned} &u_b=U\bar{u},\ v_{gb}=U\delta_b\bar{v}_g,\ \ v_{ab}=U\frac{\delta_b}{(D/L)}\bar{v}_a, \\ &w_b=U\bar{w}/(D/L), \qquad T_b=U\frac{(\sigma_H S)}{\delta_b}\bar{T}, \end{aligned} \tag{2.3.4.10a, b, c, d, e}$$

where the over-barred variables are order one. Note that we have distinguished between the geostrophic and ageostrophic cross-channel velocity components. These variables are functions of $\eta=y/\delta$ and of $x$ and $z$. They individually satisfy

$$\begin{aligned} &\bar{u}=-\bar{p}_\eta,\ \ \bar{v}_g=\bar{p}_x,\ \ \bar{v}_a=-\frac{1}{2}\bar{u}_{\eta\eta},\ \ 0=-\bar{p}_z+\frac{1}{2}\bar{w}_{\eta\eta}, \\ &\bar{w}_z=-\bar{v}_{a\eta},\ \ \bar{w}=\frac{1}{2}\bar{T}_{\eta\eta}, \end{aligned} \tag{2.3.4.11a, b, c, d, e, f}$$

and in the limit $\sigma_H S \ll E_H^{2/3}(D/L)^{2/3}$, the governing equation in the pressure is

$$\bar{p}_{\eta\eta\eta\eta\eta\eta}+4\bar{p}_{zz}=0. \tag{2.3.4.12}$$

The partial differential equation is solved in the region $\eta \geq 0, 0 \leq z \leq 1$. At the end points of this interval in $z$, the boundary conditions reflect the weakness of the Ekman pumping with slip boundary conditions and so $\bar{w}$ must vanish at those points. Vanishing

$\bar{w}$ implies that $\bar{p}_z = 0,\ z = 0, 1$. This in turn implies that the solution for $\bar{p}$ can be written as a cosine series, that is,

$$\bar{p} = Re \sum_{j=1} \cos(j\pi z) \left[ A_{1j} e^{-\gamma_j \eta} + A_{2j} e^{-\gamma_j (1/2 + i\sqrt{3}/2)\eta} + A_{3j} e^{-\gamma_j (1/2 - i\sqrt{3}/2)\eta} \right],$$

$$\gamma_j = (2j\pi)^{1/3} \tag{2.3.4.13}$$

from which, with (2.3.4.11), all the other correction functions can be found, and they are given later. Note that $\bar{p}$ and hence the horizontal velocities have a *zero vertical average* and *so the interior horizontal velocities must also have zero vertical average at the channel boundaries.* (Although $j = 0$, the barotropic mode, satisfies the boundary condition, the associated $\gamma$ is zero and does not lead to a boundary layer solution.) Thus, the structure of the Stewartson layer imposes a strong condition on the interior flow.

With the solutions to the boundary layer equations, we are now in a position to carry out the matching procedure at $y = 0$. A similar process will occur at $y = 1$, but those details can be skipped.

The matching conditions become, at $y = 0$,

$$u_I + U\bar{u} = 0,$$

$$v_I + U\left(\frac{\delta_b}{D/L}\right)\left(\bar{v}_a + \frac{D}{L} v_g\right) = 0,$$

$$w_I + U\frac{L}{D}\bar{w} = 0, \tag{2.3.4.14a, b, c, d}$$

$$T_{Iy} + U\frac{\sigma_H S}{E_H^{2/3}(D/L)^{2/3}}\bar{T}_\eta = 0.$$

$u_I$, $v_I$, and $T_I$ are order one, and $w_I$ is order $E_{\mathrm{H}}$. Recall that $v_I$ is in geostrophic balance. In the classical Stewartson layer problems involving a homogeneous fluid, the interior velocity normal to the boundary is zero, or if there is a geostrophic, order one $v_I$, then the geostrophically balanced $v_I$ must satisfy the zero conditions on its own. This, however, cannot be the case here. Since the parameter, $\frac{\sigma_H S}{E_H^{2/3}(D/L)^{2/3}}$, measuring the contribution of the boundary layer to the temperature gradient at the wall is, by hypothesis, small, *the interior temperature gradient must, to lowest order, satisfy the insulating condition on* $y = 0$. It is then impossible for the interior to satisfy that condition *and* the condition on $v$. We are forced to the conclusion then that we must choose $U$, the scale for the boundary layer correction, to achieve that balance, that is,

$$U = \frac{D/L}{\delta_b} = \left(\frac{2A_H}{fD^2}\right)^{-1/3} \gg 1. \tag{2.3.4.15}$$

This unusual constraint on the interior can be easily understood in the limit as $S$ goes to zero. Then the thermal equation is decoupled from the vertical advection, and the temperature satisfies a form of Laplace's equation (i.e., the conduction equation). It is obvious in this limit that the temperature field must directly satisfy the insulating condition directly with the interior variables.

It then follows that, to lowest order,

$$\bar{u}=0, \quad \bar{w}=0, \qquad @\eta=0 \qquad \bar{v}_a+\frac{D}{L}\bar{v}_g=-v_I. \tag{2.3.4.16a, b, c}$$

Those three conditions determine $A_{1j}$, $A_{2j}$, $A_{3j}$. Note that this implies that the flow in the boundary layer is forced by the impinging *geostrophic* flow in the interior at the wall, a conclusion that Spall (2008) has found in his numerical study. Thus, the sinking in the boundary layer is forced indirectly by the cooling as it generates a down-channel pressure gradient and a geostrophic flux toward the boundary where the sinking takes place.

The interior problem is then reduced to a solution of (2.3.3.2) subject to the condition that $T_{Iy}=0$ at $y=0,1$. That determines the baroclinic flow in the interior. The interior barotropic flow for which $T_{\mathrm{I}}$ and $w_{\mathrm{I}}$ are zero is found from the solution of (2.3.3.3 a,b).

One could think of that problem as the analogue of the Stewartson $E^{1/4}$, which, as we have seen, is not present with the slip boundary conditions at $z=0$ and 1, allowing the barotropic vorticity to diffuse into the interior. Since the boundary layer can't support a barotropic correction field, we must apply the condition

$$\int_0^1 (u_I, v_I)dz=0, \quad y=0,1. \tag{2.3.4.17}$$

### 2.3.5 An Example

To illustrate the theory, let's consider a simple example. Consider a heating function as in (2.3.2.11 a) of the form

$$H = ReH_1 e^{ikx}\cos\pi y, \qquad 0\le y\le 1. \tag{2.3.5.1}$$

(This will be a cooling if $H<0$).The form imposes a cooling on one half of the channel and a heating on the other half. In the more elaborate numerical model of Spall (2008), the downstream variation is produced naturally by the cooling that takes place as the fluid advects downstream. We short circuit that process in our linear model by specifying that variation.

We will focus our analysis on the boundary, at $y=0$, where the cooling will take place (for specified values of $x$). The form is chosen to make the satisfaction of the thermal conditions on the sidewalls, $T_{Iy}=0$, very simple. The solution to (2.3.3.2) that accomplishes that is

$$T_I = \frac{H_1}{K}\frac{\sinh Kz}{\cosh K}e^{ikx}\cos \pi y,$$
$$K^2 = (k^2+\pi^2)\frac{\sigma_v E_H}{\sigma_H E_v} \qquad (2.3.5.2a, b)$$

and the real part of this expression is understood. From the thermal wind relation, the horizontal velocities are determined up to a barotropic component, that is,

$$u_I = \pi\frac{H_1}{K^2}\frac{\cosh Kz}{\cosh K}e^{ikx}\sin \pi y + u_{I\ B}(x,y),$$
$$v_I = ik\frac{H_1}{K^2}\frac{\cosh Kz}{\cosh Kz}e^{ikx}\cos \pi y + v_{IB}(x,y). \qquad (2.3.5.3a, b)$$

Note that the thermally driven, baroclinic part of the down-channel velocity already satisfies the no-slip condition on the channel boundaries. Since the barotropic velocities must cancel the vertical mean of the baroclinic velocity on $y=0, 1$, the boundary condition for the barotropic velocities are

$$\left.\begin{aligned} u_{IB} &= 0 \\ v_{IB} &= -ik\frac{H_1}{K^3}\tanh K\cos \pi y \end{aligned}\right\} y = 0, 1. \qquad (2.3.5.4)$$

The barotropic interior velocities are in geostrophic balance and are found from the pressure field that is a solution of (2.3.3.3a, b), namely,

$$p_{IB} = e^{ikx}\left[\frac{A}{2k}y\frac{\cosh ky}{\cosh k} + \frac{B}{2k}(y-1)\frac{\cosh k(y-1)}{\cosh k} + E\frac{\sinh ky}{\sinh k} + F\frac{\sinh k(y-1)}{\sinh k}\right],$$
$$u_{IB} = -\frac{\partial p_{IB}}{\partial y},\ v_{IB} = \frac{\partial p_{IB}}{\partial x}. \qquad (2.3.5.5a, b)$$

The coefficients $A$, $B$, $E$, and $F$ are determined by applying (2.3.5.4). We obtain

$$B = A,$$
$$F = E,$$
$$E = F - \frac{A}{2k^2}\frac{\{1+1/\cosh k + k\tanh k\}}{\{\coth k + 1/\sinh k\}},$$

and (2.3.5.6a, b, c, d)

$$A = \frac{2kH_1 \tanh K/K^3}{\left\{1 - \dfrac{(1+1/\cosh k + +k\tanh)}{(\coth k + 1/\sinh k)}\right\}}$$

Now that the interior fields are determined the coefficients for the boundary layer correction functions are found using (2.3.4.16). This yields

$$A_{1j}=0,\quad A_{3j}=-A_{2j}e^{2\pi i/3},$$

and

$$A_{2j}=\frac{2ikH_1(-1)^j\tanh K}{(1-e^{2\pi i/3})(j^2\pi^2+K^2)(ik(D/L)+j\pi)}. \tag{2.3.5.7a, b, c}$$

In terms of these coefficients, the correction fields are

$$\bar{u}=\sum_{j=1}\gamma_j\Big[A_{1j}e^{-\gamma_j\eta}+(1/2+i\sqrt{3}/2)A_{2j}e^{-\gamma_j(1/2+i\sqrt{3}/2)\eta}$$
$$+(1/2-i\sqrt{3}/2)A_{3j}e^{-\gamma_j(1/2-i\sqrt{3}/2)\eta}\Big]\cos(j\pi z),$$

$$\bar{v}_g=\sum_{j=1}\left[\frac{\partial A_{1j}}{\partial x}e^{-\gamma_j\eta}+\frac{\partial A_{2j}}{\partial x}e^{-\gamma_j(1/2+i\sqrt{3}/2)\eta}+\frac{\partial A_{3j}}{\partial x}e^{-\gamma_j(1/2-i\sqrt{3}/2)\eta}\right]\cos(j\pi z),$$

$$\bar{v}_a=-\frac{1}{2}\sum_{j=1}\gamma_j^3\Big[A_{1j}e^{-\gamma_j\eta}-A_{2j}e^{-\gamma_j(1/2+i\sqrt{3}/2)\eta}-A_{3j}e^{-\gamma_j(1/2-i\sqrt{3}/2)\eta}\Big]\cos(j\pi z),$$

$$\bar{w}=\frac{1}{2}\sum_{j=1}\gamma_j^4\Big[-A_{1j}e^{-\gamma_j\eta}+(1/2+i\sqrt{3}/2)A_{2j}e^{-\gamma_j(1/2+i\sqrt{3}/2)\eta}$$
$$+(1/2-i\sqrt{3}/2)A_{3j}e^{-\gamma_j(1/2-i\sqrt{3}/2)\eta}\Big]\frac{\sin(j\pi z)}{j\pi}, \tag{2.3.5.8a, b, c, d}$$

and

$$\bar{T}_\eta=\sum_{j=1}{\gamma_j}^3\left[A_{1j}e^{-\gamma_j\eta}-\frac{1}{2}A_{2j}e^{-\gamma_j(1/2+i\sqrt{3}/2)\eta}-\frac{1}{2}A_{3j}e^{-\gamma_j(1/2-i\sqrt{3}/2)\eta}\right]\frac{\sin(j\pi z)}{j\pi}, \tag{2.3.5.9}$$

where the real parts of the preceding expressions are understood.

For the solution in the boundary layer discussed later, these series were truncated to 30 terms. Adding additional modes did not change the solution.

Figure 2.20 shows the contours of the velocity in the cross-channel direction at a position, $kx=1.25\pi$, where the flow in the upper part of the water column is being driven toward the boundary at $y=0$.

Half the channel width is shown; the solution for $v$ is antisymmetric about $y=0.5$. In $0\le y\le 0.5$, flow is being driven toward the boundary at $y=0$ where it sinks in a boundary layer of width $\delta_b=0.037$. The resulting velocity profile of the zonal velocity is shown in the half channel (the zonal velocity is *symmetric* across the channel) at $z=0.8$ in Figure 2.21a.

Note the monotonic decrease of the zonal velocity as $y=0$ is approached by the interior solution, but as we approach the boundary layer the structure of the Stewartson

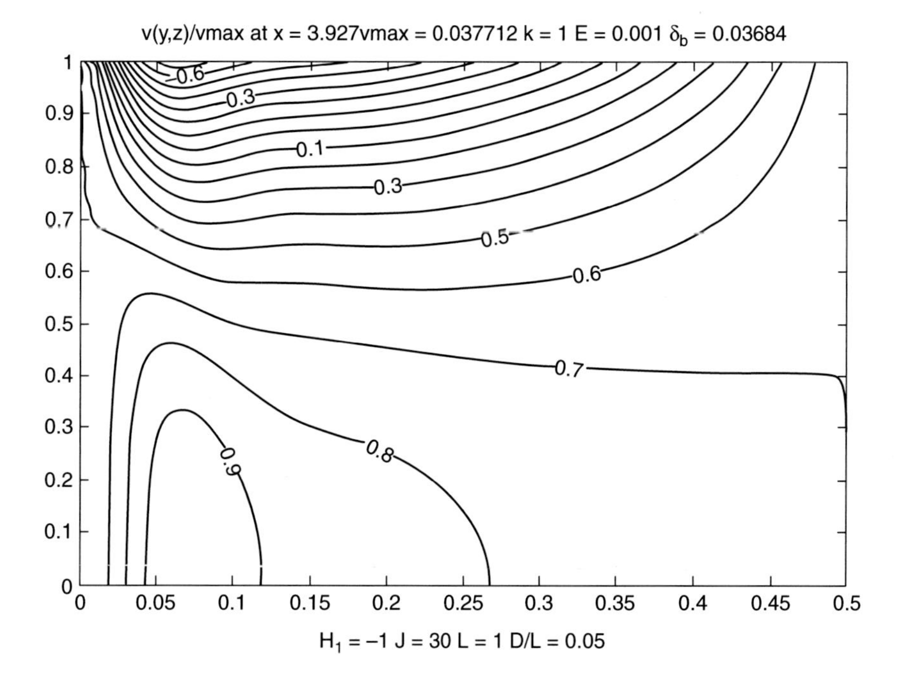


Figure 2.20. The contours of cross channel velocity, $v$, in $0 \leq y \leq 0.5$ for $E_{\mathrm{H}} = 0.001$, $D/L = 0.05$, $H_1 = -1$. The contours of $v$ are scaled with its absolute maximum of 0.0377.

layer produces a local enhancement of the down-channel velocity just as found in the numerical model of Spall (2008). We can see here that this is due entirely to the damped oscillatory behavior of the layer's structure and the relatively large amplitude of the correction driven by the need of the Stewartson layer to bring the cross-channel velocity to rest. Of course, for even smaller values of $E_{\mathrm{H}}$ as shown in Figure 2.21b, the overshoot in the along-channel velocity is even larger.

The profile of the vertical velocity, again at $z = 0.8$, is shown in Figure 2.22.

The vertical velocity is entirely limited to the sidewall boundary layer since the interior velocity, frictionally driven, is extremely small (i.e., of $O(E_H, E_v)$). As in the numerical model, the sinking is limited to the boundary region where the vorticity production due to vortex tube stretching can be balanced by viscous dissipation.

Although the motion is three dimensional, the vertical velocity is produced by the ageostrophic component of $v$ in the boundary layer, which satisfies

$$\frac{\partial v_a}{\partial y} + \frac{\partial w}{\partial z} = 0, \tag{2.3.5.10}$$

so that a streamfunction for this ageostrophic circulation can be generated. The contours of that streamfunction in the vertical plane, Figure 2.23, give a clear picture of

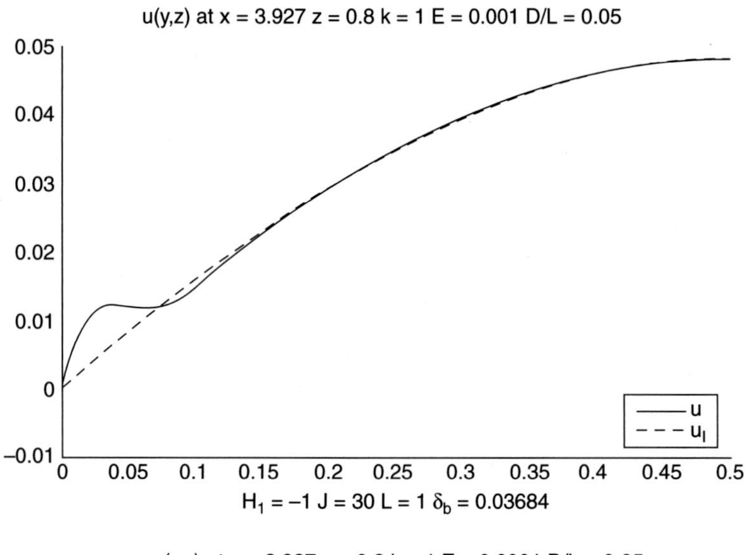


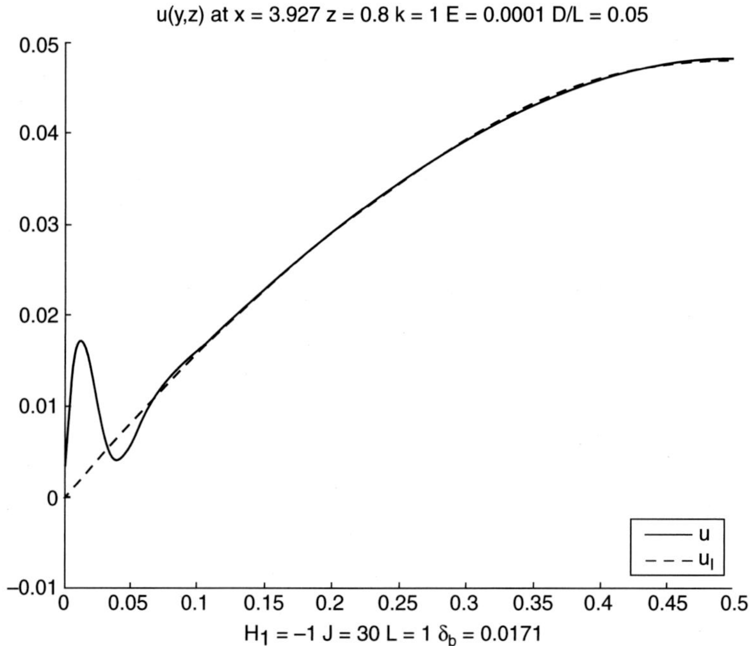


Figure 2.21 (a) The along-channel velocity, at $z = 0.8$ for the same parameters as earlier. (b) As in Figure 2.21a but now $E_H$ is 0.0001.

that part of the flow that leads to the sinking. A very similar structure is seen in Spall's nonlinear numerical calculations where the oscillatory structure of the streamfunction field in $y$ is evident. Note that the ageostrophic velocity in the boundary layer is largely driven by the *geostrophic* velocity in the interior.

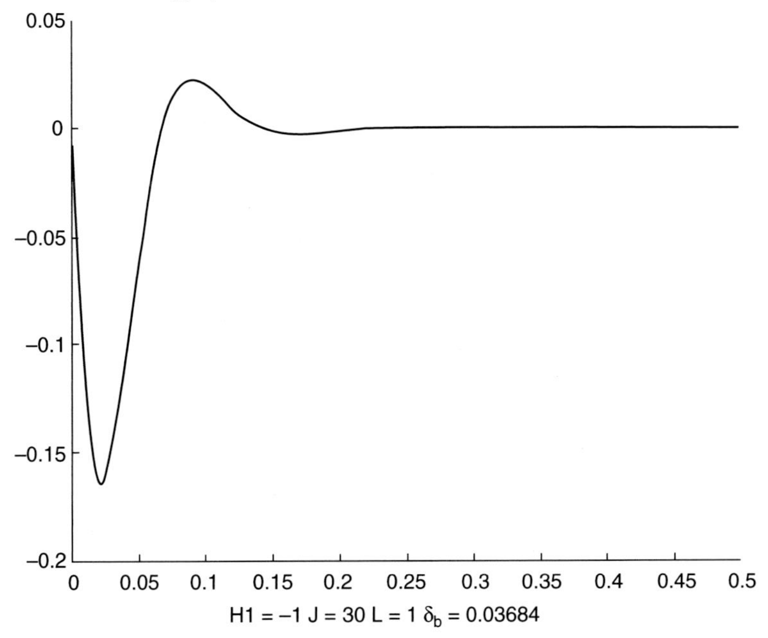


Figure 2.22. Profile of the vertical velocity at $z = 0.8$.

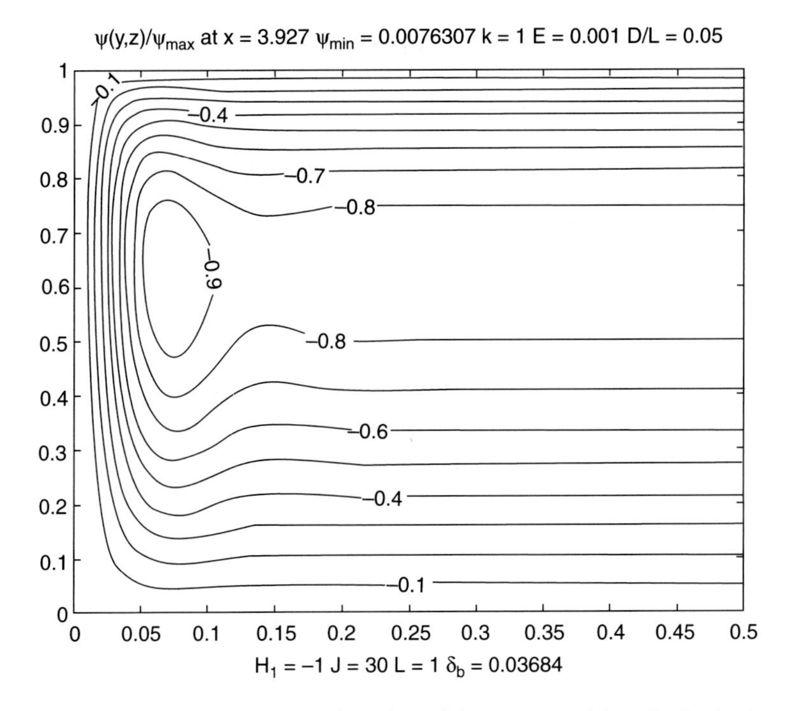


Figure 2.23. The circulation streamfunction of the ageostrophic velocity in the $y - z$ plane. All parameters are as in Figure 1. The streamfunction contours are scaled with its absolute minimum $= 0.00763$.

### *2.3.6 Discussion*

A simple linear model has been used to discuss the response of a weakly stratified layer to a nonuniform cooling of the surface. The nonuniformity in the downstream direction is imposed to mimic the downstream variation that would appear naturally if the cooling were uniform, but if, as in the nonlinear model of Spall (2008), nonlinear advection is included. In particular, the cooling has been chosen so that the geostrophic velocity in the cross-channel direction, forced by the cooling, drives an amplified response in a narrow sidewall boundary layer that is, to all intents and purposes, the same as the boundary layer introduced by Stewartson (1957). The principal difference in the treatment here is that for the layer to bring the *geostrophic* cross-channel velocity to rest at the boundary requires an amplitude for the boundary layer correction much larger than in the classical theory where the interior flow is two dimensional and the cross-channel velocity in the interior is weak and ageostrophic.

The analysis confirms the interpretation of Spall (2008) that the narrow zone to which strong vertical motion is limited in the numerical model he investigated is essentially the same as the linear Stewartson layer. This is somewhat intriguing since the numerical model is strongly nonlinear, but it does seem to imply that the basic process determining the region of sinking by cooling is robustly governed by the balance between vorticity generated by vortex stretching and the dissipation of that vorticity in narrow regions near the boundary. In particular, the net downwelling at the boundary is set by the magnitude of the geostrophic flow in the interior driven to the boundary by the along-channel pressure gradient produced by the cooling. The local boundary enhancement of the along-channel velocity is also well described by this simple analytical model and is, again, a fundamental feature of the Stewartson $E_H^{1/3}$ layer when that layer is forced by a geostrophic interior impinging flow.

If the stratification were increased enough to reverse the inequality (2.3.4.8) so that $\sigma_H S \gg E_H^{2/3}(D/L)^{2/3}$ (so that the deformation radius exceeds the width of the Stewartson layer), then the Stewartson layer would split (Barcilon and Pedlosky 1967b) into a hydrostatic layer of thickness $(\sigma_H S)^{1/2}$ and a very narrow buoyancy layer whose thickness is $(E_H D/L)^{1/2}/(\sigma_H S)^{1/4}$; however, with insulating sidewalls, this second layer is essentially absent. The vertical velocity is much reduced in the stratified hydrostatic layer, and the constraint on the interior velocity eliminates the forcing of the boundary layer by the geostrophic cross-channel flow. Thus, as might be expected, the strong downwelling seen in the model described here depends essentially on the weakness of the stratification, which in the numerical model of Spall is self-generated as the cooling-forced mixed layer.

## References

Barcilon V., and J. Pedlosky, 1967a: Linear theory of rotating stratified fluid motions, *J.Fluid Mech.* **29,** 1–16.

Barcilon V., and J. Pedlosky, 1967b: A unified linear theory of homogeneous and stratified rotating fluids, *J.Fluid Mech.* **29,** 609–621.

Greenspan, H. P., 1968: *The Theory of Rotating Fluids*. Cambridge University Press, Cambridge.

Hristova, 2009: Stability of large-scale oceanic flows and the importance of non-local effects. Ph.D. dissertation. MIT/WHOI Joint Program.

Linden, P., 1977: The flow of a stratified fluid in a rotating annulus, *J. Fluid Mech.* **79**, 435–448.

Pedlosky, J. 1965: A note on the western intensification of the oceanic circulation. *J. Marine Res.* **23**, 207–209.

Pedlosky, J. 1987: *Geophysical Fluid Dynamics.* Springer Verlag, New York.

Pedlosky, J., J. A. Whitehead, and G.Veitch, 1997: Thermally driven motions in a rotating stratified fluid: theory and experiment. *J. Fluid Mech.* **339**, 391–411.

Pedlosky, J., and M. A. Spall, 2005: Boundary intensification of vertical velocity in a $\beta$-plane basin. *J. Phys. Ocean.* **35**, 2487–2500.

Pedlosky, J., 2009: The response of a weakly stratified layer to buoyancy forcing, *J. Phys. Oecan.* **39**, 1060–1068.

Prandtl, L. 1949: *The Essentials of Fluid Mechanics* (English translation, Blackie edition, 1952), Haffner, New York, 452pp.

Spall, M. A., 2008: Buoyancy-forced downwelling in boundary currents. *J. Phys. Ocean.* **38**, 2704–2721

Stewartson, K., 1957: On almost rigid rotations. *J. Fluid Mech.* **6**, 17–26.

Stommel, H., 1948: The westward intensification of wind-driven ocean currents. *Trans. Am. Geophys. Union* **29**, 202–206.

Veronis G., 1967: Analogous behavior of rotating and stratified fluids. *Tellus* **19**, 620–634.

# 3

# Buoyancy-Forced Circulation and Downwelling in Marginal Seas

MICHAEL A. SPALL

## 3.1 Introduction

Marginal seas subject to buoyancy loss, because of their semi-enclosed geometry, are source regions for the formation of dense intermediate and bottom waters. These convective water masses generally have distinct properties relative to the open ocean and can be traced far from their formation basins. They also can transport significant amounts of heat, salt, and other tracers throughout the world ocean. The vertical circulation and meridional heat and freshwater transports are fundamental components of the oceanic circulation, and play important roles in the global climate system. Understanding how this circulation depends on the environmental parameters of the system is important if one is to better model and predict the climate system and its sensitivity to changing atmospheric conditions, such as increasing anthropogenic carbon dioxide.

The focus of this review is on the circulation and exchange resulting from surface buoyancy forcing in marginal seas. General characterisitics of the exchange between the marginal sea and the open ocean are described from eddy-resolving numerical models in idealized configurations, and the physics governing this exchange are elucidated through a combination of numerical models and simplified analytic models. Although the problems are couched in terms of marginal sea–open ocean exchange, many of the processes that emerge from this analysis are relevant to more general buoyancy-forced flows. Particular attention is paid to the dynamics involved with net vertical motions forced by surface cooling. Distinct regimes of dissipative stratified flows, weakly dissipative stratified flows, and weakly stratified flows are covered. Analytical treatments of some aspects of these flows can be found in Chapter 2 by Pedlosky (this volume). Selected results from each regime are presented here, with the objective of identifying and distinguishing the key dynamics in each of these parameter regimes. The focus here is on the processes within the marginal sea, mixing processes that lead to vertical transports and water mass modification as dense waters formed in marginal seas flow over topographic sills are not considered here (see Chapter 5 by Legg, this volume).

## 3.2 Buoyancy-Forced Circulation and Exchange

Previous studies of localized open ocean convection have been very useful for elucidation of the dynamics of convective plumes and the onset of a scale cascade from the plume scale to the mesoscale (see the review by Marshall and Schott 1999 and chapter 5 by Legg, this volume). Because those studies were forced with buoyancy loss only, they could not achieve an equilibrium since there was no source of heat to balance the cooling. Observations also indicate that the majority of the exchange between marginal seas and the open ocean is carried out in narrow boundary currents that encircle the marginal sea cyclonically. These modeling studies also neglected wind forcing, which may be an important driver of the observed cyclonic boundary current. In this section, we will build upon these previous localized convection studies to include a source of heat (to achieve equilibrium solutions), boundaries to support boundary currents, topography, and wind forcing.

The approach taken throughout this section will be to combine results from idealized configurations of primitive equation numerical models with less dynamically complete, but more dynamically revealing, conceptual and theoretical models. These simplified models are very useful because they will identify how key aspects of the circulation depend on specific environmental parameters. This generalization of the processes serves to unify our understanding of the dynamics of marginal seas and provides simple interpretations as to why various marginal seas behave differently in terms of their exchange with the open ocean.

The numerical model used in the following studies is the MIT general circulation model (Marshall et al. 1997), which solves the primitive equations on a Cartesian, staggered C-grid with level vertical coordinates. Details of the model configurations vary, but there are some aspects in common. There is generally a marginal sea separated from an adjacent basin (referred to as the open ocean) by a sill or horizontal confinement. The marginal sea is subject to buoyancy loss, and heat is provided in the open ocean to balance this cooling. All calculations discussed here use a linear relationship between temperature and density with a thermal expansion coefficient of 0.2 kg $m^{-3}C^{-1}$. The models are initialized at rest with uniform stratification and run until a statistical equilibrium is achieved, typically 20–30 years.

Heat is provided in the open ocean by restoring the model temperature toward a uniform stratification with a time scale of $O(20\text{ days})$. This approach allows the model to achieve equilibrium on the dynamic time scale, which is typically of $O(10\text{ years})$ and is set by the time it takes eddies or Rossby waves to cross the basin. The equilibration time scale for basin-scale models that have buoyancy forcing at the surface only is set by diapycnal diffision, and is typically $O(1{,}000\text{ years})$. However, the restoring approach used here does not allow the low latitudes to respond to changes in the marginal sea. The external restoring also provides potential energy to balance any loss of potential energy resulting from sinking of dense waters in the marginal sea. This is of course a gross oversimplification of what goes on to provide the stratification outside a marginal sea in the real ocean. This parameterization does, however, allow for a small computational domain, which allows for eddy resolving resolution, and removes the long time scale required to arrive at a thermodynamic equilibrium that is required when diapycnal mixing controls the stratification in the open ocean.

### *3.2.1 Influence of a Boundary*

The effects of an open ocean adjacent to the marginal sea are demonstrated by a simple extension of the localized cooling experiments reviewed by Marshall and Schott (1999). The model domain consists of a circular marginal sea connected to a small open ocean through a narrow strait (a typical domain is shown in Figure 3.1). In some cases, such as the one in Figure 3.1, there is a linearly sloping bottom topography around the perimeter of the basin. The model is forced by a repeat annual cycle consisting of cooling for a period of 2 months in the interior of the marginal sea followed by 10 months with no surface forcing. There is also a continuous restoring of temperature toward a uniform stratification in the open ocean.

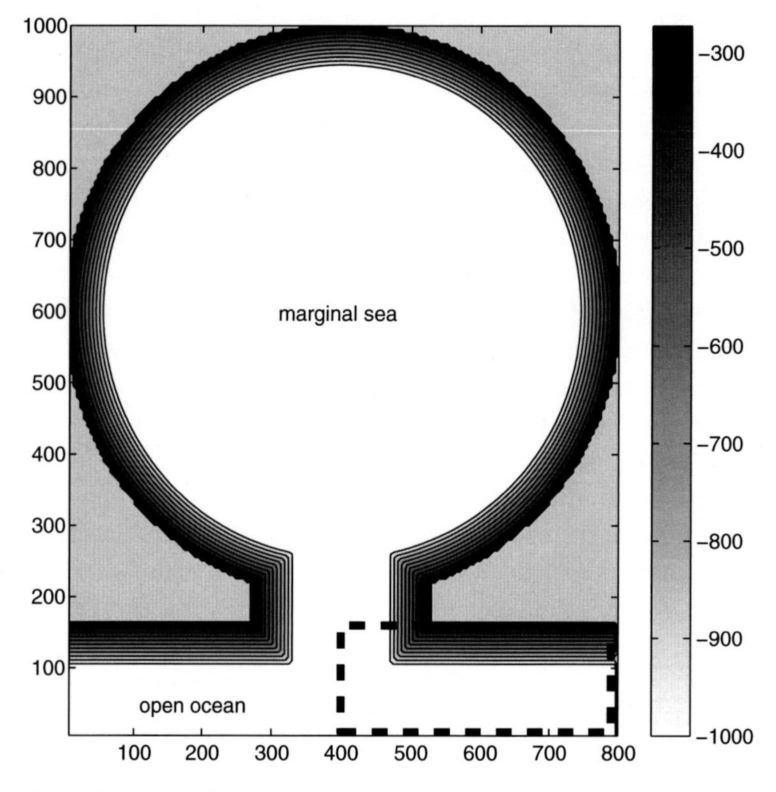


Figure 3.1. Model domain and bottom topography. Temperature is restored toward a uniform stratification within the dashed box.

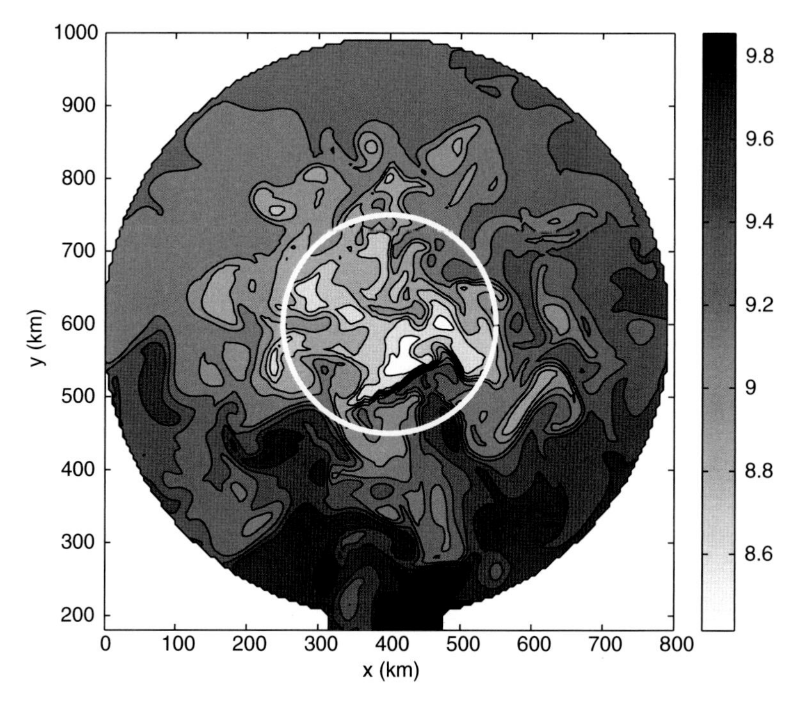


Figure 3.2. Snapshot of temperature at uppermost model level one month after cooling has ceased for a flat bottom calculation. Cooling is applied only within the white circle.

In a calculation similar to that of Spall (2003), the simplest extension of the localized open ocean convection configuration is provided by including lateral boundaries in a flat bottom ocean while keeping the heat loss localized in the interior of the basin. In this case, the stratification in the open ocean is restored to $N^2 = 4 \times 10^{-6}$ s$^{-2}$, which, with the basin depth of 1,000 m and uniform Coriolis parameter of $10^{-4}$ s$^{-1}$, gives a baroclinic deformation radius of 20 km. The horizontal grid spacing is 5 km, and there are 20 levels in the vertical with thickness 50 m.

A snapshot of the sea surface temperature 1 month after the end of the cooling period in year 24 is shown in Figure 3.2. The heat loss in the marginal sea is limited to the region inside the white circle. Cooling results in deep convection, which drives a baroclinic current and initiates baroclinic instability, much as is found in the localized cooling experiments. However, because the open ocean is maintained to be warm, a density gradient develops between the interior of the marginal sea and the open ocean. This drives a warm inflow primarily along the boundary on the right-hand side of the marginal sea (in the direction of Kelvin wave propagation). This boundary current becomes baroclinically unstable and sheds warm eddies into the basin interior. The density gradient between the interior and the boundary increases to the point where the eddies shed from the boundary transport enough heat into the interior to balance the imposed surface cooling. At this point, the model has reached a statistical

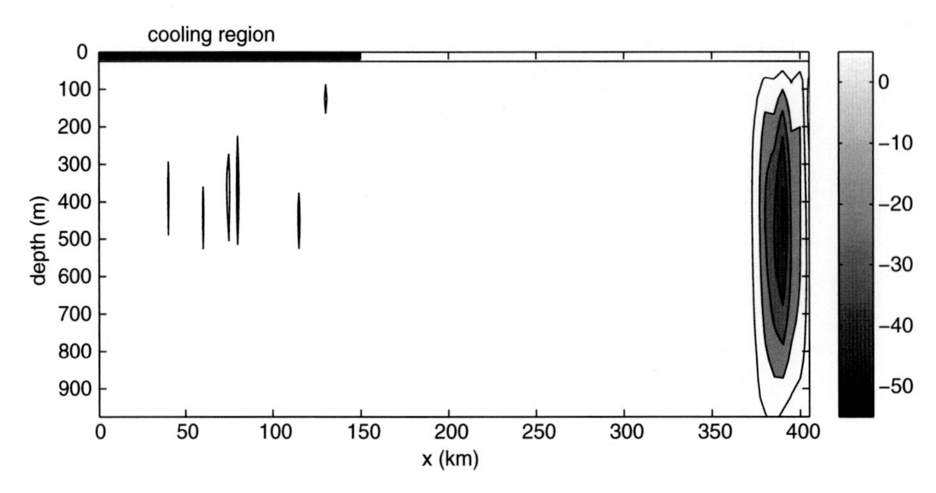


Figure 3.3. Azimuthally integrated vertical velocity as a function of depth and radius. Cooling applies only over the inner 150 km (indicated by the bold black line). Units $m^2\ s^{-1}$.

equilibrium. This balance is different from the localized cooling experiments because the eddies are shed from the boundary current, not from the edge of the cooling region.

The inflow to the marginal sea takes place in the upper ocean with a compensating outflow of dense water in the deep ocean. This implies that downwelling is taking place somewhere within the marginal sea. An azimuthal integral of the downwelling within the marginal sea averaged over the final 10 years of the 25-year integration is shown in Figure 3.3. The cooling region is indicated by the bold black line from the center of the basin out to 150 km radius. The downwelling is located adjacent to the outer boundary of the marginal sea, clearly separated from the cooling region. The net downwelling within the cooling region is essentially zero. The dynamics within the downwelling region will be discussed in Section 3.3.

This model configuration is not very realistic, but it does represent an initial step in connecting the isolated convection studies to the basin- and global-scale circulations. The important points demonstrated by this calculation are that the boundary (1) supports a warm boundary current that provides heat to balance cooling in the marginal sea and (2) permits a net vertical mass transport in the basin.

### *3.2.2 Influence of Sloping Topography*

Two aspects of this calculation that are not very realistic are the localization of surface cooling and the neglect of bottom topography. These limitations are relaxed in a model configuration that applies uniform surface cooling over the entire marginal sea and includes a sloping bottom around the perimeter of the basin (as in Figure 3.1). This section follows the results of Spall (2004); see the original study for additional details. For purposes of illustration, a calculation similar to the central calculation in Spall

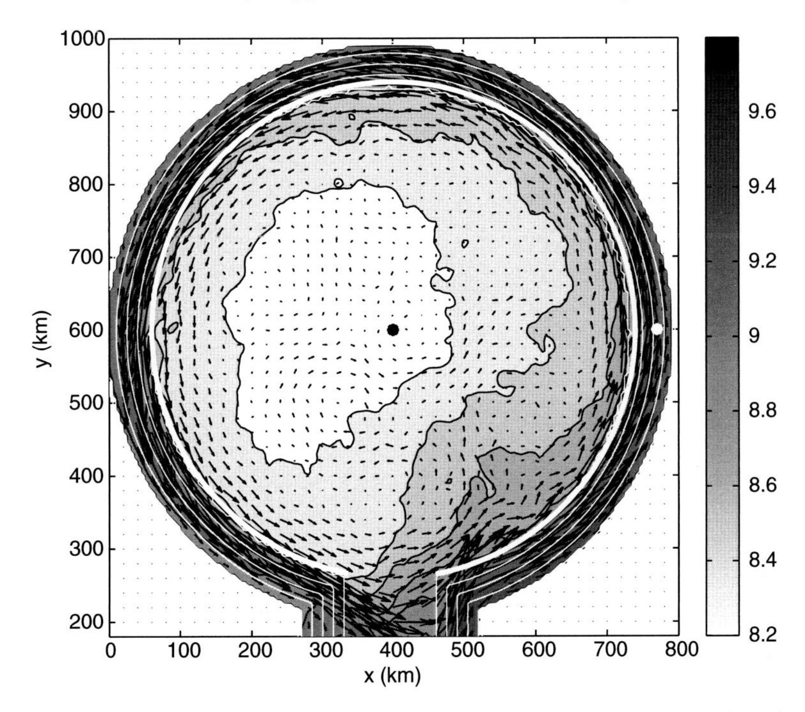


Figure 3.4. Mean temperature (°C) and horizontal velocity (every fourth grid point) at the uppermost model level. The locations of the time series in Figure 3.7 are indicated by the black dot and the white dot.

(2004) is reproduced here. The mean temperature and velocity at the uppermost level over the last 10 years of a 25-year integration are shown in Figure 3.4. As for the case with no topography and localized cooling, the heat loss in the interior drives a warm, cyclonic boundary current. The interior is cool and nearly quiescent. The boundary current is confined to the region of sloping topography (indicated by the bold white line around the perimeter of the basin) and encircles the marginal sea. The outflowing water is cooler than the inflowing water, but not as cold as the convective water mass in the interior of the basin.

A vertical section of the mean meridional velocity and temperature at $y = 600$ km are shown in Figure 3.5. The boundary current is marked by strong stratification and steeply sloping isopycnals over the bottom topography. The interior of the basin is much cooler throughout and is very weakly stratified. The maximum velocity in the boundary current is approximately 30 cm s$^{-1}$. However, unlike the case with a flat bottom, the velocity is now in the same direction throughout the water column and is very weak at the bottom. The southward flowing waters are cooler than the northward flowing waters as a result of heat loss to the north of this section. The northward mass transport along the eastern boundary is balanced by the southward transport along the western boundary. This conservation of mass transport, combined with the cooler water on outflow, requires that the southward flowing water is more barotropic than

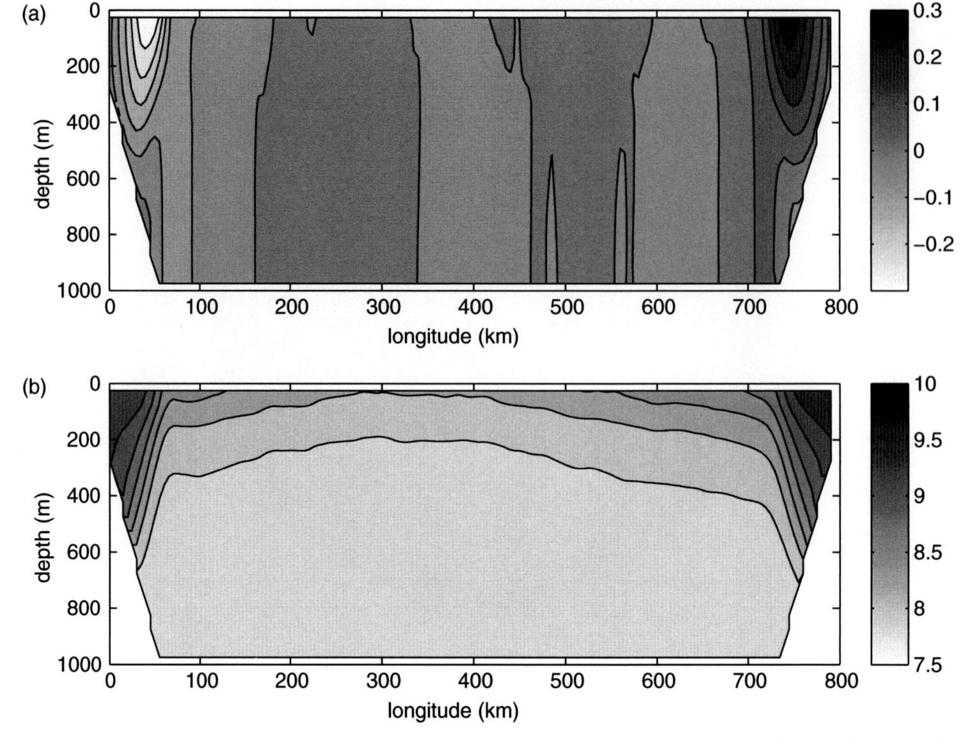


Figure 3.5. Mean vertical section of (*upper*) meridional velocity (m $s^{-1}$) and (*lower*) temperature (°C) at the midpoint of the basin ($y = 600$ km).

the northward flowing water (the maximum velocity is slightly lower, and the velocity extends slightly deeper in the water column), which in turn implies that there is a net downwelling north of this section.

The general characteristics of this velocity and hydrography section are consistent with what is observed in the upper 1,500 m of the Labrador Sea. The potential temperature and absolute velocity along the AR7W section (approximately 58°N) are shown in Figure 3.6 (from Pickart and Spall 2007).

The spinup of the circulation and hydrography are informative as to the dynamics of the equilibration. Time series of temperature at the uppermost model level at the two locations indicated by the dots in Figure 3.4 are shown in Figure 3.7. The boundary current quickly develops a seasonal cycle in which warm water is flowing into the basin over most of the year with short spikes of cooler water at the beginning of the cooling period. The interior takes much longer to equilibrate. Early in the calculation, the interior temperature cools rapidly during the forcing period and remains at a nearly constant temperature when there is no surface forcing. This cycle repeats for several years, during which time the annual mean temperature of the interior waters continues to decrease. After about 5 years, we see an increase in the interior temperature during the period when cooling is turned off, which implies an exchange with the open

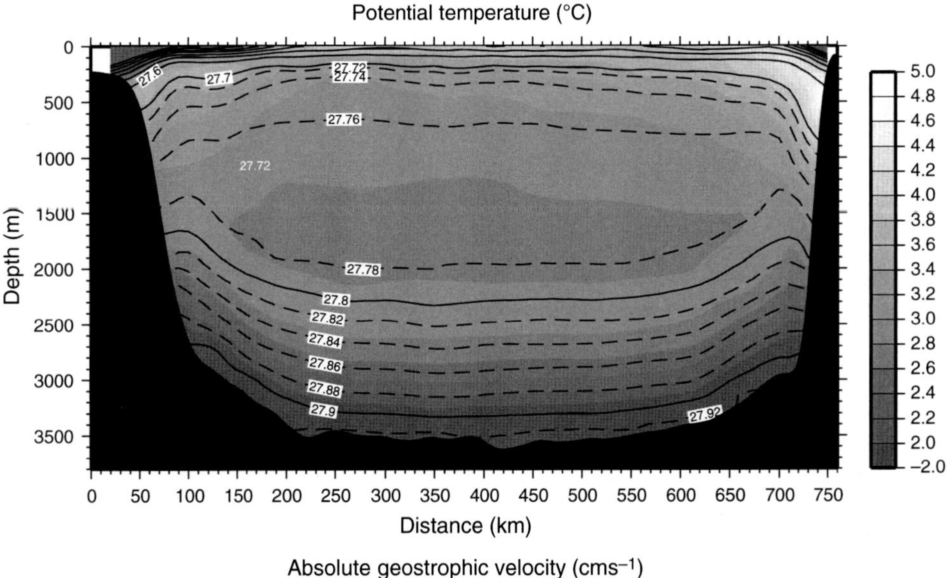


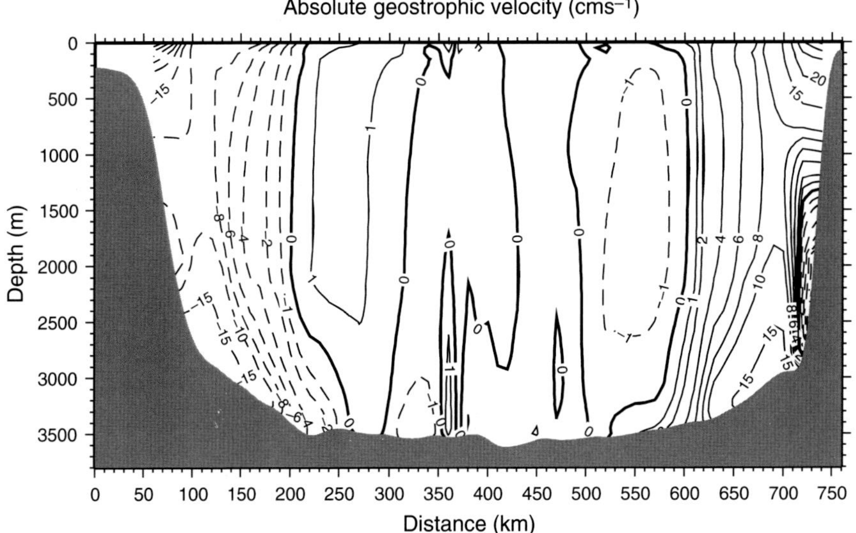


Figure 3.6. Mean (*upper*) hydrographic (shading indicates temperature, contours are potential density) and (*lower*) velocity across the AR7W section (from Pickart and Spall, 2007).

ocean via the boundary current. It takes about 15 years for the interior to achieve a statistical equilibrium, which is characterized by seasonal oscillations in temperature of $O(0.5°C)$.

The sea surface temperature and mixed layer depth at the end of the cooling period are shown in Figure 3.8. The boundary current is cooler than in the annual mean primarily as a result of direct heat loss to the atmosphere. The mixed layer is deep over most of the interior with the exception of a couple large patches of stratified

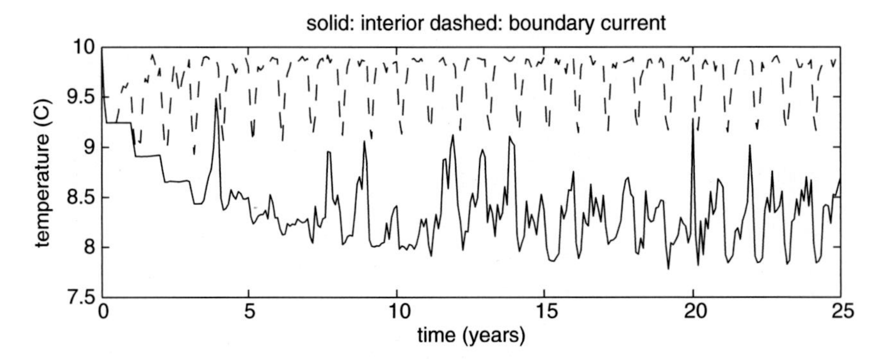


Figure 3.7. Time series of temperature at the 25 m depth for the basin interior (solid line) and inflowing boundary current (dashed line). Locations are indicated on Figure 3.4 by the black and white dots.

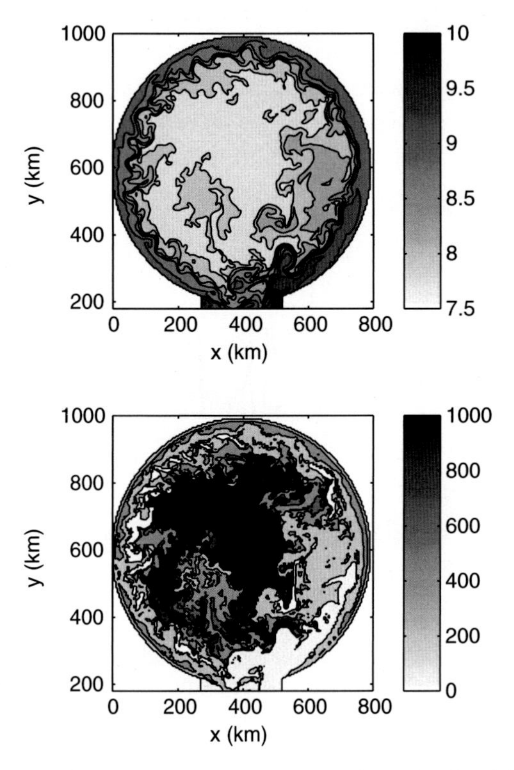


Figure 3.8. Snapshot of (*upper*) temperature and (*lower*) mixed layer depth at the end of the cooling period in year 24.

water left over from before the cooling started. We can also see warm, stratified water beginning to penetrate into the interior from the boundaries in the form of small-scale meanders and eddies. One month after the end of the cooling period, the warm water has penetrated much further into the basin and warm eddies are being shed from the

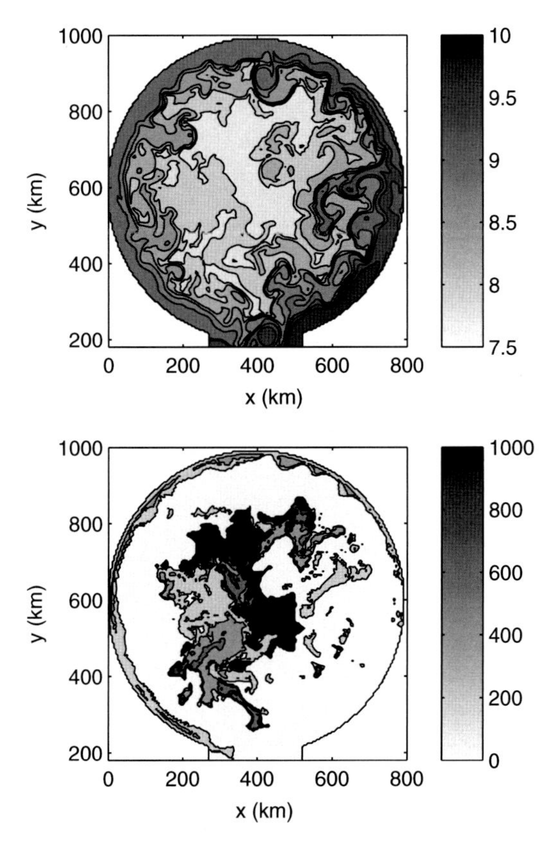


Figure 3.9. Snapshot of (*upper*) temperature and (*lower*) mixed layer depth one month after the end of the cooling period in year 24.

boundary into the interior (Figure 3.9). As a result, the mixed layer depth is less than 100 m over most of the interior.

This restratification process motivates the construction of a simpler conceptual model of the marginal sea that can be used to predict the main characteristics of the exchange between the marginal sea and the open ocean. The water masses in and around the marginal sea can be described by three water types: inflowing warm water ($T_{\text{in}}$), outflowing cool water ($T_{\text{out}}$), and the coldest water ($T_0$), which is found in the deep convection region in the interior of the basin. Several basic constraints can be reasonably imposed: mass conservation within the marginal sea, a closed heat budget within the marginal sea, and a closed heat budget within the interior of the marginal sea. It is assumed that the mean velocity field is in geostrophic balance with the mean density field. The final constraint derives from assuming that the heat loss in the interior of the marginal sea is balanced by lateral eddy heat fluxes that originate from the cyclonic boundary current.

This final constraint can be used to estimate the density of the water that is formed within the interior of the marginal sea, relative to that of the inflowing water (we get to chose a reference density that all other densities are relative to). Following Visbeck et al. (1996), a heat budget for the interior of the marginal sea can be written in terms of the mean surface heat loss $Q$ and the eddy heat flux $\overline{u'T'}$ as

$$P H_{\text{in}} \overline{u'T'} = \frac{AQ}{\rho_0 C_p}, \tag{3.1}$$

where $P$ is the perimeter of the marginal sea, $H_{\text{in}}$ is the thickness of the inflowing water, $A$ is the surface area of the interior of the marginal sea, $\rho_0$ is a typical density of seawater, the overbar indicates a time mean, primes are deviations from the mean, and $C_p$ is the specific heat of seawater. This balance implicitly assumes that the mean advection from the boundary into the interior is small. This is consistent with the model result, and with the notion that the mean flow tends to flow along geostrophic contours (bottom topography for this $f$-plane configuration).

The eddy flux is assumed to be proportional to the mean baroclinic velocity $V_{\text{in}}$ times the temperature of the boundary current relative to the interior as

$$\overline{u'T'} \propto V_{\text{in}}(T_{\text{in}} - T_0). \tag{3.2}$$

This functional relationship is supported by linear stability theory (Green 1970; Stone 1972) and in a wide range of flat bottom model calculations and laboratory experiments, as summarized in Visbeck et al. (1996). Visbeck et al. (1996) find empirically that the constant of proportionality is approximately 0.025 for a wide range of applications.

In the present configuration, the baroclinic boundary current is flowing over a sloping bottom, which we expect to alter the stability properties of the boundary current. The linear, quasi-geostrophic theory of Blumsack and Gierasch (1972) provides a simple means to incorporate the effects of a sloping bottom into (3.2), as used by Spall (2004). The Blumsack and Gierasch (1972) solution for the linear growth rate $\eta_i$ of a uniformly sheared velocity field with uniform stratification over a sloping bottom is

$$\eta_i = \left[\frac{\kappa - \tanh(\kappa)}{\tanh(\kappa)}(1-\delta) - \frac{1}{4}\left[\frac{\delta}{\tanh(\kappa)} - \kappa\right]^2\right]^{1/2}, \tag{3.3}$$

where $\kappa$ is the nondimensional wavenumber and $\delta$ is the ratio of the bottom slope to the isopycnal slope. The resulting growth rate is shown as a function of bottom slope and wavenumber in Figure 3.10a.

The flat-bottom result is recovered for $\delta = 0$. The general effect of bottom topography is to stabilize the flow, although there is a small region for $\delta$ slightly positive for which the growth rate increases relative to a flat bottom. For warm, cyclonic boundary currents, as found in convective basins, $\delta < 0$. The growth rate is strongly reduced by steeper topography and the wavelength of the most unstable waves shifts to smaller scales. The value of $\delta$ for the model calculations is approximately $-0.8$ (Figure 3.5), similar to that found for the Labrador Sea (Figure 3.6).

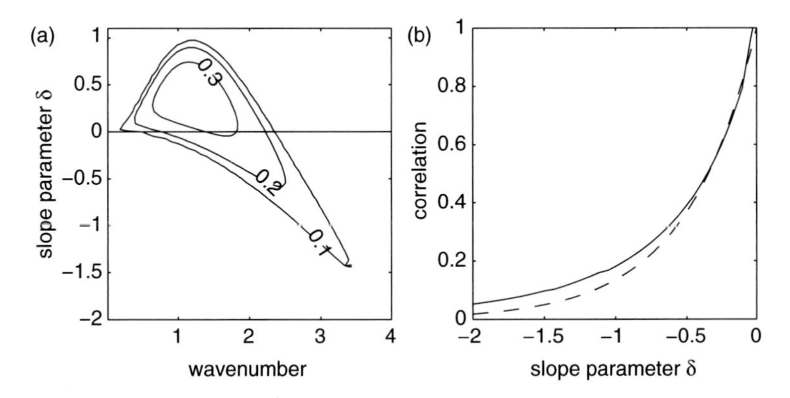


Figure 3.10. (a) Linear growth rate from (3.3) as a function of nondimensional wavenumber and $\delta$, the ratio of the topographic slope to the isopycnal slope. (b) The correlation $\overline{u'T'}$ as a function of $\delta$ for the fastest growing wave. Solid line: from the theory of Blumsack and Gierasch (1972), dashed line: $e^{2\delta}$.

Using the theory from Blumsack and Gierasch (1972), the eddy heat flux can be written as

$$\overline{u'T'} \propto \frac{\tanh(\kappa)}{\kappa - \tanh(\kappa)} \eta_i V_{\text{in}}(T_{\text{in}} - T_0) = c V_{\text{in}}(T_{\text{in}} - T_0). \tag{3.4}$$

The linear theory does not give the absolute value for the coefficient $c$, but it does provide its dependence on the slope parameter $\delta$. The value of $c$ inferred from (3.4), making use of (3.3) is shown in Figure 3.10b as a function of $\delta$ for the fastest growing wave. The value of $c$ was set arbitrarily to 1 for $\delta = 0$. The eddy heat flux is greatly reduced as the bottom slope increases relative to the isopycnal slope. For values typical of the Labrador Sea the growth rate is reduced approximately 80% relative to the flat-bottom result. The dependence of $c$ on $\delta$ is well approximated by $e^{2\delta}$, as indicated by the dashed line in Figure 3.10b. We can make use of the previous flat bottom research (Visbeck et al. 1996) to provide the constant of proportionality in (3.4) for $\delta = 0$ to be 0.025, so that the final eddy flux is related to the mean model parameters as

$$\overline{u'T'} = 0.025 e^{2\delta} V_{\text{in}}(T_{\text{in}} - T_0). \tag{3.5}$$

A similar functional relationship was tested in a series of numerical model calculations by Isachsen (2011). He found that the linear theory of Blumsack and Gierasch (1972) compared well with the nonlinear model results for the parameter range $-1 < \delta < 0$, which is the relevant range for convective basins.

It is assumed that the velocity of the inflowing water is in thermal wind balance with the density gradient between the boundary and the interior, and that the inflowing velocity is zero at the bottom, so that

$$V_{\text{in}} = \frac{\alpha g (T_{\text{in}} - T_0) H_{\text{in}}}{\rho_0 f_0 L}, \tag{3.6}$$

where $\alpha$ is the thermal expansion coefficient, $g$ is gravitational acceleration, $\rho_0$ is a reference density, and $L$ is the width of the sloping topography.

An estimate of the temperature anomaly of the interior water mass is now derived by combining (3.1), (3.5) and (3.6) to be

$$(T_{\text{in}} - T_0) = \frac{1}{H_{\text{in}}} \left( \frac{A f_0 L Q}{\alpha g P C_p c} \right)^{1/2}. \tag{3.7}$$

This result is significantly different from the isolated deep convection scaling of Visbeck et al. (1996). Its power dependence on surface heat loss and horizontal scale are different, but more importantly the temperature anomaly now depends on the rotation rate, the topographic width, and the bottom slope (through $c$). If it is assumed that the width of the boundary current is the internal deformation radius, then the scaling of Visbeck et al. is recovered (Chapman 1998). At high latitudes, the width of the sloping topography is typically much wider than the internal deformation radius, so the distinction is important. If the boundary current is to be in thermal wind balance, then this expression also determines the baroclinic shear in the boundary current through (3.6). This implies that the strength of the cyclonic boundary current is controlled by heat loss in the interior of the basin and the ability of eddies to flux heat from the boundary current into the basin interior.

The prediction for the temperature of the convective waters in the interior of the marginal sea (3.7) was tested with a numerical model by Spall (2004). A series of calculations were carried out in which the basin radius, surface heat flux, Coriolis parameter, width and amplitude of the topographic slope, and depth of the basin were varied. The characteristics of the general circulation and identification of distinct water masses in each of these calculations are similar to those in the example in Figure 3.4. The mean temperature of the water mass in the interior of the marginal sea diagnosed from each model run is compared to that predicted by (3.7) in Figure 3.11a. The temperature anomaly of the convective waters varies from approximately 0.5°C to 2°C over all the calculations. The temperature predicted by the theory compares well with the model results over the whole range of parameters tested, with a correlation of 0.95 and a least-square linear fit with slope 1.23. One of the main points from this result is that the temperature of the waters formed in the basin interior varies considerably, even when subject to the same heat loss. It is clear that the geometry of the basin and the ability of eddies to form from the boundary current control the properties of the waters formed in the basin interior.

A second quantity of interest is the temperature of the waters exported from the marginal sea ($T_{\text{out}}$). This will, in general, be warmer than the convective waters in the basin interior and cooler than the inflowing water (as required to balance the heat loss in the marginal sea). The average temperature of the outflowing waters can be estimated by requiring that the net heat flux carried into the marginal sea by the mean boundary current balances the heat loss to the atmosphere (neglect any eddy heat

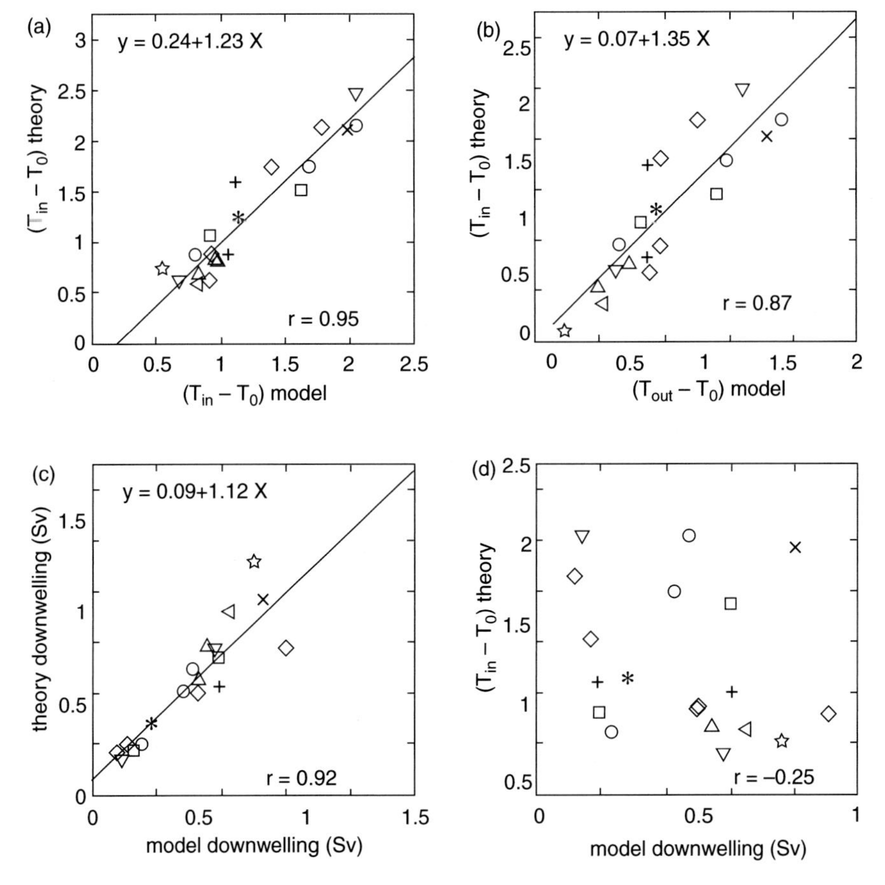


Figure 3.11. Comparison between model and theory for (a) $T_{in} - T_0$; (b) $T_{out} - T_0$; (c) net downwelling; (d) comparison between downwelling and temperature anomaly of convective waters. The least square fit to the data is given in the upper part of the panel and the correlation is given by r.

fluxes through the strait), and that the net mass flux into the marginal sea is zero. The heat budget is

$$T_{in} V_{in} H_{in} L - T_{out} V_{out} H_{out} L = \frac{AQ}{\rho_0 C_p}, \tag{3.8}$$

where the subscripts “in” and “out” refer to the inflowing and outflowing properties of the boundary current. The inflowing heat transport may be written, making use of (3.6) and (3.7) as

$$T_{in} V_{in} H_{in} L = \frac{RQL}{2\rho_0 C_p c}, \tag{3.9}$$

where $R$ is the basin radius.

The mass balance requires that $V_{in} H_{in} L = V_{out} H_{out} L$. This may be combined with (3.8) and (3.9) to produce an estimate for the outflowing temperature, relative to the interior temperature, as

$$T_{\mathrm{out}} - T_0 = (T_{\mathrm{in}} - T_0)(1 - \epsilon). \tag{3.10}$$

The factor $\epsilon = cP/L$, may be interpreted as the fraction of the inflowing waters that are fluxed into the interior by eddies (Spall 2004). For steep topography, $c \ll 1$, $\epsilon \ll 1$, and most of the water that flows into the marginal sea in the boundary current simply encircles the basin and flows back out of the marginal sea. For these cases, the temperature of the outflowing water is close to that of the inflowing water. However, for larger $cP/L$, eddies flux more heat into the basin interior and the temperature of the outflowing waters decreases. When $cP/L = 1$, all of the heat carried into the marginal sea in the boundary current is lost to the interior as it encircles the marginal sea. A slightly modified approach is required for $\epsilon > 1$ (see Spall 2004). The average temperature of the outflowing waters diagnosed from the series of model calculations is compared to that predicted by (3.10) in Figure 3.11b. Once again there is general agreement between the model and theory, with a correlation of 0.87 and slope of 1.35. For some calculations, the outflowing temperature is close to the inflowing temperature, while for others the outflowing temperature is close to the temperature of the convective water mass formed in the interior of the marginal sea.

The heat budgets of the interior of the marginal sea and the marginal sea as a whole have been used to predict the temperature of the convective and exported water masses. A related quantity of interest is the meridional overturning streamfunction, or net downwelling, in the marginal sea. The heat budgets alone do not inform us as to the means by which heat is transported into the basin. For example, the outflowing cold water could be in the upper part of the water column, in which case the heat transport would be achieved by a horizontal gyre, and there would be very little downwelling in the basin. On the other hand, the outflowing waters could be found at depth, in which case the heat transport would be carried by an overturning gyre and the downwelling would be much larger. The maximum downwelling in the marginal sea can be estimated by consideration of the thickness of the outflowing water column $H_{\mathrm{out}}$ relative to the inflowing water column $H_{\mathrm{in}}$. The downwelling transport $W$ is given simply by the product of this change in thickness times the width of the boundary current $L$ and the velocity of the boundary current $V_{\mathrm{out}}$ as

$$W = (H_{\mathrm{out}} - H_{\mathrm{in}}) V_{\mathrm{out}} L. \tag{3.11}$$

We can estimate the thickness of the outflowing water by making use of mass conservation in the marginal sea, the temperature of the outflowing water, and thermal wind balance to be

$$H_{\mathrm{out}} = \frac{V_{\mathrm{in}} H_{\mathrm{in}}}{V_{\mathrm{out}}} = \frac{H_{\mathrm{in}}}{(1-\epsilon)^{1/2}} \tag{3.12}$$

The downwelling is now derived to be

$$W = \frac{H_{\text{in}}}{\rho_0}\left(\frac{RLQ\alpha g}{2C_p c f_0}\right)^{1/2}[1-(1-\epsilon)^{1/2}] = [1-(1-\epsilon)^{1/2}]\Psi, \tag{3.13}$$

where $\Psi = V_{\text{in}} H_{\text{in}} L$ is the inflowing transport in the boundary current. The downwelling in the marginal sea is simply related to the inflowing transport by a factor of $[1-(1-\epsilon)^{1/2}]$. In the limit of $\epsilon \ll 1$,

$$W \approx \frac{\epsilon}{2}\Psi. \tag{3.14}$$

In this limit very little of the inflowing transport downwells within the marginal sea. As $\epsilon$ increases, a larger fraction of the inflowing transport downwells. For $\epsilon = 1$, all of the inflowing transport downwells within the basin. Since $\epsilon$ is a measure of the instability of the boundary current, it is clear that the net downwelling in the basin is closely related to the eddy fluxes from the boundary current into the interior.

The comparison between the maximum downwelling diagnosed from the model and that predicted by the theory (3.13) is shown in Figure 3.11c. In general, the theory compares well with the model. It is clear from these results that the downwelling is not simply related to the heat loss in the basin since for most of these calculations the heat loss is the same yet the downwelling varies by almost an order of magnitude. In fact, the downwelling transport is uncorrelated with the density anomaly of the convective water mass (Figure 3.11d). The downwelling in the basin is just what is required to balance the change in thermal wind transport from the inflowing to the outflowing boundary currents.

The relative importance of the horizontal and overturning gyres in the meridional transport of heat is also of interest. The ratio of the heat transport by the horizontal circulation to that by the overturning circulation can be estimated as

$$\frac{V_{\text{in}} H_{\text{in}} L(T_{\text{in}} - T_{\text{out}})}{W(T_{\text{in}} - T_{\text{out}})} = \frac{1}{1-(1-\epsilon)^{1/2}} \approx \frac{2}{\epsilon}. \tag{3.15}$$

The final approximation is valid for $\epsilon \ll 1$. In this limit, which is relevant for the Labrador Sea, the heat transport is dominated by the horizontal circulation, not the overturning circulation. This is consistent with estimates from repeat hydrography by Pickart and Spall (2007).

### 3.2.3 Moving Further Toward a More Realistic Configuration

Many marginal seas are separated from the open ocean by a topographic sill (e.g., the Nordic Seas, the Mediterranean Sea), while the previous configurations have only considered a sloping topography around the perimeter of the basin (which might be considered more relevant for the Labrador Sea). Another aspect of these previous calculations is that the heat loss to the atmosphere was specified, independent of the

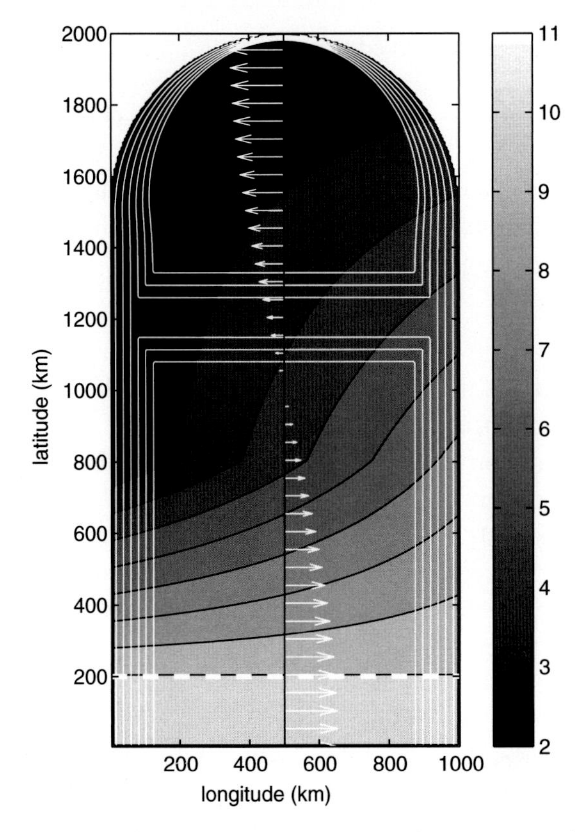


Figure 3.12. Model domain, bottom topography (white contours, c.i. = 300 m) and atmospheric temperature toward which the model sea surface temperature is restored (shaded). Temperature is restored toward a uniform stratification in the region south of the bold dashed white line at 200 km. The magnitude of the zonal wind stress is indicated by the vectors at the mid-longitude (uniform in the zonal direction).

ocean circulation. This was done to demonstrate that differences in the circulation and exchange were due only to changes in the model configuration, not due to a change in the surface forcing. However, it is expected that the heat loss to the atmosphere will depend on the ocean circulation, in particular the sensible heat flux with the atmosphere depends on the difference between the ocean temperature and the atmospheric temperature, such that a cooler ocean loses less heat to the atmosphere than a warm ocean. Finally, wind forcing has been so far neglected, but this has been done largely to simplify the problem rather than having been justified by any a priori scaling analysis. Each of these limitations are addressed in this final subsection.

The general approach is similar to what has already been discussed, and so only a brief overview will be given here. The model has been extended further to the south so that the "open ocean" is now large enough to support a wind- and buoyancy-driven circulation typical of a subpolar gyre. The domain extends $L_y = 2{,}000$ km in the meridional direction and $L_x = 1{,}000$ km in the zonal direction (Figure 3.12). The topography consists of sloping bottom around most of the perimeter, except near the southern boundary. The vertical wall at $y = 0$ prohibits topographic contours from closing around the basin and limits the excitation of a basin-scale circulation around closed contours. A sill is placed at $y = 1{,}200$ km, so that the marginal sea extends 800 km in the meridional direction. The surface heat flux is calculated by a linear relaxation of the upper model temperature toward the atmospheric temperature shown in Figure 3.12 with a relaxation constant $\Gamma$ as

$$Q = \Gamma(T_{\rm sst} - T_A) \tag{3.16}$$

where $T_{sst}$ is the sea surface temperature, $T_A$ is the atmospheric temperature, and $\Gamma$ is a restoring constant with units W $\rm m^{-2}$ $\rm C^{-1}$. The restoring temperature was chosen to represent a general warming at low latitudes and cooling at high latitudes with stronger cooling in the northwestern part of the basin. The model is forced by a wind stress as

$$\vec{\tau} = \vec{i}\tau_0^x \cos(\pi y/L_y) + \vec{j}\tau_0^y \cos(\pi x/L_x). \tag{3.17}$$

The temperature is also restored toward a uniform stratification of $N^2 = 2 \times 10^{-6}$ $\rm s^{-2}$ with a time scale of 30 days between $y = 0$ and $y = 200$ km.

The model has 5-km horizontal resolution and 20 levels in the vertical varying from 25 m in the upper 250 m to 250 m over the deepest 1,250 m. The model is initialized at rest and run for a period of 30 years. The analysis is taken over the final 5 years of integration. The inclusion of a sill is similar to what was done by Iovino et al. (2008), and some of their results will be used in the following analysis. However, the heat flux parameterization allows us to better understand what controls the meridional heat transport across the sill, a basic quantity of importance for climate, and a quantity that was specified in the studies of Spall (2004) and Iovino et al. (2008). The addition of wind forcing and a larger southern basin allow for a more complete representation of the circulation outside the marginal sea, and for the possibility of wind-driven exchange across the sill.

The mean temperature and horizontal velocity at the uppermost model level are shown in Figure 3.13. For this calculation, the restoring time scale is $\gamma = 60$ days (or $\Gamma = 20$ W $\rm m^{-2}C^{-1}$, $\gamma = h_1\rho_0 C_p/\Gamma$, where $h_1$ is the thickness of the upper model level), the Coriolis parameter at the southern limit of the model domain is $1.3 \times 10^{-4}$ $\rm s^{-1}$, and $\tau_0^x = 0.15$ N $\rm m^{-2}$, $\tau_0^y = 0$. Although clearly much more idealized than the real ocean, this calculation does represent several key aspects of the observed circulation in the North Atlantic Ocean and Nordic Seas. The mean basin-scale circulation in both the marginal sea and the northern part of the open ocean is cyclonic. Warm water is advected northward along the western boundary at low latitudes and crosses the basin to flow northward along the eastern boundary. Upon reaching the latitude of the sill, some warm water continues northward into the marginal sea and some turns toward the west along the southern flank of the ridge. The temperature of

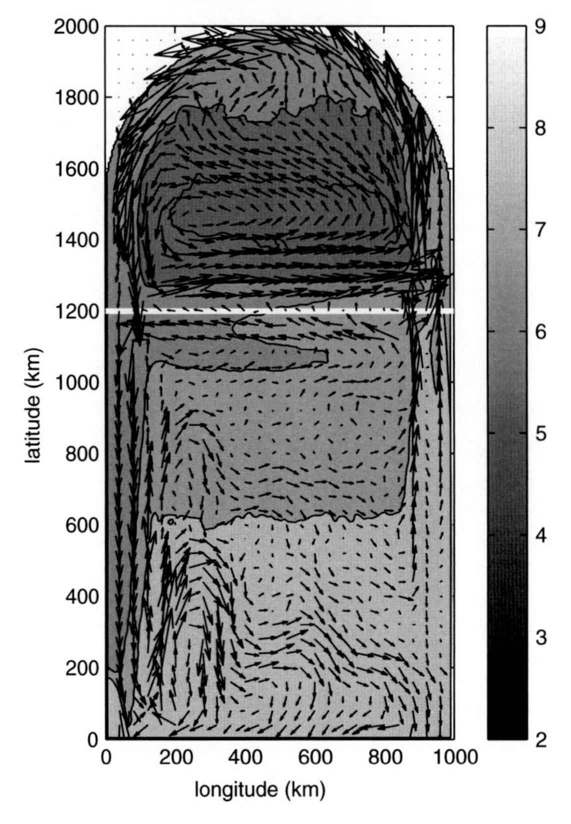


Figure 3.13. Average over the final 5 years of integration of temperature and horizontal velocity (every eighth grid vector) at the upper level. The crest of the sill is indicated by the white line at 1200 km latitude.

the water decreases along both the northward and westward pathways. The coldest waters are found in the interior of the marginal sea, while the water flowing out of the marginal sea along the western boundary is colder than the inflowing waters along the eastern boundary.

Following the same general approach as in Section 3.2.2, a heat balance in the interior of the marginal sea can be used to estimate the temperature of the convective water mass:

$$PH\overline{u'T'} = \frac{A\Gamma(T_0 - T_A)}{\rho_0 C_p}. \tag{3.18}$$

The eddy heat flux into the interior is integrated only down to the sill depth $H$ because the heat transport into the basin is confined to depths less than the sill depth (Iovino et al. 2008). The eddy heat flux is parameterized as previously in (3.5). Using thermal wind to relate the mean velocity to the temperature gradient between the boundary

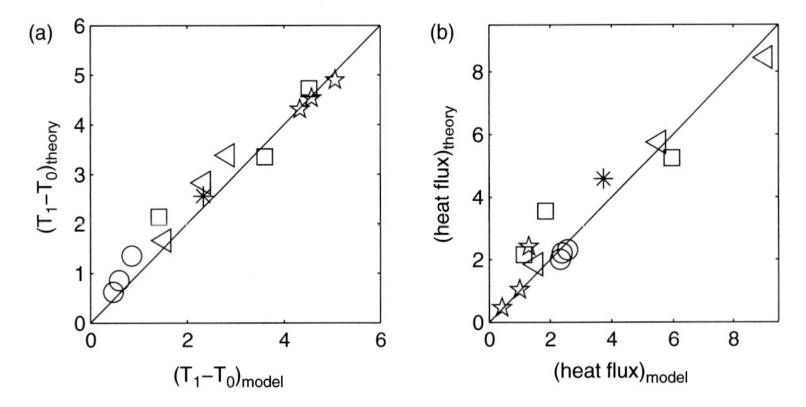


Figure 3.14. (a) Comparison of the diagnosed difference between the inflowing temperature $T_1$ and the temperature in the marginal sea interior $T_0$ from the model with that predicted by (3.7) (degrees C). The central case is indicated by the asterisk, the other symbols represent variations in the model parameters. (b) Comparison of the meridional heat flux across the sill ($10^{13}$ W) diagnosed from the model and that predicted by (3.21).

current and the interior, an estimate for the interior temperature is obtained:

$$T_{in} - T_0 = \frac{\mu}{\epsilon}\left[(1+2\epsilon/\mu)^{1/2} - 1\right](T_{in} - T_A). \tag{3.19}$$

A new nondimensional parameter has been introduced

$$\mu = \frac{A\Gamma f_0}{2\alpha g C_p H^2 (T_{in} - T_A)}. \tag{3.20}$$

The physical interpretation of $\mu$ will be discussed shortly.

Thirteen model calculations have been carried out in which the sill depth, restoring constant, and Coriolis parameter have been varied. The dignosed temperature of the convective waters (relative to the inflowing waters) is compared to that predicted by (3.19) in Figure 3.14a. The atmospheric temperature $T_A = 2.8$ degrees C used in (3.19) and (3.20) is the spatial average over the marginal sea. The central calculation in Figure 3.13 is shown by the asterisk. The comparison is quite close over the entire range of calculations. The convective water temperature is most sensitive to the sill depth, with shallow sills producing the coldest waters. The convective water density is only weakly dependent on the Coriolis parameter and the restoring time scale.

Once the sea surface temperature is known, we can readily obtain an estimate for the heat loss to the atmosphere, which in steady state is also the meridional heat transport across the sill. Using (3.16), the heat loss is

$$Q = A\Gamma[(1 + PL/A)(T_{in} - T_A) - (T_{in} - T_0)]. \tag{3.21}$$

The term containing $PL/A$ accounts for heat loss directly from the boundary current to the atmosphere. The remaining, larger terms are driven by eddy fluxes into the interior. For simplicity, it is assumed that the temperature of the boundary current is $T_{in}$, although this will produce a slight overestimate of the heat loss from the boundary current. The meridional heat flux across the sill diagnosed from the 13 model calculations is compared to the prediction (3.21) in Figure 3.14b. While there is some scatter, in general the prediction compares well with the model results. The heat flux is most sensitive to variations in the restoring constant (triangles) and sill depth (squares). It is nearly independent of the Coriolis parameter (circles). The heat flux becomes very small for shallow sills (stars).

Note that four of these calculations do not have a sill. The presence of a sill influences the solution in two ways (Iovino et al. 2008). First, it limits the depth over which water can flow into the basin, and thus influences the strength of the velocity in the boundary current through thermal wind, and the depth to which eddies can flux heat into the interior. The deeper the sill, the larger the vertical extent over which one integrates the thermal wind relation, and the stronger the velocity in the boundary current for a given temperature gradient. The second influence is that the sill blocks some of the topographic contours from extending directly from the southern basin into the marginal sea. This effectively narrows the region over which the boundary current flows, increasing the vertical shear in the velocity for a given change in temperature from the boundary current into the interor because $L$ decreases in (3.6). The same approach works for calculations with no sill as long as the primary means of heat transport into the northern basin is through the mean eastern boundary current over the sloping bottom.

The temperature in the interior of the marginal sea will fall between the temperature of the inflowing water $T_{in}$ and the atmospheric temperature $T_A$. If the eddy flux of heat from the boundary is sufficiently strong, the interior will be flooded with boundary current water before its heat can be extracted by surface cooling. This situation is found in the Lofoten Basin of the Nordic Seas, where the basin is filled by warm water of North Atlantic origin through lateral spreading via eddies formed from the boundary current. However, if the eddy flux is relatively weak, the atmospheric influence can overwhelm this eddy flux and strongly cool the ocean. This is more similar to the interior of the Greenland Sea, where deep convection occurs and the densest waters are formed within the Nordic Seas. In the former case, there is a large heat loss to the atmosphere, whereas in the latter case the heat flux is relatively small. The primary reason for this difference is that the topography along the eastern side of the Lofoten Basin is very steep compared to that along the western side of the Greenland Sea, thus increasing the eddy exchange through a smaller $L$ in $\epsilon$.

This transition between ocean influence and atmospheric influence is made clear if we consider the nondimensional form of (3.19) by dividing both sides by the temperature scale $T_{in} - T_A$. This nondimensional temperature anomaly of the convective waters depends only on the parameter $\mu/\epsilon$. A plot of the nondimensional temperature anomaly diagnosed from the model runs as a function of $\mu/\epsilon$ is shown in Figure 3.15a.

The scaled temperature anomaly predicted by (3.19) is given by the solid line. For $\mu/\epsilon \ll 1$ the sea surface temperature in the interior of the basin is close to the inflowing temperature (ocean dominates), whereas for $\mu/\epsilon \gg 1$ the sea surface temperature is close to the atmospheric temperature (atmosphere dominates). The transition between ocean dominance and atmospheric dominance occurs around $\mu/\epsilon \approx 1$.

So, what does the ratio $\mu/\epsilon$ represent? Its meaning becomes clear if we write it as the product of three terms:

$$\frac{\mu}{\epsilon} = \frac{\rho_0 f_0}{2\alpha g H^2 (T_{\text{in}} - T_A)} \frac{1}{\epsilon} \frac{A\Gamma}{\rho_0 C_p}. \tag{3.22}$$

The first term on the right-hand side is inversely proportional to the volume transport that would be carried in the boundary current if the interior of the marginal sea were at $T_A$. The second term projects this transport into an effective eddy transport into the basin interior (recall that $\epsilon$ is the fraction of the boundary current heat transport that is fluxed into the interior by eddies). The third is an effective "transport" from the marginal sea into the atmosphere based on the strength of the air–sea exchange coefficient $\Gamma$ (the units are $m^3\ s^{-1}$). The ratio $\mu/\epsilon$ thus represents the transport from the ocean to the atmosphere compared to the transport from the boundary current into the basin interior. This is consistent with the general behavior found in Figures 3.15a and 3.15b.

The temperature of the outflowing waters can be calculated using the heat budget for the entire marginal sea, as was done in the previous section for the case with a specified heat flux. Following a similar procedure using the heat budget, thermal wind, and mass conservation, the outflowing temperature anomaly relative to the inflowing,

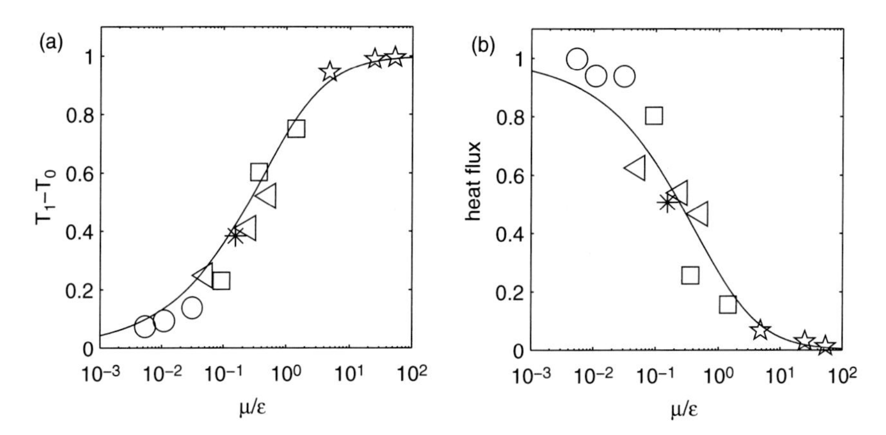


Figure 3.15. (a) The nondimensional temperature difference diagnosed from the model (symbols), from the theory (3.7) (solid line) as a function of the nondimensional parameters $\mu/\epsilon$. (b) Nondimensional meridional heat flux diagnosed from the model runs, from theory (solid line) as a function of the nondimensional parameters $\mu/\epsilon$.

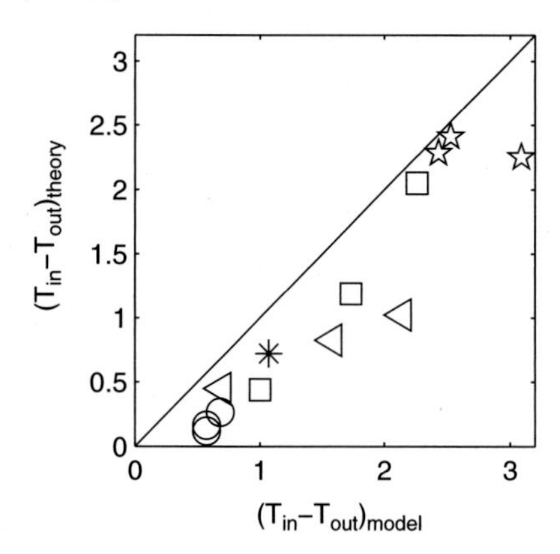


Figure 3.16. A comparison of the diagnosed temperature difference between the outflowing waters ($T_{out}$) and the inflowing waters $T_{in}$ with the theory (3.10).

$T_{in} - T_{out}$, is estimated to be

$$T_{in} - T_{out} = 2\mu\Big[\Big(\frac{T_1 - T_A}{T_1 - T_0}\Big)(1 + P_e L/A) - 1\Big](T_1 - T_A). \tag{3.23}$$

The temperature anomaly of the outflowing water calculated from the model runs is compared to that predicted by (3.23) in Figure 3.16. The general trend is reproduced, but the theory underpredicts the temperature of the outflowing water slightly. Shallow sills produce the largest difference in temperature between inflowing and outflowing waters. This is because the mass transport in the boundary current is reduced for shallow sills, thus requiring a larger temperature anomaly to balance surface cooling. The outflowing water temperature depends on both $\mu$ and the ratio $\mu/\epsilon$, so a simple comparison of the model results with a single nondimensional number is not possible.

It is also useful to calculate the outflowing temperature relative to the convective water temperature:

$$(T_{out} - T_0) = (T_{in} - T_0) - 2\mu(T_1 - T_A)\Big[\frac{T_1 - T_A}{T_{in} - T_0}(1 + PL/A) - \frac{T_1 - T_0}{T_1 - T_A}\Big]. \tag{3.24}$$

The ratio $\Theta = (T_{out} - T_0)/(T_{in} - T_0)$ is the fraction of outflowing water that is composed of inflowing water. For $\mu \ll 1$ the outflowing water is close to the inflowing water temperature, whereas for $\mu \gg 1$ the outflowing water temperature approaches the atmospheric temperature. The ratio $\Theta$ can be used to obtain an estimate of the total downwelling within the marginal sea. By requiring that the net mass transport into the marginal sea is zero, and assuming that the inflowing and outflowing currents are in thermal wind balance with the temperature gradients predicted by (3.19) and (3.24), one can obtain an estimate for the maximum downwelling in the marginal sea as

$$W = 0.5\Psi(1 - \Theta). \tag{3.25}$$

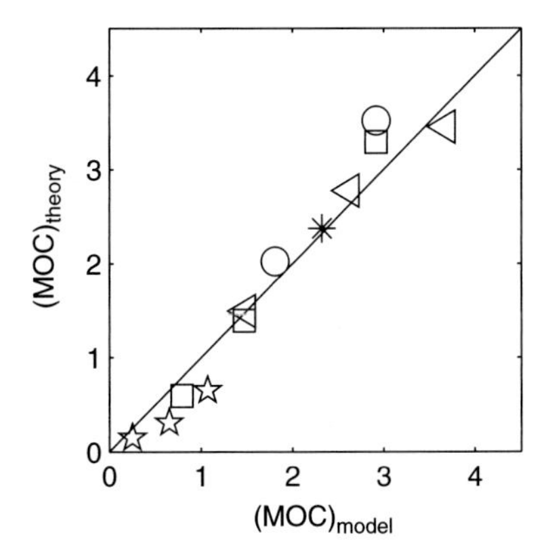


Figure 3.17. Comparison of the strength of the meridional overturning streamfunction at the sill diagnosed from the model runs with the theory (3.25).

For $\Theta \rightarrow 0$ the temperature of the outflowing water approaches the temperature of the convective water mass, whereas for $\Theta \rightarrow 1$ the outflowing water is composed of inflowing water. The transport in the boundary current is given by $\Psi$, so the downwelling is directly related to the inflowing transport through $\Theta$. The maximum overturning rate at the location of the sill diagnosed from the model is compared to that predicted by (3.25) in Figure 3.17. The theory compares well with the model results over the entire range of parameter space. The strength of the overturning varies between 0.25 and 4.12 Sv. Unlike the temperature of the water masses produced, the overturning is very sensitive to the Coriolis parameter (circles) because of its dependence on $\Psi$ which, through thermal wind, depends inversely on $f_0$. It is also sensitive to the restoring constant (triangles) and the sill depth (squares).

### *3.2.4 Influence of Wind Forcing*

The theory derived earlier does not take specific account of wind forcing. It simply defines the characteristics of the boundary current that would be required to balance surface heat loss with the assumption that all of the heat is provided through instability of the boundary current. Wind forcing was not included in the previous marginal sea studies of Spall (2003, 2004), Iovino et al. (2008), Walin et al. (2004), or Straneo (2006). In the calculations of Straneo, an idealized buoyancy-forced only model was able to reproduce the general characteristics of the observed seasonal and interannual variability in the Labrador Sea, suggesting that wind effects are secondary.

We can expect wind forcing to influence the exchange across the sill in several ways. First, if there is a zonal component to the wind, there will be a meridional Ekman transport across the sill. Another way in which the wind forcing can drive flow across the sill is through the Sverdrup-driven interior flow. Cyclonic wind stress curl will force upwelling into the Ekman layer and drive a poleward flow on a beta-plane.

The third way in which wind forcing can drive an exchange is if there is a wind stress parallel to the eastern or western boundary. For a northward wind adjacent to the eastern boundary, the Ekman transport will force a convergent flow into the boundary, which will result in an increase in sea surface height. The resulting pressure gradient will in turn drive a northward flow along the boundary. If the topographic contours extend from the southern domain directly into the marginal sea, then this will result in a wind-driven exchange.

A series of model calculations have been run in which the zonal wind stress $\tau_0^x = 0.0, 0.30$ N m$^{-2}$ with sill depths of 300, 600, 1,000, and 2,000 m. Recall that the calculations in Section 3.2.3 all had $\tau_0^x = 0.15$ N m$^{-2}$. The resulting temperature of the convective water mass, meridional heat flux across the sill, and meridional overturning at the sill are shown in Figure 3.18. On the vertical axis is the value in the absence of wind forcing. Each symbol represents a diagnosed quantity for a case with wind forcing (asterisks, $\tau_0^x = 0.15$ N m$^{-2}$, circles, $\tau_0^x = 0.30$ N m$^{-2}$). If the data points were to fall on the diagonal line, then the results would be the same as the case with no wind. It is found that the properties of the exchange across the sill are very nearly the same for all cases and for each sill depth. The influences of the Ekman-driven transport and the Sverdrup transport on the exchange are neglegible.

Calculations with a meridional wind of strength $\tau_0^y = 0.075, 0.15$ N m$^{-2}$ and $\tau_0^x = 0$ are shown by the right-facing and left-facing triangles, respectively. Each of the exchange quantities of heat transport, temperature anomaly, and meridional overturning strength are essentially the same as found for no wind forcing. There is a slight increase in the transport along the eastern boundary into the marginal sea (increasing from 8.6 Sv to 9.2 Sv to 9.7 Sv for $\tau_0^y = 0.0, 0.075, 0.15$ N m$^{-2}$), but this has little effect on the properties of the exchange.

## 3.3 Dynamics of Downwelling

A quantity of much interest for the climate system is the meridional overturning circulation. Considerable attention has been paid to where these dense waters upwell and return to the surface. The two dominant regimes are diapycnal mixing in the deep ocean (Polzin et al. 1997) and wind-driven upwelling in the Southern Ocean (Toggweiler and Samuels 1995). There has also been a lot of interest in where the poleward flowing waters in the upper limb of the thermohaline circulation become more dense, or downwell in density space (Talley 2003). Until fairly recently, little has been known about where or how these upper layer waters sink to intermediate and deep depths. It should be evident from the previous section that the net downwelling attained in marginal seas is closely related to the surface buoyancy forcing, but it should also be clear that the dynamics of the flow also influence the net downwelling rate and that the downwelling is not colocated with the region where waters are made

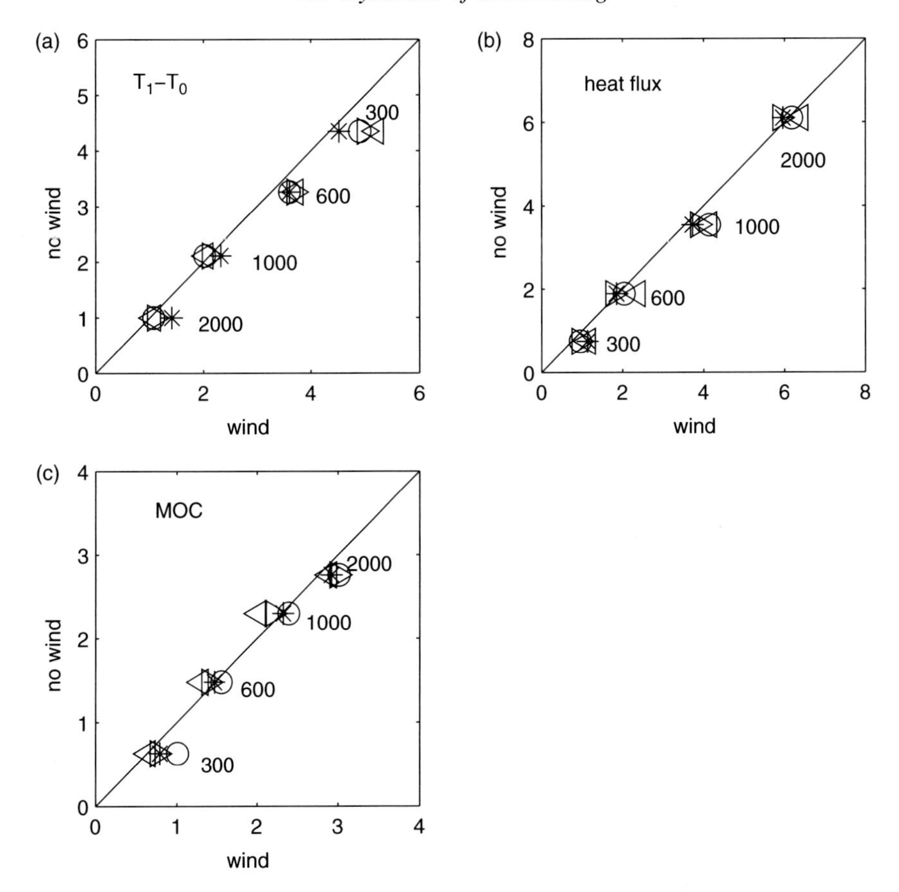


Figure 3.18. Sensitivity to wind stress for (a) temperature anomaly of convective waters, (b) meridional heat flux across the sill, and (c) meridional overturning streamfunction at the sill. Each symbol represents a calculation with a different wind forcing as: asterisk, $\tau_0^x = 0.15$ N m$^{-2}$; circle, $\tau_0^x = 0.30$ N m$^{-2}$; right-facing triangle, $\tau_0^y = 0.075$, left-facing triangle 0.15 N m$^{-2}$. The numbers next to each set of symbols are the sill depths.

more dense. In this section, the dynamical balances in several different regimes that can support net downwelling motions in buoyancy-forced basins are discussed.

### 3.3.1 Dissipative, Stratified Flows

The continuously stratified linear model described by Barcilon and Pedlosky (1967, hereafter BP67) defines two viscous boundary layers that are required to satisfy boundary conditions for the normal and tangential components of the velocity and thermal insulation. The wider of the two layers, called the hydrostatic layer, scales as $\sigma^{1/2} L_d$, where $\sigma = A_h/A_T$ is the horizontal Prandtl number and $L_d$ is the internal deformation radius. The width of this boundary layer is controlled by coupled balances in the density and vorticity equations. Horizontal diffusion of density is balanced by vertical advection of the mean stratification. This vertical advection gives rise to stretching of planetary vorticity in the vorticity equation, which is balanced by horizontal diffusion of relative vorticity into the boundary. This interplay gives rise to the dependence on the horizontal Prandtl number. Thus, subgridscale mixing plays a key role in the dynamics. The downwelling in Figure 3.3 is carried in this layer, and it was shown by Spall (2003) that the characteristics of this downwelling region follow the scaling predicted by BP67, even in these time-dependent, nonlinear numerical model calculations. There also exists a thinner, nonhydrostatic layer that does not contribute sigificantly to the net vertical mass transport, (see Chapter 2 by Pedlosky, this volume).

### 3.3.2 Weak Dissipation, Stratified Flows

Even though the boundary layer theory for downwelling in stratified flows by BP67, and its applications in Spall (2003) and Pedlosky (2003), depends on the diapycnal mixing of density due to horizontal diffusion, diapycnal mixing in the ocean is generally weak. The aim of this section is to explore downwelling in a convective basin with very small explicit diffusion of density. For a more detailed discussion, the reader is referred to Spall (2010a).

The MITgcm is configured in an elongated basin subject to cooling at the surface, which is connected to a smaller rectangular region through a strait (Figure 3.19). The domain has topography along the perimeter that slopes exponentially from the surface down to the bottom depth of 3,000 m with a horizontal e-folding scale that varies from 40 km over most of the basin to 10 km along the eastern boundary between $y = 400$ km and $y = 700$ km. The model is forced by applying 50 W m$^{-2}$ uniform cooling at the surface over the northern basin and by restoring toward a temperature profile with uniform vertical stratification of $N^2 = (g/\rho_0)\partial\rho/\partial z = 10^{-6}$ s$^{-2}$ in the southern rectangular region. This gives a first baroclinic deformation radius, based on the full ocean depth, of $L_d = NH/f_0 = 30$ km.

The vertical grid spacing is 100 m over the full depth of 3,000 m (30 levels). The model has variable horizontal grid spacing, as indicated in Figure 3.19. The grid spacing in the region along the eastern boundary is 1 km. The model was first spun up for a period of 20 years with a slightly lower resolution grid (the region of 1 km grid spacing was 2 km). The temperature and velocity fields at the end of that calculation were then interpolated onto the finer grid to initialize the final 200-day calculation used in the following analysis.

The model incorporates second-order vertical viscosity and diffusivity with coefficients $10^{-5}$ m$^2$ s$^{-1}$. The vertical diffusion is increased to 1,000 m$^2$ s$^{-1}$ for statically unstable conditions in order to represent vertical convection. Horizontal viscosity is parameterized as a second-order operator with the coefficient determined by a Smagorinsky (1963) closure as $A_h = (\nu_s/\pi)^2 L^2[(u_x - v_y)^2 + (u_y + v_x)^2]^{1/2}$, where

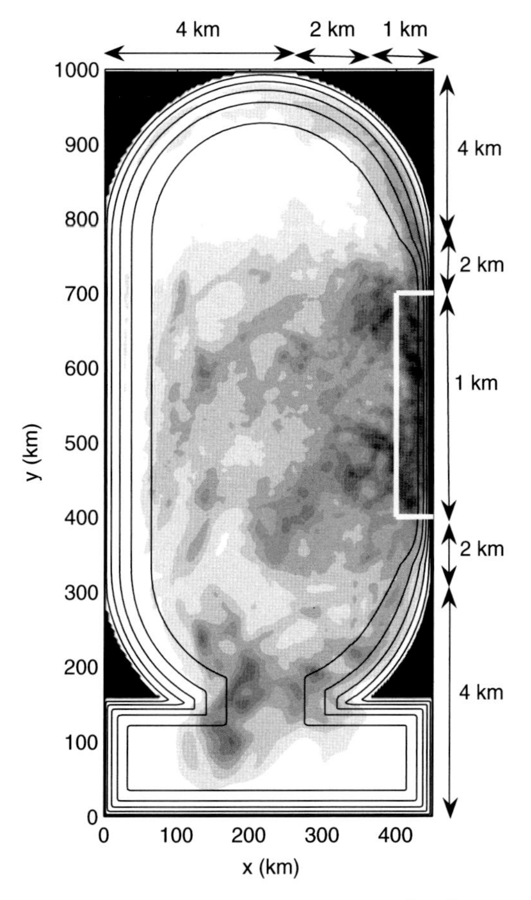


Figure 3.19. Eddy kinetic energy (maximum $\approx 500\,\text{cm}^2\,\text{s}^{-2}$, contour interval is 50 $\text{cm}^2\,\text{s}^{-2}$) over the model domain from the high-resolution calculation. Topography is indicated by the thin white contours (contour interval 500 m). The analysis region is marked by the bold white box along the eastern boundary. Zonal and meridional grid spacings are indicated along the top and right-hand sides of the figure, respectively.

$\nu_s = 2.5$ is a nondimensional coefficient, $L$ is the grid spacing, and $u$ and $v$ are the horizontal velocities (subscripts indicate partial differentiation). Temperature is advected with a third-order direct space/time flux-limiting scheme. There is no explicit horizontal diffusion of temperature.

The basin-scale circulation consists of a warm, cyclonic boundary current that enters the cooling basin along the eastern side of the strait and exits the basin along the western side of the strait, similar to the circulation discussed in Section 3.2.2. The boundary current accelerates over the region of steep topography, which results in enhanced instability and eddy formation compared to the rest of the domain where the topographic slope is weaker (Katsman et al. 2004; Wolfe and Cenedese 2006; Bracco et al. 2008; Spall 2010a, 2010b). This is reflected in the eddy kinetic energy (Figure 3.19), which is a maximum in the region of steep topography and has high

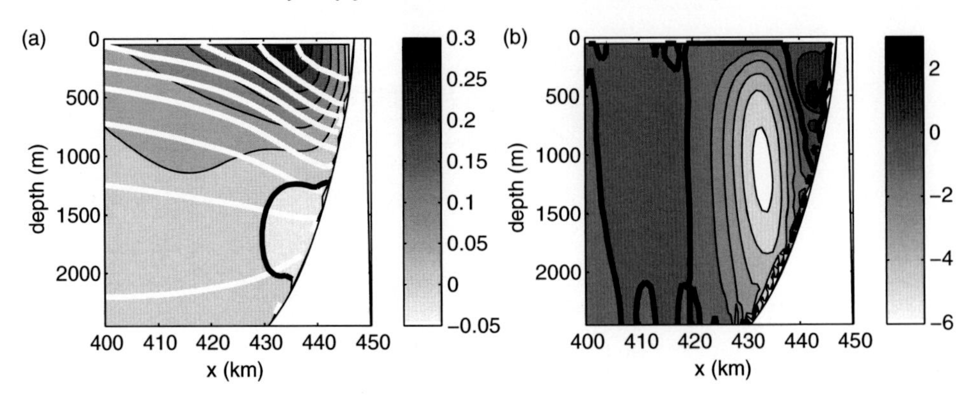


Figure 3.20. Mean $x$-depth sections averaged over 200 days and between $y = 400$ km and $y = 700$ km. (a) Meridional velocity (shaded, m s$^{-1}$) and temperature (white contours, contour interval 0.1°C); (b) vertical velocity ($10^{-4}$ m s$^{-1}$). The bold black line marks the zero contour in each panel.

values extending into the basin interior as a result of eddy shedding from the boundary current. Similar regions of enhanced eddy kinetic energy adjacent to steep topography are found along the eastern boundaries of the Labrador Sea (Prater 2002) and Lofoten Basin in the Nordic Seas (Poulain et al. 1996).

Vertical sections of the meridional average of the 200-day time-mean meridional velocity and temperature within the white box in Figure 3.19 are shown in Figure 3.20. The mean boundary current is approximately 30 km wide with a maximum velocity just over 30 cm s$^{-1}$. The mean horizontal velocity is essentially parallel to the boundary; there is very little mean flow between the boundary region and the interior.

The time-mean Eulerian vertical velocity is small everywhere in the cooling basin except along the region of steep topography and eddy activity. A meridional average of the vertical velocity in the analysis box is shown in Figure 3.20b. More than 95% of the total downwelling in the basin is concentrated in this relatively narrow band of $O(20$ km) width along the eastern boundary, where the eddies are formed. The maximum basin-integrated downwelling transport of $1.8 \times 10^6$ m$^3$ s$^{-1}$ is located at 1,100 m depth.

The importance of eddy fluxes in the region of downwelling is indicated by the flux divergence of the advective terms in the temperature tendency equation shown in Figure 3.21. The advective terms essentially balance at all depths, indicating that local surface cooling does not penetrate below 100 m in the boundary current and that diffusive effects are small. The mean advection is acting to warm the region throughout the water column. This represents the inflow of warm water in the cyclonic boundary current. Anticyclonic eddies transport warm water from the boundary current into the interior and act to cool the region at all depths (as shown in Figure 3.24). Similar warm anticyclones are observed to be formed along the eastern boundaries of the Labrador Sea and the Lofoten Basin (Lilly and Rhines 2002; Poulain et al. 1996). The vertical

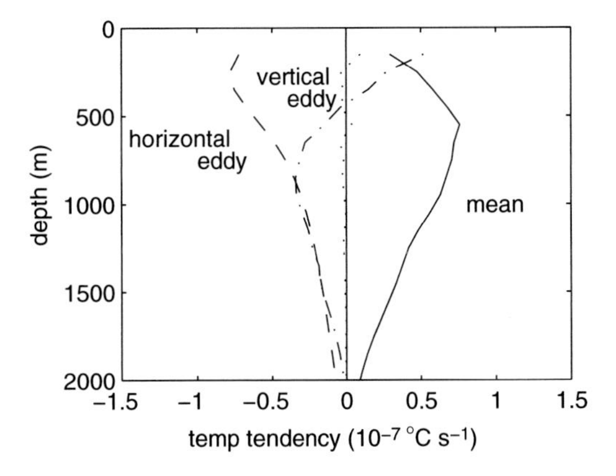


Figure 3.21. Advective influences on the temperature tendency as a function of depth averaged over the white box in Figure 3.19: solid line, mean flux divergence; dashed line, horizontal eddy flux divergence; dot-dash line, vertical eddy flux divergence; dotted line, sum of advective flux divergences.

eddy fluxes cool the deep ocean and warm the upper ocean. This is consistent with baroclinic instability, which releases potential energy by upwelling warm water and downwelling cool water (analysis of the energy conversion terms confirms baroclinic instability is active).

The mean temperature and northward transport between 400 km and the boundary is shown in Figure 3.22a. The temperature in the boundary current decreases at all depths as one moves northward, as expected since the boundary current loses heat as a result of eddy exchange with the interior. The mean along boundary transport deepens, but does not decrease, as the boundary current flows northward as a result of the downward mean vertical velocity. If eddies were not important for the heat budget, this would require an unrealistically large diapycnal mixing coefficient of $O(10^{-2}\,\mathrm{m^2\,s^{-1}})$ to balance this cross-isotherm mean advection.

The influence of eddies on density advection can be represented in compact form by considering the Transformed Eulerian Mean formulation (e.g., Andrews and McIntyre 1978). This representation includes the effects of eddy fluxes, but written in such a way that they appear as an advection acting on the mean density field. If it is assumed that the eddy advection of density is adiabatic, then these eddy-induced cross-front and vertical velocities may be written in the form of a streamfunction (where subscripts indicate partial differentiation) as

$$\tilde{u} = -\psi_z, \quad \tilde{w} = \psi_x. \tag{3.26}$$

There are numerous choices for the streamfunction $\psi$ as reviewed, for example, by Vallis (2006, pp. 304–314). For the primitive equations, a convenient choice is

$$\psi = -\frac{\overline{u'\rho'}\,\overline{\rho}_z - \overline{w'\rho'}\,\overline{\rho}_x}{\overline{\rho}_x^2 + \overline{\rho}_z^2}, \tag{3.27}$$

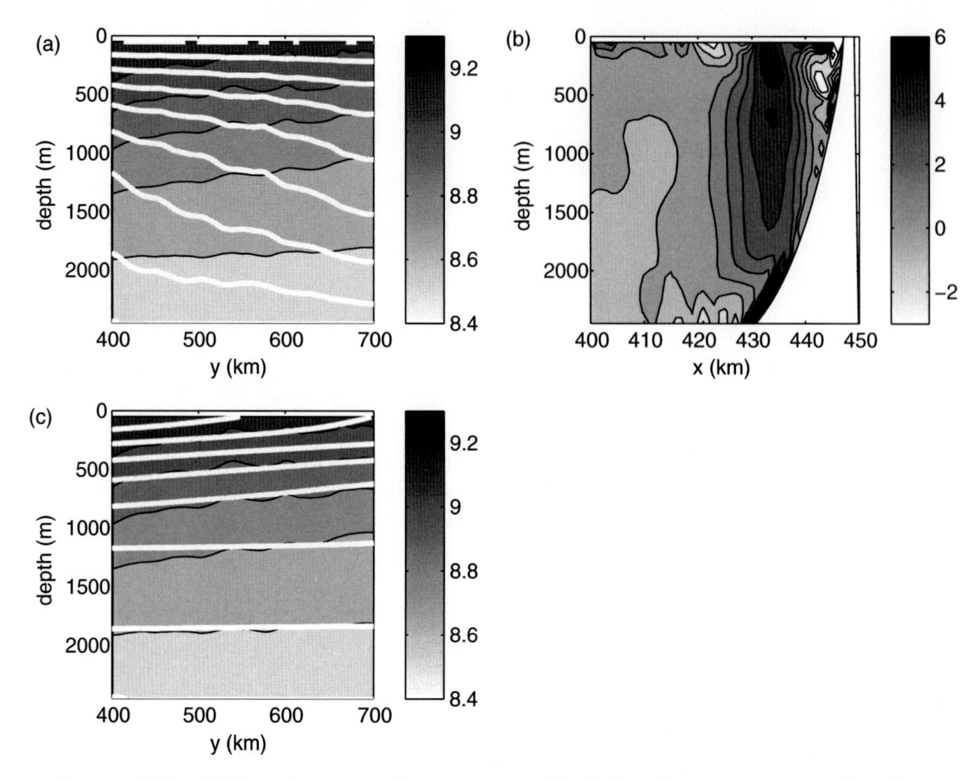


Figure 3.22. (a) Zonally averaged temperature (shaded) and transport streamfunction relative to the surface (white contours, c.i. $10^6$ m$^3$ s$^{-1}$) for the meridional flow based on the Eulerian mean vertical velocity, as a function of depth and meridional distance. (b) Eddy-induced vertical velocity ($10^{-4}$ m s$^{-1}$) calculated from (3.26) and (3.27). (c) As in Figure 3.22a except that the streamfunction is based on the transformed Eulerian mean vertical velocity.

where the overbar indicates a time average and the primed variables are deviations from the mean. The eddy flux is dominated by the zonal component, $(\overline{v'\rho'})_y \ll (\overline{u'\rho'})_x$ and $\overline{\rho}_{zy} \ll \overline{\rho}_{zx}$, so the meridional eddy flux term is neglected.

The streamfunction in (3.27) was calculated at each grid point in the analysis box and then averaged in the meridional direction to yield a two-dimensional depiction of the eddy-induced velocity. The time-averaged terms are only weakly dependent on latitude, so this averaging is used to reduce the influence of synoptic variability in $\psi$. The effective eddy-induced vertical velocity $\tilde{w}$ calculated from this averaged streamfunction via (3.26) is shown in Figure 3.22b. It is nearly equal in magnitude and opposite in sign to the Eulerian mean vertical velocity in Figure 3.20b. Such an upward eddy heat flux is consistent with Figure 3.21 and baroclinic instability. An along-boundary transport streamfunction was calculated using the zonally averaged residual vertical velocity (the sum of sections in Figures 3.20b and 3.22b). This

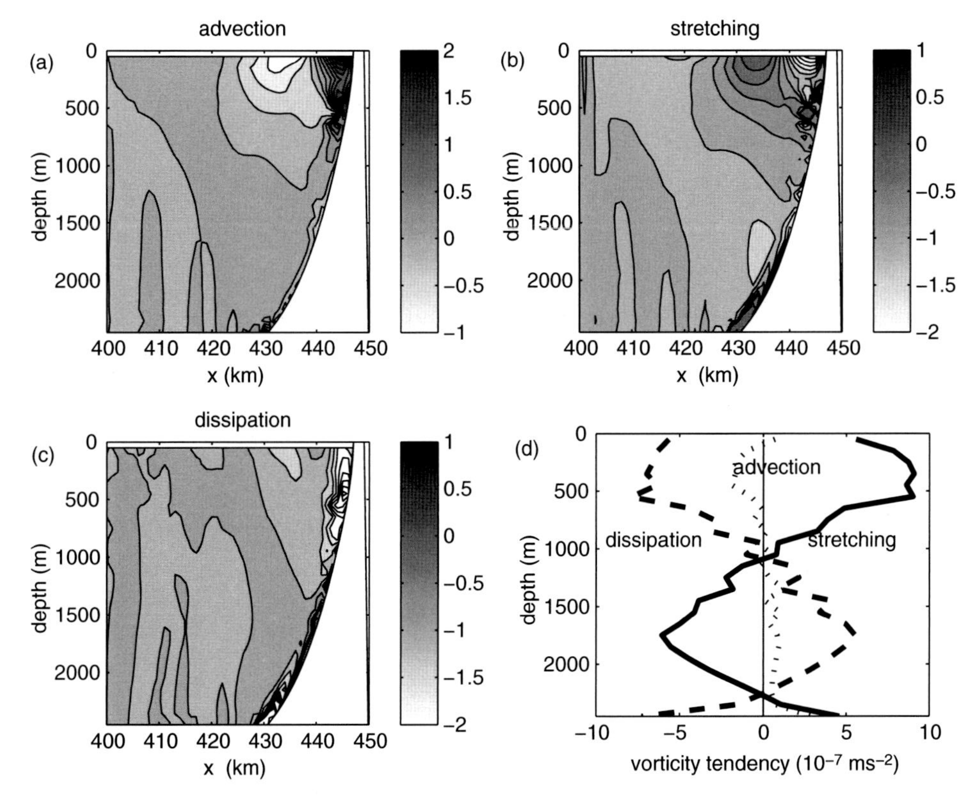


Figure 3.23. Meridionally and temporally averaged terms in the relative vorticity equation (3.28), (units $10^{-10}$ s$^{-2}$). (a) advection; (b) stretching of planetary vorticity; (c) dissipation; (d) zonal integral of the terms in Figures 3.23a–3.23c ($10^{-7}$ m s$^{-2}$).

results in upwelling that is in general agreement with the rise of the mean isopycnals (Figure 3.22c), indicating that the flow is nearly adiabatic.

The question remains as to what balances the mean Eulerian vertical velocity in the vertical component of the relative vorticity equation.

$$\zeta_t = \underbrace{-\vec{v}\cdot\nabla\zeta - w\zeta_z - \nabla w\cdot\vec{v}_z + \zeta w_z - \beta v}_{advection} + \underbrace{f w_z}_{+stretching} + \underbrace{A\nabla^2\zeta + K\zeta_{zz}}_{+dissipation} \approx 0 \quad (3.28)$$

The balance in the boundary layer theory of BP67 is between the last two terms, stretching and dissipation. All three terms have been calculated from the model output and averaged in the meridional direction to produce zonal sections (Figure 3.23). The advection term includes all advective effects, including horizontal advection of relative vorticity, vertical advection of relative vorticity, tilting of horizontal vorticity, and advection of planetary vorticity.

The strong mean Eulerian downwelling results in stretching of planetary vorticity, $f w_z$, that tends to increase the upper ocean relative vorticity in the region of maximum downwelling. This is balanced to a large degree by the an advective export of positive

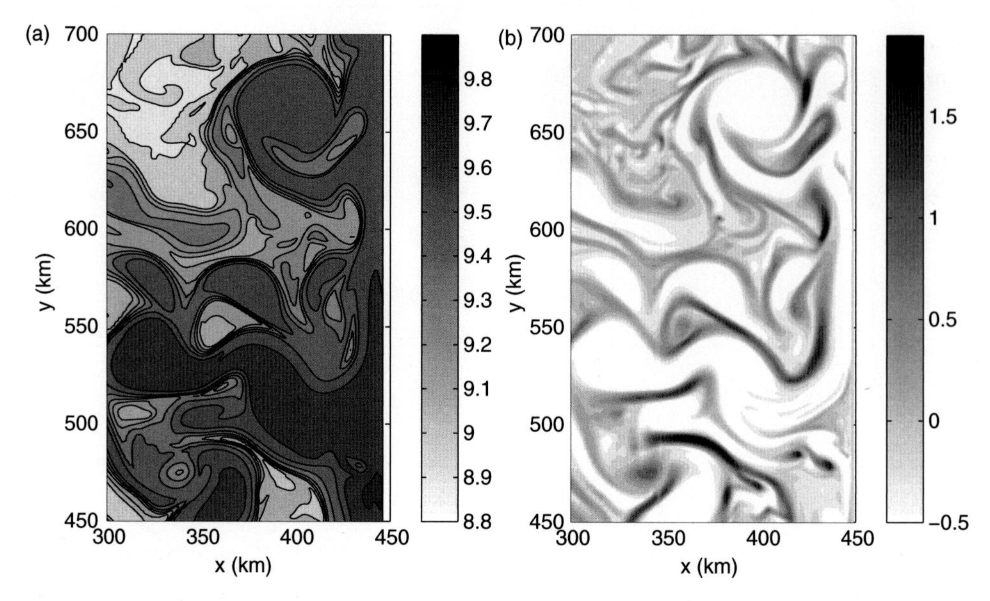


Figure 3.24. Snapshot at 50 m depth near the eastern boundary of (a) temperature (contour interval 0.1°C) and (b) relative vorticity divided by the Coriolis parameter (contour interval 0.25). The vorticity field is characterized by large, warm, anticyclonic eddies surrounded by narrow filaments of very strong cyclonic vorticity.

vorticity (primarily the horizontal eddy flux divergence of relative vorticity). The majority of the relative vorticity flux is toward the boundary, but there is also an export of positive relative vorticity into the interior (not shown). The flux toward the boundary goes to zero at the boundary because of the no-normal flow boundary condition, giving rise to a flux convergence (the positive region near the boundary in the upper 700 m) that is primarily balanced by dissipation. Thus, unlike the theory of BP67, the region of dissipation is offset from the region of downwelling, but they are connected through the lateral eddy flux term. The positive relative vorticity flux into the basin interior is somewhat surprising since the dominant form of eddies exported from the boundary current are anticyclonic, and points to the importance of submesoscale filaments of positive relative vorticity in the eddy field (Figure 3.24). The primary balance, when integrated over the whole downwelling region, is between stretching and dissipation, similar to the BP67 balance. However, the positive relative vorticity flux into the interior balances up to 20% of the stretching (Figure 3.23d).

### *3.3.3 Weakly Stratified Flows*

The idealized modeling study of Spall and Pickart (2001) found that essentially all of the downwelling at high latitudes in a low-resolution, basin-scale model occurred within the mixed layer adjacent to the boundary. Their model was non–eddy permitting and had relatively weak diapycnal mixing, so there was no means to support downwelling in the stratified part of the water column. They argued that the magnitude of the downwelling was determined by the mixed layer depth and density gradient along the boundary through thermal wind balance and a no-normal flow condition through the wall. This weakly stratified regime differs significantly from the previous theoretical analysis, which assumes that the stratification is significant. This motivates a more detailed analysis of the weakly stratified regime, and the discussion here follows closely from Spall (2008). The primary approach is numerical, although a scaling is used to identify the key boundary layer that supports the downwelling motion. For a more detailed discussion of this boundary layer and its connection to the interior flow, see Pedlosky (2009) or Chapter 2 in this volume.

A high-resolution numerical model is used to calculate the vertical motions within the mixed layer forced by cooling of a boundary current. The MITgcm is run with nonhydrostatic, Boussinesq physics. The model domain is a channel of width 20 km, length 48 km, and depth 500 m. The model horizontal grid spacing is 100 m and the vertical grid spacing is 10 m. Subgridscale mixing of momentum and temperature are parameterized by a horizontal Laplacian mixing with coefficients of 1 $m^2$ $s^{-1}$ and vertical mixing coefficients of $10^{-5}$ $m^2$ $s^{-1}$.

The initial stratification is uniform with $N^2 = 4.8 \times 10^{-6}$ $s^{-2}$. The model is forced with a specified inflowing velocity in geostrophic balance with the density field and cooled at the surface with a uniform heat flux of 500 W $m^{-2}$. The Coriolis parameter is $f_0 = 10^{-4}$ $s^{-1}$ and uniform. The inflowing velocity has a maximum value at the surface and decreases linearly to zero at 500 m depth. The model is initialized with this velocity field and a geostrophically balanced density field and sea surface height. The inflow conditions are steady in time, and the outflow boundary conditions for temperature, normal velocity, and tangential velocity are determined by an Orlanski radiation condition (Orlanski 1976). The northern and southern lateral boundary conditions are no normal flow, no-slip, and no normal heat flux.

Consider first the case of surface cooling that linearly decreases from 1,000 W $m^{-2}$ at the southern boundary to zero at the northern boundary.[1] The geostrophically balanced initial condition and inflow have a zonal velocity that is uniformly sheared in the vertical and constant in the horizontal. The maximum inflow velocity is $U = 30\,\mathrm{cm\,s^{-1}} = H_0 M^2 / f_0$, where $H_0 = 500$ is the domain depth, $M^2 = (g/\rho_0)\rho_y = b_y$ is the horizontal stratification. The temperature change across the basin is 0.6°C at all depths, giving $M^2 = 6 \times 10^{-8}$ $s^{-2}$. The average temperature between days 3 and 10 at 45 m depth is shown in Figure 3.25a. The temperature along the offshore boundary is uniform because there is no heat loss there, whereas the temperature along the southern boundary decreases by approximately 0.4°C. As a result, the total change in density across the channel is less at the outflow than at the inflow, implying a decrease

[1] The calculations are all on an $f$-plane, but, for convenience, the direction of flow will be considered towards the east, and the offshore side of the boundary current will be toward the north.

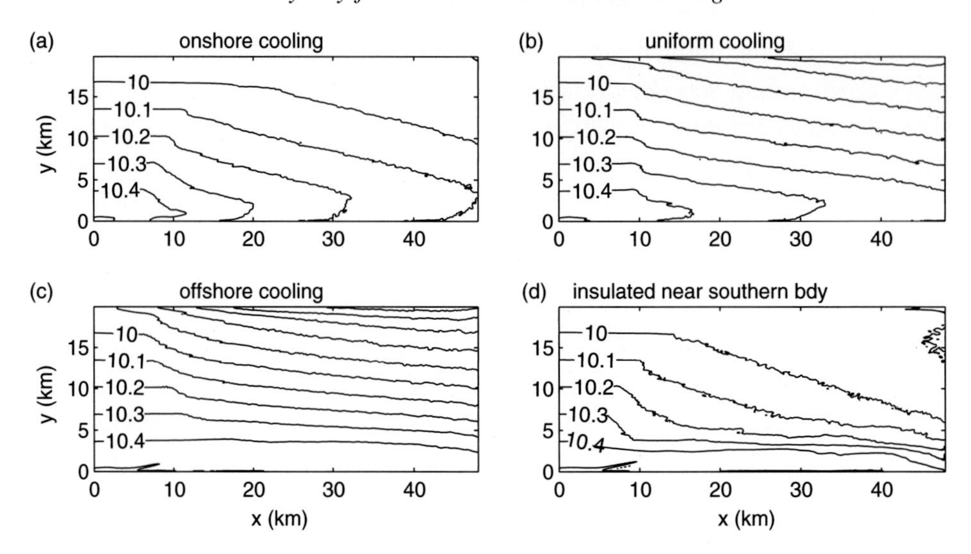


Figure 3.25. Temperature at 45 m depth averaged between days 3 and 10 for cases with uniform inflowing horizontal velocity at each depth. Heat flux: (a) linearly decays from 1,000 W m$^{-2}$ at $y = 0$ to 0 at $y = 20$ km; (b) spatially uniform at 500 W m$^{-2}$; (c) linearly increases from 0 at $y = 0$ to 1,000 W m$^{-2}$ at $y = 20$ km; (d) linearly decays from 1,176 W m$^{-2}$ at $y = 3$ km to 0 at $y = 20$ km.

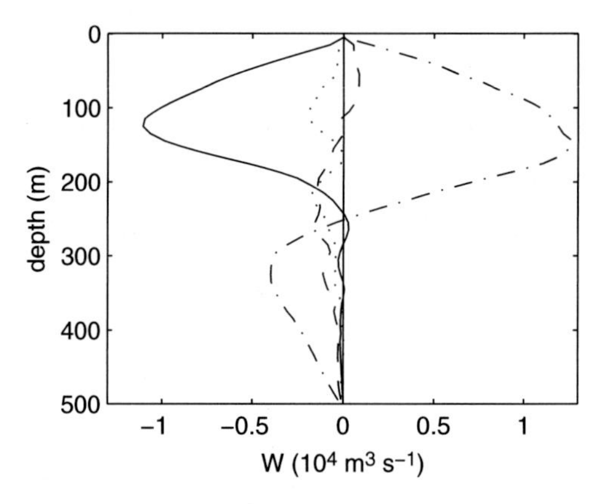


Figure 3.26. Vertical mass transport averaged between $x = 20$ km and $x = 30$ km between days 3 and 10: solid line, cooling decays offshore; dot-dashed line, cooling increases offshore; dashed line, uniform cooling; dotted line, insulated within 3 km of the southern boundary.

in the vertical shear of the geostrophic velocity at this depth. The average net vertical transport over the region $x = 20$ km to $x = 30$ km is shown in Figure 3.26 by the solid line. The net vertical motion is primarily downwelling in the upper ocean with a maximum of $1.11 \times 10^4$ m$^3$ s$^{-1}$ near 100 m depth, decreasing to zero at the surface and at the base of the mixed layer.

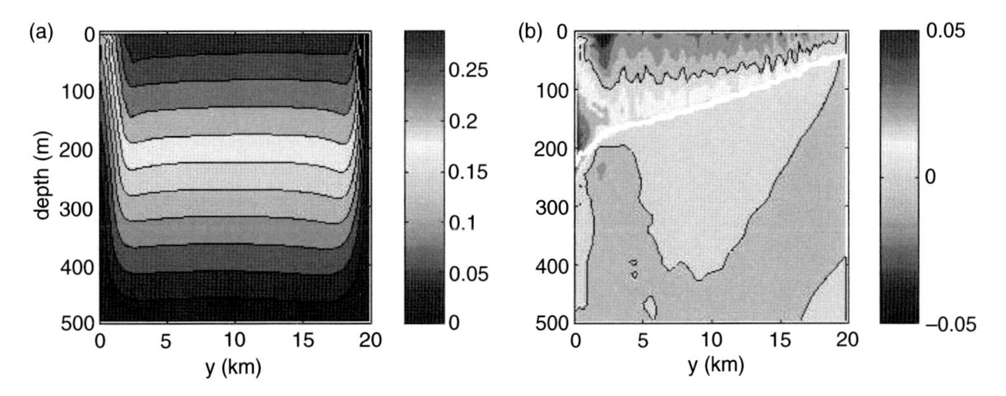


Figure 3.27. (a) The mean zonal velocity averaged between $x = 20$ km and $x = 30$ km for the calculation with surface cooling linearly decreasing from 1,000 W m$^{-2}$ at $y = 0$ to zero at $y = 20$ km. (b) The difference in the mean zonal velocity between the calculation with surface cooling and one with no surface forcing (m s$^{-1}$). The zero contour is in black and the mixed layer depth from the case with surface cooling is given by the white line.

#### *3.3.3.1 Along-Channel Evolution*

The vertical section of the average zonal velocity is shown in Figure 3.27a. The development of the no-slip boundary layers is evident along the northern and southern walls. The velocity in the interior, below the mixed layer, is nearly uniform at each depth, and close to the value at the inflow, approximately 30 cm s$^{-1}$ at the surface and decreasing linearly toward zero at the bottom. There has been some increase in the interior velocities in response to the development of the no-slip boundary layers in order to conserve mass within the domain. Within the mixed layer, however, the velocity near the surface has decreased from its inflow value and the velocity near the base of the mixed layer has increased, particularly in the southern part of the domain. These adjustments make it difficult to distinguish between changes in the zonal velocity resulting from the buoyancy forcing and changes resulting from the development of the no-slip boundary layers. A calculation was carried out that had no surface forcing at all, but was otherwise identical to this calculation. The difference between the mean zonal velocity between $x = 20$ km and $x = 30$ km for these two calculations can be attributed soley to the buoyancy forcing and is shown in Figure 3.27b. The depth of the mixed layer for the case with surface cooling is indicated by the bold white line. The zero contour is black. Most of the velocity change due to buoyancy forcing is found within the mixed layer. The zonal velocity is decreased in the upper portion of the mixed layer and increased near the base of the mixed layer, resulting in weaker vertical shear throughout most of the mixed layer. Very close to the southern boundary, the zonal velocity is increased throughout the mixed layer. This is different from the behavior in the interior and is due to the development of a very narrow boundary layer, discussed next.

### 3.3.3.2 The Nonhydrostatic Layer

There is very intense downwelling right next to the southern boundary. The width of this downwelling layer can be estimated following the approach of Stewartson (1957) (see also Pedlosky, 2009, and Chapter 2 in this volume), who found that there are two narrow boundary layers required to transition a region of interior flow driven by stress at the surface and bottom to that of an adjacent flow driven at a different speed. A boundary layer of width $E^{1/4}$ exists to allow the geostrophic flow parallel to the boundary to smoothly transition from one regime to the other, where $E = A/f_0L^2$ is the horizontal Ekman number, $A$ is the horizontal viscosity, and $L$ is a horizontal length scale. However, this transitional layer cannot support the vertical mass transport that is required to match the upper and lower Ekman layers. This is achieved in a narrower, nonhydrostatic, boundary layer of nondimensional width $E^{1/3}$, or dimensional width $(AH/f_0)^{1/3}$, where $H$ is a vertical length scale. Even though the $E^{1/4}$ layer does not exist for the present problem, where the forcing is due to an along-boundary pressure gradient and not surface and bottom Ekman layers, the $E^{1/3}$ layer that carries the vertical mass transport does. The zonal pressure gradient, which was not considered in the original solution by Stewartson, does not alter the width of this boundary layer. For the values used here ($f_0 = 10^{-4}$ s$^{-1}$, $A = 1$ m$^2$ s$^{-1}$, $H = 100$ m), the horizontal scale of the downwelling region is predicted to be $O(100$ m$)$.

The downwelling near the wall in the model is contained mostly within 1 grid cell of the boundary, so it is not well resolved with the standard 100-m grid. An identical calculation was carried out with the meridional resolution increased to 25 m between 0 and 200 m and 50 m between 200 m and 400 m from the boundary. The net vertical transport in this case is very similar to the standard resolution case, $1.24 \times 10^4$ m$^3$ s$^{-1}$ compared to $1.11 \times 10^4$ m$^3$ s$^{-1}$. The mean zonal, meridional, and vertical velocities between $x = 20$ km and $x = 30$ km near the southern boundary at 45 m depth are shown in Figure 3.28. The downwelling is still concentrated within 100 m of the wall, so the horizontal scale of this downwelling region, while only marginally resolved with the standard grid, is $O(100$ m$)$, consistent with that predicted by the $E^{1/3}$ Stewartson layer. Linear interpolation of the vertical transport indicates that 80% of the total downwelling occurs within 107 m of the boundary. A calculation with the horizontal viscosity increased to 8 m$^2$ s$^{-1}$ results in 80% of the downwelling occurring within 204 m of the boundary, in close agreement with the expected doubling of the boundary layer width for a factor of 8 increase in viscosity.

A detailed analysis of this boundary layer in the linear limit by Pedlosky (2009) verifies that the horizontal scale of the downwelling region, and the abrupt gradient in the along-boundary flow, scales as $E^{1/3}$. His analysis shows that the weak vertical stratification is key to the existence of this narrow, nonhydrostatic layer whose function is to bring the cross-channel geostrophic flow to zero at the wall. This Stewartson layer

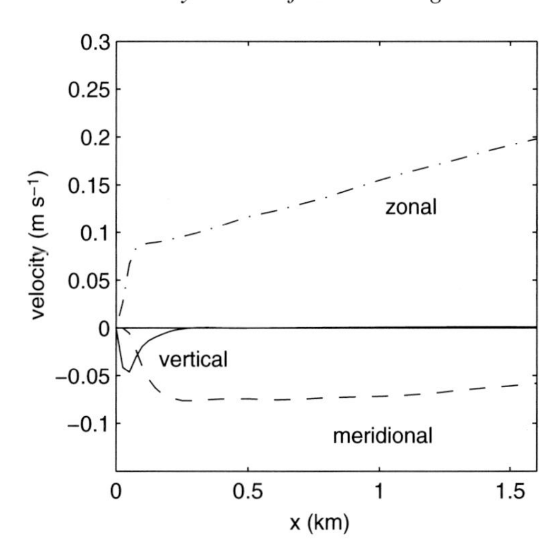


Figure 3.28. The mean horizontal and vertical velocities near the southern boundary at 45 m depth from the calculation with enhanced meridional resolution near the boundary.

is likely to be replaced by a bottom Ekman layer if the vertical wall is replaced by a sloping bottom.

### *3.3.3.3 Cooling Distribution*

Insight into what determines the net downwelling resulting from surface buoyancy loss is gained by varying the distribution of the surface cooling, while keeping the total buoyancy loss the fixed. A case with spatially uniform cooling of 500 W $m^{-2}$ and uniform horizontal and vertical stratification results in density changes along both the southern and northern boundaries (Figure 3.25b). The magnitude of the change in temperature along each boundary is similar, so that the net change in density across the channel at the outflow is very similar to the net change in density at the inflow, even though the density itself has increased. The net vertical mass transport between 20 and 30 km is very small (Figure 3.26). Even though there is a mixed-layer of $O(100$ m) depth, and mixed-layer instabilities form and drive a restratifying cross-channel circulation, they do not drive a net vertical motion when integrated across the basin. A revealing calculation is obtained with a heat loss that is zero at the southern boundary and increases linearly to 1,000 W $m^{-2}$ at the northern boundary. Now, the temperature is nearly constant along the southern boundary but decreases by approximately 0.4°C along the northern boundary (Figure 3.25c). The net vertical motion in this case is upward at about the same amplitude as the downwelling found in the case with cooling enhanced along the southern boundary (Figure 3.26).

A final calculation was carried out in which the heat loss increased from zero at the northern boundary to 1,176 W $m^{-2}$ at 3 km from the southern boundary and was zero within 3 km of the southern boundary. The sea surface temperature is constant along the northern boundary, but is now also nearly constant along the southern boundary (Figure 3.25d). Because no heat loss occurs at this point, there is no means to support a strong pressure gradient, and the along-channel velocity simply advects the isotherms downstream. The net vertical motion is also very weak (Figure 3.26). A similar sensitivity to surface insulation within 200 km of the boundary was found by Spall and Pickart (2001) for the basin-scale overturning circulation in a non-eddy-resolving climate model. The present results suggest that this process will remain important for the basin-scale thermohaline circulation, even when the lack of convection is limited to within a few kilometers of the boundary. This result is also consistent with the modeling study of Walin et al. (2004), in which a baroclinic current was cooled and formed a barotropic boundary current yet resulted in no net downwelling. The form of their surface forcing resulted in no heat loss adjacent to the boundary, and was thus unable to support a pressure gradient, or downwelling, along the boundary.

The two calculations here that do not have a change in the density gradient across the channel both have the interior ageostrophic overturning cell driven by the mixed-layer instabilities, yet neither has any appreciable net vertical motion in the basin, demonstrating that these cells play no direct role in the net sinking in the basin. Sinking is achieved when the density within the mixed layer increases along a boundary in the direction of Kelvin wave propagation; upwelling results when it decreases in the direction of Kelvin wave propagation.

#### *3.3.3.4 Parameter Dependencies*

The results in the previous section isolate the pressure gradient along the boundary as the key feature that controls net vertical motion. The pressure gradient on the boundary is related to the mixed-layer depth and the density gradient along the boundary through the hydrostatic relation. Each of these calculations had the same net surface heat loss yet demonstrated completely different net vertical motions, clearly demonstrating that there is no direct relationship between heat loss and downwelling. The key to understanding the downwelling is to understand what controls the pressure gradient along the boundary. A simple model of the temperature within the mixed layer is now formulated to provide a framework with which to understand and predict how the buoyancy-forced downwelling will vary with environmental parameters.

For simplicity, it will be assumed that the pressure does not vary along the offshore side of the boundary current. For cyclonic boundary currents that encircle marginal seas subject to buoyancy forcing, this is roughly consistent with having the offshore edge of the boundary current being defined by an isotherm. The net downwelling is then determined by the lateral, large-scale flow into the very narrow nonhydrostatic layer adjacent to the boundary. An important assumption here is that the boundary layer exists in order to satisfy the no-normal flow boundary condition and conserve mass, and that the pressure gradient is set by the flow in the boundary current just outside the narrow boundary layer. It is also assumed that all of the transport toward the boundary layer downwells within the boundary layer. This is in close agreement with the numerical results and is also supported by the linear theory of Pedlosky (2009).

Consider the buoyancy balance near the southern boundary within the mixed layer but outside the nonhydrostatic layer of width $E^{1/3}$. If the along-channel velocity at the base of the mixed layer is $U$, the mixed layer depth is $h$, the mixed layer buoyancy is $b = -g\rho/\rho_0$, and the surface buoyancy flux $B = \alpha g Q/\rho_0^2 C_p$, then the density equation within the mixed layer may be written as

$$U b_x = -\frac{B}{h}, \tag{3.29}$$

where $\alpha$ is the thermal expansion coefficient, $g$ is the gravitational acceleration, $C_p$ is the specific heat of seawater, and the variables $U, b$, and $h$ are functions of downstream distance $x$ only. This is a balance between the along-boundary advection of buoyancy and surface cooling. If it is assumed that the mixed layer is an unstratified layer overlaying a uniformly stratified region below, the depth of the mixed layer can then be related to the buoyancy as

$$h = -\frac{b}{N^2}, \tag{3.30}$$

where $N^2 = b_z$ is the Brunt-Väisälä frequency and $b$ is taken to be relative to the surface buoyancy in the absence of any cooling ($h = 0$ when $b = 0$). Combining with (3.29), the buoyancy gradient in the along-channel direction can be written as

$$b_x = \left(\frac{BN^2}{2Ux}\right)^{1/2}. \tag{3.31}$$

The downstream buoyancy gradient increases with increasing cooling, as expected. However, the buoyancy gradient also depends on the along-channel velocity because the balance is between horizontal advection of buoyancy and surface cooling. The dependence on stratification enters because the mixed layer will be shallower for stronger underlying stratification, and the buoyancy change will be larger for a shallower mixed layer.

Spall and Pickart (2001) considered the geostrophic flow within a mixed layer subject to cooling and found that, for a density that is increasing downstream, the flow will be toward the boundary in the upper half of the mixed layer and away from the boundary in the lower half of the mixed layer. Thermal wind gives a maximum downwelling at the middepth of the mixed layer, per unit along boundary distance, of

$$W = -\frac{b_x h^2}{8 f_0}. \tag{3.32}$$

This expression was found to compare well with a series of low-resolution, basin-scale wind and buoyancy-driven general circulation models.

If the mixed-layer depth were known, the downwelling rate could be derived from (3.29) and (3.32) to be

$$W = \frac{Bh}{8 f_0 U}. \tag{3.33}$$

It is clear from (3.30) that the mixed-layer depth will increase downstream as the boundary current is cooled and the buoyancy decreases. Equation (3.32) may be combined with (3.30) and (3.31) to provide an estimate of the downwelling that varies with downstream position as

$$W = \frac{1}{4 f_0}\left(\frac{B^3 x}{2N^2 U^3}\right)^{1/2}. \tag{3.34}$$

Although this expression is more complicated than (3.33), it is also more revealing regarding the competing effects that influence net downwelling. The downwelling increases with increasing cooling, as expected, but it also increases with increasing distance, decreasing velocity, decreasing stratification, and decreasing Coriolis parameter. Downwelling depends on distance because the mixed-layer depth increases with distance downstream. Less downwelling is found for stronger boundary currents because the pressure gradient is less due to stronger horizontal advection balancing the surface cooling, but it is also due to the fact that a stronger horizontal advection limits the depth of mixing (3.29). The downwelling also increases with decreasing stratification because the mixed layer will penetrate further for the same cooling rate. The increasing downwelling with decreasing Coriolis parameter is simply due to the geostrophic balance resulting in more flow toward the boundary to balance a given pressure gradient.

A series of model calculations was carried out by Spall (2008) in order to test the parameter dependencies predicted by (3.34). The model was forced with uniform horizontal and vertical stratification and a heat loss that was maximum at the southern boundary and linearly decreased to zero at the northern boundary. The magnitude of the surface cooling, vertical stratification, along-channel geostrophic flow, and Coriolis parameter were each varied (see Spall 2008 for details). Each of these model calculations was carried out with 200-m horizontal resolution and 10-m vertical resolution; however, the circulation features are very similar to the previously discussed calculation with higher resolution. The maximum net downwelling per unit along boundary distance is also similar (1.11 $m^2\ s^{-1}$ for the high-resolution calculation and 1.04 $m^2\ s^{-1}$ for the low-resolution calculation). The maximum downwelling rate was calculated, as in Figure 3.26, for each of these calculations and is compared with the theory in Figure 3.29. The velocity scale used in (3.34) has been taken to be proportional to the surface geostrophic velocity at the inflow, $U = cH_0 M^2/f_0$, where the constant $c = 0.43$ produces a least-square fit line to the data with slope 1. It is expected that $c < 1$ because the velocity decreases within the no-slip boundary layer,

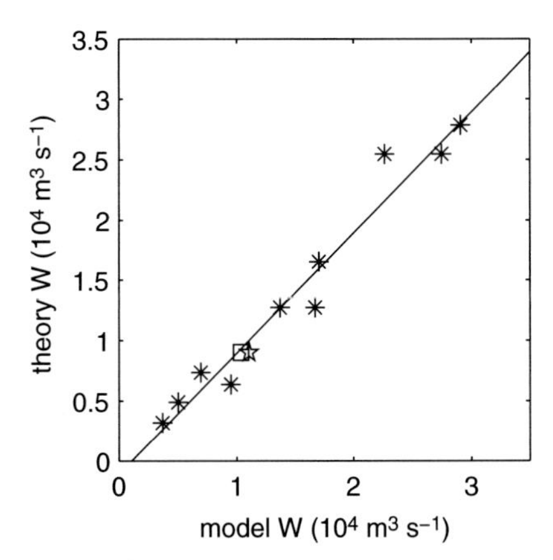


Figure 3.29. Comparison of the net downwelling rate per unit along-boundary distance calculated from the model and that predicted by the theory (3.34). The square marks the central calculation and the star marks a calculation with the same parameter settings and a horizonal resolution of 100 m.

which is much wider than the downwelling layer. Nonetheless, the geostrophic scaling allows for a systematic means to estimate the influence of the horizontal velocity on the downwelling and makes clear the dependence on the controlling parameters $M^2$ and $f_0$.

The agreement between the downwelling diagnosed from the model and that predicted by the simple theory is quite good. The central calculation is indicated by the square (low resolution); and star (high resolution); the two are nearly-indistinguishable on this scale. It is interesting that all but three of these calculations are subject to the same heat loss at the surface yet the net downwelling varies by a factor of 6.

It is somewhat counterintuitive that the simple theory (3.34) indicates that the total downwelling does not depend on the offshore extent of the boundary current or the amplitude or pattern of cooling away from this near-boundary region. To demonstrate this independence, the model was run with a maximum heat loss of 1,000 W m$^{-2}$ at the southern boundary that linearly decreased to zero at the northern boundary in a channel 40 km wide, twice as wide as in the standard case. The horizontal velocity, horizontal stratification, and vertical stratification were the same as the standard case, but due to the wider domain the total heat loss was twice as large. The total net downwelling in this case was $1.08 \times 10^4$ m$^3$ s$^{-1}$, essentially the same as for the 20-km-wide channel. The downwelling is independent of the current width provided that the current transports enough heat to balance the surface cooling and maintain the along-boundary baroclinic pressure gradient.

It does not appear to be necessary to resolve the nonhydrostatic physics and convective plumes explicitly. A hydrostatic calculation with horizontal viscosity and diffusivity increased to 5 m$^2$ s$^{-1}$, and with vertical convection parameterized by increasing the vertical diffusivity to 1,000 m$^2$ s$^{-1}$ for unstable profiles, results in a net downwelling of $1.10 \times 10^4$ m$^3$ s$^{-1}$, close to the standard calculation. The subgridscale mixing is sufficiently large that the symmetric instabilities are supressed, but all other aspects of the zonally averaged flow are similar to the nonhydrostatic result. The density within the mixed layer is essentially uniform with depth. The mean cross-channel flow in the interior is now toward the boundary in the upper mixed layer and away from the boundary in the lower mixed layer, as expected from geostrophy (Spall and Pickart 2001).

The redistribution of mass in the vertical is achieved by a geostrophic flow toward the boundary in the upper mixed layer, downwelling very close to the boundary, and a return flow away from the boundary below the mixed layer. Thus, while the acceleration at the base of the mixed layer is physically very close to the deceleration at the surface, the water parcels had to make a relatively long traverse all the way to the narrow boundary layer in order to sink to the deeper depth.

## 3.4 Summary

Buoyancy-forced marginal seas are the source regions for many of the dominant water masses in the world ocean. While relatively small in geographic extent, these basins are the locations of most regions of deep convection and produce water masses whose signature can be seen far from their source. The properties of the product waters produced in such marginal seas are of great importance for the general oceanic circulation, for understanding its sensitivities to changes in atmospheric forcing and climate, and for our ability to properly represent these processes in large-scale climate models. In Section 3.2, simple dynamic and thermodynamic constraints were combined to provide analytic estimates of the key properties of the exchange between semienclosed marginal seas and the open ocean. The key dynamical components are a mean boundary current that exchanges mass and tracers between the marginal sea and the open ocean and baroclinic instability that controls the exchange of buoyancy between the boundary current and the interior of the marginal sea. It was shown that the boundary current arises because of the buoyancy forcing in the marginal sea, and its strength is determined by the eddy exchange dynamics. These two conditions, along with mass conservation, buoyancy budgets, geostrophic momentum balances, and an eddy flux parameterization, allow for direct estimates of the density of the convective waters, density of the exported waters, exchange rate between the marginal sea and the open ocean, net downwelling in the marginal sea, and in some cases the heat flux into the marginal sea. A key aspect of these solutions is that the mean flow through the interior of the marginal sea is neglegible. This is no longer true if topographic contours directly connect the interior of the marginal sea with the open ocean, in which case eddy fluxes are much less important (e.g., Spall 2005).

The dynamics that support net vertical motions that result from buoyancy loss at the surface were discussed in Section 3.3. Three distinct parameter regimes can be identified: dissipative, stratified; weakly dissipative, stratified; unstratified. Although the detailed dynamical balances in each of these regimes vary, there are some common elements. The net vertical motion resulting from buoyancy loss in the basin is downward and is located near the lateral boundaries. The magnitude of the downward mass transport is just what is required to maintain a thermal wind balance in the along-boundary geostrophic flow. Because buoyancy has been extracted from the basin, the horizontal density gradient is reduced, and the transport in the boundary current is shifted from the baroclinic mode to the barotropic mode as one moves cyclonically around the marginal sea. This vertical motion introduces stretching into the vertical vorticity equation, which is ultimately balanced by lateral dissipation into the side boundary. The details of these viscous lateral boundary layers vary, depending on stratification and resolved physics. In all cases, the region of net downward motion is distinct from the region where buoyancy is extracted from the fluid and where there is a net diapycnal mass transport from lighter to denser density surfaces.

The dynamics of downwelling motions present some challenges for both the observational and large-scale climate modeling communities. The vertical velocities are highly variable in space and time, making direct observations of net vertical mass transport difficult. Perhaps the best approach would be to measure inflowing and outflowing transports along a segment of a boundary current that is subject to buoyancy loss. An initial attempt in this direction was carried out by Pickart and Spall (2007). However, synoptic variability is likely to be large relative to the net transports, so long-time and high-resolution measurements are required. Global-scale climate models will not be able to directly resolve the processes that control buoyancy-forced downwelling for some time. However, there is hope in that the net downwelling is determined by the deformation-scale flow, and the details of the downwelling dynamics do not need to be resolved. So, progress will rely on getting the mean boundary currents correct (velocity, mixed layer depth, along-boundary variations in density). Since much of the heat balance within the boundary currents involves eddy fluxes with the interior, this will require an accurate eddy flux parameterization. The eddy-resolving calculations presented here suggest that parameterizations should be adiabatic; identify regions of enhanced eddy fluxes (including the role of topography); and transport tracers far from the boundary current. This last constraint will be difficult to achieve since it requires a nonlocal eddy flux parameterization and knowledge of how eddies lose their heat to the atmosphere and surrounding ocean.

## Acknowledgments

The work summarized in this paper was supported by the National Science Foundation under Grants OCE-0240978, OCE-0726339, and OCE-0850416, and by the Office of Naval Research under Grants N00014-01-1-0165 and N00014-03-1-0338.

Joe Pedlosky, Fiamma Straneo, and Bob Pickart are acknowledged for numerous fruitful discussions on a wide range of topics related to this work. I would like to thank the organizers for the opportunity to present this work at the Alpine Summer School on Buoyancy-Driven Flows at the Gran Paradiso National Park in the Aosta Valley, Italy, and several of the students for careful reading and useful comments on the first draft of this chapter.

## References

Andrews, D. G., and M. E. McIntyre, 1978: Generalized Eliassen-Palm and Charney-Drazin theorems for waves on axisymmetric mean flows in compressible atmospheres. *J. Atmos. Sci.***35,** 175–185.

Barcilon, V., and J. Pedlosky, 1967: Unified linear theory of homogeneous and stratified rotating fluids. *J. Fluid Mech.* **29,** 609–621.

Blumsack, S. L., and P. J. Gierasch, 1972: Mars: The effects of topography on baroclinic instability. *J. Atmos. Sci.* **29,** 1081–1089.

Bracco, A., J. Pedlosky, and R. S. Pickart, 2008: Eddy formation near the west coast of Greenland. *J. Phys. Oceanogr.* **38,** 1992–2002.

Chapman, D. C., 1998: Setting the scales of the ocean response to isolated convection. *J. Phys. Oceanogr.* **28,** 606–620.

Green, J. S., 1970: Transfer properties of the large-scale eddies and general circulation of the atmosphere. *Quart. J. Roy. Meteor. Soc.* **96,** 157–185.

Iovino, D., F. Straneo, and M. A. Spall, 2008: On the effect of a sill on dense water formation in a marginal sea. *J. Mar. Res.* **66,** 325–345.

Isachsen, P. E., 2011: Baroclinic instability and eddy tracer transport across sloping bottom topography: How well does a modified Eady model do in primitive equation simulations? *Ocean Modelling* **39**, 183–199.

Katsman, C. A., M. A. Spall, and R. S. Pickart, 2004: Boundary current eddies and their role in the restratification of the Labrador Sea. *J. Phys. Oceanogr.* **34,** 1967–1983.

Lilly, J. M., and P. B. Rhines, 2002: Coherent eddies in the Labrador Sea observed from a mooring. *J. Phys. Oceanogr.* **32,** 585–598.

Marshall, J., and F. Schott, 1999: Open–ocean convection: Observations, theory, and models. *Rev. Geophys.* **37,** 1–64.

Marshall, J., C. Hill, L. Perelman, and A. Adcroft, 1997: Hydrostatic, quasi–hydrostatic, and non-hydrostatic ocean modeling. *J. Geophys. Res.* **102,** 5733–5752.

Orlanski, I., 1976: A simple boundary condition for unbounded hyperbolic flows. *J. Comput. Phys.* **21,** 251–269.

Pedlosky, J., 2003: Thermally driven circulations in small ocean basins. *J. Phys. Oceanogr.* **33,** 2333–2340.

Pedlosky, J., 2009: The response of a weakly stratified layer to buoyancy forcing. *J. Phys. Oceanogr.* **39,** 1060–1068.

Pickart, R. S., and M. A. Spall, 2007: Impact of Labrador Sea convection on the North Atlantic meridional overturning circulation. *J. Phys. Oceanogr.* **37,** 2207–2227.

Polzin, K. L., J. M. Toole, J. R. Ledwell, and R. W. Schmitt, 1997: Spatial variability of turbulent mixing in the abyssal ocean. *Science* **276,** 93–96.

Poulain, P.-M., A. Warn-Varas, and P. P. Niiler, 1996: Near surface circulation of the Nordic Seas as measured by Lagrangian drifters. *J. Geophys. Res.* **101,** 18237–18258.

Prater, M., 2002: Eddies in the Labrador Sea as observed by profiling RAFOS floats and remote sensing. *J. Phys. Oceanogr.* **32,** 411–427.

Smagorinsky, J. 1963: General circulation experiments with the primitive equations: I. The basic experiment. *Mon. Wea. Rev.* **91,** 99–164.
Spall, M. A., 2003: On the thermohaline circulation in flat bottom marginal seas. *J. Mar. Res.* **61,** 1–25.
Spall, M. A., 2004: Boundary currents and water mass transformation in marginal seas. *J. Phys. Oceanogr.* **34,** 1197–1213.
Spall, M. A., 2005: Buoyancy-forced circulations in shallow marginal seas. *J. Mar. Res.* **63,** 729–752.
Spall, M. A., 2008: Buoyancy-forced downwelling in boundary currents. *J. Phys. Oceanogr.* **38,** 2704–2721.
Spall, M. A., 2010a: Dynamics of downwelling in an eddy resolving convective basin. *J. Phys. Oceanogr.* **40**, 2341–2347.
Spall, M. A., 2010b: Non-local topographic influences on deep convection: An idealized model for the Nordic Seas. *Ocean Modelling.* **32,** 72–85.
Spall, M. A., and R. S. Pickart, 2001: Where does dense water sink? A subpolar gyre example. *J. Phys. Oceanogr.* **31,** 810–826.
Stewartson, K., 1957: On almost rigid motions. *J. Fluid Mech.* **3,** 17–26.
Stone, P., 1972: A simplified radiative-dynamical model for the static stability of rotating atmospheres. *J. Atmos. Sci.* **29,** 405–418.
Straneo, F., 2006: Heat and freshwater transport through the central Labrador Sea. *J. Phys. Oceanogr.* **36,** 606–628.
Talley, L., 2003: Shallow, intermediate, and deep overturning components of the global heat budget. *J. Phys. Oceanogr.* **33,** 530–560.
Toggweiler, J. R., and B. L. Samuels, 1995: Effect of Drake Passage on the global thermohaline circulation. *Deep–Sea Res.* **42,** 477–500.
Vallis, G. K., 2006: *Atmospheric and Oceanic Fluid Dynamics*, Cambridge University Press, Cambridge.
Visbeck, M., J. Marshall, and H. Jones, 1996: Dynamics of isolated convective regions in the ocean. *J. Phys. Oceanogr.* **26,** 1721–1734.
Walin, G., G. Broström, J. Nilsson, and O. Dahl, 2004: Baroclinic boundary currents with downstream decreasing buoyancy: A study of an idealized Nordic Seas system. *J. Mar. Res.* **62,** 517–543.
Whitehead, J. A., and J. Pedlosky, 2000: Circulation and boundary layers in differentially heated rotating stratified fluid. *Dyn. Atmos. Oceans* **31,** 1–21.
Wolfe, C. L., and C. Cenedese, 2006: Laboratory experiments on eddy generation by a buoyancy coastal current flowing over variable bathymetry. *J. Phys. Oceanogr.* **36,** 395–411.

# 4

# Buoyant Coastal Currents

STEVE LENTZ

## 4.1 Introduction

Relatively fresh river or estuarine water entering the coastal ocean forms a buoyant plume that often turns anticyclonically (to the right in the Northern Hemisphere) and forms a buoyant gravity current that can flow large distances along the coast before dispersing (e.g., Mork 1981; Munchow and Garvine 1993a; Rennie et al. 1999; Royer 1981). The tendency for the buoyant water to turn and flow along the coast as a relatively narrow current is a consequence of Earth's rotation. The focus here is on two aspects relevant to buoyant gravity currents in the ocean: (1) determining the characteristics of buoyant coastal currents flowing along a sloping bottom and (2) determining the influence of wind forcing on buoyant coastal currents.

Buoyant coastal currents are important components of the circulation on most continental shelves (e.g., Simpson 1982; Hill 1998). Buoyant coastal currents also transport constituents, such as sediment, marine organisms, nutrients, and chemical pollutants large distances from their river or estuarine sources. Therefore, determining the ultimate distribution and fate of these constituents depends on understanding buoyant coastal currents and their alongshore range of influence (e.g., Wiseman et al. 1997; Epifanio et al. 1989). Two examples of societal problems where buoyant coastal currents play an important role are hypoxia and abrupt climate change.

Hypoxia is dissolved oxygen concentrations that are reduced to a level that is detrimental to marine organisms. Hypoxia associated with nutrient transport from rivers to the coastal ocean is a global problem (Diaz 2001). As an example, there is a large hypoxic region in the Gulf of Mexico associated with the Mississippi River (Wiseman et al. 1997). The Mississippi River, which drains most of the central United States, transports large quantities of nutrients from fertilizers used in agriculture. The nutrients enter the Gulf of Mexico in the Mississippi River plume and fuel rapid biological productivity. The resulting organic matter settles to the bottom where it decomposes and in the process uses up the available oxygen. At the same time, the layer of buoyant water associated with the Mississippi River plume isolates the subsurface water from the oxygen-rich surface waters by inhibiting vertical mixing. The consequence is a large region where the subsurface waters are so depleted in oxygen that it is harmful or fatal to the marine organisms. In this case, the buoyant plume from the Mississippi River plays two key roles: (1) spreading the nutrients along shore and (2) providing the stratification that isolates the subsurface waters from the oxygen-rich surface waters.

The second example concerns the cause of a dramatic cooling event that occurred about 8,000 years ago. One hypothesis for the cause of this cooling event is the sudden drainage of a large lake (Lake Agassiz) that is thought to have covered much of Canada 8,000 years ago (Barber et al. 1999). This lake was contained by a large glacial ice sheet. The ice dam may have failed suddenly, leading to a huge volume of fresh water draining into the Labrador Sea (Clarke et al. 2003). The hypothesized scenario is that the resulting lens of fresh water reduced or eliminated deep water formation in the northern North Atlantic, resulting in a reduction in the Meridional Overturning Circulation (MOC), which in turn resulted in global cooling. A key element of this scenario is that the fresh water spread out over the northern North Atlantic. Numerical modeling studies (Renssen et al. 2002; Manabe and Stouffer 1997; Seidov and Maslin 1999) suggest that the volume of fresh water from Lake Agassiz spread over portions of the North Atlantic is sufficient to reduce or shutdown the MOC and hence could account for the well- documented abrupt cooling and associated climate changes about 8,000 years ago (Alley et al. 1997). However, in these "hosing" studies, the fresh water is spread over the surface of some portion of the North Atlantic, often the Labrador Sea where deep convection occurs, and these studies suggest that the results are sensitive to where the fresh water is placed. Recent numerical simulations show that fresh water from a massive drainage event will not simply spread out over the Labrador Sea, but rather will form a buoyant coastal current that carries the water southward along shore before it gets exported offshore (Condron and Winsor, personal communication). Thus, it is important to understand the dynamics of the buoyant coastal current and the resulting pathway of the freshwater if we want to understand this climate change event.

The dynamics of buoyant gravity currents along a wall over a flat bottom is a classical problem that is relatively well understood from laboratory, theoretical, and numerical model studies (see Chapter 1 by Linden, in this volume). Buoyant gravity currents in the ocean are often strongly influenced by Earth's rotation. Experimental studies and theory have provided estimates of the characteristics of rotating buoyant gravity currents flowing along a wall in terms of the buoyant current transport $Q$, the Coriolis parameter $f$, and reduced gravity $g' = g\Delta\rho/\rho_o$, where $\Delta\rho$ is the density anomaly of the buoyant water relative to the ambient water having density $\rho_o$ (Stern 1982; Griffiths and Hopfinger 1983; Griffiths 1986; Helfrich et al. 1999). Specifically, the thickness of the buoyant coastal current at the wall is $h_w = (2Qf/g')$, the width $W \sim 0.4L_R$ scales with the deformation radius ($L_R = (g'\ h_w)^{1/2}/f$), and the nose propagates with a speed $c_w \sim 1.4\ (g'h_w)^{1/2}$.

These classical ideas and specifically the derived scales have been applied to buoyant coastal currents in the ocean. However, in the ocean, buoyant coastal currents flow along sloping continental shelves. This raises the question of how relevant the vertical wall results are to oceanic buoyant coastal currents and more generally how a bottom slope modifies the characteristics of buoyant coastal currents. Some hints were evident from laboratory (Whitehead and Chapman 1986) and numerical modeling (Chao 1988; Kourafalou et al. 1996; Garvine 1999) studies that found buoyant currents along a sloping bottom are more stable, wider, and propagate more slowly than buoyant currents along a vertical wall.

## 4.2 A Simple Model for Buoyant Coastal Currents over a Sloping Bottom

A simple model for buoyant gravity currents flowing along a slope, based on a sequence of studies by Chapman and Lentz (1994), Yankovsky and Chapman (1997), and Lentz and Helfrich (2002), is presented. Consider a buoyant gravity current with density $\rho_o$-$\Delta\rho$ flowing along a bottom with slope $\alpha$ in an ocean with density $\rho_o$ (Figure 4.1). Let $h(y)$ be the position of the interface between the buoyant current and the ambient water. Let $W_\alpha$ be the offshore location where the interface intersects the bottom and $W_p = W_\alpha + W_w$, where the plume interface intersects the surface.

Assume the flow is hydrostatic $P_z = -g\rho$ and the alongshelf flow is geostrophic $fu = -P_y/\rho_o$. This implies that the shear in the alongshelf flow is in thermal wind balance with the cross-shelf density gradient; that is,

$$f\frac{\Delta u}{\Delta z} = \frac{g}{\rho_o}\frac{\Delta\rho}{\Delta y}. \tag{4.1}$$

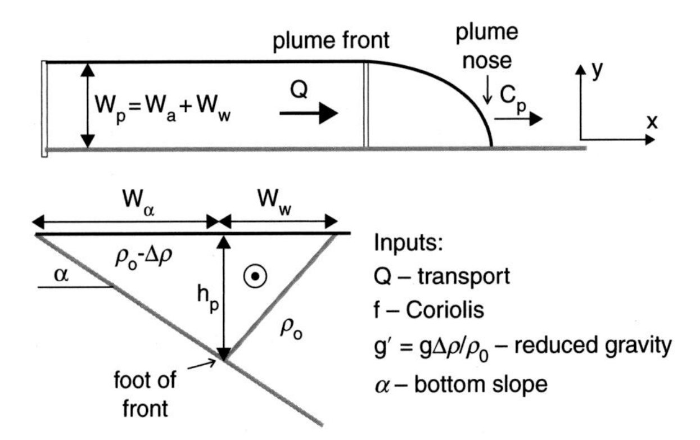


Figure 4.1. Schematic of buoyant coastal current showing structure and scales in plan view (*top*) and cross section (*bottom left*).

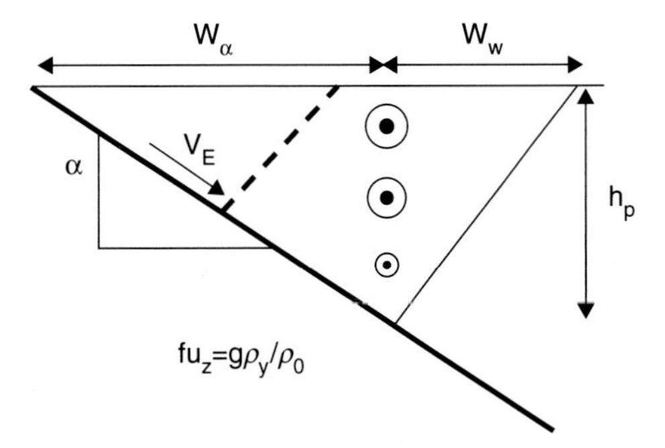


Figure 4.2. Schematic of plume cross-section showing influence of offshore bottom Ekman transport on position of plume front.

Note that the only place where there is a change in density is at the interface, so the thermal wind balance implies there is a velocity jump at the interface given by

$$\Delta u = \frac{g\Delta\rho}{f\rho_o}\frac{\Delta z}{\Delta y} = \frac{g\Delta\rho}{f\rho_o}h_y, \tag{4.2}$$

where we have taken the limit as $\Delta y$ and $\Delta z$ go to zero and $h_y$ is the interface slope. To determine the total geostrophic alongshore velocity we need to know the near-bottom flow. A nonzero, near-bottom alongshelf flow results in a bottom stress that drives a cross-slope Ekman transport in the bottom boundary layer (Ekman 1905) that will move the foot of the front (Figure 4.2). Therefore, to reach an equilibrium in which the front does not move, the bottom stress, and hence the near-bottom flow, must be zero. Away from the front we expect the near-bottom velocity to go to zero because of bottom drag.

If the near-bottom velocity is zero everywhere, then the alongshelf velocity is zero everywhere, except above the plume interface (i.e., $W_\alpha \leq y \leq W_p$ and $z \geq h$). The resulting plume transport, $Q$, is

$$Q = \int_{W_\alpha}^{W_p}\int_{-h}^{0} u\,dz\,dy = \int_{W_\alpha}^{W_p} \frac{g\Delta\rho}{2f\rho_o}\frac{d(h^2)}{dy}dy = \frac{g'h_p^2}{2f}, \tag{4.3}$$

where $g' = g\Delta\rho/\rho_o$ is reduced gravity and $h_p$ is the equilibrium water depth at the foot of the plume front. Solving for the equilibrium depth of the foot of the front given the buoyant gravity current transport yields

$$h_p = \sqrt{\frac{2Qf}{g'}}. \tag{4.4}$$

Note that this thickness does not depend on the bottom slope and is the same value as for a vertical wall. That is because there are the same two basic assumptions in either case: the flow is geostrophic, and the alongshelf flow is zero at the base of the plume.

Knowing $h_p$ and the bottom slope provides an estimate of the distance from the coast to the foot of the front

$$W_\alpha = \frac{h_p}{\alpha}, \tag{4.5}$$

where the bottom slope $\alpha$ is assumed small so that $\tan(\alpha) \sim \alpha$.

To estimate the total width of the plume at the surface, $W_p$, we need to estimate the offshore distance from the foot of the front to the offshore edge of the plume at the surface, $W_w$. Assuming the plume front adjusts to a geostrophic balance, the scale of the width of the frontal region is the baroclinic deformation radius based on $h_p$,

$$W_w \approx \frac{\sqrt{g' h_p}}{f} = \frac{c_w}{f}, \tag{4.6}$$

where $c_w = \sqrt{g' h_p}$ is the internal wave speed. This is also the width scale for a buoyant gravity current along a vertical wall noted previously. One can make a more detailed estimate of the frontal structure and width through a formal analysis of the geostrophic adjustment of a front over a slope (Hsueh and Cushman-Roisin 1983), remembering that $h_p = \sqrt{2Qf/g'}$ and noting that $W_\alpha \approx \frac{h_p}{\alpha} = \frac{c_w}{f}\frac{c_w}{c_\alpha}$ where $c_\alpha = \frac{\alpha g'}{f}$ is a topographic speed scale. Thus, the total width of the plume is

$$W_p = W_\alpha + W_w \approx \frac{c_w}{f}\left(1 + \frac{c_w}{c_\alpha}\right). \tag{4.7}$$

Note that the width scale for a buoyant coastal current over a sloping bottom is wider than for a buoyant current along a wall (given by $W_w$ as noted previously) by the factor $(1 + c_w/c_\alpha)$.

The propagation speed of the nose of the buoyancy current $c_p$ may be estimated by assuming that the ageostrophic nose region has a finite extent, that the nose shape is steady as it propagates along the coast, and that there is minimal mixing of the gravity current fluid with the ambient fluid. Mixing and detrainment at the nose of a gravity current propagating along a vertical wall is observed in laboratory experiments (Stern et al. 1982; Griffiths and Hopfinger 1983; Kubokawa and Hanawa 1984). However, Whitehead and Chapman (1986) found that mixing at the nose was nearly eliminated over a slope. With these assumptions, volume conservation at the nose of the buoyant current implies $Q\Delta t = Ac_p\Delta t$, where the cross-sectional area of the plume is approximated as $A \sim h_p W_p/2$ and $\Delta t$ is a small-time increment. Solving for the propagation speed using the expressions for $h_p$ and $W_p$ yields

$$c_p \sim \frac{Q}{A} \sim \frac{c_w}{1 + c_w/c_\alpha} = \frac{c_w c_\alpha}{c_w + c_\alpha}. \tag{4.8}$$

Note that the scale for the plume propagation speed is always slower than the propagation speed scale along a wall by the factor $1/(1 + c_w/c_\alpha)$ and the propagation speed $c_p$ tends to scale with the slower of the two speeds $c_w$ or $c_\alpha$.

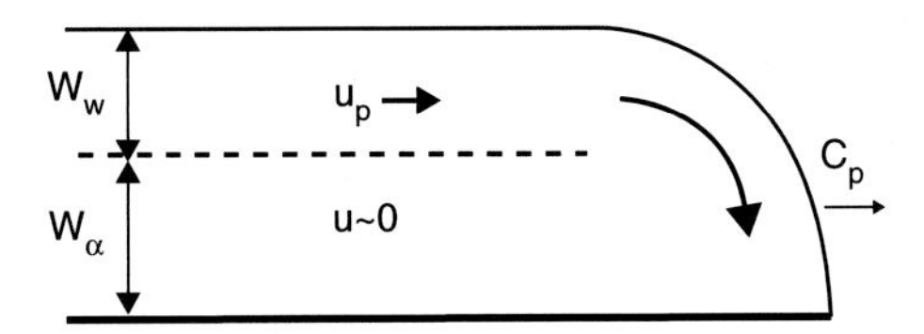


Figure 4.3. Schematic of flow field in buoyant coastal current.

The character of the associated flow field also depends on $c_w/c_\alpha$. Onshore of the foot of the front, the flow should be weak (Figure 4.3; Chapman and Lentz 1994). This follows from the assumption that the alongshore flow is geostrophic and, since there is no cross-slope density gradient onshore of the front, the geostrophic flow is vertically uniform and equal to the near-bottom flow. However the near-bottom flow must be zero since there is no bottom stress in equilibrium. If the transport $Q$ is confined to the frontal region, then taking the area of the frontal region to be approximately $A_f \sim W_w h_p/2$, and using the expressions for $W_w$ (4.6) and $h_p$ (4.4), conservation of volume transport implies that the average velocity is

$$u_p \sim \frac{Q}{A_f} = \frac{2Q}{W_w h_p} \sim c_w. \tag{4.9}$$

This provides a scale for the average velocity but does not provide any insight into the structure of the flow within the frontal region. Assume that behind the gravity current nose the flow is divided into a region near the front moving at an average velocity $u_p \sim c_w$ and an onshore quiescent region.

Since

$$\frac{u_p}{c_p} = 1 + c_w\big/c_\alpha \geq 1,$$

the approaching flow must turn shoreward within the ageostrophic nose region and come to rest to fill the quiescent onshore region with buoyant water. In a frame propagating with the nose, the flow approaches the nose in the offshore region and flows back upstream near the coast. The flow $u_p$ is confined near the front by geostrophy and is independent of the bottom slope. In the wall limit, the entire gravity current is moving at an average flow speed of $u_p$ (Figure 4.4 left schematic). Since there is no quiescent onshore region, the nose propagates at the average flow speed (i.e., $c_p = u_p$). Over a sloping bottom, $c_p$ must be less than $u_p$ because as the flow approaches the nose, it takes time to fill in the quiescent onshore region (Figure 4.4 right). As the bottom slope decreases, the cross-sectional area of the quiescent onshore region increases since $h_p$ is independent of the bottom slope and $W_\alpha = h_p/\alpha$ (see Figure 4.5). As the cross-sectional area of the quiescent region onshore of the front increases, more time is required to fill in this region, and $c_p$ must decrease. Thus, this flow structure provides a simple kinematic reason for the reduction in the nose propagation speed over a sloping bottom.

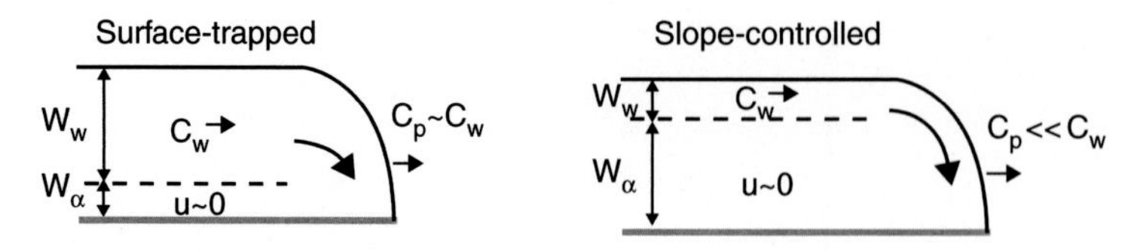


Figure 4.4. Schematic of flow field for a surface-trapped (*left*) and slope-controlled (*right*) buoyant coastal current.

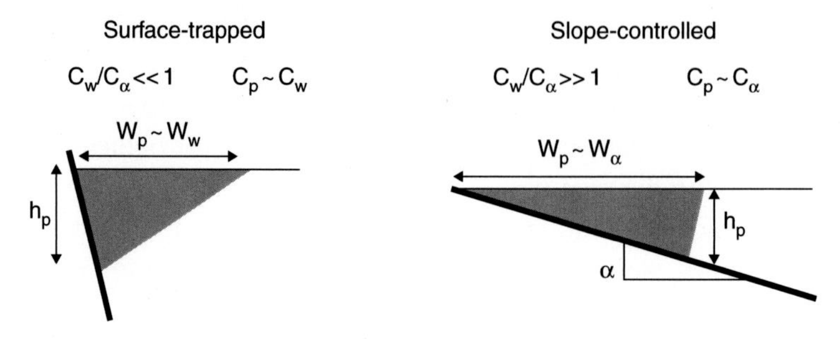


Figure 4.5. Summary of buoyant coastal current width and propagation speed in surface-trapped and slope-controlled limits.

The key nondimensional parameter in this model of a buoyant gravity current along a slope is the ratio

$$\frac{c_w}{c_\alpha} = \frac{W_\alpha}{W_w} = \frac{(2Q)^{1/4} f^{5/4}}{\alpha g'^{3/4}}$$

$$c_p \approx \frac{c_w}{1 + c_w/c_\alpha} \qquad W_p = \frac{c_w}{f}\left(1 + \frac{c_w}{c_\alpha}\right) \tag{4.10}$$

Note that $c_w/c_\alpha$ is also the scale of the ratio of the isopycnal slope $h_p/W_w$ to the bottom slope $\alpha$. From the expressions for $c_p$ and $W_p$, if $c_w/c_\alpha \ll 1$, then $c_p \sim c_w$, $W_p \sim W_w$, and the buoyancy current is surface trapped (Figure 4.5 left and Figure 4.4 left). In this limit, the gravity current is independent of $\alpha$, and the scaling theory recovers the previous results for gravity currents along a wall. If $c_w/c_\alpha \gg 1$, then $c_p \sim c_\alpha$, $W_p \sim W_\alpha$, and the buoyancy current is slope controlled (Figure 4.5 right). This slope-controlled limit will occur for buoyancy currents with large transports (though the 1/4 power dependence is weak) at faster rotation rates (higher latitudes), smaller density anomalies, and smaller bottom slopes. This analysis can be extended to consider finite-width fronts with continuous density variations (Pimenta, personal communication).

## 4.3 Evaluating the Buoyant Coastal Current Model

The proposed model of a buoyant coastal current over a sloping bottom is based on a number of simplifying assumptions. This raises the question of whether the model captures the fundamental features of buoyant coastal currents and how well it represents those features. The model is evaluated by comparing predictions from the model to laboratory models of buoyant plumes (Lentz and Helfrich 2002), more complex numerical models (Garvine 1999), and oceanographic observations (Lentz et al. 2003). Each of these comparisons has advantages and disadvantages. Laboratory models deal with a real fluid and can be controlled to isolate and evaluate specific parameter dependencies, but it is often difficult to design small-scale laboratory studies to represent the appropriate parameter space for a geophysical problem. Numerical models can be set up to represent the scales and physics of the geophysical flows of interest and can also be designed to isolate specific parameters. However, numerical models necessarily include parameterizations of unresolved processes. Finally, oceanographic observations capture the actual process of interest, but the observations tend to be sparse in space and time and are subject to measurement errors. Geophysical flows also generally can't be controlled, for example, to test parameter dependence and are subject to a wide range of complexities associated with other processes. In the following sections, all three approaches will be used to evaluate the proposed model of buoyant coastal currents flowing along a sloping bottom.

### 4.3.1 Laboratory model

To test the proposed model, a sequence of 28 laboratory experiments in a rotating tank were conducted by Lentz and Helfrich (2002).

A counterclockwise rotating tank with a sloping bottom was filled with salt water (Figure 4.6). Buoyant (less salty) water was then pumped into the tank at a fixed rate Q. The 28 experiments spanned a range of the nondimensional parameter $c_w/c_\alpha$ from 0.1 (surface trapped) to 13 (slope-controlled). In 16 of the runs, the buoyant fluid was

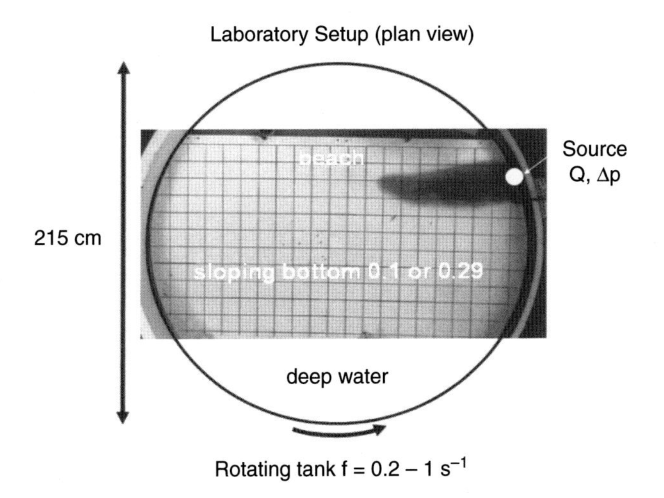


Figure 4.6. Schematic of laboratory setup (Lentz and Helfrich 2002).

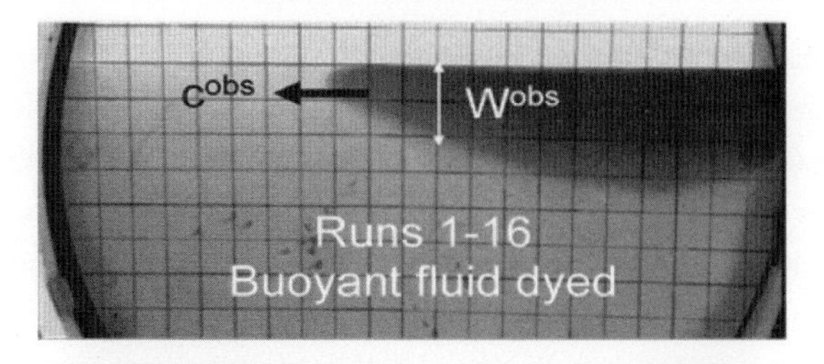


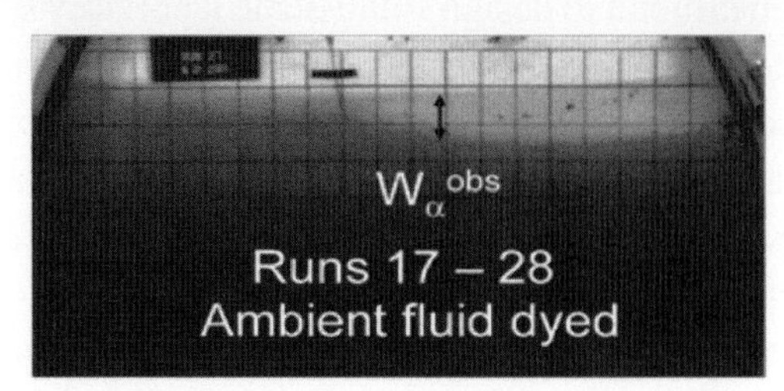


Figure 4.7. Example of laboratory runs with plume water dyed (upper) and with ambient fluid dyed (lower).

dyed, allowing a clear estimate of the plume width (Figure 4.7 upper panel). In the remaining 12 runs, the ambient fluid was dyed to estimate the location where the foot of the front intersected the bottom (Figure 4.7 lower panel).

Plume widths as a function of time were estimated at the midpoint of the tank. Plume propagation speeds were estimated as the difference of the positions of the nose of the plume in sequential images divided by the elapsed time. Observations of the flow field were collected by tracking 4-mm buoyant spheres.

Time series of the offshore position of the foot of the front ($W_\alpha^{\mathrm{obs}}$) for the 12 experiments with the ambient fluid dyed initially increase rapidly and then asymptote to a relatively constant value ranging from 2 to 27 cm (Figure 4.8 upper panel). To evaluate the laboratory results in the context of the model, time is normalized by an estimate of the time it takes the bottom Ekman transport to move the foot of the front offshore to its equilibrium position $W_\alpha$

$$t_{\mathrm{adj}} = \frac{2W_\alpha}{c_w} = \frac{2c_w}{f c_\alpha} \tag{4.11}$$

(see Lentz and Helfrich, 2002) and $W_\alpha^{\mathrm{obs}}$ by the model estimate $W_\alpha$. These scalings tend to collapse the laboratory observations (Figure 4.8 lower panel). The normalized offshore position of the foot of the front asymptotes to a width that is 70–90% of the scale width $W_\alpha$. Linear regression of scale width versus the observed width averaged over the last five observations of each run for runs that appear to reach steady state yields a correlation of 0.96.

A similar comparison for the plume width at the surface, $W_p^{\mathrm{obs}}$, shows that the model scaling again tends to collapse the observations (Figure 4.9; correlation 0.97 for average widths over time period $1 < t/t_p < 3$, where $t_p = W_p/c_p$). However,

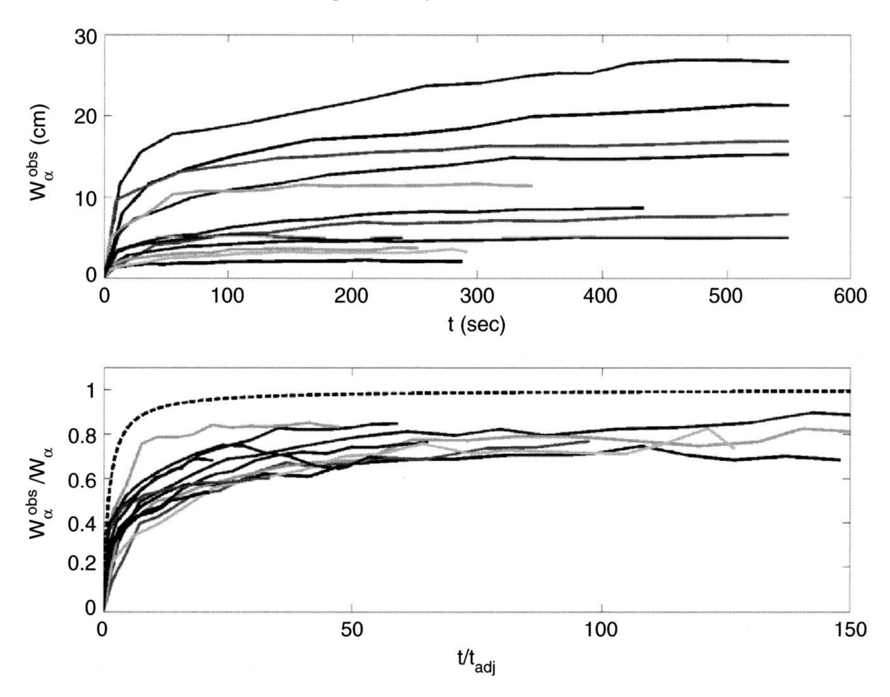


Figure 4.8. Offshore distance to where the plume front intersects the bottom ($W_\alpha^{obs}$) as a function of normalized time for model runs with $c_w/c_\alpha$ ranging from 0.1 to 13 (*top*) and $W_\alpha^{obs}$ normalized by prediction as a function of normalized time (*bottom*). Dashed line in lower panel is model prediction.

in this case the plume widths from the laboratory runs do not reach equilibrium as predicted by the simple model but instead continue to widen at a rate proportional to $(t/t_p)^{1/2}$. Lentz and Helfrich (2002) hypothesized that the continual increase in the plume width (at the surface) may be due to interfacial drag along the plume front and they showed the time dependence is consistent with this hypothesis (dashed lines in lower panel Figure 4.9). The proposed scale $c_p$ given by equation (4.8) similarly collapses the propagation speeds estimated from the laboratory runs (not shown; see Lentz and Helfrich 2002, correlation 0.94). The simple model also collapses plume width and propagation speeds from the previous laboratory study of Whitehead and Chapman (1986).

The particle tracks from the laboratory runs indicate a variation in the flow structure from surface trapped to slope controlled that is consistent with the model prediction. In a surface-trapped case ($c_w/c_\alpha = 0.17$, Figure 4.10 left panels) the maximum flow is near the coast. In a reference frame moving with the plume nose (Figure 4.10 lower panels) the current near the wall flows toward the plume nose, and farther offshore the flow turns offshore and is moving away from the nose. This is consistent with the flow pattern in previous studies of buoyant currents along a wall. In the slope-controlled case ($c_w/c_\alpha = 3.8$, Figure 4.10 right panels), the fastest flow is farther offshore, away

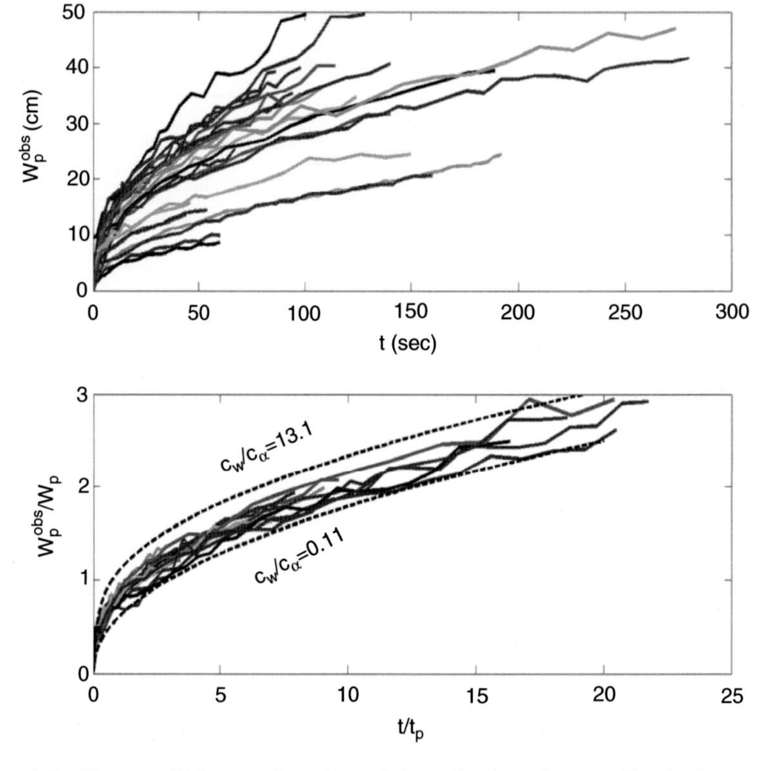


Figure 4.9. Plume width as a function of time (*top*) and normalized plume width as a function of normalized time (*bottom*). Dashed lines indicate predicted time dependence due to interfacial drag ($t^{1/2}$) for $c_w/c_\alpha = 13.1$ (upper dashed line) and $c_w/c_\alpha = 0.11$ (lower dashed line).

from the wall. In a reference frame moving with the nose, the circulation pattern is the opposite of the surface-trapped case, with flow toward the nose farther offshore, turning onshore near the nose and flowing away from the nose near the wall.

### 4.3.2 Numerical Model

Another opportunity to evaluate the simple model is provided by a sequence of 66 numerical model runs conducted by Garvine (1999) to study the processes controlling the alongshore extent of buoyant plumes. Garvine varied inflow transport and density, the width of the source, bottom slope on the shelf, and latitude (Coriolis frequency) and reported plume widths half way between the source and the end of the plume (where the density anomaly was less than 5% of the initial anomaly).

The buoyant plumes decay downstream from the source because of mixing, which reduces the density anomaly and hence $g'$ (Figure 4.11). To determine whether the proposed model collapses Garvine's numerical model results, his estimates are compared to the simple model estimates of $W_p$ based on the input transport $Q_i$, density anomaly

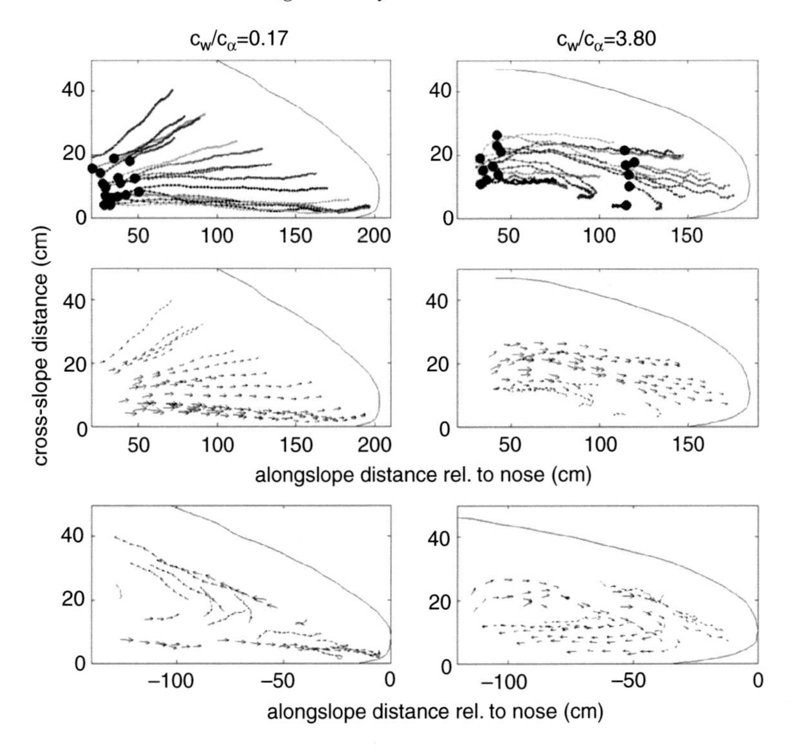


Figure 4.10. Particle tracks (*top*), velocity vectors inferred from particle tracks (*middle*) and velocities in a reference frame moving with the nose of the plume (*bottom*) for a run with $c_w/c_\alpha = 0.17$ (*left*) and $c_w/c_\alpha = 3.8$ (*right*).

at the source $\Delta\rho_i$, bottom slope and Coriolis frequency. The numerical model widths are proportional to $W_p$, with the notable exception of the model runs at low latitudes (small Coriolis frequency) (Figure 4.12). Excluding the low-latitude ($\leq 25^o$) runs, the correlation is 0.98. The regression slope is about 4, which presumably accounts for the relationship between the source transport and density anomaly and the local values where the width is measured. The failure of the scaling at low latitudes is almost certainly due to the tendency for an increasing fraction of the inflow transport to be trapped in a growing recirculation cell (bulge) near the source (see Figure 4.11, bottom), rather than in a buoyant coastal current. At the equator, where $f = 0$, there is no buoyant coastal current in the numerical model, only a growing bulge.

### *4.3.3 Ocean Observations – The Chesapeake Bay Buoyant Coastal Current*

Observations of a buoyant coastal current that emanates from Chesapeake Bay are used to further evaluate the simple model. Relatively fresh water piles up in Chesapeake Bay during northward wind stresses and is then released as a buoyant coastal current when the winds relax or reverse (Rennie et al. 1999). The released fresh water propagates alongshore as a narrow buoyant coastal current, often extending along shore more than 100 km. Figure 4.13 shows a Synthetic Aperature Radar (SAR) image of a buoyant

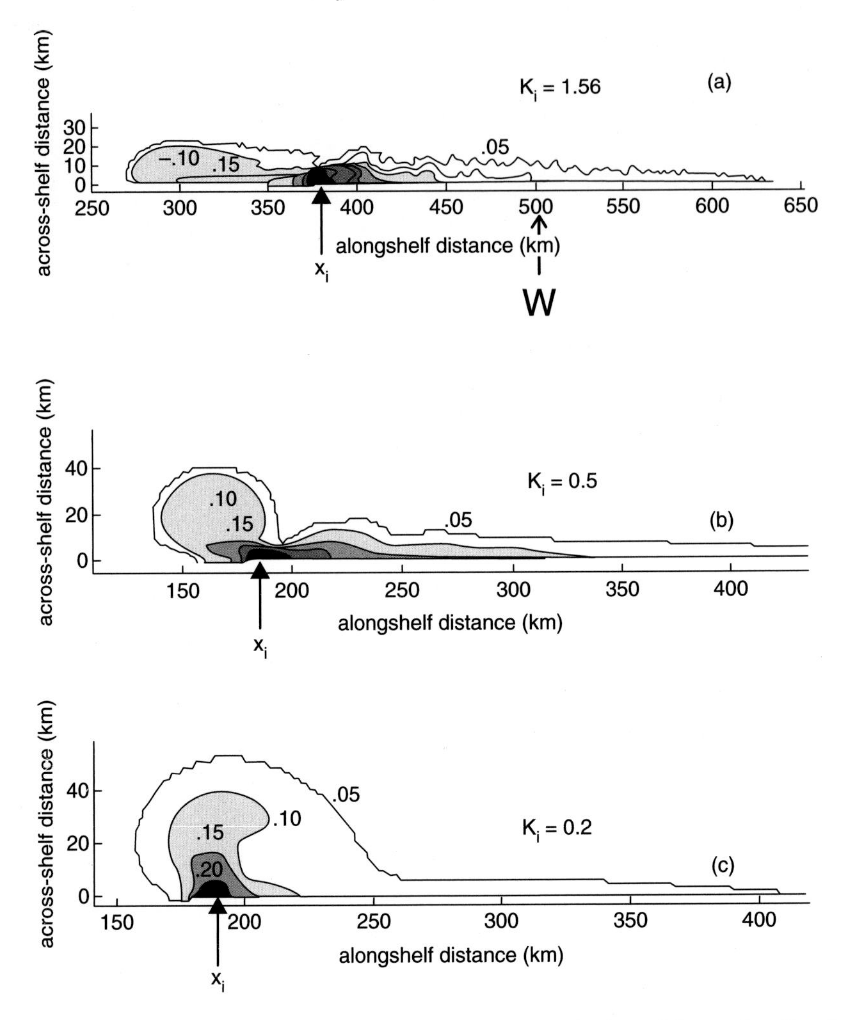


Figure 4.11. Examples of buoyant plumes from numerical model runs by Garvine (1999) with different inlet Kelvin numbers (inlet width divided by baroclinic deformation radius at the inlet). Plume widths were reported halfway between the source and the end of the plume.

coastal current from Chesapeake Bay (Donato and Marmorino 2002). A number of studies have made observations of the buoyant coastal current from Chesapeake Bay (Boicourt 1973; Rennie et al. 1999; Johnson et al. 2001; Hallock and Marmorino 2002; Donato and Marmorino 2002).

Here the focus is on observations collected as part of a study of the influence of winds and waves on the inner-shelf circulation off North Carolina. Three sets of measurements from this study are examined: (1) ship and small-boat surveys that provide salinity sections of the cross-shelf structure of the plume, (2) an alongshore array of salinity sensors that provide estimates of the propagation speed of the nose of the plume, and (3) moored-current observations on a cross-shelf transect that provide

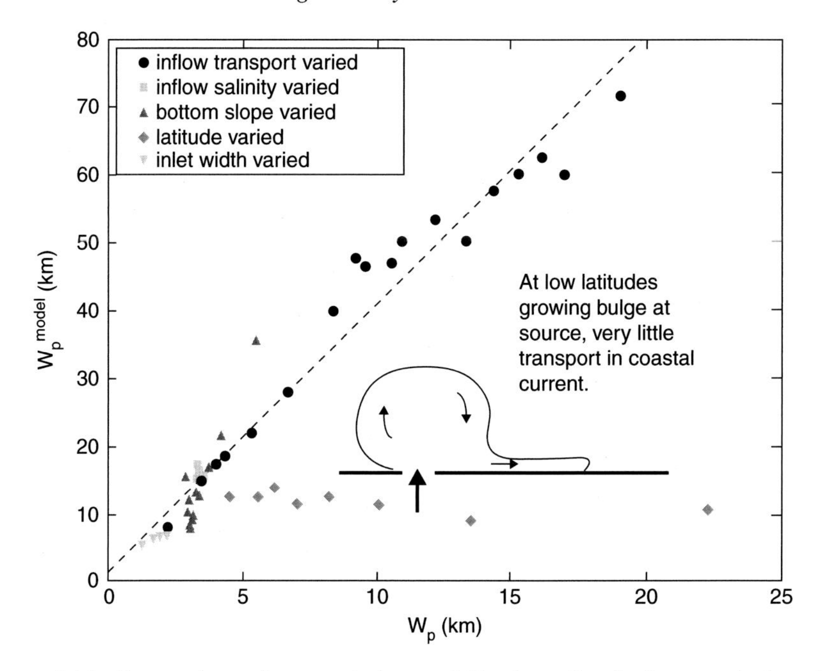


Figure 4.12. Comparison of reported plume widths from Garvine's numerical model runs with predicted scale widths $W_p$. There is a linear relationship with a slope of about 4, except for runs at low latitude dominated by formation of a bulge at the source.

information on the velocity structure of the buoyant coastal current as it propagates southward past the array.

The alongshelf propagation of the buoyant water released from Chesapeake Bay can be clearly seen in the salinity time series from the alongshelf array of sensors (Figure 4.14). The time series show distinct low-salinity pulses with more delayed arrival times with distance southward alongshore. Note also that the low-salinity pulses become saltier with distance along shore, consistent with entrainment of the saltier shelf water. The average propagation speed of the buoyant plume between any two sites may be estimated as the separation between two sites divided by the time it takes the pulse to travel from one site to the next. The predicted propagation speed $c_p$ can be determined from estimates of $g'$ using the density jump associated with the plume arrival, the bottom slope, the Coriolis frequency, and either the plume thickness $h_p$ (taken to be 8 m) or equivalently a rough estimate of the average plume transport. In general, the predicted plume propagation speeds are not correlated with the observed propagation speeds (Figure 4.15 all symbols) unless one only considers the time when winds are weak (Figure 4.15 solid symbols).

Lentz and Largier (2006) showed that much of the discrepancy is due to the ambient shelf current driven by the wind stress. Comparisons of plume thickness and width also show a strong dependence on wind stress (Fong et al. 1997; Hallock and Marmorino

Figure 4.13. A Synthetic Aperture Radar image showing buoyant plume propagating southward along shore from Chesapeake Bay (Donato and Marmorino 2002).

2002; Lentz and Largier 2006). These results emphasize the importance of wind forcing to the response of buoyant coastal currents. One aspect of the response of buoyant coastal currents to wind forcing, offshore dispersal by upwelling favorable winds, is the focus of the next section.

The moored observations may also be used to examine the structure of the flow field near the leading edge (nose) of the buoyant coastal current as it propagates past the mooring sites (following Lentz et al. 2003; see also Donato and Marmorino 2002; Melton et al. 2009; Woodson et al. 2009). The basic question of interest is whether the flow field associated with the Chesapeake buoyant coastal current resembles the flow field observed in laboratory studies of flow along a wall, with maximum alongshore flow near the wall and an offshore turning of the flow in the nose region (Figure 4.16 left panels), or whether it resembles the proposed flow field for a buoyant coastal current flowing along a sloping bottom with maximum alongshore flow offshore near

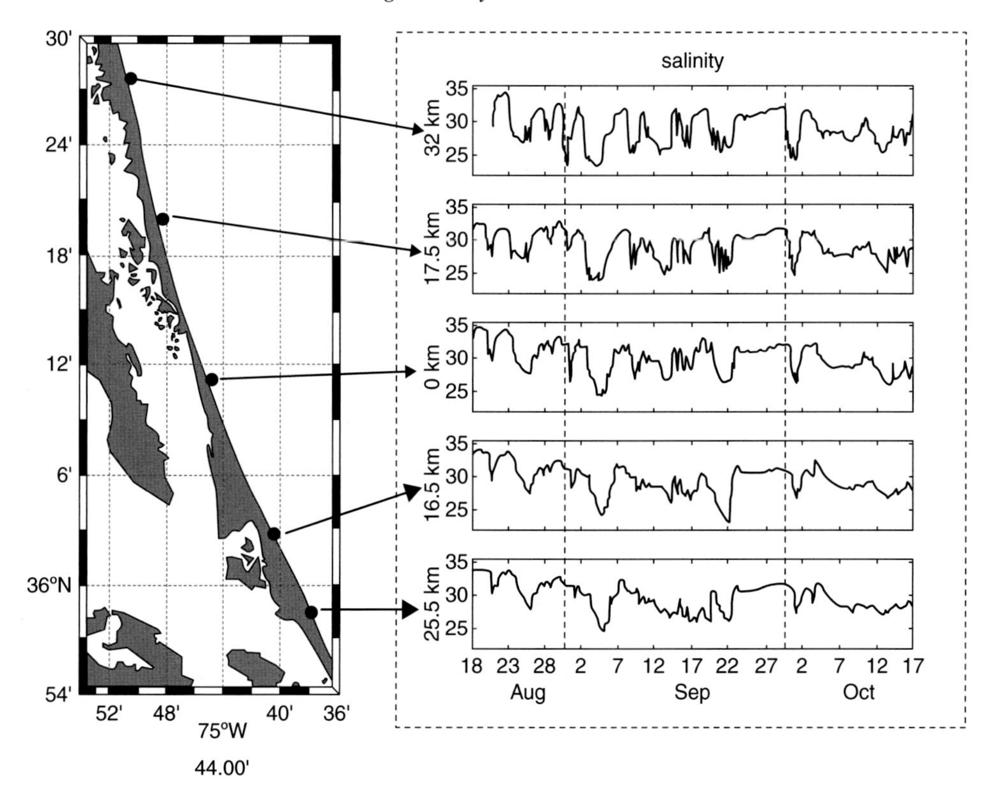


Figure 4.14. Salinity time series (*right*) from five alongshore locations south of Chesapeake Bay showing the alongshore propagation of low-salinity pulses. Dashed lines mark the time of two low-salinity plume arrivals at the northernmost site. Note time lag at arrivals at sites farther to the south consistent with alongshore propagation.

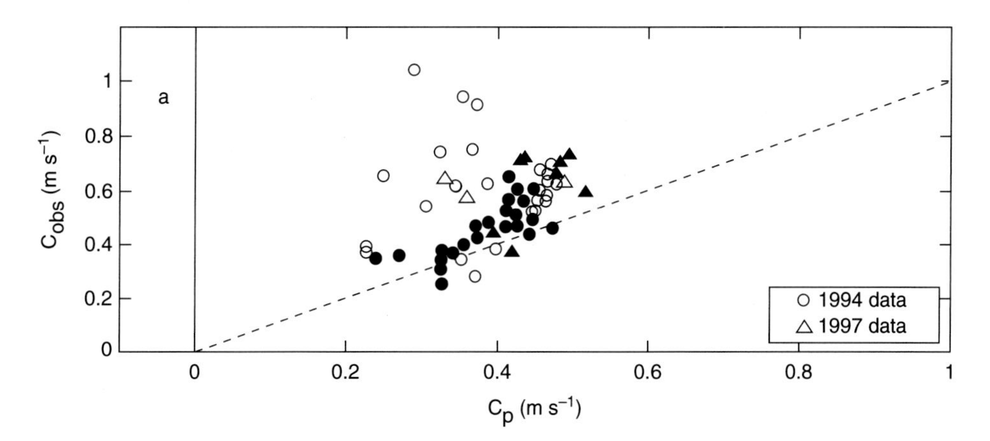


Figure 4.15. Observed propagation speeds as a function of model-predicted propagation speeds *cp*. There is a correlation between the observed and predicted speeds only during low-wind stresses ($|\tau^s| < 0.02$ N/m$^2$, solid symbols).

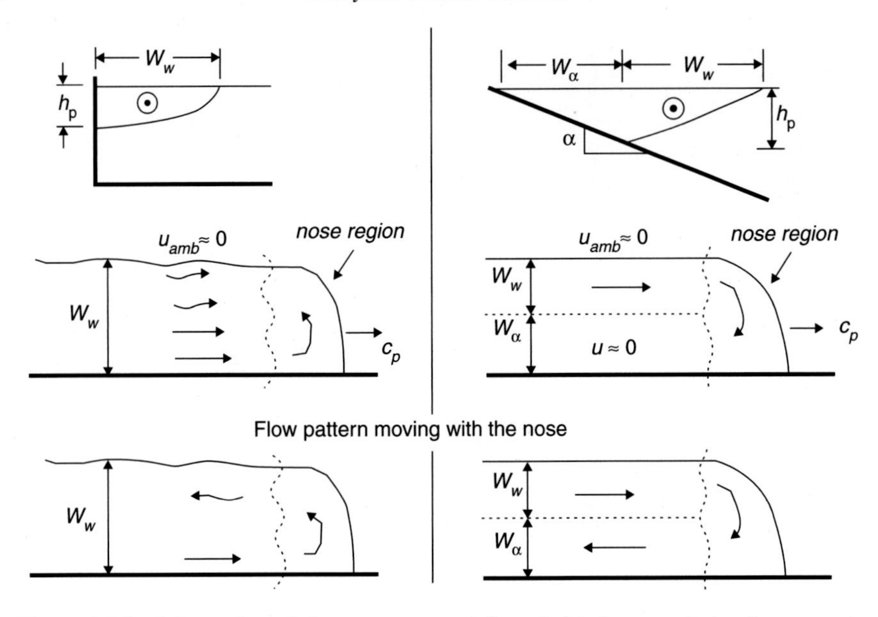


Figure 4.16. Schematics of the structure and flow field characteristics for a gravity current along a wall (*left*) and along a sloping bottom (*right*).

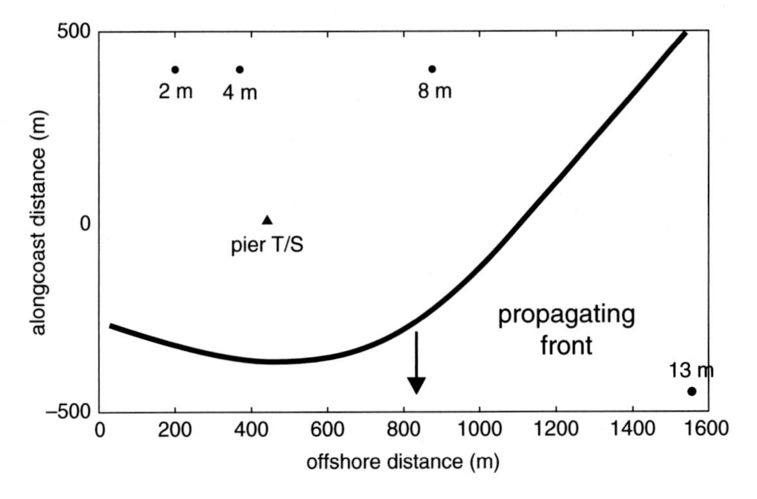


Figure 4.17. Schematic showing mooring locations and schematic of propagating plume front from the Chesapeake Bay (Lentz et al. 2003).

the front, weaker alongshore flow near the coast, and a tendency for the flow to turn onshore near the nose (Figure 4.16 right panels).

To examine the structure of the flow near the nose of the Chesapeake plume, the time series from the fixed moored observations are used to infer the alongshelf structure of the front by assuming the structure does not evolve in time as the front propagates

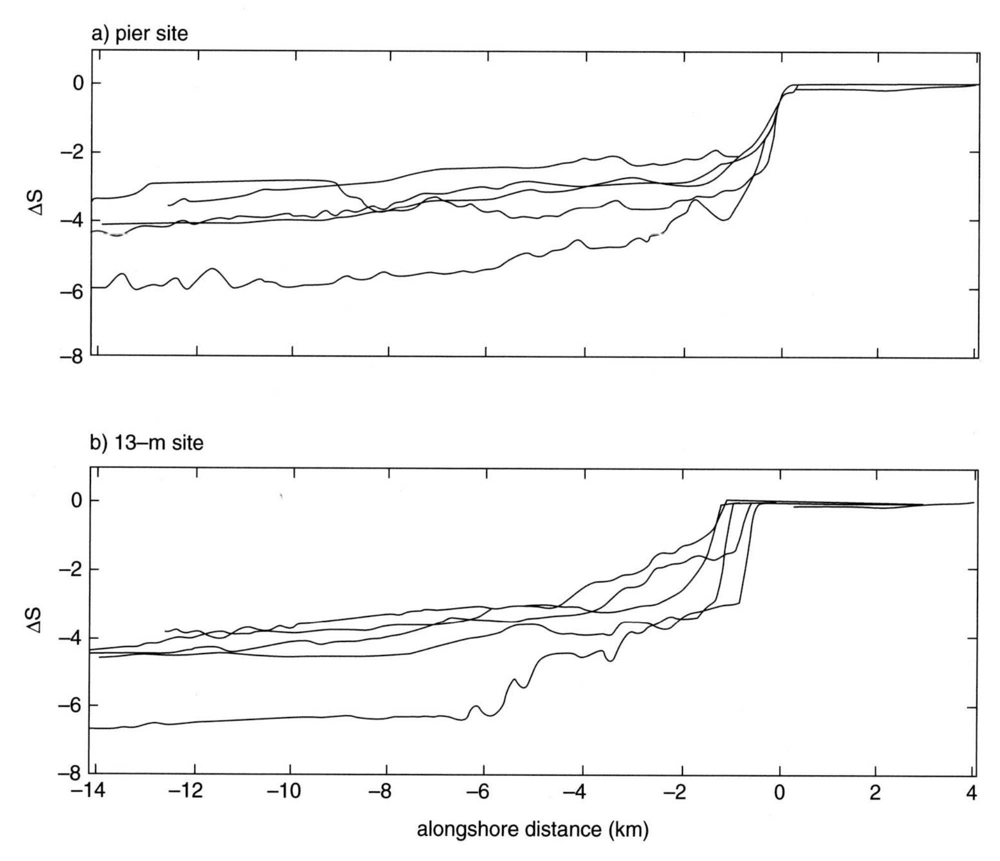


Figure 4.18. Relative salinity $\Delta S$ versus alongshore distance for five plume events passing the (a) pier site and (b) 13-m sites. Alongshore variation is estimated from observed temporal variations by assuming advection at the front propagation speed. (Time increases from right to left.) For each event, the origin (alongshore distance = 0) is defined by the arrival of the salinity front at the pier site, and the salinities are relative to values ~4 km downstream from the front. The salinity drop at the 13-m site lags the drop at the pier site by about 1 km as depicted schematically by the curve labeled "propagating front" in Figure 4.17.

past the moorings (Figure 4.17). Then temporal changes are transformed to alongshore spatial variations by multiplying the time by the propagation speed of the front (~0.5 m/s). The instrument sample rate is 4 minutes, which implies an alongshelf displacement resolution of 120 m.

Five buoyant coastal current events were chosen when the wind stress was weak, to focus on the coastal current flow field in the absence of wind forcing. Figure 4.18 shows salinity time series from two sites, one about 0.4 km offshore (pier site) and the other about 1.6 km offshore (13-m site), tranformed to alongshore distance (1 hr, ~2 km). There is a salinity jump at both sites as the nose of the buoyant plume passes the mooring sites. The time when the density jump reaches the pier site has been set to zero, so that alongshore distances are relative to the nose of the plume front (with positive values ahead of the front). There is a consistent salinity structure for all five events. Relative to the ambient salinity just ahead of the plume nose, there is an abrupt salinity decrease of 2–4 psu (salinities are reported in practical salinity units) . This front is a few hundred meters wide. The front arrival at the 13-m site lags the pier site by about 1 km, consistent with curvature near the nose of the front (Figure 4.17).

Figure 4.19 shows contour plots of an average composite for the five events of the vertical and alongshore structure of the alongshore current relative to the ambient flow 4 km ahead of the front. The alongshore flow increases from 0 to about 20 cm/s over a distance of about 4 km prior to the arrival of the front. There is a velocity jump at the front of 30–50 cm/s at the 8-m site to a velocity in excess of the nose propagation speed ($c_{obs} \sim$ 50 cm/s; Figure 4.19b). The velocity jump is smaller both onshore and offshore of the 8-m site. At the 8-m site the region of super critical flow where the velocity exceeds the nose propagation speed extends about 5 km behind the front. At the 13-m site there is a return flow toward Chesapeake Bay of about 10 cm/s below the plume (Figure 4.19c).

The corresponding average cross-shelf flow structure shows offshore flow ($\sim$10 cm/s) slightly ahead of and below the front (Figures 4.20b and 4.20c). This is consistent with the ambient fluid being displaced offshore as the buoyant plume propagates along shore. There is a sharp gradient at the front interface with onshore flow ($-$10 cm/s) behind the front at the 8-m site (Figure 4.20b), weaker onshore flow at the 13-m site (Figure 4.20c), and near-zero cross-shelf flow at the 4-m (and 2-m not shown) sites (Figure 4.20a). This flow pattern behind the plume front, notably the strong onshore flow behind the front, is consistent with the proposed flow pattern for a buoyant coastal current over a sloping bottom.

In summary, laboratory and numerical modeling studies support the proposed model for a buoyant coastal current over a sloping bottom. Some features of the buoyant coastal current emanating from Chesapeake Bay are also consistent with the proposed model. Observations of a buoyant coastal current in a broad range of parameter space are needed to evaluate the proposed model. Finally, wind forcing can have a profound impact on buoyant coastal currents, and one aspect of that is discussed in the next section.

## 4.4 Response of Buoyant Coastal Currents to Wind Forcing

The previous sections focused on buoyant coastal currents in isolation. In the ocean, buoyant coastal currents are influenced by a range of other processes, most notably wind forcing. Buoyant coastal currents may be strongly influenced by wind forcing because the momentum imparted by the wind is typically trapped in the thin

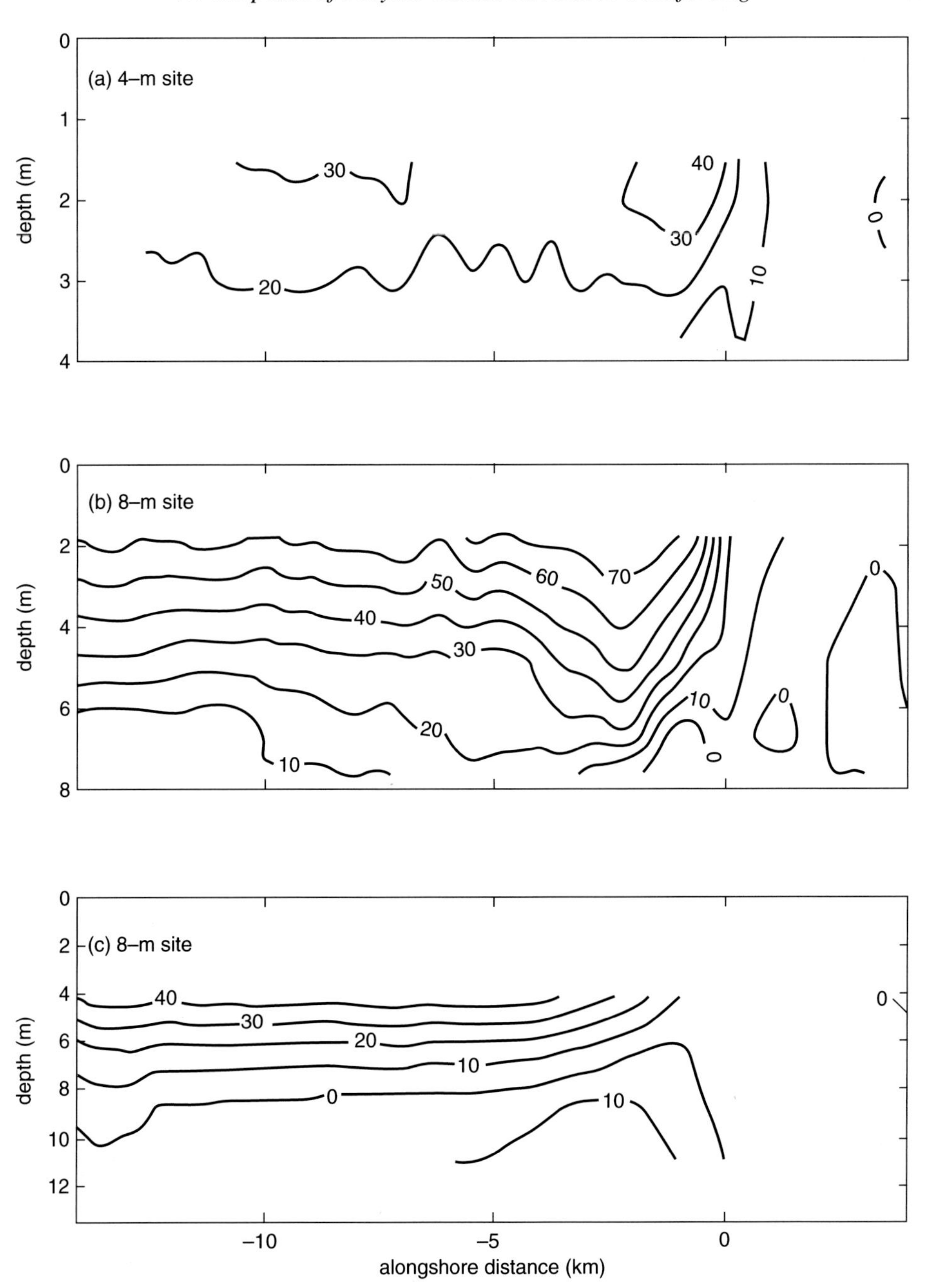


Figure 4.19. Contours of alongshore velocity (cm/s) averaged over five events (positive flow is southward) as a function of depth (distance below the surface) and alongshore distance at the (a) 4-, (b) 8-, and (c) 13-m sites. Positive alongshore velocities are southward. For each event, the origin (alongshore distance = 0) is defined by the arrival of the salinity front at the pier site (7-m depth), advection at $c_{obs}$ is assumed, and the velocities are relative to values ∼4 km downstream from the front. (Time increases from right to left.)

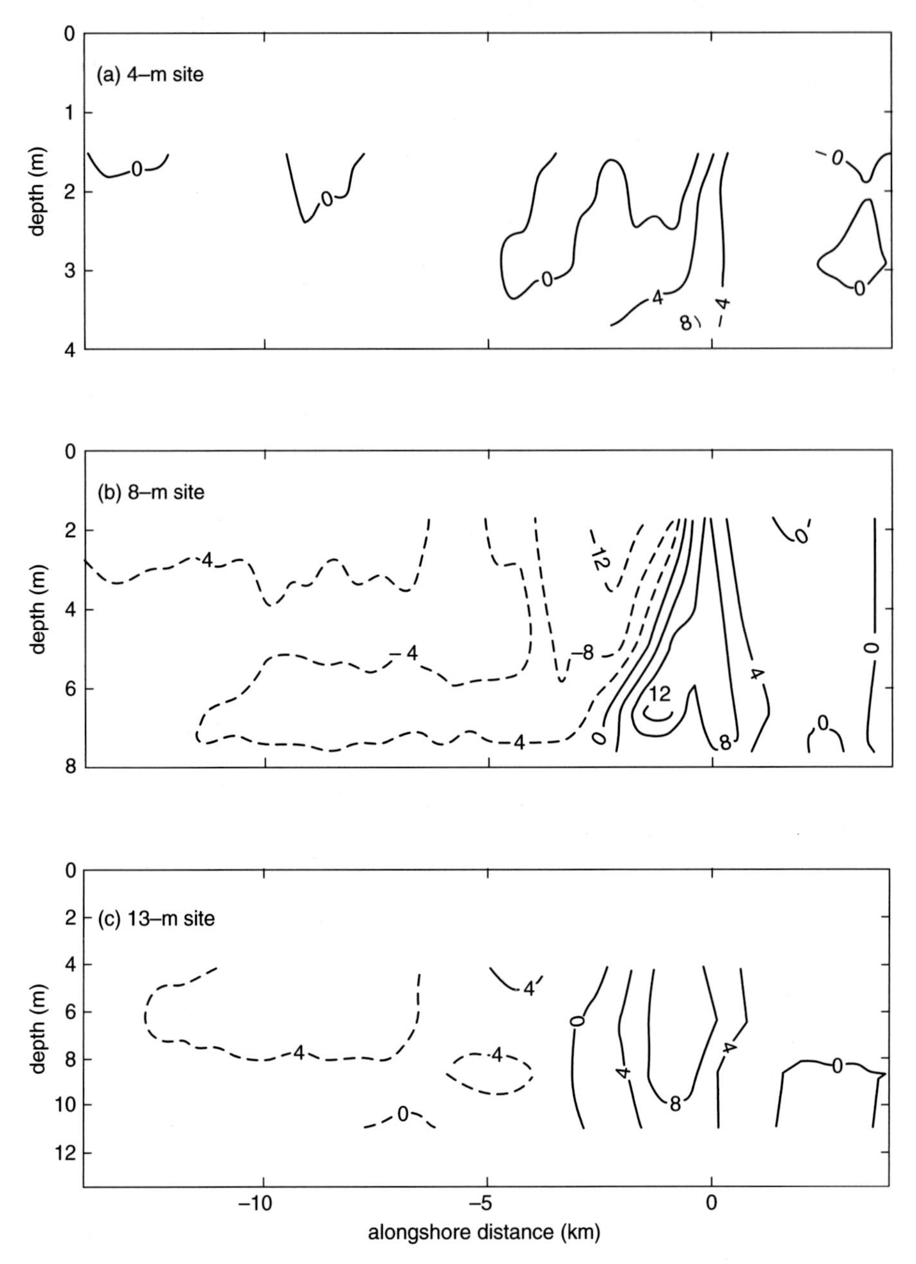


Figure 4.20. Contours of cross-shore velocity (cm/s) averaged over five events (positive flow is southward) as a function of depth (distance below the surface) and alongshore distance at the (a) 4-, (b) 8-, and (c) 13-m sites. Positive cross-shore velocities are offshore. For each event, the origin (alongshore distance = 0) is defined by the arrival of the salinity front at the pier site (7 m depth), advection at $c_{obs}$ is assumed, and the velocities are relative to values ~4 km downstream from the front. (Time increases from right to left.)

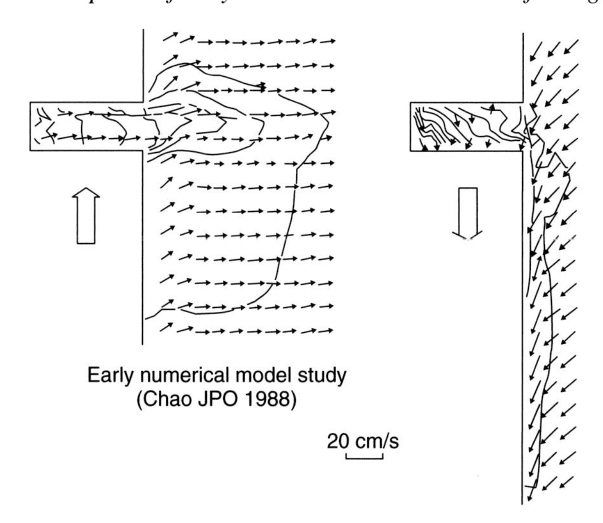


Figure 4.21. Plume geometry and flow field in response to upwelling (*left*) and downwelling (*right*) favorable wind stress (from Chao 1988).

buoyant plume that is isolated from the underlying water by strong density gradients. Consequently, even light winds can have a strong influence on buoyant coastal currents.

An early modeling study of wind-forced buoyant plumes by Chao (1988) showed a marked asymmetry in the response to alongshore winds that are either in the direction of the plume propagation (downwelling favorable winds, Figure 4.21 right panel) or in the opposite direction (upwelling favorable winds, Figure 4.21 left panel).

Upwelling winds cause the buoyant plume to widen and thin, and tend to reduce the alongshore extent of the plume. Downwelling winds cause the plume to narrow and deepen and to increase the alongshore extent of the plume. Subsequent studies using more sophisticated numerical models with much better spatial and temporal resolution show the same qualitative picture of the response to wind forcing (e.g., Choi and Wilkin 2007).

This response is consistent with simple Ekman dynamics. An upwelling-favorable wind stress advects the plume offshore, while the subsurface return flow moves the foot of the front onshore into shallower water (Figure 4.22a). Consequently the plume spreads offshore and thins. A moderate downwelling wind stress steepens the front, moving the near-surface portion of the front onshore and the foot of the front offshore (Figure 4.22b). This results in a plume that is more tightly confined to the coast (narrower) and thicker. A third possibility can occur if the downwelling-favorable winds are strong enough to rapidly mix the plume water with the underlying ambient water (Figure 4.22c; see Blanton et al. 1989). This results in a plume having a broader front with vertical isopycnals.

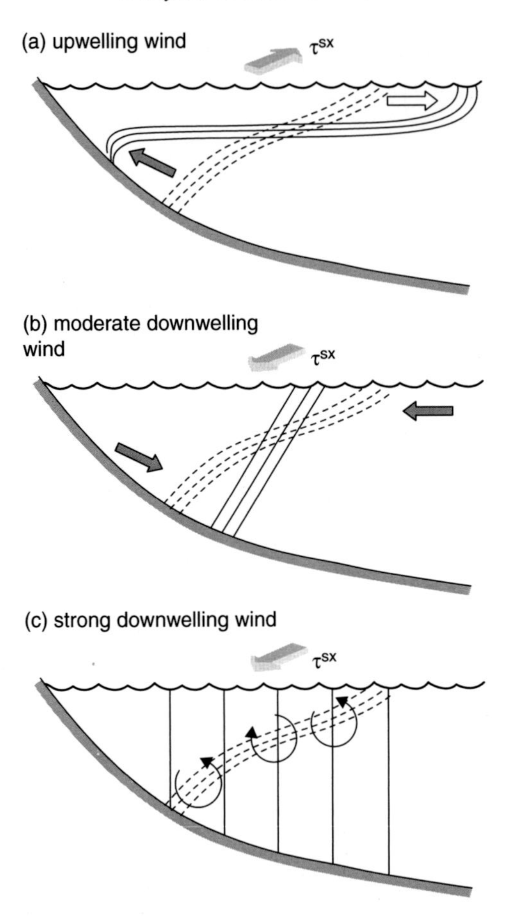


Figure 4.22. Schematic of buoyant plume response to different alongshelf wind forcing. (a) Upwelling winds flatten the plume front, causing the plume to thin and widen. (b) Moderate downwelling winds steepen the front, causing the plume to thicken and narrow. (c) Strong downwelling winds force vertical mixing that widens the plume front but causes little change in the plume width.

Observations of the Chesapeake Bay buoyant coastal current response to wind forcing exhibit examples of these responses. The first example is a transition to strong downwelling-favorable wind stress (Figure 4.23). The left panels show time series of the wind stress (top panel) and salinities from three mooring sites 1.5, 5.4, and 16.5 km offshore. There are between two and four salinity time series from different depths ranging from a few meters below the surface (blue) to near the bottom (in all cases salinity increases with depth). The dashed lines in the left panels indicate the times of two cross-shelf salinity sections (Figure 4.23 right panels): one just prior to the onset of strong downwelling winds and the second two days later. Prior to the onset of the wind stress, there is a buoyant coastal current that is 5–10 m thick and extends about 15 km offshore. There are large vertical salinity gradients, with relatively fresh

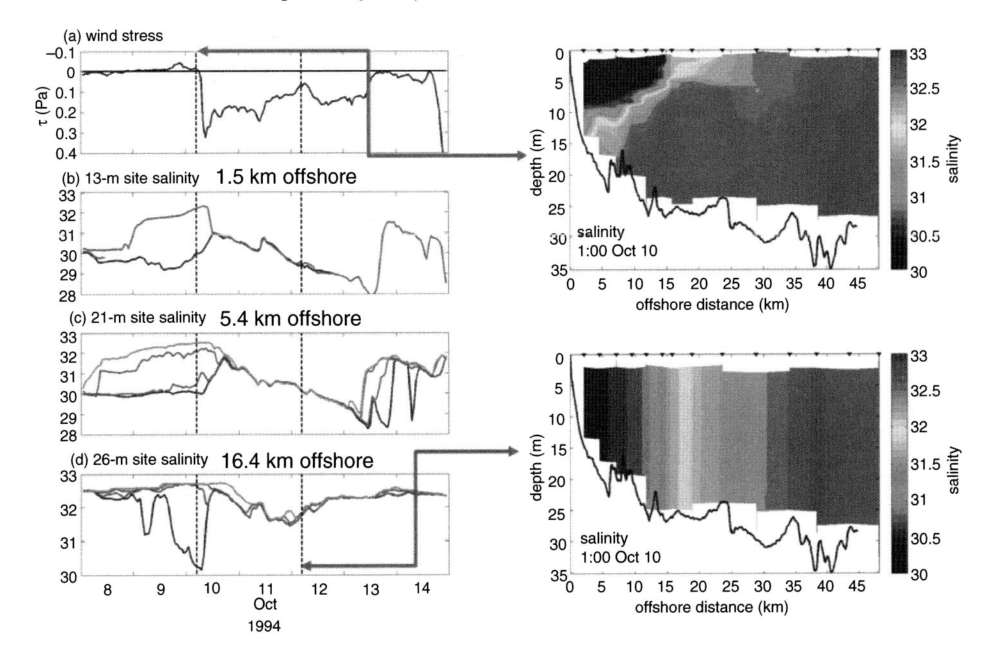


Figure 4.23. Time series of alongshelf wind stress, and salinity at different depths 1.5, 5.4, and 16.4 km offshore (*left*) and corresponding cross-shelf salinity sections prior to and after a strong downwelling wind stress event (*right*). For a color version of this figure please see the color plate section.

plume water (29–30 psu) overlying saltier ambient shelf water (32–33 psu), at all three mooring sites. Shortly after the onset of the wind forcing, the water column becomes well mixed at all three mooring sites. This is also clearly evident in the subsequent salinity section where the plume width remains about the same, but the isopycnals are vertical. Note also that at the shallower mooring sites the salinity decreases steadily during the downwelling winds, presumably due to the more rapid alongshelf flow of plume water from Chesapeake Bay that is subject to less entrainment.

The second example shows the onset of a light upwelling-favorable wind stress (Figure 4.24). In this case, prior to the onset of upwelling favorable winds, the plume is about 10 m deep and about 5 km wide. With the onset of light upwelling favorable winds, the plume thins and begins to spread offshore and detach from the coast. The salinity at the offshore mooring site (15.6 km offshore) remains high, $\geq$32 psu (indicating the plume does not extend this far offshore) until August 27, when the near-surface salinity suddenly decreases as the plume arrives at this site. At the same time, salinities at the other two sites closer to shore are increasing and are actually saltier than the offshore site as the plume detaches from the coast.

Notice that not only do the upwelling winds detach the low-salinity plume from the the coast, but also the salinity of the plume is clearly increasing, suggesting that

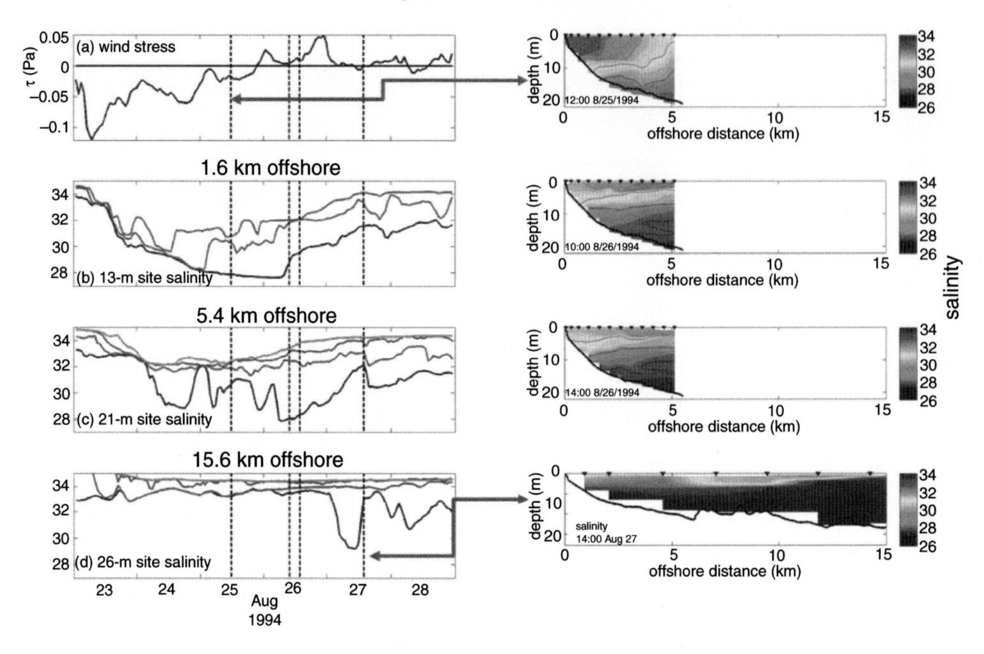


Figure 4.24. Time series of alongshelf wind stress, and salinity at different depths 1.5, 5.4, and 16.4 km offshore (*left*) and corresponding cross-shelf salinity sections prior to and after a weak upwelling wind stress event (*right*). For a color version of this figure please see the color plate section.

the plume is entraining ambient water (Figure 4.24 right panels). This suggests that upwelling-favorable winds are potentially a very effective mechanism for dispersing buoyant coastal currents. To properly understand this, we need to understand what controls the structure of the buoyant coastal current as it responds to upwelling-favorable winds and what controls the entrainment process.

To gain insight into the dynamics of the upwelling response of a buoyant coastal current, Fong and Geyer (2001) carried out a sequence of numerical modeling experiments and developed a simple model of the response based on the insights they gained from the numerical model results. The numerical model consisted of a long ($\geq$400-km) channel with a buoyant source near the upstream end of the channel (Figure 4.25). They applied an inflow of buoyant water at the source for 36 days to form a steady coastal current and then applied an upwelling-favorable wind stress for 3 days and tracked the response of the plume. They carried out 20 numerical model runs in which they varied the wind forcing, the source transport, and the water depth at the source. The latter two parameters resulted in variations in the initial buoyant plume thickness and density. The following two figures show plan views (Figure 4.25) and vertical cross-sections (Figure 4.26) of the buoyant plume response to an upwelling-favorable wind stress of 0.1 Pa from one of the numerical model runs. As expected, the plume

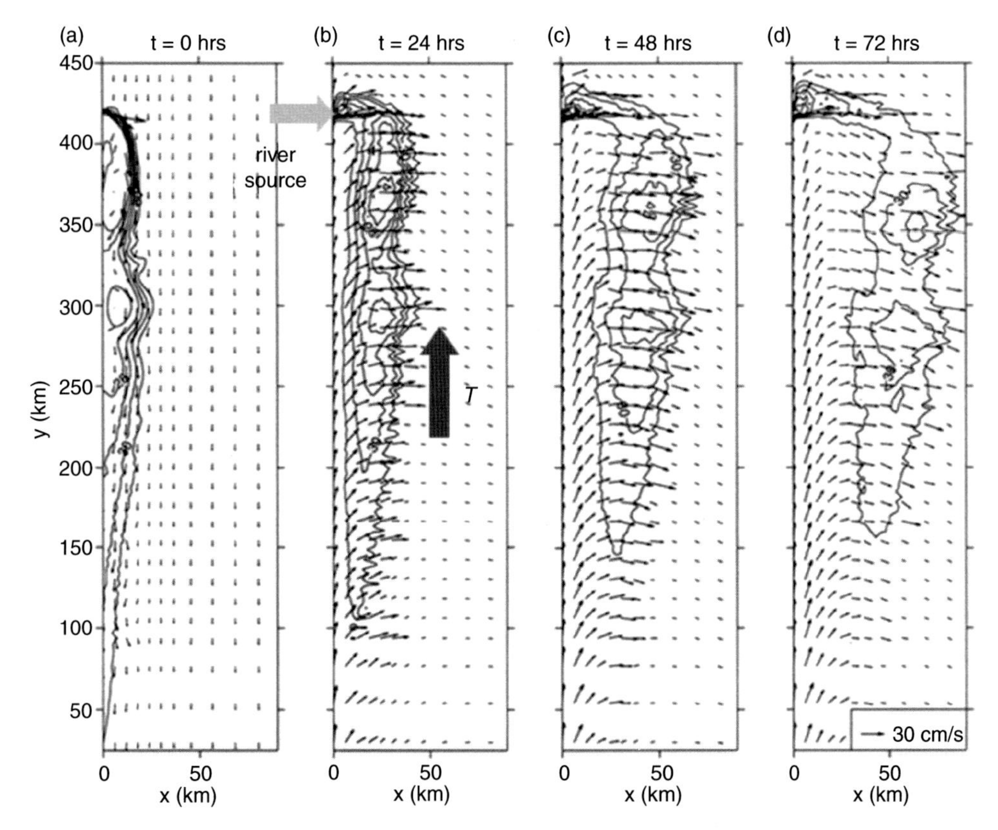


Figure 4.25. Sequence of plan views of salinity (contours) and currents from a numerical model run prior to ($t = 0$) and after the onset of upwelling favorable winds (from Fong and Geyer 2001). (Note in this case $x$ is offshore and $y$ is alongshore.)

moves offshore and widens, and the density anomaly of the plume decreases because of entrainment.

Fong and Geyer proposed a simple model to explain the evolution of the buoyant plume thickness. They assumed that the offshore edge of the plume thickens until there is a balance between buoyancy and shear-generated mixing as measured by a critical value for the bulk Richardson number (e.g., Pollard et al. 1973), that is,

$$R_b = \frac{g \Delta \rho h_o}{\rho_o \Delta v^2}, \tag{4.12}$$

where $\Delta v = \tau^{sx} / \rho_o f h_o$ is the velocity jump at the base of the plume based on the Ekman transport. Substituting this expression for $\Delta v$ into the expression for $R_b$ (4.12) and solving for $h_o$ yields

$$h_o = \left\{ \frac{R_b (\tau^{sx} / \rho_o f)^2}{g \Delta \rho / \rho_o} \right\}^{1/3}. \tag{4.13}$$

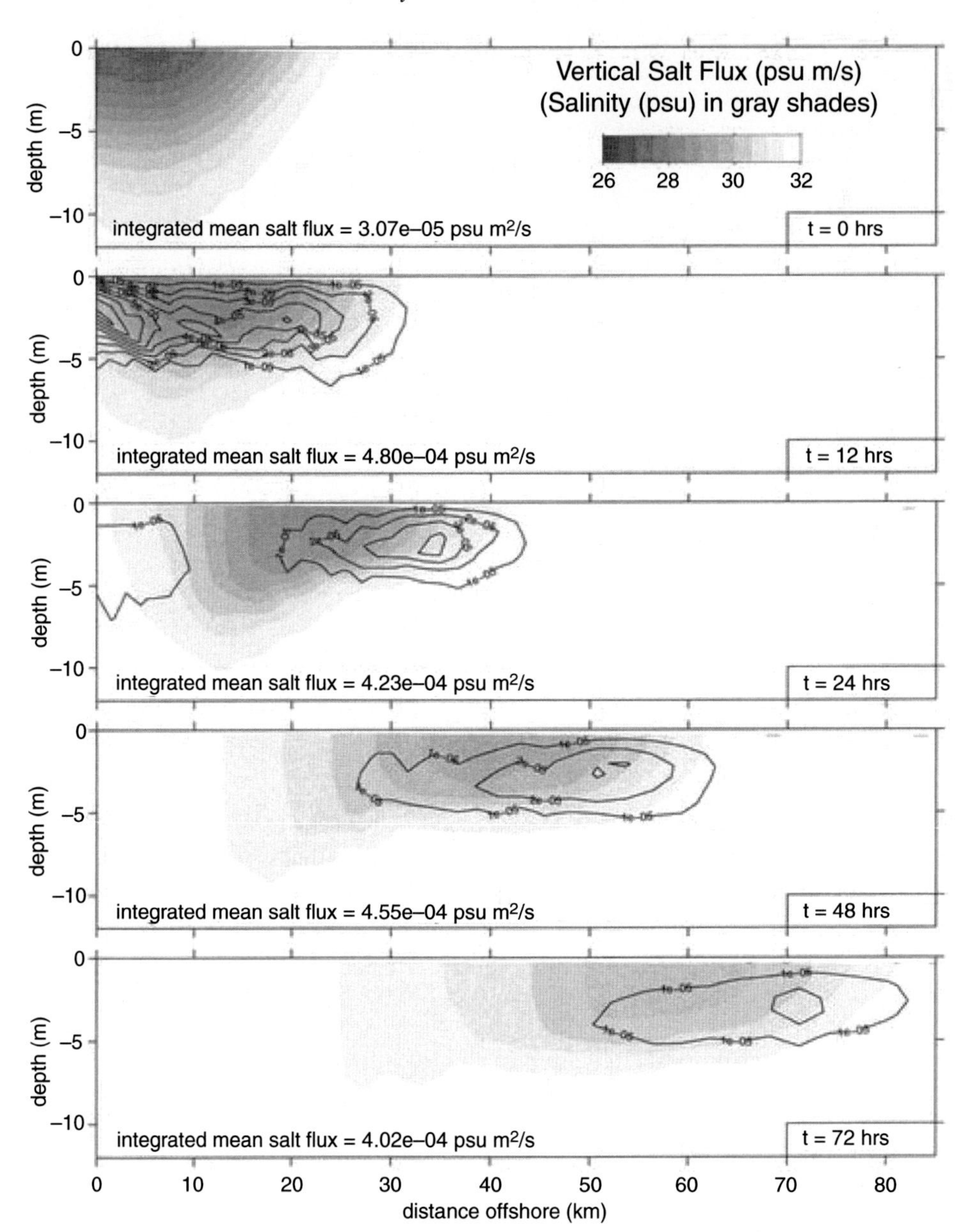


Figure 4.26. Sequence of salinity sections (gray scale) just prior to and after the start of upwelling winds. Vertical salt flux is shown in contours (from Fong and Geyer 2001).

The plume is dragged offshore maintaining a constant thickness $h_o$ over the outer portion of the plume (Figure 4.27). Given the original cross-sectional area of the plume $A_i$ and assuming volume is conserved, one can also estimate the final width of the plume.

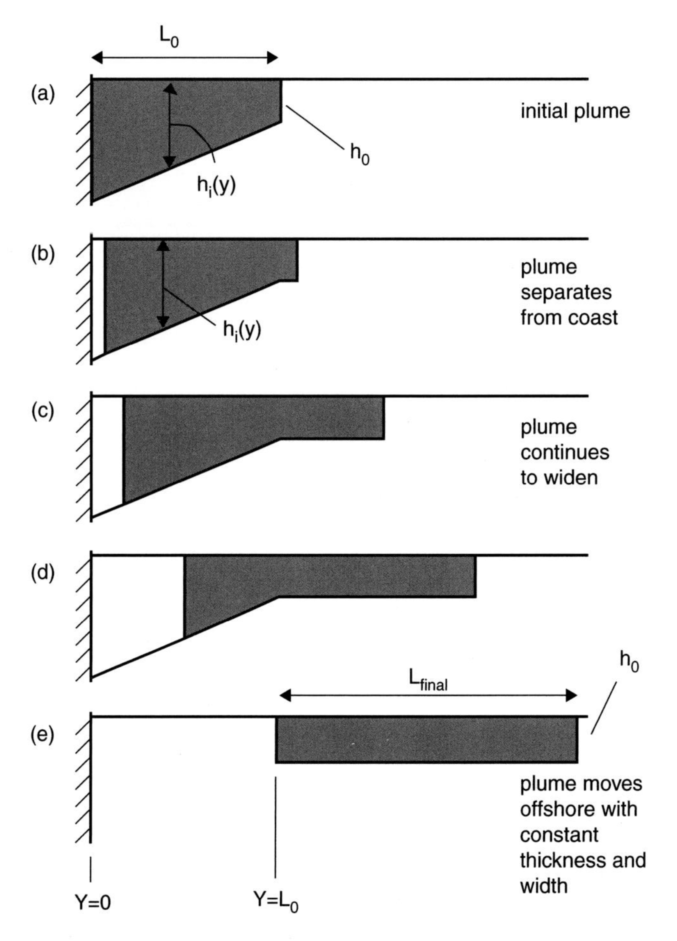


Figure 4.27. Schematic of a simple model for the response of a buoyant coastal current to upwelling winds (from Fong and Geyer 2001).

Fong and Geyer found excellent agreement between the plume thickness estimated from the bulk Richardson number criteria and the plume thickness from the numerical model results (Figure 4.28).

However, as noted by Fong and Geyer, their model ignores entrainment that will change both the density anomaly of the plume (as also seen in the observations above) and the plume volume. This motivated development of a simple model that included entrainment (Lentz 2004). Assume a two-dimensional plume flowing along a vertical wall with initial density anomaly $\Delta\rho_i$, thickness at the wall $h_i$, and width $W_i$ prior to the onset of upwelling-favorable winds (Figure 4.29). Note the following can be extended to the case of a sloping bottom, but for simplicity and clarity only flow along a vertical wall is considered here. Following Fong and Geyer, at the onset of the wind forcing, assume the portion of the plume shallower than $h_o$ mixes with the ambient shelf water until there is a balance between buoyancy and shear-generated turbulence

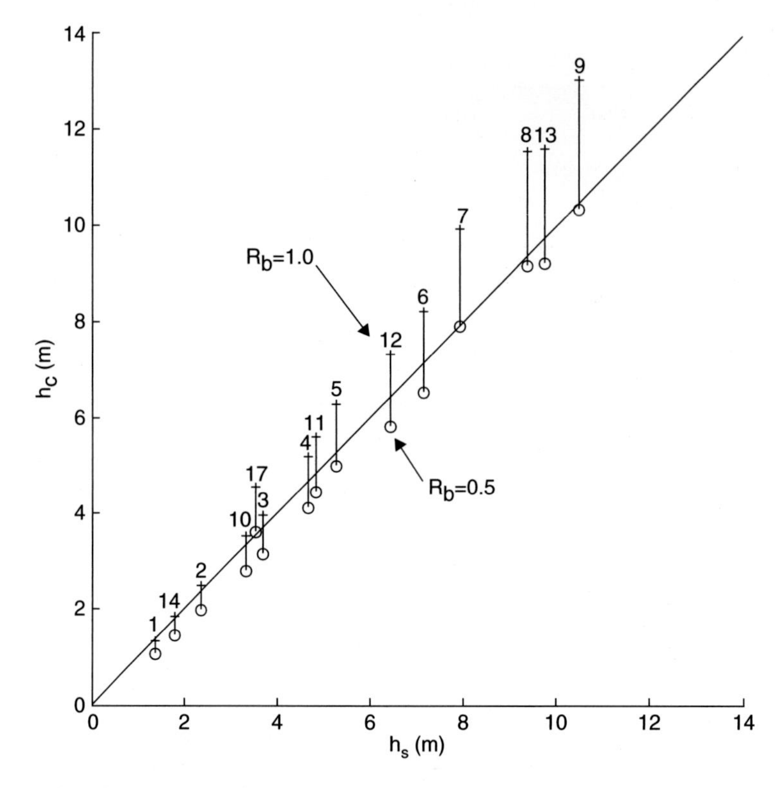


Figure 4.28. Comparison of the plume thicknesses from the numerical model ($h_s$) and predicted ($h_c$) from the simple model proposed by Fong and Geyer for two different choices of the critical value for $R_b$.

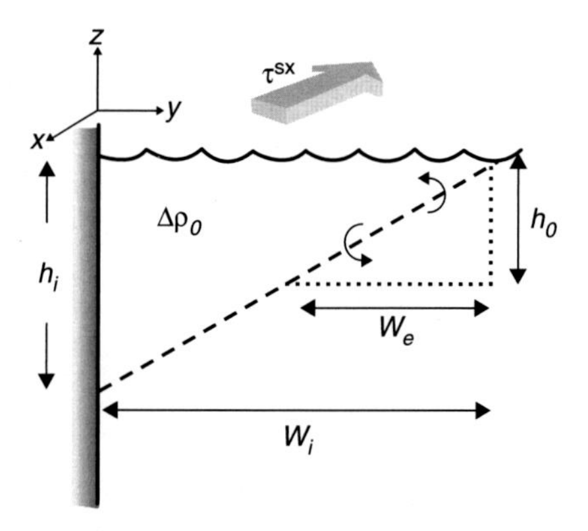


Figure 4.29. Schematic of entrainment at the onset of the wind forcing.

that satisfies a bulk Richardson number criteria that yields, as before,

$$h_o = \left\{ \frac{R_b V_E^2}{g\Delta\rho_o/\rho_{\text{amb}}} \right\}^{1/3} \quad \text{where } V_E = \frac{\tau^{sx}}{\rho_o f}. \tag{4.14}$$

But in this case, we do not know $\Delta\rho_o$. To solve for $\Delta\rho_o$, we invoke conservation of buoyancy,

$$\Delta\rho_o(A_i + A_e) = \Delta\rho_i A_i, \tag{4.15}$$

where $A_i = W_i h_i/2$, $A_e = W_e h_o/2$, and $W_e = W_i h_o/h_i$. Note that for simplicity we've assumed a triangular geometry for the plume. First solving for $\Delta\rho_o$ in terms of $h_o$ yields

$$\Delta\tilde{\rho}_o = \begin{cases} (1+\tilde{h}_o^2)^{-1} & \tilde{h}_o \leq 1, \\ (2\tilde{h}_o)^{-1} & \tilde{h}_o > 1, \end{cases} \tag{4.16}$$

where $\Delta\tilde{\rho}_o = \Delta\rho_o/\Delta\rho_i$ and $\tilde{h}_o = h_o/h_i$ are density anomaly and plume thickness normalized by the initial values prior to the wind forcing. The case where $\tilde{h}_o > 1$ corresponds to winds that are strong enough to mix the entire plume (see Lentz 2004 for a discussion of this limit). Substituting the expressions in (4.16) for $\Delta\rho_o$ into the expression for $h_o$ (4.14) yields

$$\begin{array}{ll} \tilde{h}_o^3 - \gamma^2(\tilde{h}_o^2+1)/2 = 0 & \gamma \leq 1, \\ \tilde{h}_o = \gamma & \gamma > 1, \end{array} \tag{4.17}$$

where

$$\gamma = \left(\frac{2R_b}{g_i' h_i}\right)^{\frac{1}{2}} \frac{|\tau^{sx}|}{\rho_{\text{amb}} f h_i} = \sqrt{2R_b}\frac{v_E}{c_i},$$

where $v_E = V_E/h_i$, $c_i = \sqrt{g_i' h_i}$, and $g_i' = g\Delta\rho_i/\rho_{\text{amb}}$.

These expressions provide estimates of the plume density anomaly and thickness at the offshore edge of the plume at the onset of the wind forcing.

As the wind continues to drag the plume offshore, the numerical model results of Fong and Geyer indicate there continues to be entrainment concentrated at the offshore edge of the plume (see the vertical salt fluxes in Figure 4.26). To estimate this entrainment at the offshore edge of the plume, imagine a two-step process. First, there is a geostrophic adjustment that results in the slumping of the isopycnals over a time scale of $1/f$. This slumping region at the offshore edge of the plume has a width that scales with the local baroclinic deformation radius $W_e$ (Figure 4.30). The slumping region is shallower than the equilibrium thickness of the plume and hence is subject to entrainment. The second step is wind-driven vertical mixing that again satisfies a bulk Richardson number criteria (as in equation (4.14)). Given these two steps we can make an estimate for the change in area of the plume due to entrainment over a time scale of $1/f$ as

$$\frac{\partial A}{\partial t} = \frac{W_e h}{2} f = \frac{1}{2}\sqrt{g' h^3} = \frac{1}{2}\sqrt{Ri_b} V_E. \tag{4.18}$$

In reality these two processes are, of course, happening simultaneously. Integrating (4.18) in time yields

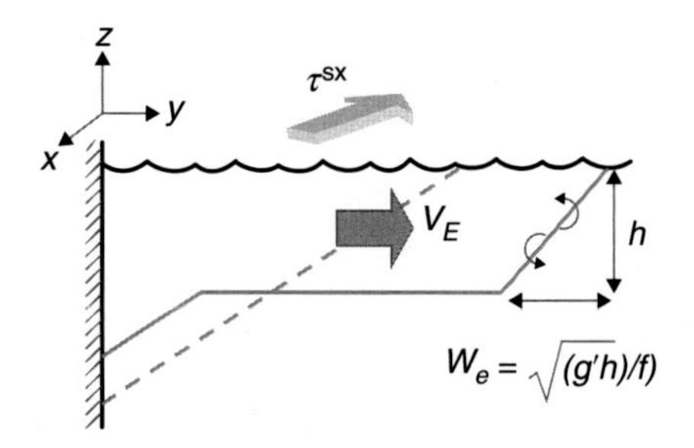


Figure 4.30. Schematic of wind-driven entrainment at the offshore edge of the buoyant plume. The entrainment is assumed to be the result of a competition between geostrophic adjustment associated with the buoyancy forcing and wind-driven mixing in the region where the isopycnals slope upward to intersect the surface. The width of this region is assumed to scale with the baroclinic deformation radius. The dashed line shows the initial shape of the plume.

$$A = A_o(1+\hat{t}), \tag{4.19}$$

where $\hat{t} = t/t_s$ and $t_s = 2A_o/\sqrt{Ri_b}V_E$ is the time it takes entrainment to double the initial cross-sectional area of the plume. From buoyancy conservation

$$\Delta\rho A = \Delta\rho_o A_o \rightarrow \Delta\rho = \Delta\rho_o(1+\hat{t})^{-1}, \tag{4.20}$$

and from the bulk Richardson number constraint on the plume thickness

$$h = \left\{\frac{Ri_b V_E^2}{g\Delta\rho/\rho_{\text{amb}}}\right\}^{1/3} \rightarrow h = h_o(1+\hat{t})^{1/3}. \tag{4.21}$$

We can estimate the offshore displacement of the plume edge from the Ekman transport, the plume thickness (4.21), and the initial plume width $W_i$ as

$$Y = W_i + \int_0^t \frac{V_E}{h}dt' = W_i + \frac{3A_o}{\sqrt{Ri_b}h_o}[(1+\hat{t})^{2/3} - 1]. \tag{4.22}$$

The plume width $W$ can also be estimated. Prior to the plume separating from the coast, the plume width is just $Y$, the distance offshore of the outer edge of the plume. After the plume separates from the coast, the plume width is just $A/h$ for the simple rectangular geometry that yields from (4.19) and (4.21)

$$W = \frac{A}{h} = \frac{A_o}{h_o}(1+\hat{t})^{2/3}. \tag{4.23}$$

The time when the plume separates from the coast, $t_{\text{sep}}$, can be found by solving for the time at which $Yh = A$ (see Lentz 2004).

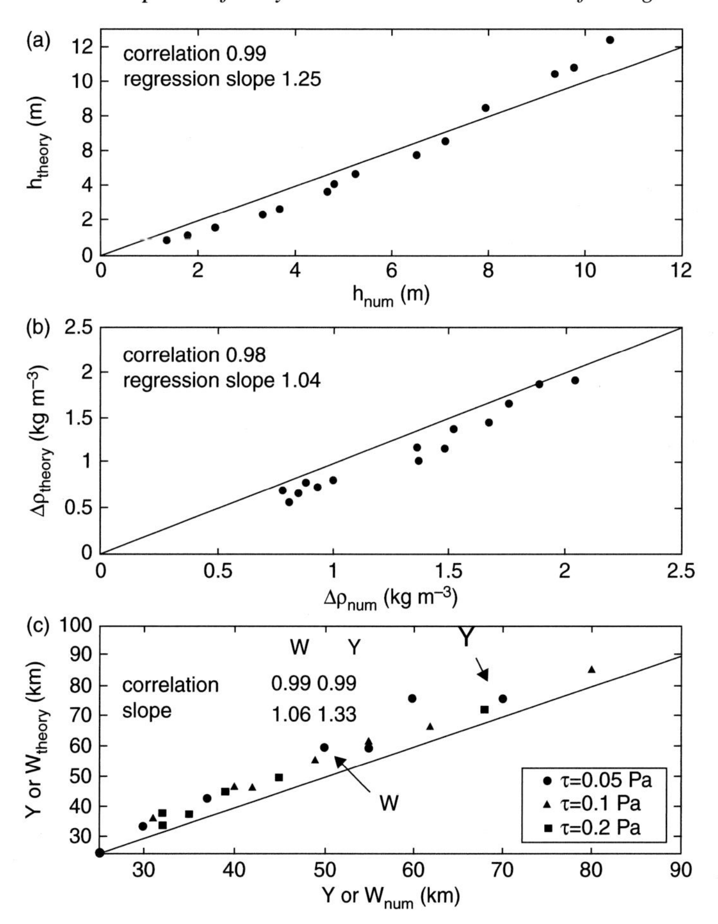


Figure 4.31. Comparisons of numerical model estimates with theoretical estimates of (a) plume thickness $h$ estimated from simple model, (b) density anomaly $\Delta\rho$ estimated from model, and (c) width $W$ estimated from model (triangles) and the offshore edge of plume $Y$ estimated from model (circles). Results in Figure 4.31c are for three numerical model runs with different wind stress magnitudes. For reference a line with a slope of 1.0 is shown in each frame.

This simple model provides estimates of the plume thickness, density, width, and offshore position. An obvious question is to what extent this simple model reproduces the numerical model results of Fong and Geyer. Fong and Geyer report 15 estimates of plume thickness and 14 estimates of plume density anomaly for model runs in which they varied the wind stress and initial plume characteristics. They also show the time evolution of plume cross-section (as previously) for three model runs with different wind stress magnitudes (0.05, 0.1, and 0.2 Pa). The comparisons suggest the simple model reproduces the numerical model results quite well (Figure 4.31). This suggests the simple model is capturing the fundamental physics determining the plume characteristics in the numerical model.

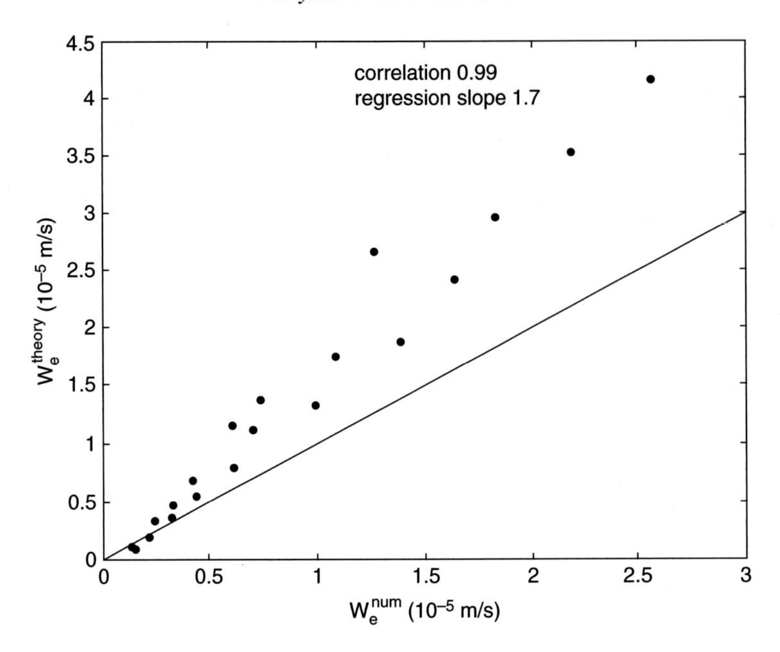


Figure 4.32. Comparisons of plume-average entrainment rates 72 hours after the onset of the wind forcing estimated from the 20 numerical model runs of Fong and Geyer (2001), $w_e^{\mathrm{num}}$, and from simple model $w_e^{\mathrm{theory}}$. A line with a slope of 1.0 is shown for reference.

Another, more severe test is whether the simple model provides reasonable estimates of the entrainment rate. The average entrainment velocity from the simple model can be estimated (4.18) and (4.23) as $\bar{w}_e = \frac{\partial A/\partial t}{W}$. Comparison with estimates of the average entrainment rate in the plume reported by Fong and Geyer indicates a very high correlation, but the regression coefficient indicates the simple theory entrainment rate is 1.7 times larger than the numerical model (Figure 4.32). The factor of 1.7 is probably due to the oversimplified geometry used in the simple model, and the particular choice of a critical bulk $R_i$ number ($R_b = 1$ used here). The agreement supports the idea that the entrainment is due to wind-driven mixing near the offshore edge of the plume.

What about the ocean? Does this simple model provide useful estimates of the response of buoyant coastal currents to upwelling winds? In general, it is difficult to find detailed enough observations of buoyant plumes and simple enough conditions to test these ideas. However, one example from the Chesapeake plume suggests this simple model may be relevant to the ocean (Lentz 2004). We look at a period in August 1994 when there was initially a period of no wind and a buoyant coastal current was present on August 11 (Figure 4.33). This was followed by a 12-day period when winds were more or less continuously upwelling favorable. Over this period, there were eight salinity sections collected from a ship or small boat on a cross-shelf transect located about 100 km south of Chesapeake Bay.

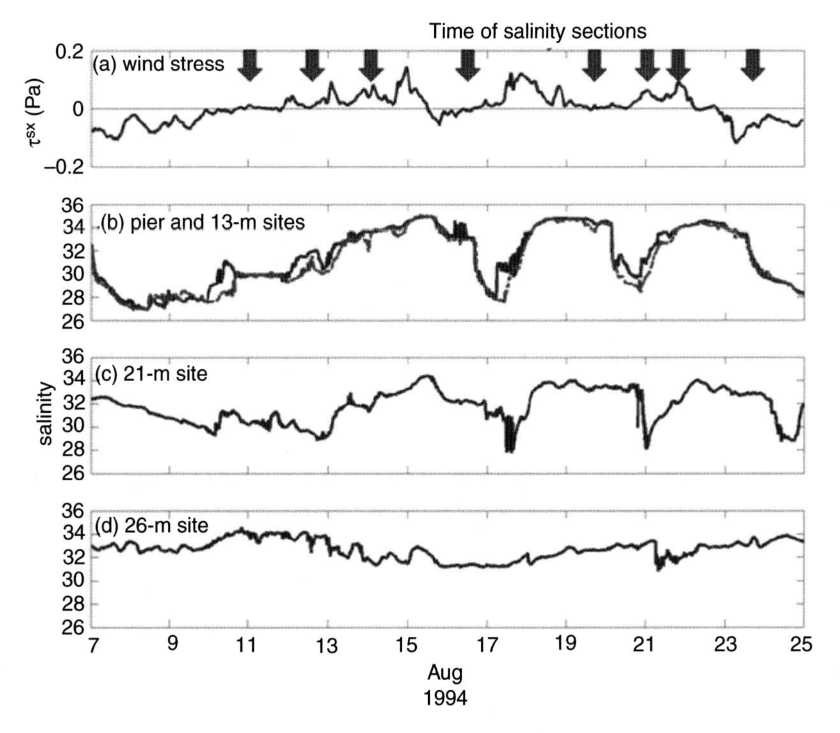


Figure 4.33. Time series from August 7–25, 1994, of (a) wind stress, and near-surface salinities at mooring sites (b) 1.3 km offshore, (c) 5.6 km offshore, and (d) 16 km offshore. Times of shipboard transects shown in Figure 4.34 are noted by arrows at top of figure. Buoyant coastal plume events are evident in Figure 4.33b on August 7–11, 16–17, and 20.

The resulting sections provide a partial view of the evolution of the buoyant plume over this period (Figure 4.34). On August 11 winds are nearly zero, and there is a buoyant plume at the site that is about 5 km wide and 10 m thick. After the onset of weak upwelling winds on August 12, the plume thins to about 5 m and spreads to almost 20 km offshore. By August 14 and 16, after stronger upwelling-favorable wind stresses, the plume has thickened, the salinity anomaly has decreased, and the plume width extends beyond the outer edge of these short surveys (more than 20 km offshore). By August 19, the plume has thickened to a maximum depth of 40 m, the salinity anomaly has continued to decrease, and the plume is more than 40 km wide. On August 21 a new plume from Chesapeake Bay has arrived at the site, and it subsequently entrains shelf water, deepens, and widens. By August 23, when the winds have reversed to downwelling favorable, the original plume has moved back on shore. There is reasonable agreement between estimates of the plume thickness, density anomaly, width, and offshore extent from the salinity sections with estimates from the simple model (Figure 4.35).

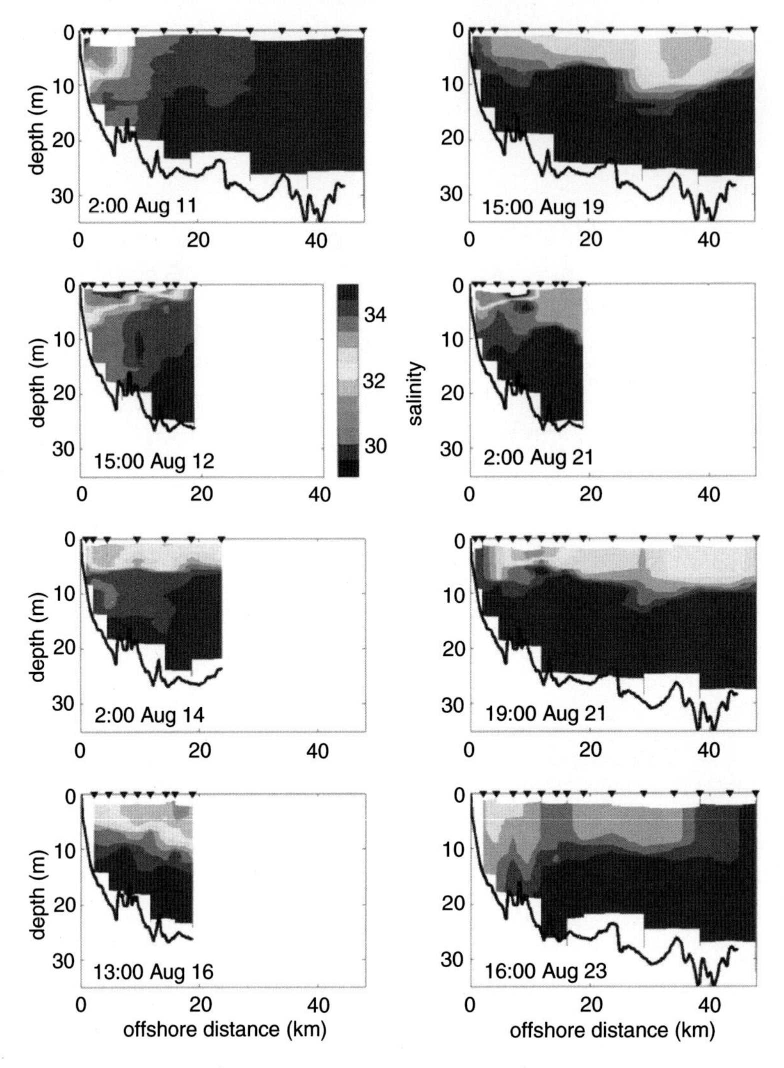


Figure 4.34. Sequence of salinity sections from shipboard hydrographic transects in August 1994. The salinity scale is shown next to the August 12 panel. For a color version of this figure please see the color plate section.

It is interesting to contrast the theoretical estimates for the Chesapeake plume with estimates from a larger plume such as the Mississippi River plume. For a moderate, upwelling wind stress of 0.1 Pa and the initial conditions used for the Chesapeake plume, $h_i = 8$ m, $W_i = 7$ km, and $\Delta\rho_I = 5$ kg m$^{-3}$, the resulting time scales for entrainment to double the size of the plume and for the plume to separate from the coast are $t_s = 18$ hours and $t_{sep} = 5$ h, respectively. In contrast, for the same wind stress, using $h_i = 15$ m, $W_i = 40$ km, and $\Delta\rho_I = 5$ kg m$^{-3}$ for the Mississippi River (from Wiseman et al. 1997) yields $t_s = 7$ days and $t_{sep} = 4$ days. While the characteristic plume thickness scales are similar for the two plumes because the wind stress and

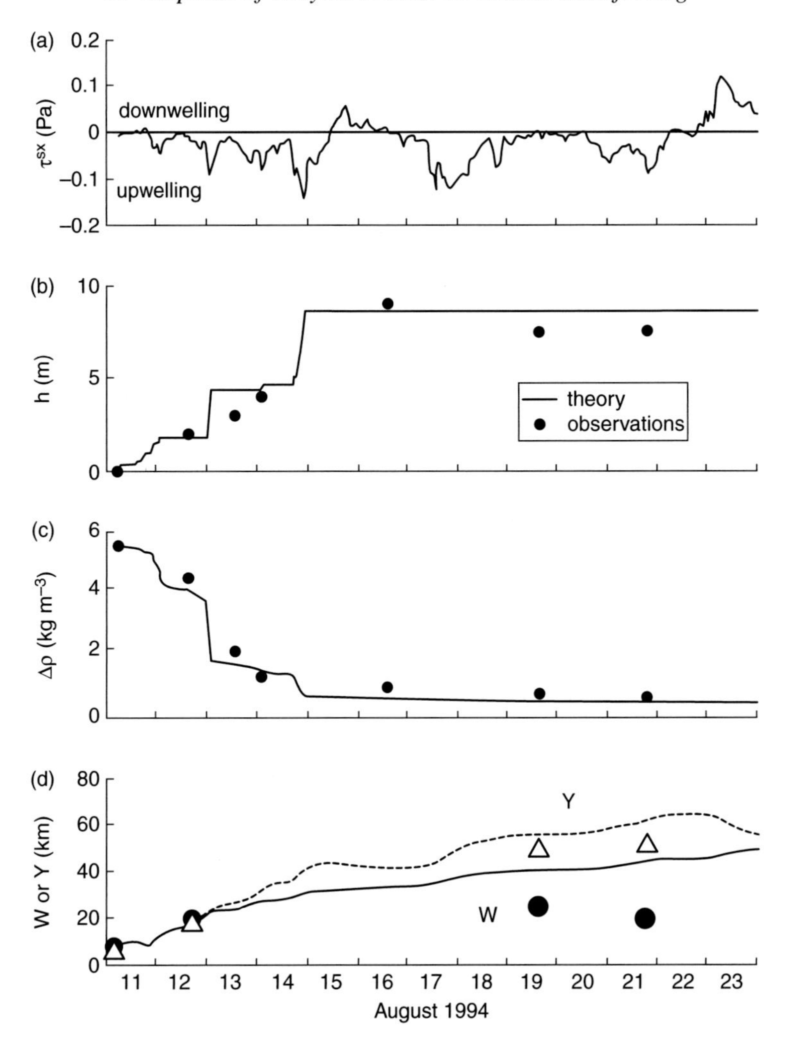


Figure 4.35. Comparisons of theoretical estimates (lines) and observations (symbols) of (b) buoyant plume thickness $h$, (c) density anomaly $\Delta\rho$, and (d) width $W$ (solid line and circles) and offshore position $Y$ (dashed line and triangles) for a period in August when wind stresses, shown in (a), were primarily upwelling favorable (positive values).

density anomalies are the same, the time scales differ by almost an order of magnitude because of the difference in cross-sectional areas. For the Chesapeake, both $t_s$ and $t_{\text{sep}}$ are similar to the time scale of a typical wind event. This implies that a single wind event can cause the Chesapeake plume to separate from the coast and the associated entrainment can double the plume volume. Since the density anomaly scales as $(1 + t/t_s)^{-1}$, it takes a 0.1 Pa upwelling wind stress about a week to decrease the Chesapeake plume density to 10% of its original value. For the Mississippi plume, both $t_s$ and $t_{\text{sep}}$ are longer than a typical wind event, which may explain why the Mississippi plume persists as a buoyant coastal plume for much of the year, except in mid to late summer when persistent upwelling-favorable winds move the low-salinity water toward the east (Cochrane and Kelly 1986; Wiseman et al. 1997; Li et al. 1997). Furthermore, the theoretical estimate suggests it would take months for an upwelling wind stress of 0.1 Pa to reduce the density anomaly of the Mississippi plume to 10% of its initial value. This comparison suggests that buoyant plumes similar to or smaller than the Chesapeake plume should be strongly influenced by wind forcing on time scales of a day or two, while much larger plumes, such as the Mississippi plume, should be more strongly influenced by wind forcing on monthly to seasonal time scales.

In summary, the simple model for the response of a buoyant coastal current to upwelling-favorable winds provides insight into the cross-shelf dispersal of plume water. The key feature of the model is wind-driven entrainment concentrated at the offshore edge of the plume. The model reproduces the basic plume characteristics from a more complex primitive equation numerical model and thus provides insight into the underlying physics. The model also reproduces the observed response of the Chesapeake plume for one period of upwelling winds.

## Acknowledgments

I thank Hannah Hiester, Robert McEwan, Travis Miles, Madeline Miller, and Romain Pennel for their contributions in editing this chapter.

## References

Alley, R. B., et al., 1997: Holocene climatic instability: A prominent, widespread event 8200 yr ago. *Geology* **25**, 483–486.

Barber, D. C., et al., 1999: Forcing of the cold event of 8,200 years ago by catastrophic drainage of Laurentide lakes. *Nature* **400**, 344–348.

Blanton, J. O., L.-Y. Oey, J. Amft, and T. N. Lee, 1989: Advection of momentum and buoyancy in a coastal frontal zone. *J. Phys. Oceanogr.* **19**, 98–115.

Boicourt, W. C., 1973: The circulation of water on the continental shelf from Chesapeake Bay to Cape Hatteras. Ph.D. thesis. The Johns Hopkins University.

Chao, S. Y., 1988: Wind-driven motion of estuarine plumes. *J. Phys. Oceanogr.* **18**, 1144–1166.

Chapman, D. C., and S. J. Lentz, 1994: Trapping of a coastal density front by the bottom boundary layer. *J. Phys. Oceanogr.* **24**, 1464–1479.

Choi, B.-J., and J. L. Wilkin, 2007: The effect of wind on the dispersal of the Hudson River plume. *J. Phys. Oceanogr.* **37**, 1878–1897.

Clarke, G., D. Leverington, J. Teller, and A. Dyke, 2003: Superlakes, megafloods, and abrupt climate change. *Science* **301**, 922–923.

Cochrane, J. D., and F. J. Kelly, 1986: Low-frequency circulation on the Texas-Louisiana continental shelf. *J. Geophys. Res.* **91**, 10645–10659.

Diaz, R. J., 2001: Overview of hypoxia around the world. *J. Env. Quality* **30**, 275–281.

Donato, T. F., and G. O. Marmorino, 2002: The surface morphology of a coastal gravity current. *Cont. Shelf Res.* **22**, 131–146.

Epifanio, C. E., A. K. Masse, and R. W. Garvine, 1989: Transport of blue crab larvae by surface currents off Delaware Bay, USA. *Marine Ecology Prog. Series* **54**, 35–41.

Fong, D. A., and W. R. Geyer, 2001: Response of a river plume during an upwelling favorable wind event. *J. Geophys. Res.* **106**, 1067–1084.

Fong, D. A., W. R. Geyer, and R. P. Signell, 1997: The wind-forced response of a buoyant coastal current: Observations of the western Gulf of Maine plume. *J. Mar. Syst.* **12**, 69–81.

Garvine, R. W., 1999: Penetration of buoyant coastal discharge onto the continental shelf: A numerical model experiment. *J. Phys. Oceanogr.* **29**, 1892–1909.

Griffiths, R. W., 1986: Gravity currents in rotating systems. *Annu. Rev. Fluid Mech.* **18**, 59–89.

Griffiths, R. W. , and E. J. Hopfinger, 1983: Gravity currents moving along a lateral boundary in a rotating frame. *J. Fluid Mech.* **134**, 357–399.

Hallock, Z. R., and G. O. Marmorino, 2002: Observations of the response of a buoyant estuarine plume to upwelling favorable winds. *J. Geophys. Res.* **107**, 3066, doi:10.1029/2000JC000698.

Helfrich, K. R., A. C. Kuo, and L. J. Pratt, 1999: Nonlinear Ross by adjustment in a channel. *J. Fluid Mech.* **390**, 187–222.

Hill, A. E., 1998: Buoyancy effects in coastal and shelf sea. K. H. Brink and A. R. Robinson (eds.) *The Sea,* Vol. 10, John Wiley & Sons. pp 21–62.

Hsueh, Y. , and B. Cushman-Roisin, 1983: On the formation of surface to bottom fronts over steep topography. *J. Geophys. Res.* **88**, 743–750.

Johnson, D. R., A. Weidemann, R. Arnone, and C. O. Davis, 2001: Chesapeake Bay outflow plume and coastal upwelling events: Physical and optical properties. *J. Geophys. Res.* **106**, 11613–11622.

Kourafalou, V. H., L.-Y. Oey, J. D. Wang, and T. N. Lee, 1996: The fate of river discharge on the continental shelf, 1, Modeling the river plume and the inner shelf coastal current. *J. Geophys. Res.* **101(C2),** 3415–3434.

Kubokawa, A., and K. Hanawa, 1984: A theory of semigeostrophic gravity waves and its application to the intrusion of a density current along a coast. Part 2. Intrusion of a density current along a coast in a rotating fluid. *J. Oceanogr. Soc. Japan* **40**, 260–270.

Lentz, S. J., 2004: The response of buoyant coastal plumes to upwelling-favorable winds. *J. Phys. Oceanogr.* **34**, 2458–2469.

Lentz, S. J., and K. R. Helfrich, 2002: Buoyant gravity currents along a sloping bottom in a rotating frame. *J. Fluid Mech.* **464**, 251–278.

Lentz, S. J., and J. Largier, 2006: The influence of wind forcing on the Chesapeake Bay buoyant coastal current. *J. Phys. Oceanogr.* **36**, 1305–1316.

Lentz, S. J., S. Elgar, and R. T. Guza, 2003: Observations of the flow field near the nose of a buoyant coastal current. *J. Phys. Oceanogr.* **33**, 933–943.

Li, Y., D. Nowlin, Jr., and R. O. Reid, 1997: Mean hydrographic fields and their interannual variability over the Texas-Louisiana continental shelf in spring, summer, and fall. *J.Geophys. Res.* **102**, 1027–1049.

Manabe, S., and R. J. Stouffer, 1997: Coupled ocean-atmosphere model response to freshwater input: Comparison to Younger Dryas event. *Paleoceanography* **12**, 321–336.

Melton, C., L. Washburn, and C. Gotschalk, 2009: Wind relaxations and poleward flow events in a coastal upwelling system on the central California current. *J. Geophys. Res.* **114**, doi:10.1029/2009JC005397.

Mork, M., 1981: Circulation phenomena and frontal dynamics of the Norwegian coastal current. *Philos. Trans. R. Soc. London Ser. A* **302**, 635–647.

Munchow, A., and R. W. Garvine, 1993a: Dynamical properties of a buoyancy-driven coastal current. *J. Geophys. Res.* **98**, 20063–20077.

Munchow, A., and R. W. Garvine, 1993b: Buoyancy and wind forcing of a coastal current. *J. Mar. Res.* **51**, 293–322.

Pollard, R. T., P. B. Rhines, and R. O. R. Y. Thompson, 1973: The deepening of the wing-mixed layer. *Geophys. Fluid Dyn.* **3**, 381–404.

Rennie, S., J. L. Largier, and S. J. Lentz, 1999: Observations of low-salinity coastal current pulses downstream of Chesapeake Bay. *J. Geophys. Res.* **104**, 18227–18240.

Renssen, H., H. Goosse, and T. Fichefet, 2002: Modeling the effect of freshwater pulses on the early Holocene climate: The influence of high-frequency climate variability. *Paleoceanography* **17**, Art. No. 1020.

Royer, T. C., 1981: Baroclinic transport in the Gulf of Alaska: a fresh water driven coastal current. *J. Mar. Res.* **39**, 251–266.

Seidov, D. , and M. Maslin, 1999: North Atlantic deep water circulation collapse during Heinrich events. *Geology* **27**, 23–26.

Simpson, J. E., 1982: Gravity currents in the laboratory, atmosphere, and ocean. *Annu. Rev. Fluid Mech.* **14**, 213–234.

Stern, M. E., J. A. Whitehead, and B. L. Hua, 1982: The intrusion of the head of a gravity current along the coast of a rotating fluid. *J. Fluid Mech.* **123**, 237–266.

Whitehead, J. A., and D. C. Chapman, 1986: Laboratory observations of a gravity current on a sloping bottom: The generation of shelf waves. *J. Fluid Mech.* **172**, 373–399.

Wiseman, W. J., N. N. Rabelais, R. E. Turner, S. P. Dinnel, and A. MacNaughton, 1997: Seasonal and interannual variability with the Louisiana coastal current: Stratification and hypoxia. *J. Mar. Syst.* **12**, 237–248.

Woodson, C. B., L. Washburn, J. A. Barth, D. J. Hoover, A. R. Kirincich, M. A. McManus, J. P. Ryan, and J. Tyburczy, 2009: Northern Monterey Bay upwelling shadow front: Observations of a coastally and surface-trapped buoyant plume. *J. Geophys. Res.* **114**, doi:10.1029/2009JC005623.

Yankovsky, A. E., and D. C. Chapman, 1997: A simple theory for the fate of buoyant coastal discharges. *J. Phys. Oceanogr.* **27**, 1386–1401.

# 5

# Overflows and Convectively Driven Flows

SONYA LEGG

## 5.1 Introduction to Overflows

### *5.1.1 What Are Dense Overflows?*

Dense water formed in semi-enclosed seas often has to flow through narrow straits or down continental slopes before it reaches the open ocean. These regions of dense water flowing over topography are known as dense overflows. The dense water has been formed through a variety of processes including surface cooling, the addition of salt in the form of brine from freezing pack ice in high-latitude seas, and evaporation in enclosed subtropical seas. The dense overflows are regions of significant mixing, which modifies the temperature and salinity signal of the dense water. Many of the deep water-masses of the ocean originate in these overflows and have their properties set by the mixing that occurs therein. For example, the Nordic overflows occurring in gaps in the Greenland-Iceland-Scotland Ridge (e.g., the Denmark Straits and the Faroe Bank Channel) are the source of most of the North Atlantic Deep Water (NADW), whereas Antarctic Bottom Water (AABW) is replenished by dense overflows from the Weddell and Ross seas in the Antarctic. Together these two deep water-masses are responsible for most of the deep branches of the meridional overturning circulation (MOC). Other overflows, such as the Red Sea overflow and Mediterranean outflow, contribute to important saline waters at intermediate depths. The properties of the deep and intermediate water-masses covering much of the abyssal ocean are therefore determined to a large extent by the mixing that takes place in the overflow, and hence these localized mixing regions play a significant role in influencing the large-scale ocean circulation. Here we will examine the observations of transport of dense water, as well as mixing and entrainment for the major overflows shown in Figure 5.1, highlighting their similarities and differences. Most differences stem from the different locations and involve different topography, tidal influence, Coriolis, and surrounding water-masses and currents.

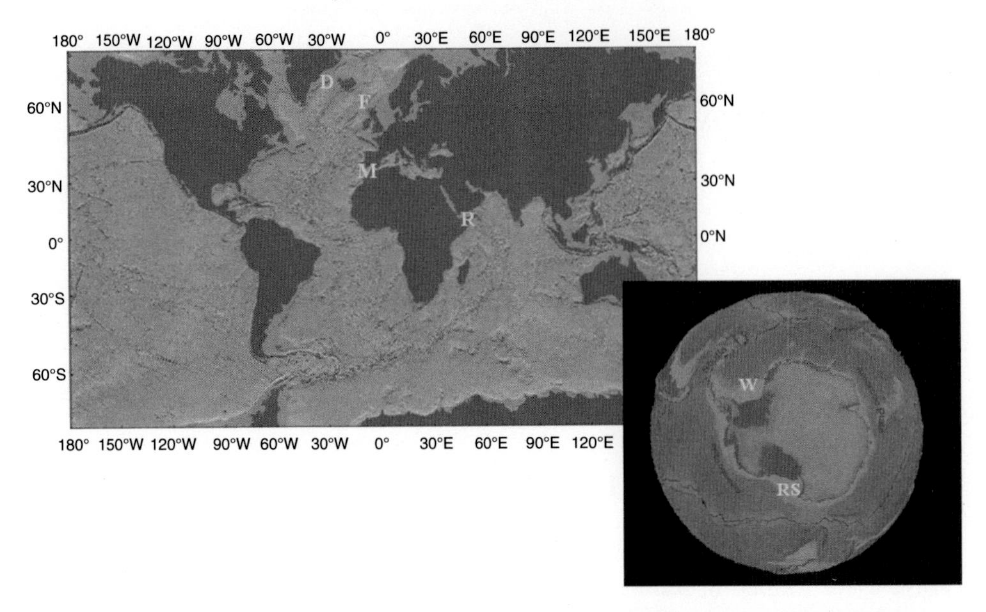


Figure 5.1. Location of the major overflows: the Denmark Straits overflow (D), the Faroe Bank Channel overflow (F), the Mediterranean outflow (M), the Red Sea overflow (R), and the Weddell (W) and Ross sea (RS) overflows. This figure is derived from a figure in Legg et al. (2009) originally created by Arnold Gordon.

### *5.1.2 Denmark Straits Overflow*

The Denmark Straits overflow, between Greenland and Iceland, carries dense water from the Greenland-Iceland-Norwegian (GIN) seas into the North Atlantic (Figure 5.2a). The channel is relatively wide ($\approx$ 100 km) relative to the small deformation radius [the scale on which rotation is important, approximately $L = \sqrt{g'h}/f$, where $g'$ is the buoyancy anomaly or reduced gravity of the overflow water ($g' = -g\Delta\rho/\rho_0$ where $g$ is the gravitational acceleration, $\Delta\rho$ is the density anomaly, and $\rho_0$ is the reference density), $h$ is the thickness of the overflow layer, and $f$ is the Coriolis parameter] of about 5 km. Unlike many other overflows, the Denmark Straits overflow has flow in the same direction at most depths across the channel (Girton et al. 2001). The compensating inflow of light water into these seas occurs on the eastern side of the ridge separating the North Atlantic from the GIN seas, not at the location of the outflow. Key parameters for the Denmark Straits and other major overflows are summarized in Table 5.1.

Cross-sections at the sill reveal dense water banked against the right-hand boundary (facing downstream), as expected for a geostrophic flow in a wide channel (Macrander et al. 2005). The width of the dense water outflow is only about 15–20 km. The barotropic flow means that the southward flow out of the Greenland sea extends well above the dense layer. At the sill, there is approximate hydraulic control, that is, the

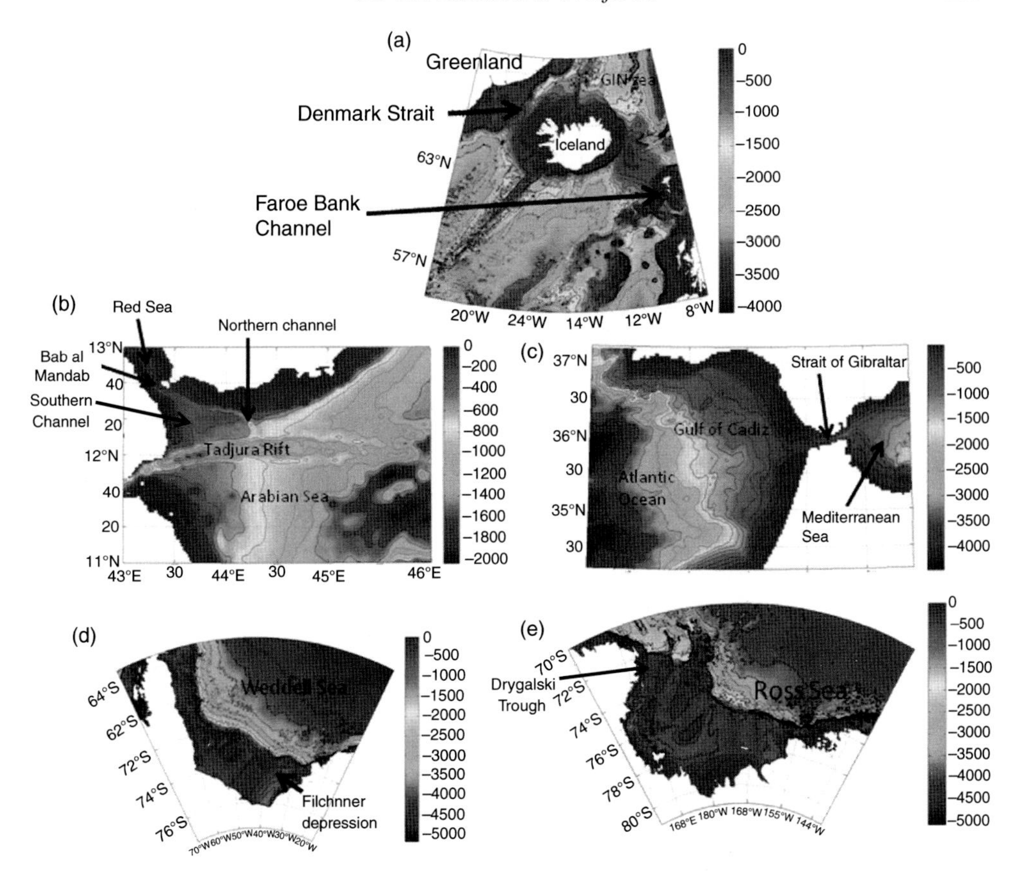


Figure 5.2. The topography of the major overflow regions: (a) Denmark Strait and Faroe Bank Channel overflows; (b) Red Sea outflow; (c) Mediterranean Outflow; (d) Weddell Sea overflow; (e) Ross Sea overflow. Color scale shows the depth in meters and is different for each location. The contour spacing is 200 m for the first 2,000 m depth, and 500 m thereafter. These images were created from the ETOPO5 database, with the assistance of Mehmet Ilicak. For a color version of this figure please see the color plate section.

flow makes a transition from a subcritical flow (speed less than long internal wave speed) to a supercritical flow (speed greater than the long internal wave speed). For such a control point at the sill, the transport $Q$ is given by

$$Q = \frac{g' h_u^2}{2f}, \tag{5.1}$$

where $h_u$ is the upstream height of the dense water layer, as expected for a wide channel in a rotating system (Whitehead 1998). Changes in upstream density structure are therefore reflected in changes in the magnitude of transport through the straits (Macrander et al. 2007).

Table 5.1. *Table of Observations, Summarizing the Major Similarities and Differences Between the Major Overflows*[a]

| | Faroe Bank | Denmark Straits | Ross Sea | Weddell Sea | Red Sea | Med Sea |
|---|---|---|---|---|---|---|
| **Source Water** | | | | | | |
| Potential temp. (°C) | 0 | 0.25 | −1.9 | −1.9 | 22.8 | 14.0 |
| Salinity | 34.92 | 34.81 | 34.8 | 34.67 | 39.8 | 38.4 |
| $\sigma_0$ | 28.07 | 27.94 | 27.9 | 27.8 | 27.7 | 28.95 |
| Sill depth (m) | 800 | 500 | 600 | 500 | 200 | 300 |
| **Product Water** | | | | | | |
| Potential temp. (°C) | 3.3 | 2.1 | −1.0 | −1.0 | 21.7 | 11.8 |
| Salinity | 35.1 | 34.84 | 34.72 | 34.67 | 39.2 | 36.4 |
| $\sigma_0$ | 27.9 | 27.85 | 27.85 | 27.75 | 27.48 | 27.6 |
| Depth (m) | 3,000 | 1,600 | >3,000 | 2,000 | 750 | 850 |
| **Transport (Sv)** | | | | | | |
| Source | 1.8 | 2.9 | 0.6 | 1.0 | 0.3 | 0.8 |
| Product | 3.3 | 5.2 | 2 | 5 | 0.55 | 2.3 |
| Tidal current (m/s) | 0.2 | 0.1 | 0.3 | 0.2 | 0.8 | 0.1 |
| Froude number, $Fr = U/\sqrt{g'H}$ | 1 | 0.3–1.2 | 0.9–1.1 | 1 | 0.6–1.3 | 1 |
| Entrainment rate, $W_e/U$ | $5\times10^{-4}$ | $1\times10^{-3}$ | $6\times10^{-3}$ | | $2\times10^{-4}$ | 5-20$\times10^{-4}$ |
| Coriolis parameter, $f$ ($s^{-1}$) | $1.3\times10^{-4}$ | $1\times10^{-4}$ | $1.3\times10^{-4}$ | $1.3\times10^{-4}$ | $1.3\times10^{-5}$ | $3.1\times10^{-5}$ |
| Deformation radius, (km) $L_R = \sqrt{g'H}/f$ | 30 | 5 | 7 | 7 | 40 | 100 |

[a]A more complete version of this table is shown in Legg et al. (2009), and the full table is available at http://www.usclivar.org/gceobs.php. Data are taken from the field campaigns described in the following publications: Faroe Bank, Mauritzen et al. (2005); Denmark Straits, Girton et al. (2003); Ross Sea, Gordon et al. (2004); Weddell Sea, Foldvik et al. (2004); Red Sea, Peters et al. (2005a); and Mediterranean Sea, Price et al. (1993).

The dense water descends the slope from the sill depth at 600 m to about 2,500 m at Cape Farewell. As the overflow water descends, its density decreases because of the mixing with ambient water. The dense water does not proceed directly down the direction of maximum slope but rather closely follows the isobaths, only gradually moving deeper downstream (Girton and Sanford 2003). A purely geostrophic flow would follow isobaths exactly; the cross-isobath flow is an indicator of frictional and eddy processes.

Along with the decrease in density, there is an increase in transport as the plume proceeds downstream (Girton and Sanford 2003). This is an indication of "entrainment" – the mixing in of ambient fluid, which both dilutes the plume density and increases its volume.

As the plume moves down the slope, the dense water breaks into blobs or "boluses," associated with eddying behavior. These eddies also have a significant cyclonic surface signature because of the barotropic nature of the flow through the straits (Bruce 1995). The mechanism of cyclogenesis in the presence of this barotropic flow is discussed in Spall and Price (1998). The flow becomes less barotropic downstream of the straits (Kase et al. 2003).

### *5.1.3 Faroe Bank Channel Overflow*

The Faroe Bank Channel, which connects the Norwegian Sea and Iceland Basin south of the Faroe Islands (Figure 5.2a), is deep (a sill depth of 850 m) and very narrow (15 km across). However, this deep channel extends only up to about 200 m depth, so the surface waters are not constrained. Unlike the flow in the Denmark Straits, the flow in the channel is not barotropic (i.e., unidirectional with depth). At the sill, dense water is banked against the right boundary (looking downstream) because of the effect of Coriolis (Mauritzen et al. 2005). There is some cross-channel variation in the interface stratification between light and dense waters due to the combined effects of mixing and cross-channel circulation. In the narrow channel the flow is accelerated up to 1 m/s. About 50 km downstream of the sill, the topographic slope increases, the flow accelerates further, and the flow becomes supercritical (Girton et al. 2006) (i.e., the flow speed $U$ exceeds the long-wave phase speed given approximately by $\sqrt{g'h}$, so that the Froude number $Fr = U/\sqrt{g'h} > 1$). The channel opens out onto a broad slope, and when the dense flow encounters this slope, it spreads out suddenly, perhaps through a transverse rotating hydraulic jump (Pratt et al. 2007) (i.e., a sudden transition from a narrow supercritical flow to a wide subcritical flow). As the dense water moves downstream, the plume becomes warmer and saltier, and hence less dense, because of the entrainment of ambient water, and the total transport of dense water is approximately doubled (Mauritzen et al. 2005). Measurements of dissipation show that mixing occurs both at the interface between the dense water and ambient water and in the frictional bottom boundary layer (Fer et al. 2004). The highest values of entrainment are found at the region where increased slope leads to an increase in the dense flow velocity.

On the broad slope, the flow is highly variable, with oscillations with a period of about 3–4 days associated with cold core eddies or waves (Geyer et al. 2006).

### *5.1.4 Red Sea Overflow*

The Red Sea overflow is characterized by two narrow channels that are only about 5 km wide and that continue for about 100 km (Figure 5.2b). The path of the dense water is therefore topographically constrained throughout its descent from the strait known as Bab al Mandab until the channels reach the edge of the deep Tadjura rift. Because the channel widths are small compared to the local deformation radius (which is large at this near-equatorial location) the flow in the channels is not affected by rotation. There are therefore no eddy processes modifying the structure of the overflow as it descends (Peters et al. 2005a).

At Bab al Mandab the sill is only 150 m deep, so mixing can entrain seasonally modified near-surface waters. The nature of the exchange flow at the strait is modified on seasonal time scales and by the tides; however, the tidal influence on the dense overflow waters decays within about 25 km downstream (Peters et al. 2005a).

The overflow waters in the two channels show slightly different properties. The flow in the southern channel is weaker and has lower salinity due to a sill separating the main flow from the entrance to the southern channel. Both channels show seasonal changes in the overflow water, with much weaker summer outflows. The dense overflow is more episodic than the overflow in many other regions because the transport out of the straits is strongly influenced by seasonal and intraseasonal variations in wind forcing.

The structure of the overflow within the channel reveals a thick, interfacial layer separating the warm, salty, dense water from the overlying ambient water, generated by mixing between the overflow and the environment (Peters et al. 2005b). In this layer, temperature and salinity both gradually decrease with height, stable stratification, and strongly sheared velocity. Below this layer, the core of the overflow remains relatively unmodified by mixing. Frictional processes lead to a sheared boundary layer at the bottom, associated with vertically homogenized tracers. The narrow channels therefore protect the densest water in the overflow from dilution, and most of the doubling in transport occurs in the thick interfacial layer (Matt and Johns 2007).

Whereas the Faroe Bank Channel and Denmark Strait overflows are the source of the densest water in the northernmost North Atlantic (AABW being the densest water in other parts of the Atlantic), and so do not separate from the boundary at a neutral buoyancy level, the Red Sea overflow water has usually reached neutral buoyancy just before the narrow channels reach the edge of the much deeper Tadjura rift. However, on occasion the Red Sea overflow water must be dense enough to fall to the bottom of this strait, as indicated by the deep homogeneous salty layer (Bower et al. 2005).

### *5.1.5 Mediterranean Overflow*

The Mediterranean overflow originates at the 14-km-wide Strait of Gibraltar connecting the Atlantic Ocean and the Mediterranean Sea (Figure 5.2c). At the strait, there is an exchange flow; (on average) Atlantic water enters the Mediterranean at the surface and saltier Mediterranean water outflows below (Baringer and Price 1997). The shallowest point, the Camarinal Sill, is only 280 m deep, and the topography rapidly deepens and opens out in the Gulf of Cadiz.

At the sill, the flow is strongly influenced by the tides. Much of the mixing in the vicinity of the straits, in the form of Kelvin-Helmoltz billows and hydraulic jumps, occurs when the tides are flowing out of the Mediterranean. When the tides are reversed, the dense water outflow may be temporarily arrested (Wesson and Gregg 1994).

Subsurface mixing farther down the slope is less dependent on the tides. As the outflow moves gradually downward, following topographic contours to the right under the influence of rotation, entrainment in regions of high Froude number causes a dilution of the salinity and an increase in the total transport (Baringer and Price 1997). The outflow develops two distinct cores: one farther up the slope and another, denser core lower down. These cores are generated by a combination of small-scale topography and slightly differing mixing rates in different parts of the initial outflow (Borenas et al. 2002).

Ultimately, the mixing reduces the density of the overflow water sufficiently that it reaches a neutral buoyancy level at middepths, leading to the formation of an intermediate water mass characterized by high salinity. Eddies formed on the slope, influenced by variations in topography, migrate offshore as lenses of salty water (known as "meddies") at the neutral buoyancy level (Serra and Ambar 2002). Modeling of the Mediterranean overflow is discussed by Treguier et al. in Chapter 7 of this volume.

### 5.1.6 Antarctic Overflows

The Antarctic overflows are the least studied overflows because of the difficulty in making measurements in the Southern Ocean. In contrast to the overflows discussed earlier, dense water gravity currents in the Antarctic do not originate at narrow straits, but rather on wide areas of the continental shelf, such as the Ross and Weddell shelves. High-salinity shelf water is a product of complex circulations driven by brine rejection under freezing ice. (Since sea-ice has a lower salinity than the salt water from which it originates, the salt becomes concentrated under the freezing ice.) In the Weddell Sea (Figure 5.2d), the Filchner Depression is one source of dense water, modified by circulation in the cavity under the ice shelf (Foldvik et al. 2004). As this dense water moves down the continental slope, it encounters two ridges that steer some of the water steeply down the slope, while the rest moves at a shallow angle of descent. Darelius et al. (2009) showed that beyond the ridges, the dense layer is warmer and thicker than before. There is significant temporal variability in the overflow at several different subinertial frequencies, suggesting mesoscale eddies or topographic waves. In the Ross Sea (Figure 5.2e), dense water from the Drygalski Trough moves in a broad current with a shallow angle of descent governed by a balance between Coriolis effects and friction (Gordon et al. 2009). However, recent observations in the Ross Sea have shown that intermittent cascades of dense water can run directly down the slope, probably connected with tidally generated spills of dense water over the shelf edge (Gordon et al. 2004). Tides in this region have large amplitudes near the shelf break and can cause a cross-shelf displacement of 20 km, carrying shelf water over the sill and down the slope (Padman et al. 2009). As for other overflows, there appears to be a turbulent layer in the sheared region at the top of the overflow and another frictionally driven turbulent layer adjacent to the bottom boundary (Visbeck and Thurnherr 2009). The mixing in these overflows is strongly influenced by the large effect of thermobaricity (i.e., the nonlinear dependence of density on pressure in the equation of state) at these temperatures. Entrainment leads to an approximate doubling in transport.

### 5.1.7 Midocean Ridge Overflows

As dense water moves through the world abyssal ocean, it continues to encounter topographic features, including narrow canyons and sills; examples in the Atlantic Ocean include the Romanche Fracture Zone (Ferron et al. 1998; Tregier et al., Chapter 7, this volume), the Vema Fracture Zone, and the myriad of small canyons in the Mid-Atlantic Ridge. When a dense layer moving through a canyon encounters a sill, it may be locally accelerated, which increases the Froude number and causes mixing on the downstream side of the sill, as evidenced by decreases in density and increases in transport and small-scale overturning (St Laurent and Thurnherr 2007). (The relationship between Froude number and mixing will be discussed in Section 5.2.) Unlike the overflows moving down the continental slope, the dense water does not continue to accelerate down sloping topography, and so the enhanced mixing is highly localized. Nonetheless, as the bottom water encounters many successive narrow canyons and sills, this localized entrainment can have a significant cumulative effect.

### 5.1.8 Common Features of Overflows

All the overflows described so far have some common features, as summarized in Figure 5.3. A dense water-mass moves down topography, accelerating under gravity. This acceleration leads to the development of strong shear between the dense water and the lighter overlying water-masses. There is also strong shear in the frictional layer at the bottom boundary. These two shear layers may lead to turbulence-generating instability. In the case of the interfacial shear layer, turbulence leads to the mixing in of ambient water, which dilutes the dense current while increasing its transport. Turbulence in the frictional shear layer at the bottom boundary leads to the homogenization of dense water properties in this layer.

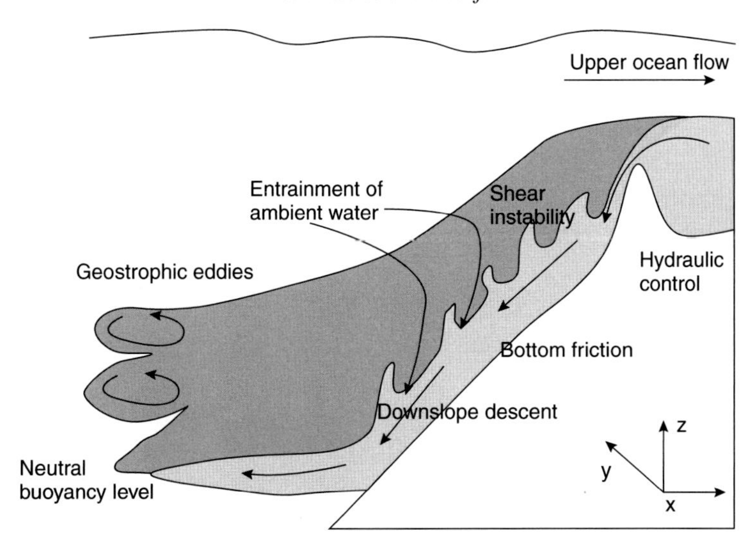


Figure 5.3. A schematic representation of the processes common to most oceanic overflows. This figure is derived from a figure in Legg et al. (2009).

The overflows, however, differ in many respects, including the following:

1. *The topography*: Some overflows are constrained by narrow canyons from start to finish (e.g., the Red Sea outflow), whereas others originate at narrow straits that open into wider slopes (e.g., the Faroe Bank Channel and Mediterranean overflows). Narrow canyons constrain the flow laterally, restricting the role of eddies and lateral jumps in spreading and mixing. Overflows on slopes may also be influenced by smaller-scale topographic features such as canyons and ridges (e.g., Antarctic overflows).
2. *The tides*: Some overflows (e.g., the Mediterranean and Antarctic overflows) are strongly influenced by tides, particularly at the sill. Tides enhance mixing and force dense water out over the continental slope.
3. *Rotation*: Overflows near the equator in narrow channels (e.g., the Red Sea outflow) are relatively unaffected by rotation. In other overflows (e.g., the Denmark Straits and Antarctic overflows), rotation constrains the flow to approximately follow topographic contours and allows the formation of mesoscale eddies.
4. *The overlying flow*: In some overflows the flow is approximately barotropic (e.g., the Denmark Straits overflow), whereas in others the overlying flow is in the opposite direction (e.g., the Mediterranean overflow). The overlying flow direction is important in influencing the eddy behavior of the overflow, that is, whether the eddy formation is due to baroclinic instability or due to downstream increases in the barotropic flow (see Spall and Price 1998).

Table 5.1 Shows a comparison of key parameters for the different major overflows of Figure 5.1.

## 5.2 Overflow Processes: Focus on Entrainment

### 5.2.1 The Entrainment Concept

A fundamental property of most overflows is that their transport increases downstream while their tracer properties become diluted (e.g., Mauritzen et al. 2005; Matt and Johns 2007). This can be explained as the result of **entrainment**, the mixing of ambient water into the descending dense current.

One conceptual representation of overflows is as an entraining stream tube (Price and Baringer 1994). In this model, shown schematically in Figure 5.4, the dense overflow water descends the slope under the influence of gravity, modified by rotation and friction, in a steady state (i.e., at any particular location, properties are independent of time). Entrainment modifies both the tracer properties of the overflow (i.e., by dilution) and the total transport (as more fluid is mixed into the descending flow). A bulk view of this entrainment assumes that any entrained ambient fluid is mixed throughout the overflow water and ignores variations in density and velocity within the overflow except in the downstream direction. We can then write conservation equations for the changes in total transport and in overflow plume tracer properties.

$$\frac{d}{dx}\int_L Uhdy = \int_L w_e dy \; ; \; \frac{d}{dx}\int_L Uqdy = \int_L w_e q_e dy, \tag{5.2}$$

where $x$ is the downstream direction, $y$ is the cross-stream direction, $h$ is the thickness of the plume, $L$ is the total cross-sectional width of the plume, $U$ is the downstream velocity, and $w_e$ is the **entrainment velocity**. $q$ is a conserved tracer (such as temperature or salinity), and $q_e$ is the tracer value in the ambient fluid adjacent to the plume boundary. With the addition of a momentum equation (which may involve gravitational acceleration, bottom friction/drag, and rotation), an assumption of the plume shape or self-similarity (i.e., that the plume retains the same shape and form, no matter how much it expands), an equation of state relating tracers and density anomaly, and a parameterization of the entrainment velocity $w_e$ in terms of the other quantities,

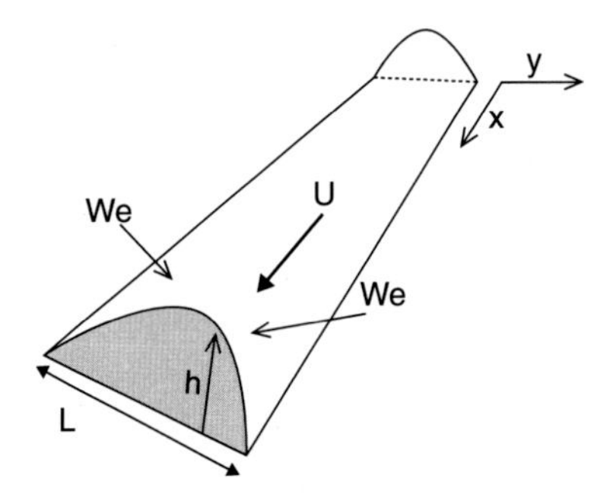


Figure 5.4. The entraining stream-tube model of a descending overflow, of width $L$, thickness $h$, and downstream velocity $U$. Transport increases downstream because of entrainment, where $W_e$ is the entrainment velocity.

solutions for the downstream overflow plume evolution can be found. Being able to parameterize the entrainment velocity is therefore important. An obvious parameterization, in keeping with this bulk view of the plume, is to assume a relationship between $w_e$ and $U$ of the form

$$w_e = UE, \tag{5.3}$$

where $E$ is a nondimensional entrainment coefficient (Turner 1986). A full parameterization requires an understanding of the factors that control $E$.

Equations (5.2) and (5.3) can be rearranged to give two different ways to estimate the entrainment coefficient: the **transport method** of diagnosing entrainment,

$$E = \frac{w_e}{U} = \frac{\frac{d}{dx}\int_L hU dy}{UL}, \tag{5.4}$$

and the **tracer method** of diagnosing entrainment,

$$E = \frac{w_e}{U} = \frac{h}{(q_e - q)}\frac{dq}{dx}. \tag{5.5}$$

The former requires good sampling across the plume to capture the full transport, whereas the latter requires sampling only in the center of the plume. The tracer method is therefore more commonly used to diagnose entrainment from observations (e.g., Girton and Sanford 2003); in numerical simulations and laboratory experiments, the transport method can also be used (e.g., Riemenschneider and Legg 2007).

### 5.2.2 *Causes of Entrainment*

The causes of entrainment are most completely understood in the case of turbulent, nonrotating, dense currents on a uniform slope and in a uniform environment. In this scenario, the dense water accelerates down the slope because of the influence of gravity, and the resultant shear between the overflow and the quiescent overlying water leads to Kelvin-Helmholtz instability at the interface separating the dense water from the lighter ambient water. The instability tends to pull lighter water under denser water, and subsequent turbulence mixes this lighter water into the dense water plume. The mixture must be denser than the surrounding ambient fluid, and so the plume continues to accelerate even as it is diluted. The development of shear-driven instability in a dense current on a slope can be seen in both two- and three-dimensional simulations (e.g., Ozgokmen and Chassignet 2002; Ozgokmen et al. 2004) and in laboratory experiments (e.g., Pawlak and Armi 2000).

From linear stability theory, shear instability is possible when the gradient Richardson number $Ri = N^2/(dU/dz)^2 < 1/4$ (the Miles-Howard criterion). This criterion can be derived on energetic grounds: When the kinetic energy extracted from the mean flow is sufficient to provide the increase in potential energy needed to mix parcels in the vertical, instability is possible. Shear instability therefore derives energy from the mean flow. However, using the stream tube model and ignoring entrainment and friction, we can show that the kinetic energy in the mean flow is buoyancy driven (i.e., a parcel loses potential energy by moving down the slope and gains kinetic energy). The mixing in this buoyancy-driven flow therefore derives its energy source from the kinetic energy of the large-scale flow, but that kinetic energy is in turn derived from the potential energy of the dense water originating high on the slope. The kinetic energy of the mean downslope flow is therefore converted to potential energy through mixing or lost to dissipation through frictional processes.

This (simple) picture is considerably complicated by the addition of bottom friction and rotation. With rotation, in addition to the gravitational force $g'$ directed downward (where $g'$ is the reduced gravity of the dense current), there is the Coriolis force $fU$ directed in the horizontal, perpendicular to the flow, where $U$ is the flow speed, and $f$ is the Coriolis parameter, as shown in Figure 5.5. (We are assuming here that rotation is aligned with the vertical.) A balance is achieved when the component of these two forces in the direction across the slope is equal and opposite:

$$g' \sin\theta = fU \cos\theta, \tag{5.6}$$

where $\theta$ is the angle of the slope relative to the horizontal. This balance is achieved when the along-slope velocity is given by

$$U = \frac{g' \nabla D}{f}, \tag{5.7}$$

where $\nabla D = \tan\theta$ is the bottom slope, and the dense current will then flow along isobaths, with the shallower region to the right in the Northern Hemisphere. The acceleration of the dense current is therefore limited by rotation.

Friction allows this geostrophic balance to be broken so that dense water can cross isobaths. When the influence of bottom friction is large [as expressed by a large Ekman number, $Ek = (\delta/h)$ where the Ekman thickness $\delta = \sqrt{2\nu/f}$, and $\nu$ is the viscosity (for laminar flows, this would be the molecular viscosity), $f$ is the Coriolis parameter, and $h$ is the dense current thickness] and Froude number is small (i.e., $Fr = U/\sqrt{g'h} < 1$), the dense fluid moves down slope in a thin laminar frictional Ekman layer, with little

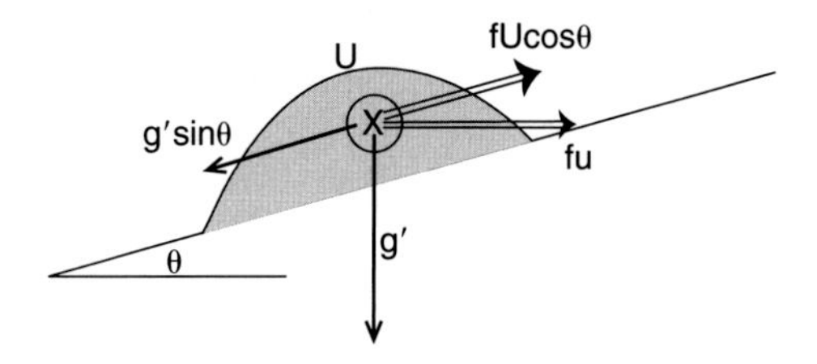


Figure 5.5. A schematic of the balance of forces operating on a steady dense inviscid overflow on a slope of angle $\theta$, with along-slope velocity $U$ in the presence of rotation. The rotation vector is assumed to be vertical.

mixing. For large Ekman numbers combined with moderate to large Froude numbers, the dense overflow develops roll-like waves, which break and cause mixing (Cenedese et al. 2004).

For lower $Ek$, large $f$ flows, the geostrophic flow develops eddies (Cenedese et al. 2004; Lane-Serff and Baines 1998) through barotropic or baroclinic instability (depending on the nature of the flow in the overlying layer), in which mixing may occur primarily through lateral stirring processes. Observations of the Denmark Straits overflow indicate that entrainment by mesoscale eddies dominates from about 200 m downstream of the sill (Voet and Quadfasel 2010). Entrainment by these lateral eddy processes, which is probably the least understood aspect of overflows, will not be examined further in this chapter. Figure 5.6 summarizes the different regimes for rotating overflows, as identified by Cenedese et al. (2004).

Changes in bottom topography also influence entrainment. In particular, if the bottom slope changes abruptly, the dense current may transition from a supercritical state (a thinner, faster current with $Fr > 1$) to a subcritical state (a thicker, slower current with $Fr < 1$) through a hydraulic jump, as shown in Figure 5.7, leading to locally enhanced entrainment. This type of localized entrainment is seen especially in flow over sills in canyons in the Mid-Atlantic Ridge (Ferron et al. 1998) and in the Faroe Bank Channel, where the hydraulic jump is strongly influenced by rotation (Girton et al. 2006).

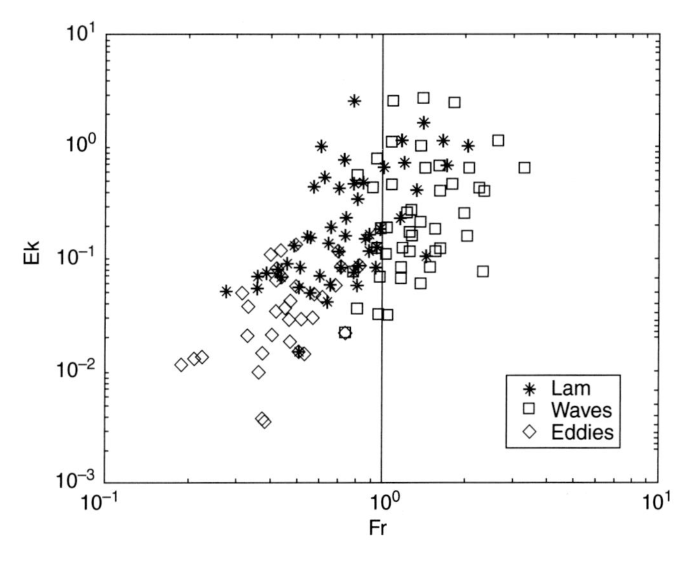


Figure 5.6. The regime diagram for rotating gravity currents, as determined by Cenedese et al. (2004), showing the laminar ("lam"), roll-wave ("waves"), and eddy regimes ("eddies") as a function of Froude and Ekman numbers. This figure is figure 11 in Cenedese, C., J. A. Whitehead, T. A. Ascarelli and M. Ohiwa, 2004: A dense current flowing down a sloping bottom in a rotating fluid. *J. Phys. Oceanogr.* **34**, 181–203. © Americal Meteorological Society. Reprinted with permission.

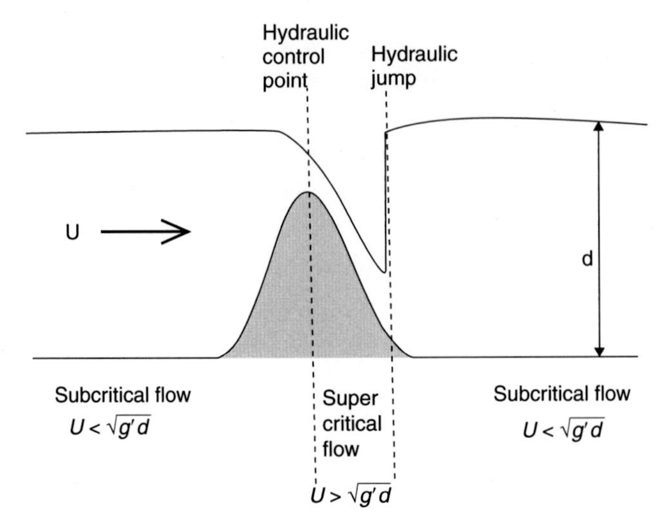


Figure 5.7. A schematic showing a hydraulic jump downstream of a sill, where flow transitions from a supercritical to a subcritical state, which may be a location of enhanced entrainment.

Bottom roughness can influence entrainment by increasing the length scale of mixing so that it penetrates the full depth of the dense layer (Ozgokmen and Fischer 2008). With rotation, small-scale topography can have an even more complicated effect.

### *5.2.3 Parameterizing Entrainment*

If we could parameterize the entrainment coefficient $E$ in terms of the bulk properties of the flow, then provided that the bulk view of the overflow plume is a good representation, the downstream evolution of plume properties due to entrainment would be predictable.

Entrainment has often been parameterized in terms of the bulk Froude number

$$E = E(Fr) \text{ where } Fr = \frac{U}{\sqrt{g'h}} \tag{5.8}$$

where $U$ is the plume velocity relative to its surroundings, $g'$ is the plume buoyancy anomaly, and $h$ is the plume thickness, or the related bulk Richardson number

$$Ri_b = 1/Fr^2. \tag{5.9}$$

Experiments by Ellison and Turner (1959) give an empirical relationship

$$E = \frac{0.08Fr^2 - 0.1}{Fr^2 + 5} \text{ for } Fr^2 \geq 1.25\ ;\ E = 0 \text{ for } Fr^2 < 1.25. \tag{5.10}$$

An important feature of this parameterization is a cut-off Froude number below which no entrainment occurs. This cut-off Froude number is equivalent to a cut-off bulk Richardson number of $Ri_c = 0.8$, above which no entrainment occurs.

However, recent studies have shown that entrainment depends on the Reynolds number $Re = U/(\nu L)$, where $L$ is the length scale of the motion, and on the phenomenological regime of the downslope flow. In particular, for rotating overflows, the plume may be in a laminar regime (little mixing); a roll-wave regime, in which waves may or may not be breaking; an eddying regime; or a fully turbulent regime (Cenedese et al. 2004; Cenedese and Adduce 2008). For the roll-wave regime, entrainment has been found to occur well below the cut-off Froude number identified by Ellison and Turner.

A new parameterization (Cenedese and Adduce 2010) that takes into account the dependence of entrainment on Reynolds number and the entrainment at low Froude number is

$$E = \frac{E_{\min} + AFr^{\alpha}}{1 + AC_{\inf}(Fr + Fr_0)^{\alpha}}, \tag{5.11}$$

where

$$C_{\inf} = \frac{1}{E_{\max}} + \left(\frac{B}{Re^{\beta}}\right) \tag{5.12}$$

where the nondimensional constants $A = 3.4 \times 10^{-3}$, $F_0 = 0.51$, $\alpha = 7.18$, $B = 243.52$, and $\beta = 0.5$ are determined by nonlinear regression to the available data from many different laboratory studies and field observations. $E_{\min}$ is a minimum value of entrainment for very low Froude numbers, and $E_{\max}$ is the maximum value for large Froude numbers. This parameterization of entrainment is shown as a function of $Fr$ for different $Re$ in Figure 5.8.

These parameterizations of entrainment rely on the bulk view of the overflow, assuming that the velocity is uniform over the dense current and that any ambient fluid is rapidly mixed throughout the dense layer. However, observations and simulations reveal vertical structure in both the tracers and velocity fields, as shown schematically in Figure 5.9. Therefore, an alternative view point is to express local entrainment in terms of the local gradient Richardson number (i.e., $E = f(Ri_g)$), where the gradient Richardson number is defined as $Ri_g = N^2/(dU/dz)^2$, which may vary over the width and height of the overflow plume. This type of parameterization is also more amenable to inclusion in numerical models (Hallberg 2000).

The entrainment velocity can be thought of as a diapycnal velocity, moving fluid across isopycnals from one density class to another. In this sense, entrainment is particularly well suited to numerical models that use density as their vertical coordinate. The HYCOM model, a hybrid coordinate model that at depth approaches isopycnal coordinates, has a parameterization of entrainment based upon high-resolution, numerical simulations of overflows. It is difficult to directly measure the dependence of entrainment from one fluid density class to another on the local Richardson number. The

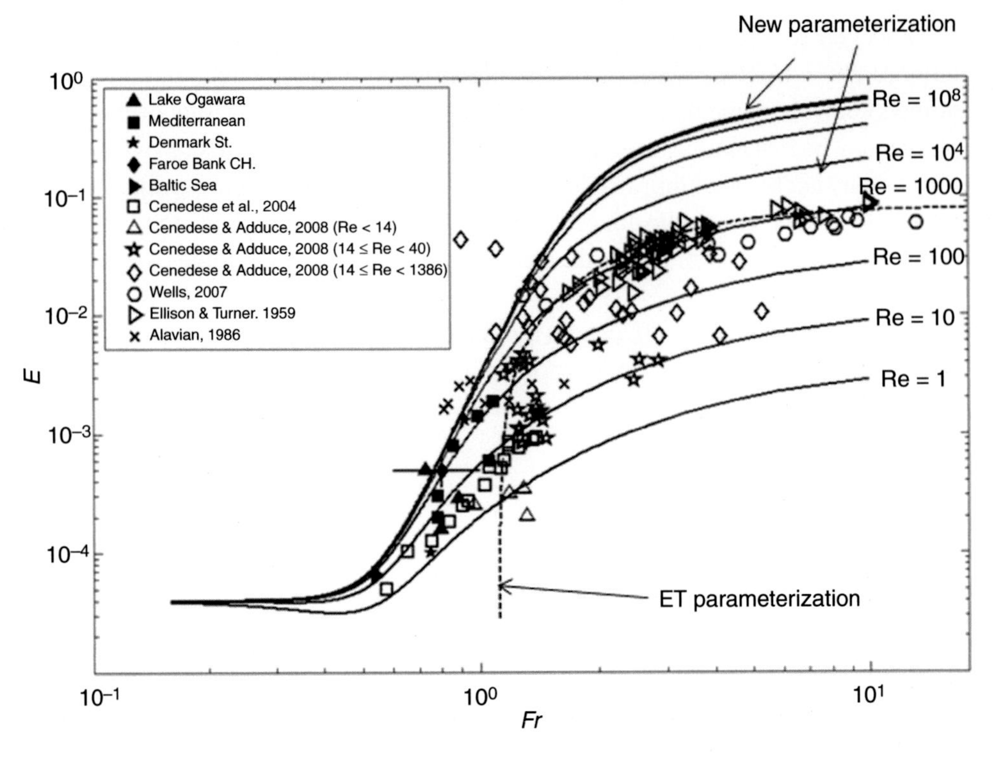


Figure 5.8. The new entrainment parameterization developed by Cenedese and Adduce (2010), shown as a function of the Froude number for different Reynolds numbers and compared with laboratory and field data. Also shown is the Ellison and Turner (1959) parameterization. This is figure 2 from Cenedese, C. and C. Adduce, 2010: A new parameterization for entrainment in overflows. *J. Phys. Oceanogr.* **40**, 1835–1850. © American Meteorological Society. Reprinted with permission.

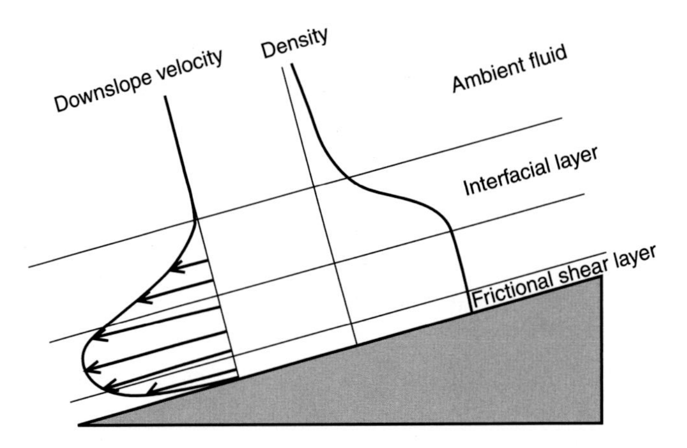


Figure 5.9. A schematic of the vertical structure of an overflow, showing downstream velocity and density profiles.

development of the parameterization therefore focuses on calibrating a physically reasonable parameterization to give the best agreement between the overflows simulated by the high-resolution model (the "truth") and the low-resolution hydrostatic model incorporating the parameterization. Using this approach, Xu et al. (2006) developed an entrainment parameterization of form

$$E = E_0(1-(Ri/Ri_c)) \text{ for } Ri < Ri_c \text{ ; } E = 0 \text{ for } Ri > Ri_c, \qquad (5.13)$$

where $E_0 = 0.2$ and $Ri_c = 0.25$, as determined from the comparison between parameterization and "truth" simulation. In this parameterization, $Ri$, the layer Richardson number, is calculated by

$$Ri_k = \frac{\Delta \rho_k g h_k}{\rho_k \Delta U_k^2}, \qquad (5.14)$$

where $\rho_k$ is the density of the layer, $h_k$ is the layer thickness, and $U_k$ is the layer horizontal velocity.

Another recent parameterization of the shear-driven mixing in overflows is the Jackson et al. (2008) parameterization. This is written in terms of a diffusivity, making it easy to implement in height-coordinate models as well as in isopycnal models:

$$\frac{\partial^2 \kappa}{\partial z^2} - \frac{\kappa}{L_B^2} = SF(Ri_g), \qquad (5.15)$$

where $\kappa$ is the diapycnal diffusivity, $L_B$ is the buoyancy length scale at which stratification suppresses turbulence, and $S = dU/dz$ is the vertical shear of the horizontal flow. $F(Ri_g)$ is a nondimensional function of the gradient Richardson number. In the currently implemented version of this parameterization,

$$Fr(Ri_g) = \frac{0.15(1 - Ri_g/Ri_c)}{1 - 0.9Ri_g/Ri_c}, \qquad (5.16)$$

where $Ri_c$ is a critical Richardson number. In many ways, this is similar to the previous parameterization, with the dimensions of entrainment determined by the shear and the magnitude dependent on a nondimensional function of Richardson number. The main difference comes from the middle term in equation (5.15), which in practice is small in overflows and significant only in other shear-driven mixing regions. Comparison between this new scheme, other entrainment parameterizations, and direct numerical simulation of shear-driven turbulence is shown in Figure 5.10.

Neither of these shear-driven mixing parameterizations include the low-Froude-number (i.e., high-Richardson-number) mixing found by Cenedese et al. (2004) in the roll-wave regime, nor do they attempt to capture any mixing associated with eddy processes. Turbulence closure models (e.g., so-called 'k-$\epsilon$ models' employing two additional prognostic equations for turbulent kinetic energy and dissipation) have also been shown to perform well in simulating the entrainment for most overflows (Ilicak

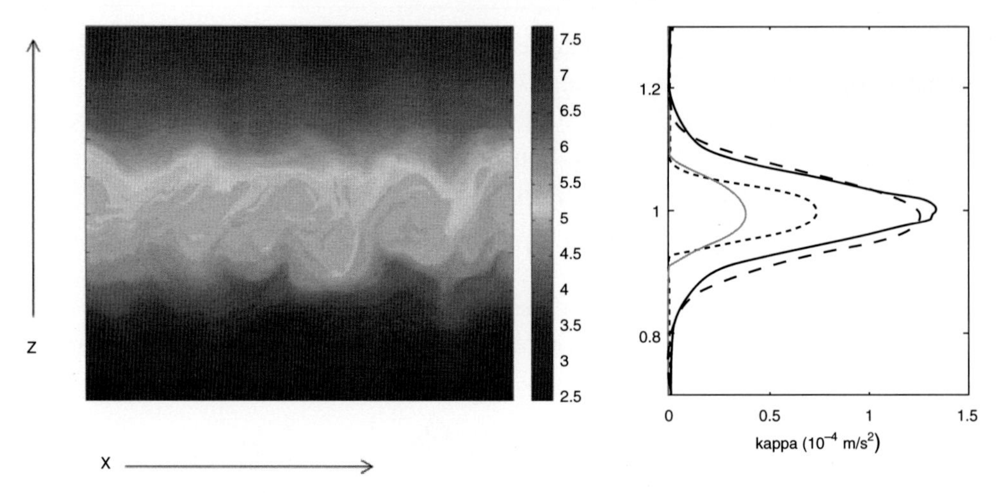


Figure 5.10. The Jackson et al. (2008) parameterization of shear-driven mixing. The left panel shows a snapshot of the temperature field from a three-dimensional, high-resolution numerical simulation; the right panel shows the diffusivity diagnosed from this simulation (solid) and compared with several different parameterizations: the new parameterization (dashed), the Ellison-Turner parameterization (gray), and the Xu et al. (2006) parameterization (dot-dashed).

et al. 2008), although their structure functions may need further modification to capture the low-Froude-number mixing regime.

Both the Xu et al. (2006) parameterization and the Jackson et al. (2008) parameterization were developed by the multi-institutional Gravity Current Entrainment Climate Process Team, whose goal was to develop parameterizations of overflow entrainment for climate models. A limitation of both schemes is that they parameterize the entrainment in terms of the resolved shear. Unresolved shear, for example that generated by small-scale topographic features, may enhance the entrainment and is the focus of ongoing investigations (e.g., Ilicak et al. 2011).

### *5.2.4 Detrainment*

When a dense current on a slope moves through a stratified environment, it is possible for fluid to detrain from the dense current (i.e., leave the boundary and flow into the interior at constant depth or following isopycnals). When ambient fluid has been mixed into the dense current, and that intermediate-density mixture reaches a depth where the ambient fluid is of the same density, the overflow water no longer experiences any buoyancy-driven acceleration downslope relative to the environment. Instead, the overflow can then flow out into the interior, leaving the slope.

A bulk view of detrainment follows from the bulk view of entrainment given previously. If all the overflow water is well mixed and any entrained ambient fluid is rapidly mixed through the whole depth of the overflow plume, then all the overflow

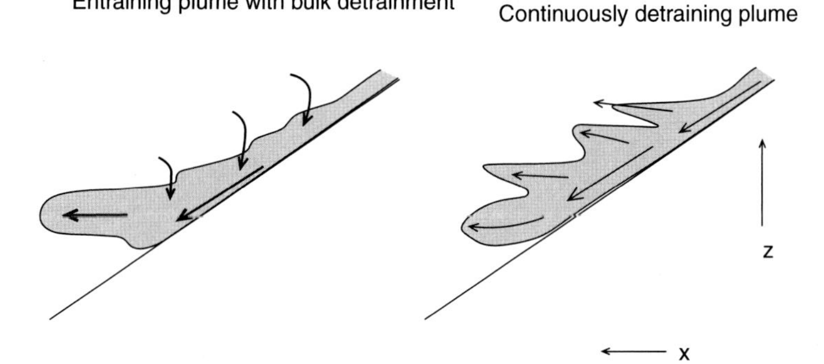


Figure 5.11. A schematic showing detrainment from descending overflows. (*Left*) entrainment over most of the slope is followed by bulk detrainment at the neutral buoyancy level. (*Right*) Detrainment of small amounts of fluid occurs continuously over most of the slope, as seen in the experiments of Baines (2001).

fluid will leave the boundary at the same density level and flow into the interior, as shown schematically in the left panel of Figure 5.11. This neutral buoyancy level will be a function of the initial density of the overflow, the entrainment that has taken place, and the environmental stratification. Diagnosing the detrainment depth can therefore be a useful method to estimate the entrainment, provided the incoming buoyancy flux and ambient stratification are known. Simulations and laboratory experiments have shown that the total entrainment, and hence the neutral buoyancy level, are approximately independent of the slope steepness (Ozgokmen et al. 2006). Steeper slopes might be expected to have a higher local entrainment, due to the higher velocities and hence lower Richardson number, but the horizontal distance traveled is greater for a shallower slope. The entrainment rate per unit of vertical descent is therefore approximately constant. The neutral buoyancy level can therefore be estimated for a two-dimensional, nonrotating entraining plume with constant entrainment rate per unit depth $E$ as

$$Z_{\text{eq}} \sim E^{-1/3} \frac{B^{1/3}}{N}, \tag{5.17}$$

where $B$ is the buoyancy flux per unit length in the 2-D plume (with units of buoyancy $\times$ velocity $\times$ height $= \text{m}^3\text{s}^{-3}$). Laboratory experiments (Wells and Nadarajah 2009) support this scaling for the neutral buoyancy level. Bulk detrainment is seen in the Red Sea outflow and Mediterranean outflow.

Alternatively if the structure of the plume varies over its depth, ambient fluid is not completely mixed into the plume. Then, as a parcel of environmental fluid is mixed into the top of the plume, it is initially denser than the surroundings and moves down with the plume. Farther down the slope, the same parcel may be neutrally buoyant and will separate from the plume. This partial mixing can therefore lead to a continuous detrainment from the plume (Baines 2001), as shown schematically in the right panel of Figure 5.11. Detrainment is particularly important if momentum loss due to bottom drag is sufficient to balance the increase in plume momentum due to buoyancy-driven acceleration (Baines 2005). Mauritzen et al. (2001) suggested some continuous detrainment may occur in the Mediterranean outflow plume, but this type of detrainment is more difficult to detect in observations.

### 5.2.5 The Frictional Bottom Boundary Layer

In the presence of rotation, bottom friction plays an important role in allowing down-slope flow in a thin Ekman layer. Even without rotation, bottom friction modifies the plume structure. In particular, in the frictionally controlled boundary layer, the down-slope velocity is reduced, leading to strong shear (as shown in Figure 5.9). In this region, tracer properties are homogenized by mixing, in contrast to the interfacial shear layer where a vertical gradient in tracer properties persists. This bottom layer can therefore be a significant sink of kinetic energy, while having less effect on mixing than the interfacial layer (especially if the two zones are separated by a thick overflow layer, as in the Red Sea overflow, Peters et al. 2005b).

It is important to properly capture the homogenization of tracers induced by the mixing driven by frictionally generated shear. One such parameterization, developed by Legg et al. (2006) as part of the gravity current Climate Process Team (CPT), estimates the energy lost to bottom drag and then uses a fraction of that energy to raise the potential layer adjacent to the boundary by mixing. Without this parameterization in an isopycnal model, little mixing would occur near the bottom of a thick overflow, and the overflow would unphysically split into a dense unmixed layer continuing down to the bottom and an interfacial zone modified by mixing, detraining at an intermediate depth. This parameterization has an important impact on the ability of an isopycnal global circulation model to represent the Mediterranean overflow. Without including the frictional bottom boundary layer mixing, the model Mediterranean overflow splits into two branches, including a spurious plume that descends to the bottom of the Atlantic with little mixing. With the new parameterization, this spurious bottom plume is eliminated (Legg et al. 2009).

### 5.2.6 Inhomogeneities Across the Overflow Plume

Entrainment, especially in a rotating overflow, can be significantly modified by lateral variations across the plume, coupled with circulation in the cross-channel direction. For dense currents of small thickness relative to the Ekman layer thickness, in a channel of width near the deformation radius and low bulk Froude number, Umlauf et al. (2010) found that fluid is initially entrained into the interfacial layer, but it is then advected by the transverse circulation into the deeper part of the plume. A lateral stratification exists on the left-hand side of the plume (looking upstream) and the dense fluid layer is tilted in this direction because of rotation. An upslope, frictionally driven circulation exists in the left-hand boundary layer. The net entrainment therefore depends not only on the gradient Richardson number at the interface, but also on the effectiveness of the lateral circulation in removing mixed fluid from the interfacial region. The importance of lateral variations in entrainment and transverse circulation is also seen in observations from the Faroe Bank Channel (Fer et al. 2010).

### *5.2.7 Summary*

For nonrotating plumes, the concept of bulk entrainment, depending on the bulk Froude number, has proven to be a useful predictor of the increase in transport and dilution of tracer properties in a dense current descending a slope. However, with the addition of rotation, entrainment becomes more complicated, particularly through the influence of Ekman layers and mesoscale eddies. Different regimes then exist in which the entrainment mechanisms vary (e.g., breaking roll waves, lateral mixing by eddies). Several parameterizations of entrainment exist, but currently no parameterization includes all the physics, particularly for low Froude numbers, or captures the full vertical structure of the plume.

## 5.3 Convectively Driven Ocean Flows

### *5.3.1 Convective Plumes*

An overflow, specifically one that can be described as a dense turbulent current descending a slope, is one example of an entraining plume in oceanography. Another example of entraining plumes is the individual coherent structures of open-ocean convection forced by surface buoyancy loss. Here we will examine these vertically oriented entraining plumes and consider how their structure depends on the size of the forcing area, the influence of rotation, and the influence of ambient stratification.

Oceanic convection is initiated by heat loss to the atmosphere in regions where cold wintertime air blows over an ice-free ocean. The deepest ocean convective layers are found in regions where stratification is relatively weak, and water is trapped within a cyclonic gyre-scale circulation. Examples include the Labrador Sea, the Irminger Sea, and the Greenland Sea (Marshall and Schott 1999; Pickart et al. 2002). Deep convection can also occur at lower latitudes when density is modified by evaporation as well as heat loss (e.g., in the Mediterranean). This chapter focuses on the smaller scales of oceanic convection; Chapter 3 by Spall (this volume) considers the interaction between convective circulation and the gyre scale.

Convective plumes are initiated at the surface thermal boundary layer (the thin layer where diffusive processes dominate advective processes). If we ignore the details of the boundary layer flow and consider how a dense parcel of fluid develops as it descends, then we can write the following entraining parcel equations (Turner 1986):

$$\begin{aligned} \frac{d}{dt}V &= V_e S && \text{conservation of mass,} \\ \frac{d}{dt}V\overline{W} &= V\overline{g'} && \text{conservation of momentum,} \\ \frac{d}{dt}V\overline{T} &= V_e T_E S && \text{conservation of temperature,} \end{aligned} \tag{5.18}$$

where $V$ is the volume of the blob, $S$ is the surface area, $W$ is the vertical velocity, $g'$ is the buoyancy anomaly or reduced gravity of the thermal, $T$ is the temperature, and $T_E$ is the temperature of entrained fluid. $V_e$ is the entrainment velocity, the velocity at which the boundary of the thermal expands outward because of entrainment of quiescent fluid into the turbulent region. $\overline{q}$ represents the average of a quantity $q$ over the whole turbulent region. We will assume a linear equation of state so that $g' = g\epsilon T'$, where $\epsilon$ is the coefficient of thermal expansion and $T'$ is the temperature anomaly.

For parcels descending vertically, entrainment does not have to overcome the effect of gravity (e.g., lighter ambient fluid may be entrained from the side or from below). The entrainment rate is therefore not dependent on Richardson number/Froude number, as it was in the case of the gravity current (where entrainment always pulls lighter fluid downward). The entrainment velocity is therefore parameterized as

$$V_e = \alpha W, \tag{5.19}$$

where the nondimensional entrainment coefficient $\alpha$ is assumed constant (and has been found from observations and laboratory experiments to have a value of about 0.1–0.2).

If we assume the parcel has a self-similar spherical shape (i.e., does not change its shape as it descends) and descends through a homogeneous environment, as shown in Figure 5.12, left panel, then for an initial buoyancy $V\overline{g'} = Q$, the solutions to the preceding equations are of the form

$$r \sim \alpha z\,; \quad W \sim \left(\frac{Q}{\alpha^3 z^2}\right)^{1/2}\,; \quad g' \sim \frac{Q}{\alpha^3 z^3}. \tag{5.20}$$

If, instead of an isolated parcel, we have a steady plume, supplied with constant buoyancy flux, the governing entraining plume equations are

$$\begin{aligned} \frac{d}{dz}WA &= V_e l && \text{conservation of mass,} \\ \frac{d}{dz}A\overline{W}^2 &= A\overline{g'} && \text{conservation of momentum,} \\ \frac{d}{dz}WA\overline{T} &= V_e T_E l && \text{conservation of temperature,} \end{aligned} \tag{5.21}$$

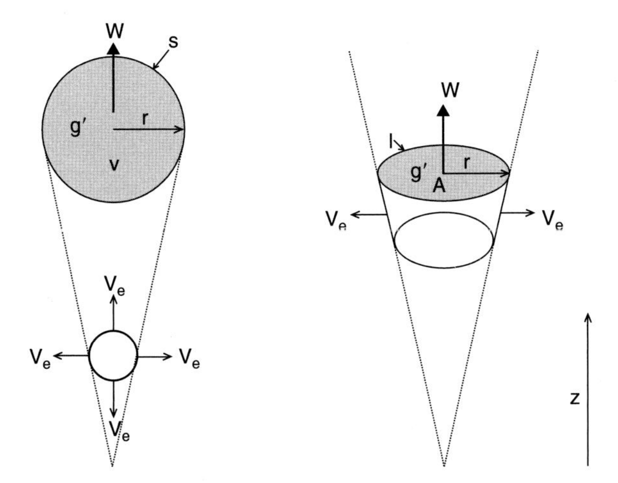


Figure 5.12. A schematic showing the classical entraining thermal (*left*) and steady entraining plume (*right*) described in detail in Turner (1986). The thermal, assumed to be spherical with radius $r$, entrains fluid with entrainment velocity $V_e$ over the surface $S$. The circular plume with radius $r$ entrains over the perimeter length $l$. Both have vertical velocity $W$ and buoyancy anomaly $g'$.

where $A$ is the cross-sectional area of the plume, and $l$ is the length of the bounding perimeter of the plume (see Figure 5.12, right panel). For a circular plume forced by a point source with total buoyancy flux $F_0$, (which has units of buoyancy × velocity × area, i.e., $\mathrm{m^4\,s^{-3}}$) solutions are of the form

$$r \sim \alpha z\,;\; W \sim \left(\frac{F_0}{\alpha^2 z}\right)^{1/3}\;;\; g' \sim \left(\frac{F_0^2}{\alpha^4 z^5}\right)^{1/3}, \tag{5.22}$$

where $r$ is the radius of the thermal, and $z$ is the vertical distance the thermal has traveled.

These classical plume solutions, dating from more than 50 years ago, have been widely compared against laboratory experiments and observations of plumes (e.g., smoke plumes) (Turner 1986).

In the ocean, of course, the source of buoyancy is not a point but rather is distributed over a finite area. If surface buoyancy flux varies on length scales much larger than the depth of the convecting layer, then that convecting area will be occupied by many plumes. The relevant forcing parameter is then $B_0$, the surface buoyancy flux per unit area (units of buoyancy × velocity, i.e., $\mathrm{m^2\,s^{-3}}$). For a plume that extends down to a distance $h$, the buoyancy flux entering that plume is therefore $F_0 \sim B_0\alpha^2 h^2$, as shown in Figure 5.13, and so the similarity solutions for a convecting layer of depth $h$ are of the form

$$W \sim (B_0 h)^{1/3}\;;\; g' \sim \left(\frac{B_0^2}{h}\right)^{1/3}. \tag{5.23}$$

This assumes that the plumes are always able to expand through entrainment as they descend. Numerous laboratory and numerical simulations have shown that in a rotating system, lateral expansion through entrainment is limited once the plume Rossby number becomes small (Jones and Marshall 1993; Maxworthy and Narimousa 1994). The depth $h_{\mathrm{rot}}$ at which this occurs can be estimated by estimating the Rossby number for a plume at depth $h$

$$Ro(h) \sim \frac{W(h)}{fr(h)} \sim \left(\frac{B_0}{f^3h^2}\right)^{1/3}. \tag{5.24}$$

When $Ro(h) = 1$, then

$$h = h_{\mathrm{rot}} \sim \left(\frac{B_0}{f^3}\right)^{1/2}. \tag{5.25}$$

The plume is then prevented from expanding further beyond a width of $l_{\mathrm{rot}} \sim h_{\mathrm{rot}}$ as it descends, as indicated schematically in Figure 5.13, and the velocity scale will then have the form

$$W_{\mathrm{rot}} \sim (B_0 h_{\mathrm{rot}})^{1/3} \sim \left(\frac{B_0}{f}\right)^{1/2} \tag{5.26}$$

This rotational control on the plume width will occur only if $h_{\mathrm{rot}} < H$, where $H$ is the depth of the fluid, or the depth of the convecting layer. This criterion for rotational control can be written as $Ro_c < 1$, where $Ro_c$, the convective Rossby number, is given by

$$Ro_c = \left(\frac{B_0}{f^3H^2}\right)^{1/3}. \tag{5.27}$$

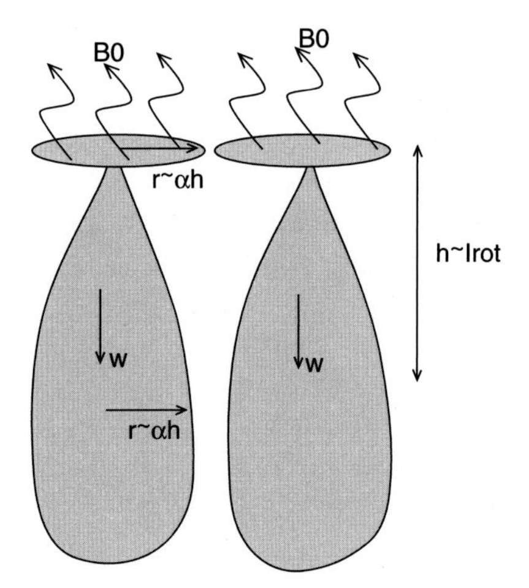


Figure 5.13. A schematic showing multiple plumes generated from a distributed surface buoyancy loss $B_0$. Plume expansion through entrainment occurs until a depth $h_{rot}$ where the local Rossby number is of order 1.

$Ro_c$ can be considered as the ratio between the Coriolis time scale and the convective time scale. Rotation is important if the Coriolis time scale is less than the time taken for a convective element to cross the convecting layer.

Laboratory and numerical experiments have verified the scaling for $h_{rot}$ and $W$ given above (Jones and Marshall 1993; Maxworthy and Narimousa 1994).

For oceanic parameters, we can calculate $h_{rot}$ as follows: If $B_0 = 10^{-7}\text{m}^2\text{s}^{-3}$, and $f = 10^{-4}\text{s}^{-1}$, then $h_{rot} \sim 10^{2.5}$ m. Oceanic convection is therefore rotationally controlled when the convective layer depth is more than several hundred meters. This occurs only in a few regions (e.g., the Labrador Sea, the Greenland Sea, and the Gulf of Lyons in the Mediterranean). Shallow ocean convection (e.g., on shelves or in shallow mixed layers) is not rotationally controlled (due to small $H$) and neither is atmospheric convection (due to large $h_{rot}$ resulting from large $B_0$).

The rotational constraints on plume expansion can be explained on energetic and angular momentum conservation grounds: With rotation, plumes spin-up cyclonic vorticity through the convergence of the surface layer fluid into the plume (Julien et al. 1996a). For a large radius plume, the energy required to spin-up this vortical circulation exceeds the energy available through conversion of available potential energy. Hence, the plume is energetically constrained to have a maximum width. The cessation of expansion as a plume descends could imply no further entrainment and hence no further change in $g'$. Alternatively, it could imply entrainment balanced by detrainment, leading to a continuing change in $g'$ as the plume descends. Analysis of numerical simulations for Rayleigh-Benard convection shows that exchange – entrainment balanced by detrainment – dominates in rotating convection (Julien et al. 1999). A probable cause of the detrainment is the vortical motion associated with the plumes, which leads to filaments being drawn off the plumes and mixed with ambient fluid.

In an oceanic scenario the plumes will descend until either they encounter the ocean floor, or (more likely) they reach neutral buoyancy (which occurs if the convective layer is bounded below by a stably stratified region). When the plume passes through its neutral buoyancy level, it will have finite momentum and therefore continue downward. However, it will now experience a deceleration, not an acceleration, as it is now more buoyant than the surroundings. At some finite depth below the level of neutral buoyancy, the plume will no longer descend. In this region of overshoot, plumes can sweep fluid from below into the convective layer; this entrainment of denser fluid will result in a negative buoyancy flux. When convection driven by a surface buoyancy loss into a stably stratified fluid results in a negative buoyancy flux at the base of the convecting layer, this is termed "penetrative convection." The negative buoyancy flux is possible only because of the coherent structures accelerating over the depth of the convective layer. For nonpenetrative convection, we can predict the increase in the depth $h$ of the convective layer with time $t$ on energetic grounds.

$$\frac{dh}{dt} = \frac{B_0}{N^2 h} \rightarrow h \sim \left(\frac{2B_0 t}{N^2}\right)^{1/2} \tag{5.28}$$

where $B_0$ is the buoyancy flux and $N^2 = -(g/\rho_0)\partial\rho/\partial z = \partial b/\partial z$, the buoyancy gradient, and $b$ is the buoyancy. For penetrative convection, the deepening will occur at a faster rate.

Ultimately, for oceanic convection, we are interested in the cumulative effect of many plumes on features such as the heat flux and the temperature/salinity/buoyancy stratification. Within a large convective area, there are many downwelling plumes, separated by more diffuse upwelling, so there is no net vertical mass flux due to convective plumes. To estimate the total heat flux and tracer properties, we should sum over the plume ensemble and the ambient upwelling fluid. This method is the basis of several atmospheric convection parameterizations (Arakawa and Schubert 1974; Fritsch and Chappell 1980; Donner 1993), and recent ocean convection parameterizations (Paluzkiewisz and Romea 1997; Canuto et al. 2007).

A typical nonrotating convective layer, shown in Figure 5.14, has (1) a thermal boundary layer at the surface in which the convective buoyancy flux is zero, and the diffusive buoyancy flux is large; (2) a deep unstratified region in which the convective heat flux is large and decreases approximately linearly with height; (3) an overshoot region in which stratification is stable and enhanced over the preexisting stratification value, and the buoyancy flux reverses sign, with a maximum magnitude of about 0.2 the surface value; and (4) the stably stratified region below as yet unmodified by convection.

With the addition of rotation, the convective layer has a mean negative vertical buoyancy gradient so it is marginally unstably stratified, instead of being unstratified as for nonrotating convection. This weak negative buoyancy gradient is a consequence of the reduced vertical transport of buoyancy in the presence of rotation, due to the

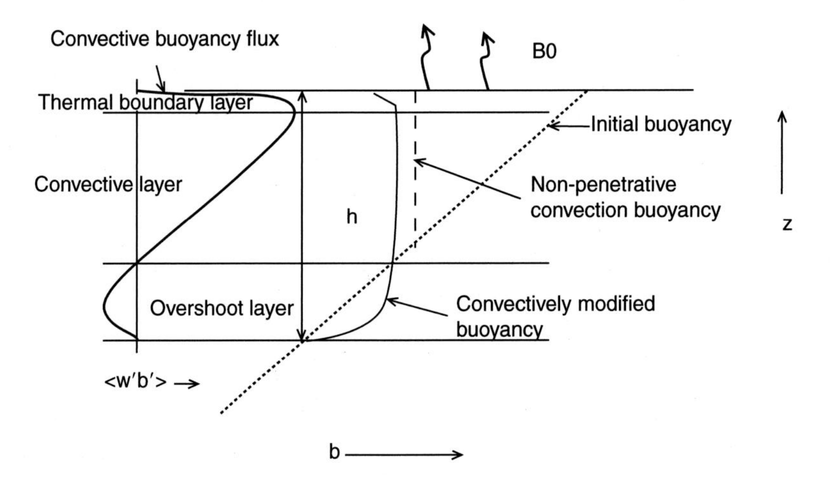


Figure 5.14. A schematic of the buoyancy profiles generated by penetrative and nonpenetrative convection, and the vertical convective buoyancy flux associated with penetrative convection.

detrainment from the plumes. The negative buoyancy flux in the overshoot zone is also weaker for rotating convection (Levi and Fernando 2002), a consequence of reduced kinetic energy transport, due to increased dissipation in the upper convective layer (Julien et al. 1996b). Numerical simulations (Wang 2006) have shown that the negative buoyancy flux scales with convective Rossby number $Ro$ as follows:

$$B = B_{nr} Ro^{0.4}, \tag{5.29}$$

where $B_{nr}$ is the negative buoyancy flux in the nonrotating case.

When the convective layer has no buoyancy gradient, as for nonrotating convection, then the buoyancy flux (which is positive over this layer) is not proportional to the local buoyancy gradient. Hence, this nonlocal buoyancy flux cannot be parameterized by a simple, down-gradient eddy diffusion model

$$\kappa = -\frac{\overline{w'b'}}{\partial b/\partial z} \to \infty \text{ if } \frac{\partial b}{\partial z} \to 0, \tag{5.30}$$

where $\kappa$ is the eddy diffusivity, $w'$ and $b'$ are the vertical velocity and buoyancy anomalies, relative to the horizontal means, and $\overline{q}$ represents an average of quantity $q$ over a large horizontal scale. This nonlocal nature of convective buoyancy fluxes is a key feature of many convective layer parameterizations (e.g., KPP, Large et al. 1994) and is a result of the coherent plumes that cross the convective layer from top to bottom. However, in the rotating case, where the vertical transports are less efficient and a weak negative buoyancy gradient persists in the convective layer, an eddy diffusivity can be used to parameterize convective mixing.

### *5.3.2 Horizontal Inhomogeneities in Convective Flows*

To this point, we have considered a horizontally homogeneous ensemble of convective plumes. Now we will consider how this ensemble behaves when there are horizontal inhomogeneities caused by either the surface forcing or the preexisting stratification.

#### *5.3.2.1 Localized Buoyancy Forcing*

If buoyancy forcing is localized but has a horizontal length scale considerably larger than the plume scale, then within the region of forcing the plumes are largely unaffected by the forcing inhomogeneity. The buoyancy forcing is communicated to the interior fluid so that a horizontal density gradient between the interior and the uncooled exterior fluid becomes established. The nature of the response to this density gradient is dependent on whether or not rotation is important. Without rotation, a gravity current would carry the denser fluid radially outward, either as an intrusion at the neutral buoyancy level or at the bottom for weak stratification. With rotation, a baroclinic

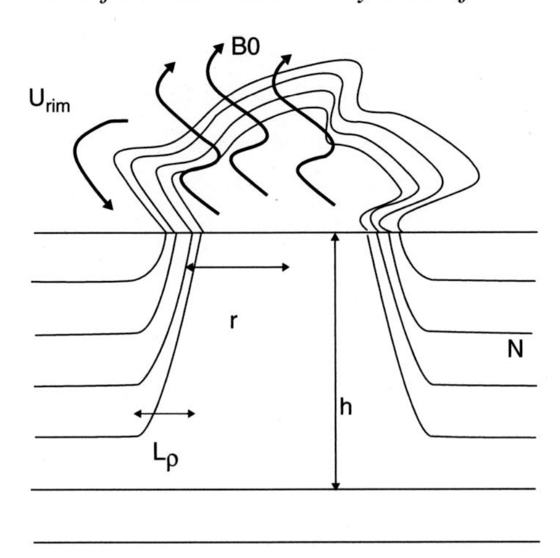


Figure 5.15. A schematic of convective mixing and geostrophic circulation generated by buoyancy loss $B_0$ confined to a localized region of radius $r$. Within the region beneath the cooling, the initial stratification $N$ is eroded, with a deepening convective layer of depth $h$.

geostrophic current is established, as shown in Figure 5.15, with current magnitude

$$U_{\text{rim}} \sim \frac{gh}{f\rho_0}\frac{\Delta\rho}{L_\rho} \tag{5.31}$$

from thermal wind balance, where $\Delta\rho$ is the density anomaly of the cooled region, and $L_\rho$, the deformation radius, is given by

$$L_\rho \sim \frac{1}{f}\left(\frac{g\Delta\rho}{\rho_0}h\right)^{1/2}. \tag{5.32}$$

Lateral transfer of dense fluid can then occur through instabilities of this current (Legg and Marshall 1993; Maxworthy and Narimousa 1995). The growth rate of the instability is a function of the density gradient, which increases as the buoyancy forcing progresses. However, eventually an equilibrium state can be established in which the lateral fluxes of buoyancy due to baroclinic instability balance the surface fluxes; then the density anomaly of the cooled region no longer increases. In this steady state, we have

$$\int_{A_1} \frac{B_0\rho_0}{g} dA = \int_h \int_{l_1} u'\rho' dl dz, \tag{5.33}$$

where $A_1$ is the area of the cooling region, $l_1$ is the bounding perimeter of the cooling region, and $u'$ and $\rho'$ are the velocity and density anomalies with respect to the mean around the perimeter.

If we assume that $< u'\rho' >$ can be parameterized as $CU_{\text{rim}}\Delta\rho$, where $C$ is a nondimensional constant, and $U_{\text{rim}}$ is the geostrophic rim velocity, then we can solve for the density anomaly $\Delta\rho_{eq}$ at which an equilibrium is reached. For localized convection forced by surface buoyancy loss over a disc of radius $r$ into a stratified fluid with buoyancy frequency $N$, we then have

$$h_{\text{max}} \sim \frac{(B_0 r)^{1/3}}{N}; \Delta\rho_{\text{eq}} \sim \frac{\rho_0 N}{g}(B_0 r)^{1/3}\ ;\ L_\rho \sim \frac{(B_0 r)^{1/3}}{f} \tag{5.34}$$

(provided $h_{\text{max}}$ is less than the total water depth). For convection into an unstratified fluid of depth $H$ we have

$$\Delta\rho_{\text{eq}} \sim \frac{\rho_0}{gH}(B_0 r)^{2/3}\ ;\ L_\rho \sim \frac{(B_0 r)^{1/3}}{f} \tag{5.35}$$

A characteristic feature of localized deep convection is the development of rim current eddies, comprising newly convected water, on a scale $L_\rho$. These eddies serve to export dense water out of the convecting patch, and the return flow of less dense water restratifies the convecting region (Legg et al. 1996; Visbeck et al. 1996).

#### *5.3.2.2 Convection in the Presence of Lateral Buoyancy Gradients*

In a rotating flow, a horizontal buoyancy gradient $\partial b/\partial y = \alpha$ always implies vertical shear through the thermal wind relation: $\partial U/\partial z = -\alpha/f$. The meridional pressure gradient is in balance with the zonal flow. For displacements that preserve the zonal nature of the flow, the absolute zonal momentum $m = U - fy$ is a conserved quantity. If this scenario is forced by a convective buoyancy loss, it will be subject to instability if either $N^2 < 0$, along constant absolute momentum surfaces, or $f + \zeta < 0$, normal to constant buoyancy (i.e., isopycnal) surfaces, are satisfied independently. (Note that surfaces of constant absolute momentum are not vertical if $\alpha \neq 0$).

These criteria can be combined to give a new criterion for instability:

$$Q = (f + \zeta).\nabla b < 0, \tag{5.36}$$

where $Q$ is the potential vorticity and $\zeta$ is the relative vorticity. Therefore, for a convective forcing into an ocean with a horizontal buoyancy gradient, the mixing does not occur in the vertical direction but rather in the direction aligned with angular momentum. Isopycnals become aligned with angular momentum surfaces, not with the vertical, as shown in Figure 5.16. Hence, finite $db/dz > 0$ persists, even though $\overline{w'b'} > 0$, and hence fluxes are upgradient. The convectively modified region is also deeper than it would be if mixing occurred only in the vertical direction.

These predictions for slantwise convection are valid for two-dimensional simulations (Straneo et al. 2002) and for the initial convection in three-dimensional simulations (Legg et al. 1998). However, for three-dimensional simulations, the buoyancy surfaces do not remain aligned with angular momentum surfaces over time, as a result of the breaking of zonal or azimuthal symmetry. If variations in

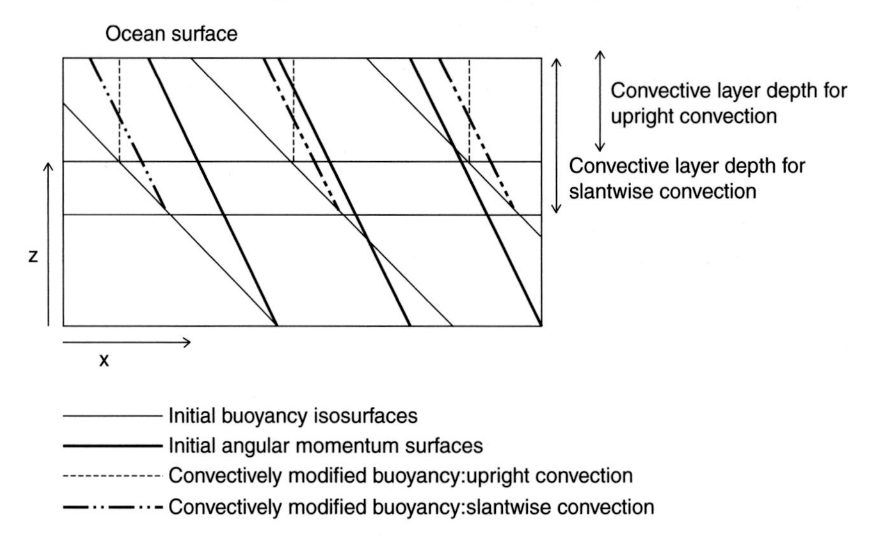


Figure 5.16. Schematic comparing buoyancy profiles generated by vertical mixing (upright convection) and mixing aligned with angular momentum surfaces (slantwise convection).

the zonal/azimuthal direction occur (e.g., through baroclinic instability), then the angular momentum constraints no longer apply, and isopycnals may slump to a position more horizontal than the angular momentum surfaces. Scenarios in which an initial, horizontally varying, stably stratified density field is convectively forced therefore proceed through an initial stage of mixing along angular momentum surfaces, followed by the development of an instability made possible by the erosion of the stratification in near-surface layers. This type of instability, known as a mixed-layer instability, differs from traditional baroclinic instability in that it involves the unstratified convectively mixed layer and exists in a parameter range of $O(1)$ Richardson number and Rossby number. Motion is highly ageostrophic: There is significant vertical motion, and eddy length scales are typically much smaller than those of more geostrophic baroclinic instability (Boccaletti et al. 2007). Buoyancy loss applied to geostrophic, mesoscale eddy fields therefore facilitates a transfer of energy to the submesoscale (Thomas et al. 2008). This mixed-layer instability can quickly restratify a convecting region. More weakly stratified regions (which may be associated with cyclonic eddies) are more susceptible to mixed-layer instabilities, and so convection tends to destroy cylonic eddies, whereas anticyclonic eddies persist (Legg and McWilliams 2001).

Since the mesoscale and submesoscale eddies play an important role in restratifying convective regions, good parameterizations of these processes are important for correctly representing convective regions in ocean models. One such parameterization (Fox-Kemper et al. 2008) is now included in several ocean-climate models.

Figure 1.1. A dust storm created by cold air flowing out from under a thunderstorm.

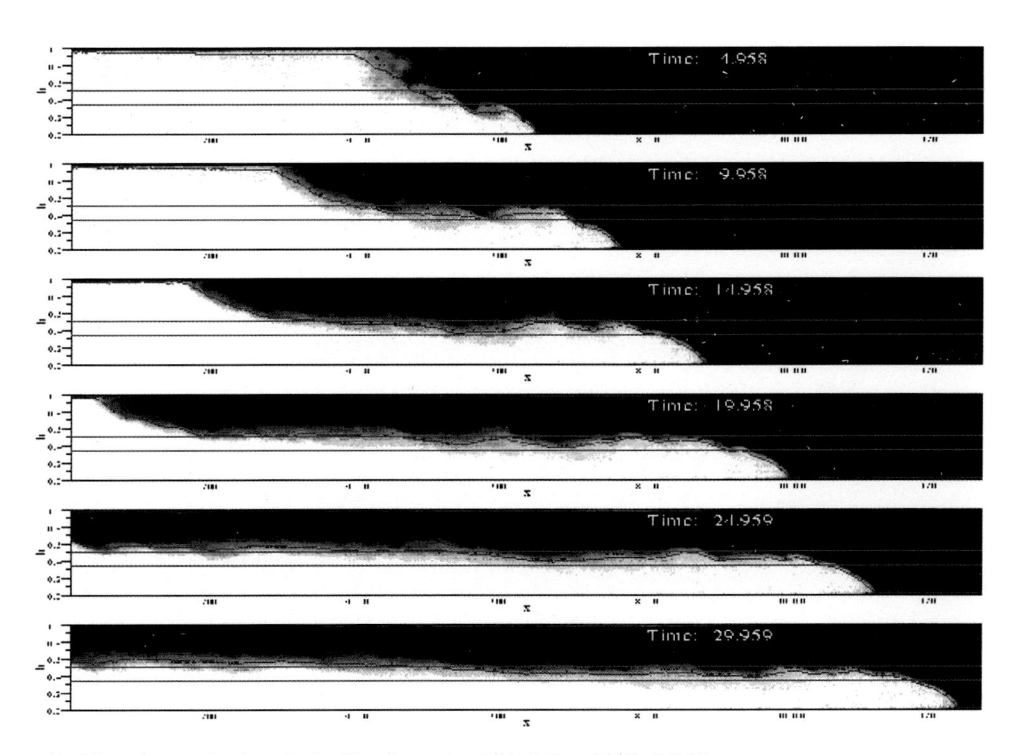


Red line shows effective depth. Blue lines give $h/H = 0.5$ and $h/H = 0.347$.

Figure 1.13. The effective depth $\overline{h}$ (Shin et al., 2004).

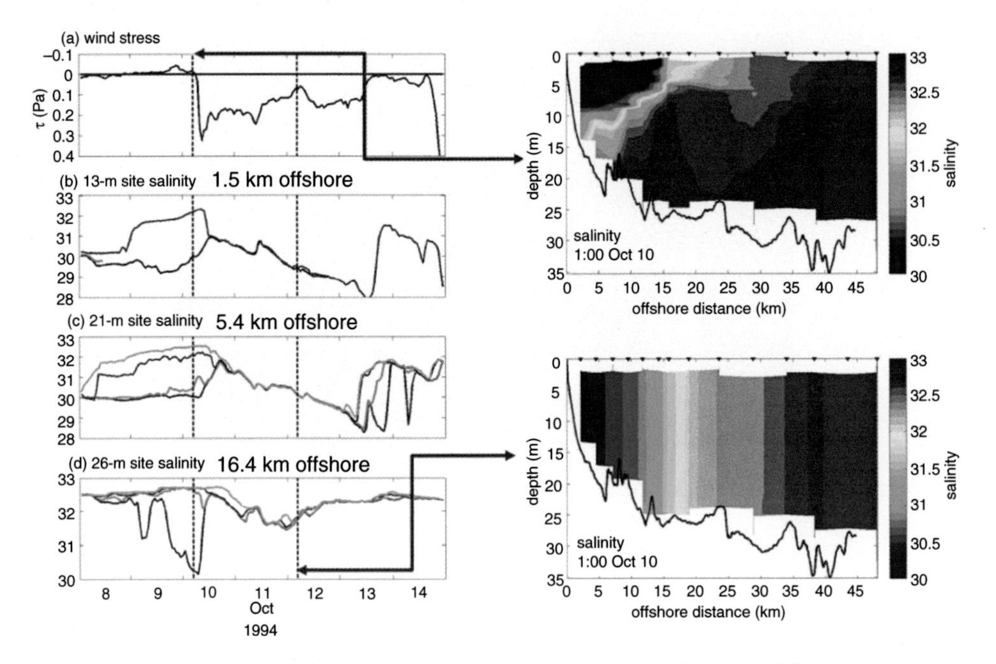


Figure 4.23. Time series of alongshelf wind stress, and salinity at different depths 1.5, 5.4, and 16.4 km offshore (*left*) and corresponding cross-shelf salinity sections prior to and after a strong downwelling wind stress event (*right*).

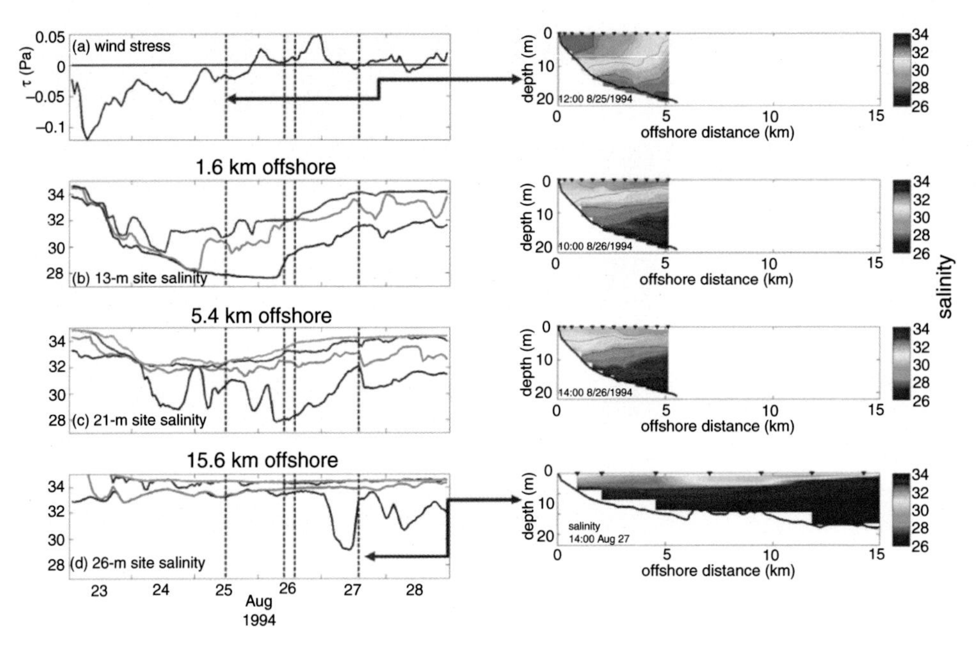


Figure 4.24. Time series of alongshelf wind stress, and salinity at different depths 1.5, 5.4, and 16.4 km offshore (*left*) and corresponding cross-shelf salinity sections prior to and after a weak upwelling wind stress event (*right*).

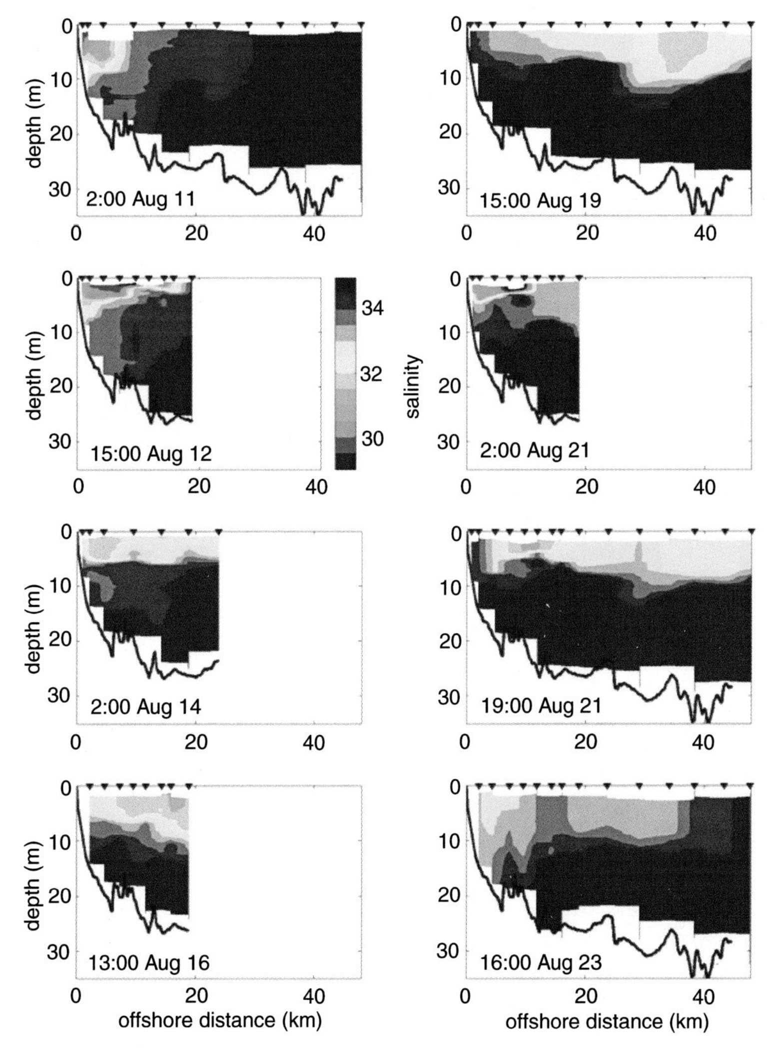


Figure 4.34. Sequence of salinity sections from shipboard hydrographic transects in August 1994. The salinity scale is shown next to the August 12 panel.

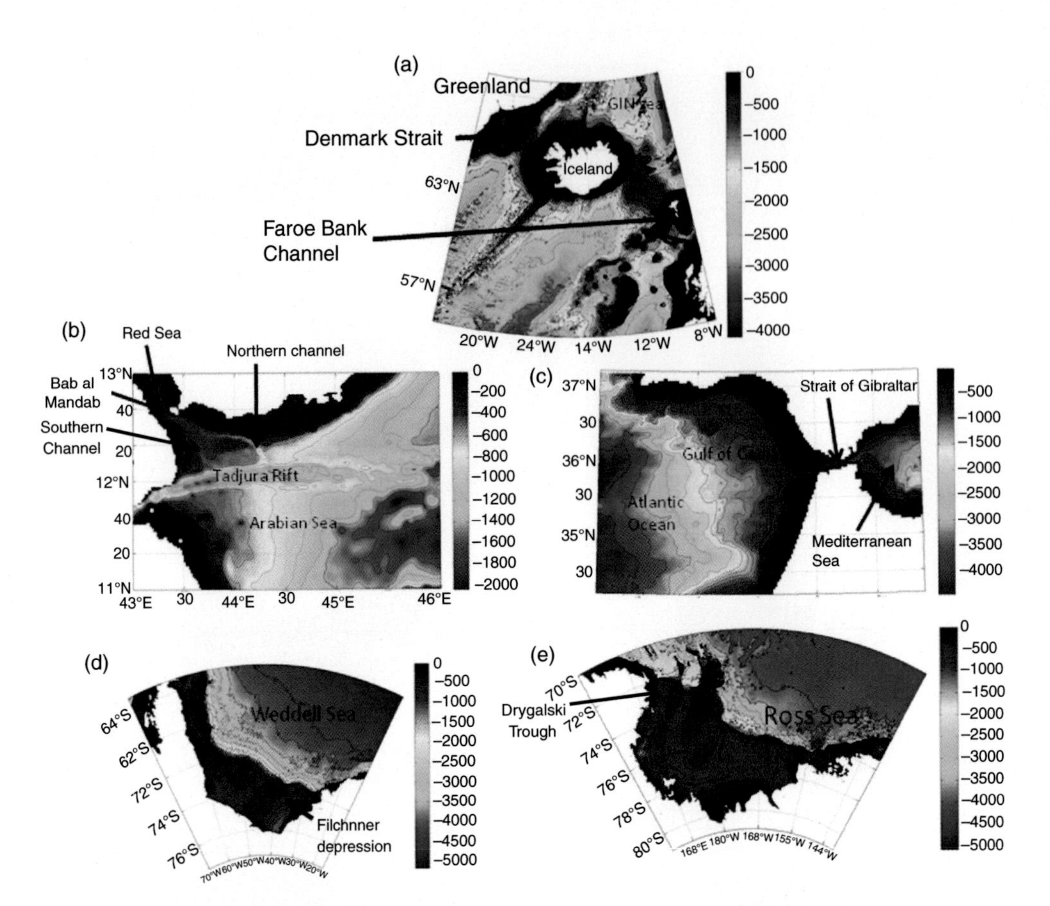


Figure 5.2. The topography of the major overflow regions: (a) Denmark Strait and Faroe Bank Channel overflows; (b) Red Sea outflow; (c) Mediterranean Outflow; (d) Weddell Sea overflow; (e) Ross Sea overflow. Color scale shows the depth in meters and is different for each location. The contour spacing is 200 m for the first 2,000 m depth, and 500 m thereafter. These images were created from the ETOPO5 database, with the assistance of Mehmet Ilicak.

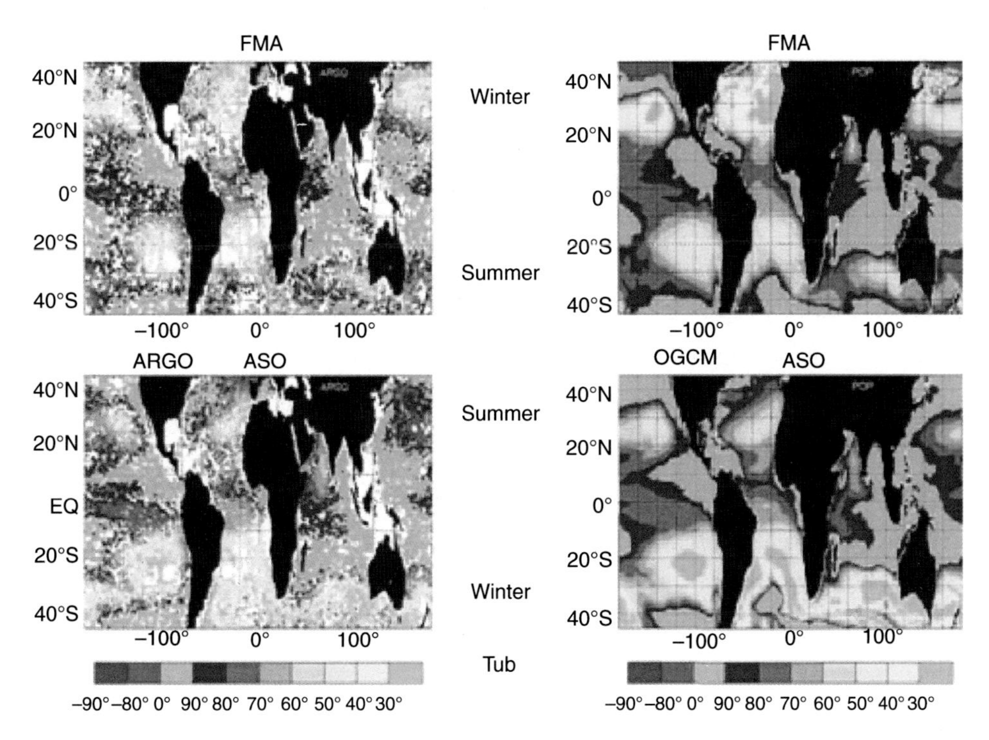


Figure 6.15. Seasonal bulk Turner angles from winter and summer in both hemispheres from an OGCM (*right*) and from individual ARGO floats (*left*). Adapted from Yeager and Large (2007).

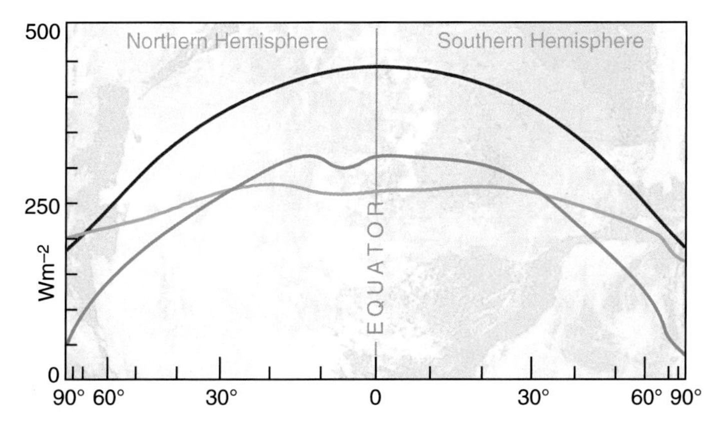


Figure 8.2. Zonally averaged radiative budget of the system Earth/Atmosphere. The red curve is the mean solar input at the top of the atmosphere. The blue curve shows the mean radiation absorbed by the system (the difference between the red and the blue curve is due to the reflection of the solar radiation). The green curve shows what is emitted back into space. The radiative budget between the blue and the green curve shows excess in the tropical regions and deficits in the midlatitudes and polar regions. From Gill (1982).

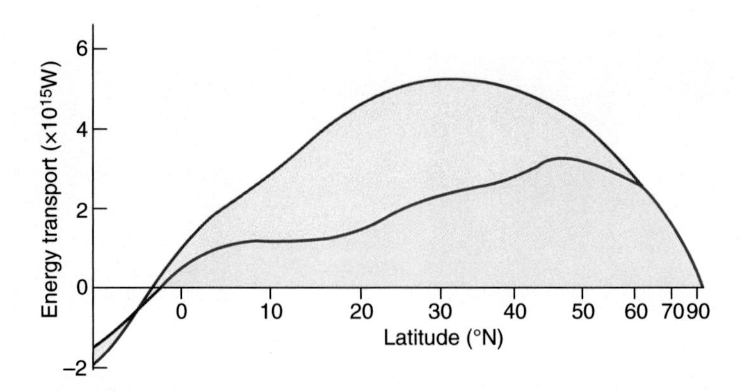


Figure 8.3. Equator/pole energy transport by the geophysical fluids in the Northern Hemisphere. The surface under the red curve is proportional to the total transport of energy necessary to balance the meridional differential heating. The green shaded area is the part transported by the oceans, and the blue shaded area is the part transported by the atmosphere. From Gill (1982).

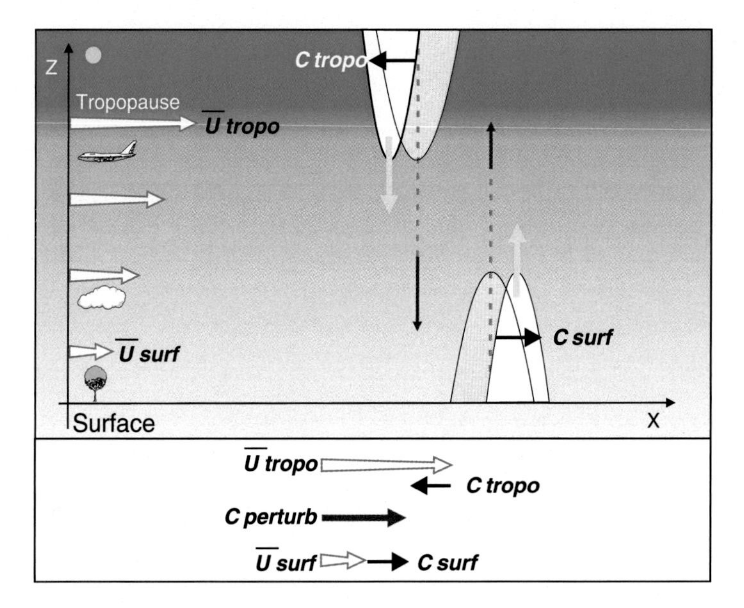


Figure 8.7. Positive baroclinic interaction between a cyclonic anomaly at the tropopause and a cyclonic anomaly at the surface. The difference of zonal advection of the system near the surface and at the tropopause due to the vertical wind shear is compensated by the propagation of the anomalies in the opposite direction. The full system propagates westward (with a characteristic speed of about 10 m/s). From Malardel (2009).

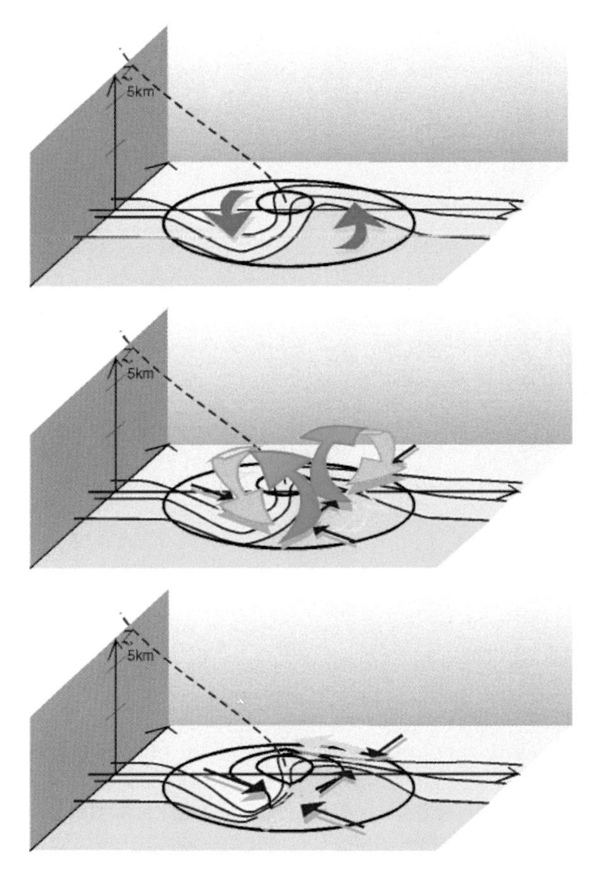


Figure 8.9. Schematic representation of the retroaction of the ageostrophic circulation on a thermal gradient in both a cold and a warm front. (*top*) The deformation of the meridional thermal gradient of the baroclinic zone starts with the advection by the quasi-geostrophic circulation around a growing cyclonic vortex (baroclinic development), (*middle*) a vertical and cross-frontal ageostrophic circulation is necessary to maintain the thermal wind balance, (*bottom*) the convergence of the ageostrophic wind reinforces the thermal gradient. However, a stronger convergent agestrophic circulation is necessary to maintain the thermal wind balance, which reinforces the gradient. Thermal quasi-discontinuities form to the southwest and northeast of the initial cyclone: the cold and warm fronts. From Malardel (2009).

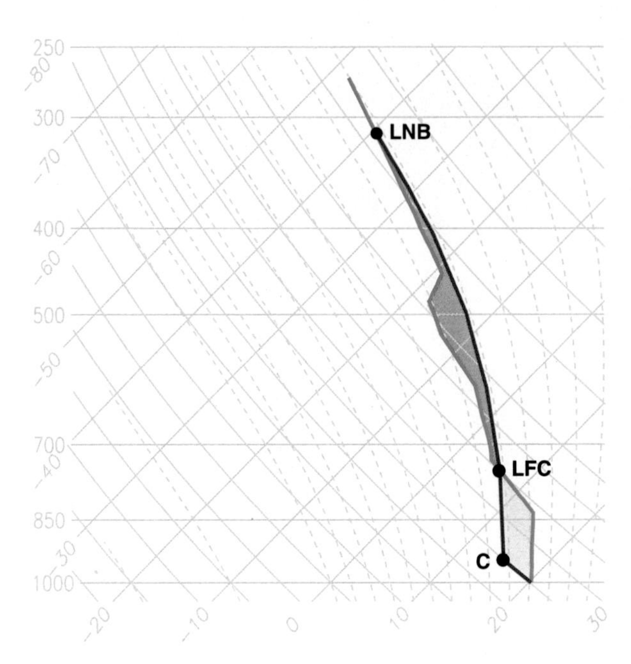


Figure 8.10. Representation of convective instability on a *p*-*T* diagram (horizontal brown lines are isobars; diagonal brown lines are isotherms). A parcel of air is lifted from an initial level of 1,000 hPa (red line). Until its level of condensation (C), the parcel follows a dry adiabat (solid green curves). Above the level of condensation, the parcel follows a saturated adiabat (dashed green curves). The temperature of the parcel decreases more slowly then because of latent heat release. Below the level of free convection (LFC), the parcel is cooler than the environment (blue line). The surface between the temperature profile of the parcel (red line) and the temperature profile of the environment (blue line) is proportional to the convective inhibition (yellow shading). Above the LFC, the saturated adiabat is warmer than the environment; the parcel is buoyant and accelerated upward until it reaches its level of neutral buoyancy (LNB). The surface between the temperature profile of the parcel and the temperature profile of the environment (orange shading) is proportional to the convective available potential energy (CAPE). From Malardel (2009).

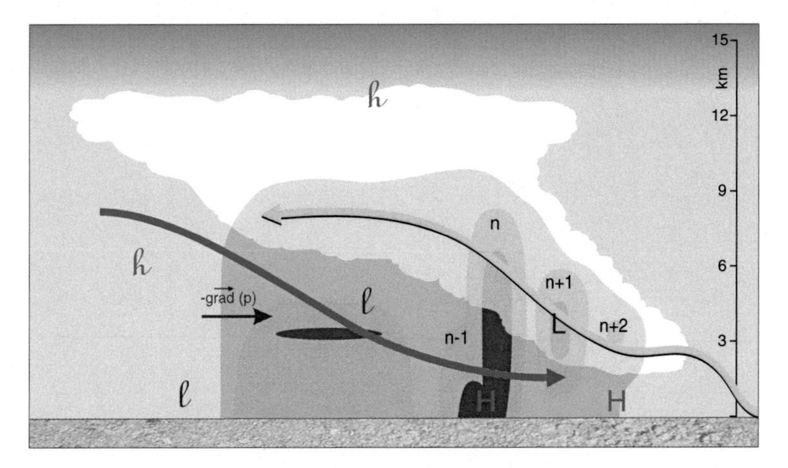


Figure 8.14. Conceptual model of a squall line. Cross-section perpendicular to the squall line. From Weisman and Przybylinski (1999).

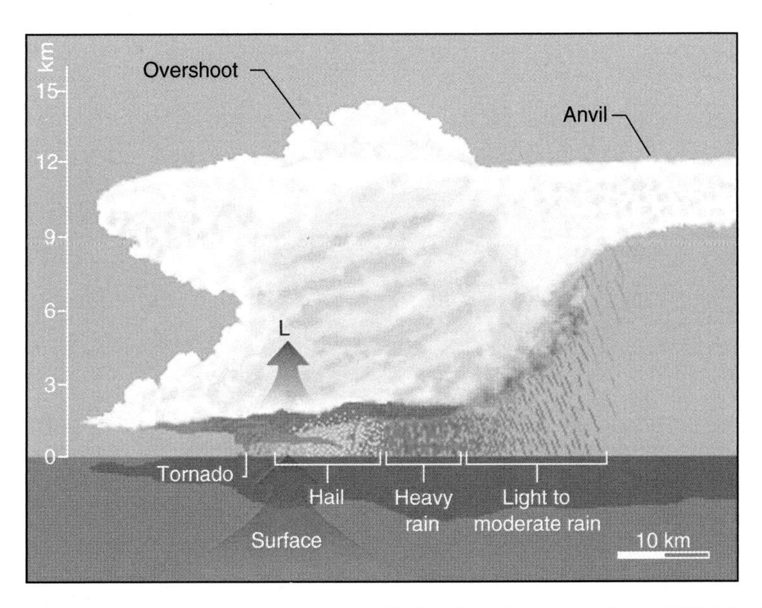


Figure 8.13. Conceptual model of a supercellular thunderstorm. From COMET (1996).

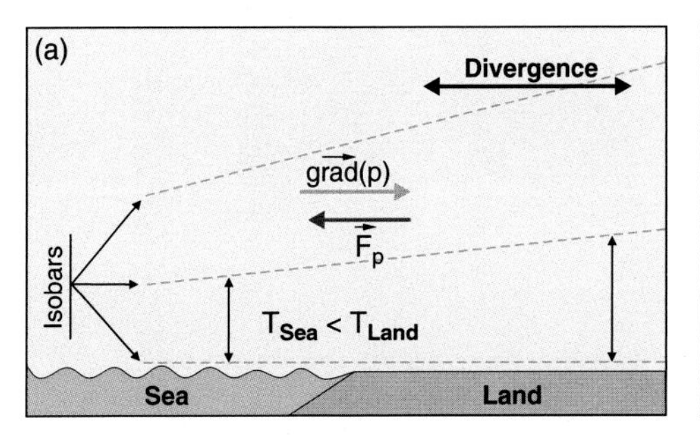


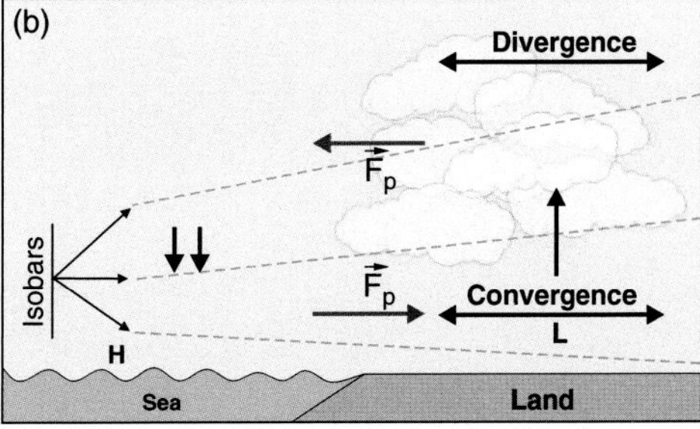


Figure 8.15. Initiation (a) and installation (b) of the land/sea breeze circulation during the day. As the air above the land surface becomes warmer than the air above the sea, the thickness between isobars above the land surface deepens. The pressure gradient force $\vec{F}_p$ is pointing toward the sea at the top of the boundary layer. Mass is then moved from the land to the sea at the top of the BL. The surface (hydrostatic) pressure in the air columns above land decreases. The pressure gradient force near the surface points toward the land, and it accelerates the air from the sea toward the land in a low-level wind called a *sea breeze*. From Malardel (2009).

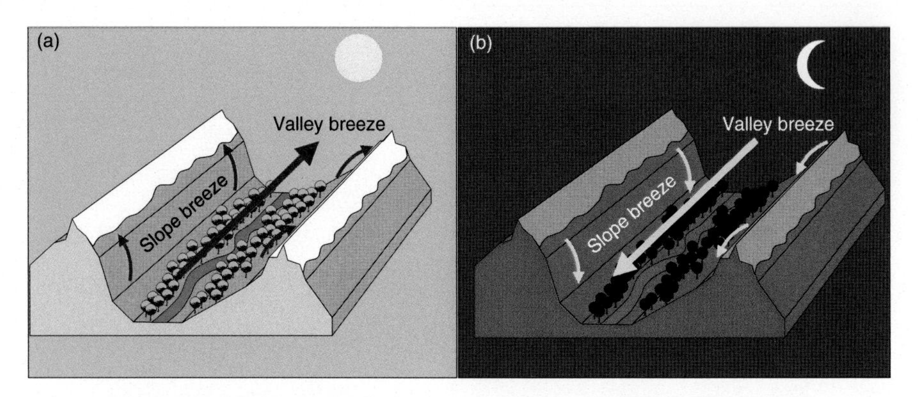


Figure 8.16. Valley and slope breezes (a) anabatic circulations during the day (b) katabatic circulations during the night. From Malardel (2009).

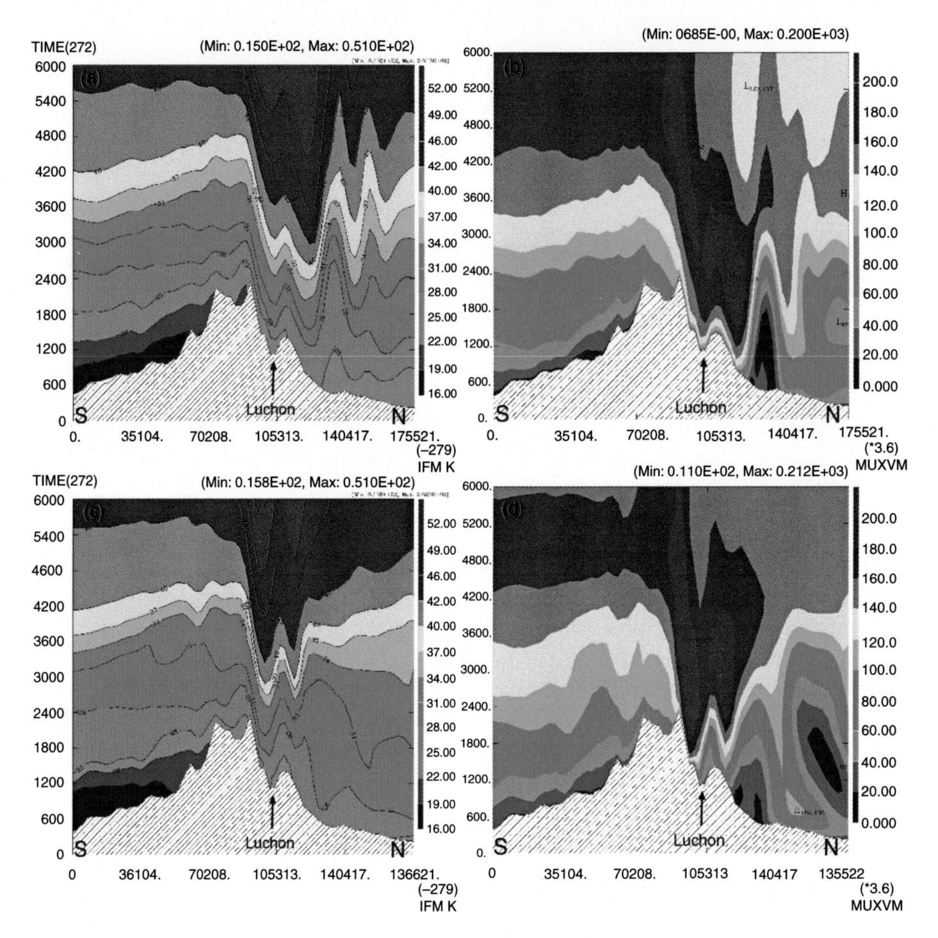


Figure 8.19. Orographic gravity waves simulated by the operational LAM Arome (Meteo-France) on 27 February 2010 on the north side of the Pyrénées. Potential temperature along a south/north cross-section of the Pyrénées: (a) nonhydrostatic simulation and (c) hydrostatic simulation. The trapped lee waves (shorter wave lengths) are filtered by the hydrostatic version. The stronger surface winds of the NH version are much more realistic. Wind intensity: (b) nonhydrostatic simulation and (d) hydrostatic simulation.

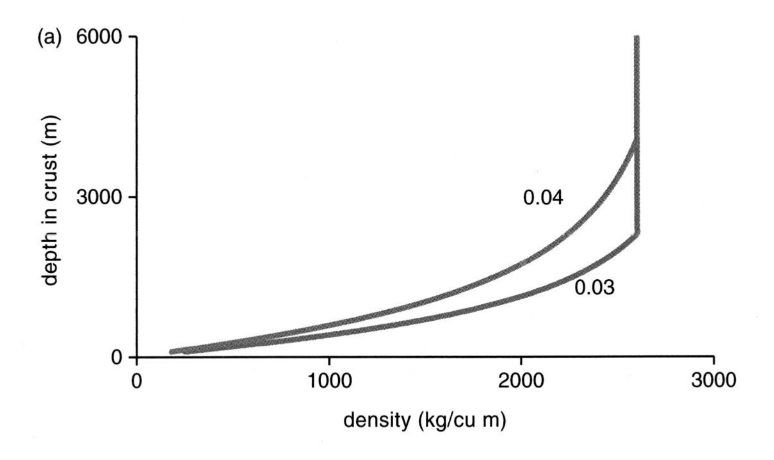


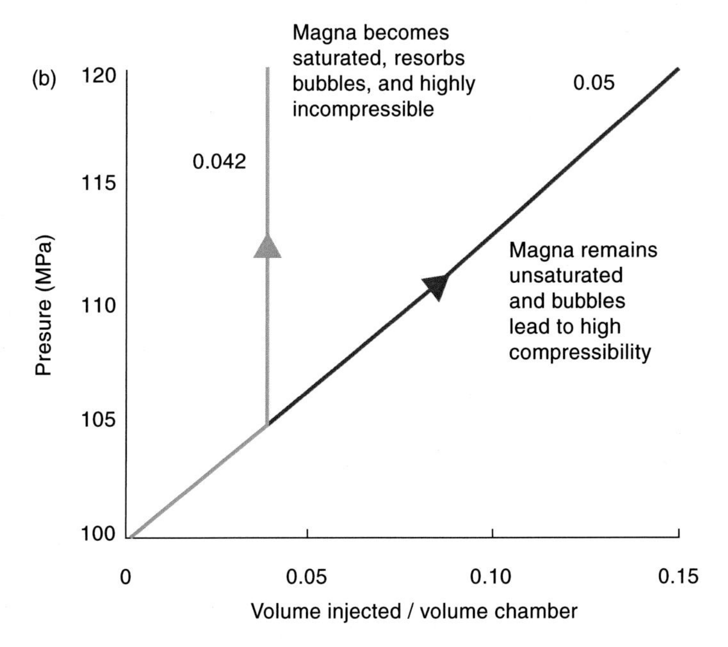


Figure 9.2. (a) Density of magma as a function of depth in the crust, and hence pressure, for magma with gas mass fraction 0.03 and 0.04, and (b) illustration of the increase in pressure of the magma reservoir as a function of the fractional increase in volume associated with injection of new magma into the magma reservoir, with the blue line showing the increase in pressure in a magma, which has no exsolved gas (total gas mass fraction 0.042) for pressures in excess of 105 MPa, where the blue line becomes nearly vertical, and the red line showing the increase pressure in a magma which still contains exsolved gas and hence is much more compressible (total gas mass fraction 0.05).

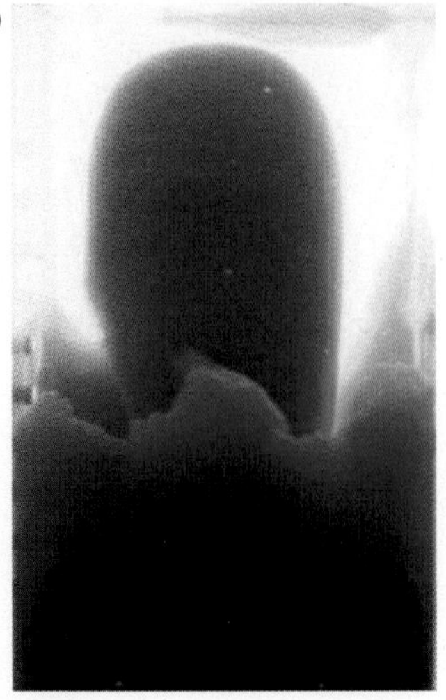

Figure 9.10. (a) The red experiment shows plumes migrating through the interface as a result of bubbles accumulating at the interface. (b) The blue experiment shows large-scale overturn as a result of bubbles in the lower layer leading to a reversal in the sign of the relative density of the two layers (adopted from Woods and Phillips 2002). (c) Illustration of a pumice erupted from Tolbachek in Kamchatka, showing both brown and white pumice intermingled in the clast.

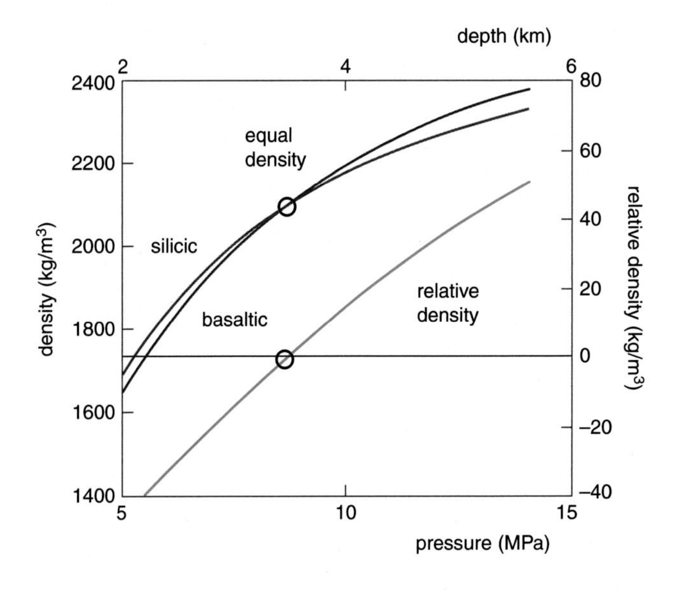


Figure 9.11. Illustration of the variation of the density of magma as a result of the decompression of a magma chamber during eruption, which can lead to overturn of the chamber. The blue and red lines show the bulk density of each of the layers (left-hand side vertical axis), whereas the green line shows the difference between the two densities (right-hand side vertical axis).

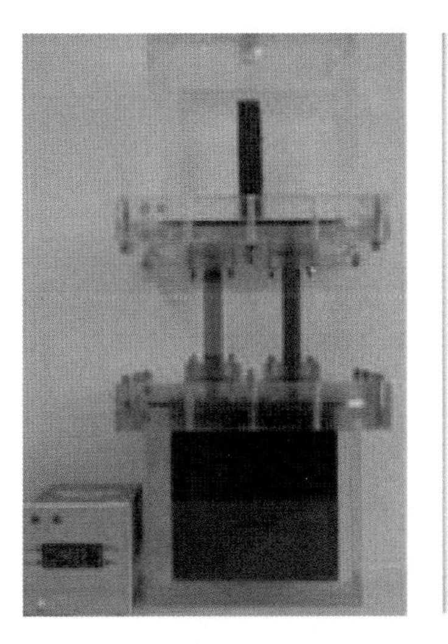

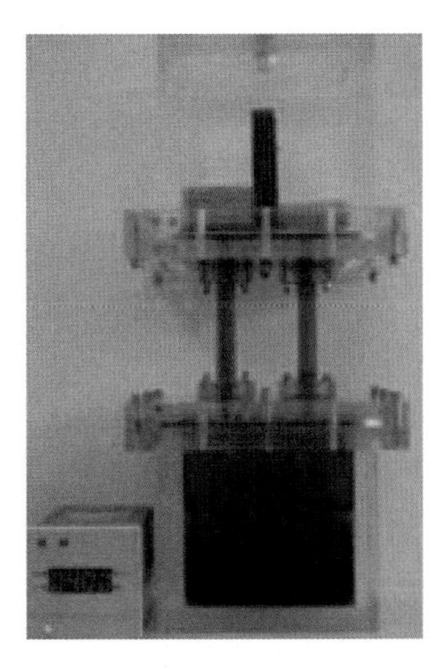

Figure 9.12. Illustration of the overturn of a magma reservoir during eruption, which leads to a change in the erupting material. In this experiment, the conduit on the right is erupting the lower layer initially (red), but following the overturn in the reservoir, the blue volatile poor layer sinks to the base of the chamber and erupts from this vent. Conversely, the conduit on the left issues the shallow fluid from the system, and therefore undergoes a transition from erupting blue to red fluid owing to the overturn.

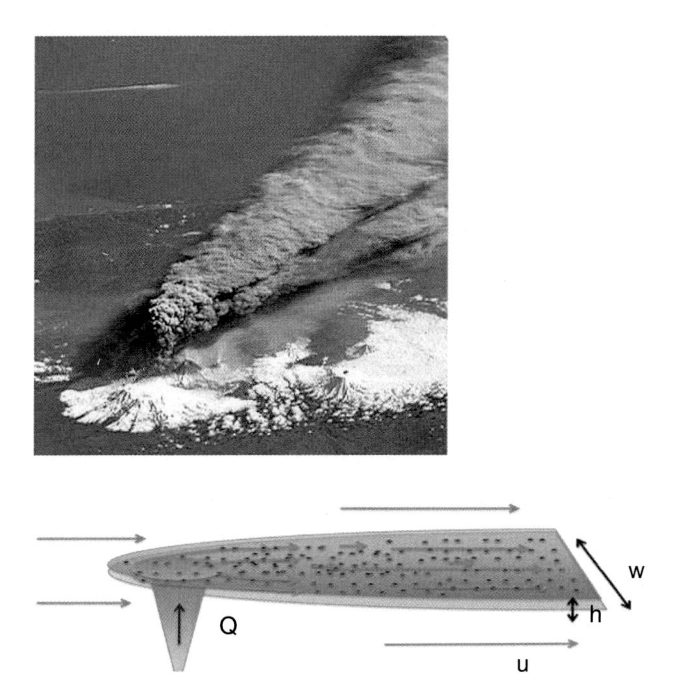


Figure 9.14. (a) Photograph from Mt. Cook, Alaska, and (b) a schematic of the dispersal of ash downwind of the eruption column.

Figure 9.16. Illustration of a small developing ash flow at the Soufriere Hills Volcano, Montserrat.

Figure 9.20. Image of two neighboring black smoker plumes, showing the turbulent nature of the particle laden, hot water.

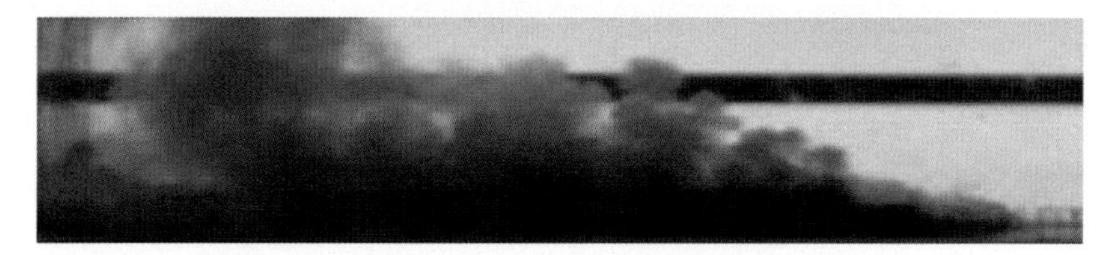

Figure 9.23. Experimental model of a $CO_2$ lake eruption triggered by a landslide. Salt solution with salt crystals is released in a tank of water containing a thin layer of lemonade at the base of the tank. On mixing with salt, the lemonade releases bubbles, and a series of convective bubbly thermal clouds rise from the flow of dense saltwater, which continues to run along the base of the tank, as a model avalanche.

Figure 10.1. (a) Debris-flow deposit in the Ravin-des-Sables watershed (France); the bucket gives a scale of the deposit thickness. (b) Debris flow on the road to la-Chapelle-en-Valgaudemar (France).

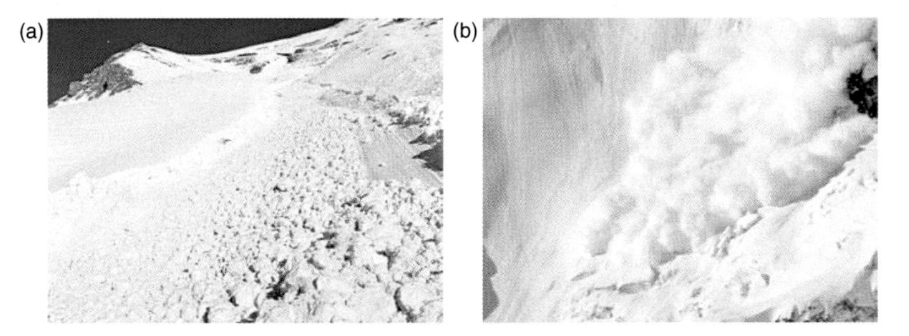


Figure 10.2. (a) Wet-snow avalanche deposit in the southern face of Grammont (Switzerland); the snowballs are approximately 10 cm in diameter. (b) Powder-snow avalanche in the northern face of Dolent (Switzerland); the typical flow depth is 20 m.

Figure 10.6. (a) Crosssection of the Malleval stream after a debris flow in August 1999 (Hautes-Alpes, France). (b) Levees left by a debris flow in the Dunant river in July 2006 (Valais, Switzerland); courtesy of Alain Delalune.

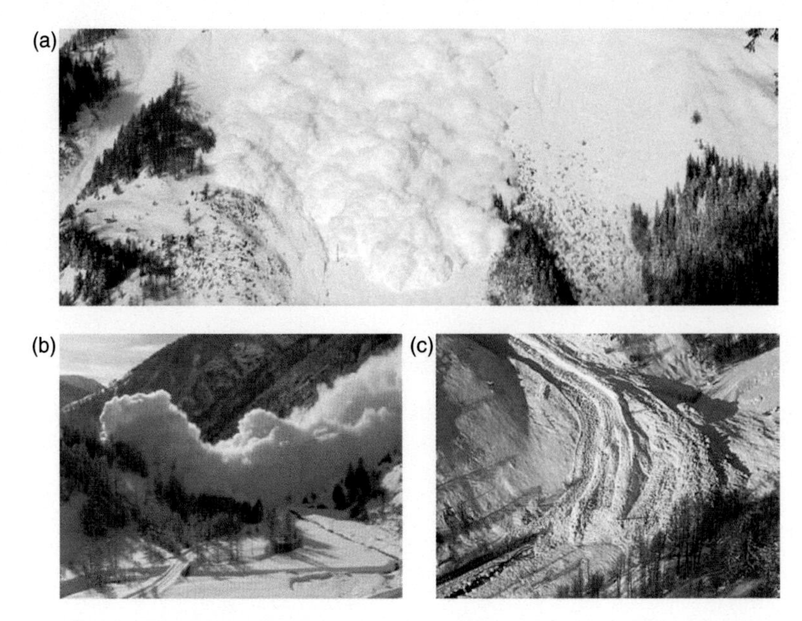


Figure 10.8. (a) Powder-snow avalanche in a flowing regime. Photograph taken in the Vallée-de-la-Sionne field site (Switzerland) in January 2004; courtesy of François Dufour, SLF. (b) Runup of a cloud of a powder-snow avalanche in a runout phase. Photograph taken at le Roux-d'Abriès, France in January 2004; courtesy of Maurice Chave. (c) Deposit of the dense core for the same avalanche; courtesy of Hervé Wadier.

Figure 10.12. Different types of snow observed in avalanche deposits. (a) Block of wet snow (size: 1 m). (b) Slurry of dry snow including weak snowballs formed during the course of the avalanche (the heap height was approximately 2 m). (c) Ice balls involved in a huge avalanche coming from the north face of the Mont Blanc (France); the typical diameter was 10 cm. (d) Sintered snow forming broken slabs (typical length: 40 cm, typical thickness 10 cm).

### 5.3.3 Summary: Contrasting Convection and Overflows

Overflows and surface convection have much in common. Both types of flows derive their energy from the initial available potential energy. In the case of overflows this comes from dense fluid initially high up on a slope, whereas for surface convection dense fluid is initially in the surface boundary layer. In both cases, the negatively buoyant structure emanating from this forcing can be described as an entraining plume, a structure that accelerates downward because of its negative buoyancy anomaly, becomes turbulent, and entrains ambient fluid as it goes, expanding and diluting as a result. However, the size of these plumes is very different. Convective plumes are limited in scale by the depth of the mixed layer [e.g., $O(1\,\text{km})$], whereas overflow plumes have an initial size set by the straits from which they come, or in the case of wide straits, by the local deformation radius. Convective plumes are transient and exist in an ensemble, leading to interactions between plumes; overflow plumes consist of a single, large, quasi-steady structure.

Convective plumes and overflow plumes experience different types of entrainment. Overflows entrain lighter fluid only from above, and so entrainment rates are very sensitive to stratification and are a function of the Richardson number. Convective plumes entrain lighter fluid sideways and from below, and so entrainment does not have to work against gravity. The entrainment is therefore approximately constant (in the absence of rotational control).

Both types of plumes are strongly influenced by ambient stratification, eventually reaching a neutral buoyancy level. In the case of overflow plumes, when the neutral buoyancy level is reached, the plume can flow horizontally into the interior. In the case of convective plumes, when the neutral buoyancy level is reached they have finite momentum in a direction approximately perpendicular to the isopycnals and so overshoot and transfer some of their kinetic energy to potential energy by entraining fluid from below into the mixed layer. The entrainment in this region at the base of the convective layer is similar in some respects to entrainment into overflows, as dense fluid is mixed upward against gravity. However, the shear generating this entrainment occurs on small spatial scales, unlike the larger scale overflow shear.

Overflow plumes are strongly controlled by topography, which steers the flow, constrains it horizontally, and provides an additional source of mixing in the bottom frictional layer. In contrast, surface convective plumes are more strongly influenced by preexisting mesoscale circulation, which is of larger scale than the plumes. Surface convective plumes may also be influenced by other processes occurring in the mixed layer (e.g., wind-driven mixing), which have not been discussed here.

Both overflows and convective flows may be strongly influenced by rotation, but the nature of that influence differs because of the different scales of the flows. Overflows, being of larger scale, are geostrophic structures, so rotation constrains the dense flow to approximately follow isopycnals and generates a shear structure that is in geostrophic balance with the density anomaly. Convective plumes are smaller-scale ageostrophic, nonhydrostatic features, but rotation can have an influence by modifying the scale of the structures, the vorticity field, and hence the kinetic energy and tracer variance dissipation.

Mesoscale eddies are important in both overflows and convectively driven flows. Eddies develop as a result of the density differences between the overflow or the convecting region and the surrounding fluid, through baroclinic instability. Vorticity may also be generated through vortex stretching in both overflows (by the descending water column) and convection (by the convergence into the plume). For overflows, this barotropic vorticity can also contribute to eddy-generating instability. The role of mesoscale eddies in contributing to lateral mixing and restratification in convecting regions is now well established. The role of mesoscale eddies in determining the final properties of the dense overflow water mass is less well understood. For convective regions, the interaction between mesoscale, submesoscale eddies, and the surface mixed layer must be considered. An analogue in overflows is the interaction between the frictionally driven mixing at the bottom boundary and the mesoscale eddies.

## References

Arakawa, A., and and W. H. Schubert, 1974: Interaction of a cumulus cloud ensemble with the large scale environment. Part I. *J. Atmos. Sci.* **31**, 674–701.

Baines, P. G., 2001: Mixing in flows down gentle slopes into stratified environments. *J. Fluid Mech.* **443**, 237–270.

Baines, P. G., 2005: Mixing regimes for the flow of dense fluid down slopes into stratified environments. *J. Fluid Mech.* **538**, 245–267.

Baringer, M. O., and J. F. Price, 1997: Mixing and spreading of the Mediterranean outflow. *J. Phys. Oceanogr.* **27**, 1654–1677.

Bocaletti, G., R. Ferrari, and B. Fox-Kemper, 2007: Mixed layer instabilities and restratification. *J. Phys. Oceanogr.* **37**, 2228–2250.

Borenas, K. M., A. K. Wahlin, I. Ambar, and N. Serra, 2002: Mediterranean outflow splitting–a comparison between theoretical models and CANIGO data. *Deep-Sea Res. II* **49**, 4195–4205.

Bower, A. M., W. E. Johns, D. M. Fratantoni, and H. Peters, 2005: Equilibration and circulation of Red Sea outflow water in the western Gulf of Aden. *J. Phys. Oceanogr.* **35**, 1963–1985.

Bruce, J., 1995: Eddies southwest of the Denmark Strait. *Deep-Sea Res.* **42**, 13–29.

Canuto, V. M., Y. Cheng, and A. M. Howard, 2007: Non-local ocean mixing model and a new plume model for deep convection. *Ocean Modelling* **16**, 28–46.

Cenedese, C., J. A. Whitehead, T. A. Ascarelli, and M. Ohiwa, 2004: A dense current flowing down a sloping bottom in a rotating fluid. *J. Phys. Oceanogr.* **34**, 188–203.

Cenedese, C., and C. Adduce, 2008: Mixing in a density-driven current flowing down a slope in a rotating fluid. *J. Fluid Mech.* **604**, 369–388.

Cenedese, C., and C. Adduce, 2010: A new parameterization for entrainment in overflows. *J. Phys. Oceanogr.* **40**, 1835–1850.

Darelius, E., L. H. Smedsrud, S. Osterhus, and A. Foldvik, 2009: Structure and variability of the Filchner overflow plume. *Tellus* **61A**, 446–464.

Donner, L. J., 1993: A cumulus parameterization including mass fluxes, vertical momentum dynamics and mesoscale effects. *J. Atmos. Sci.* **50**, 889–906.

Ellison, T., and J. Turner, 1959: Turbulent entrainment in stratified flows. *J. Fluid Mech.* **6**, 423–448.

Fer, I., R. Skogseth, and P. M. Haugan, 2004: Mixing of the Storfjorden overflow (Svalbard Archipelago) inferred from density overturns. *J. Geophys. Res.* **109**, doi:10.1029/2003JC001968.

Fer, I., G. Voet, K. S. Seim, B. Rudels, and K. Latarius, 2010: Intense mixing of the Faroe Bank Channel overflow. *Geophys. Res. Lett.* **37**, doi:10.1029/2009GL041924.

Ferron, B., H. Mercier, K. Speer, A. Gargett, and K. Polzin, 1998: Mixing in the Romanche Fracture Zone. *J. Phys. Oceanogr.* **28**, 1929–1945.

Foldvik, A., T. Gammelsrod, S. Osterhus, E. Fahrbach, G. Rohardt, M. Schroder, K. W. Micholls, L. Padman, and R. A. Woodgate, 2004: Ice shelf water overflow and bottom water formation in the southern Weddell Sea. *J. Geophys. Res.* **109**, doi:10.1029/2003JC002008.

Fox-Kemper, B., R. Ferrari, and R. Hallberg, 2008: Parameterization of mixed layer eddies, Part I: Theory and diagnosis. *J. Phys. Oceanogr.* **38**, 1145–1165.

Fritsch, J. M., and C. F. Chappell, 1980: Numerical prediction of convectively driven mesoscale pressure systems. Part I: convective parameterization. *J. Atmos. Sci.* **37**, 1722–1733.

Geyer, F., S. Osterhus, B. Hansen, and D. Quadfasel, 2006: Observations of highly regular oscillations in the overflow plume downstream of the Faroe Bank Channel. *J. Geophys. Res.* **111**, doi:10.1029/2006JC003693.

Girton, J. B., T. B. Sanford, and R. H. Kase, 2001: Synoptic sections of the Denmark Strait overflow. *Geophys. Res. Lett.* **28**, 1619–1622.

Girton, J. B. and T. B. Sanford, 2003: Descent and modification of the overflow plume in the Denmark Strait. *J. Phys. Oceanogr.* **33**, 1351–1364.

Girton, J. B., L. J. Pratt, D. A. Sutherland, and J. F. Price, 2006: Is the Faroe Bank Channel overflow hydraulically controlled? *J. Phys. Oceanogr.* **36**, 2340–2349.

Gordon, A. L., E. Zambianchi, A. Orsi, M. Visbeck, C. F. Giulivi, T. Whitworth, III, and G. Spezie, 2004: Energetic plumes over the western Ross Sea continental slope. *Geophys. Res. Lett.* **31**, doi:10.1029/2004GL020785.

Gordon, A. L., A. H. Orsi, R. Muench, B. A. Huber, E. Zambianchi, and M. Visbeck, 2009: Western Ross Sea continental slope gravity currents. *Deep-Sea Res. II*, **56**, 796–817.

Hallberg, R., 2000: Time integration of diapycnal diffusion and Richardson number-dependent mixing in isopycnal coordinate ocean models. *Mon. Wea. Rev.* **128**, 1402–1419.

Ilicak, M., T. M. Ozgokmen, E. Ozsoy, and P. F. Fischer, 2008: Performance of two-equation turbulence closures in three-dimensional simulations of the Red Sea overflow. *Ocean Modelling* **24**, 122–139.

Ilicak, M., S. Legg, A. Adcroft, and R. Hallberg, 2011: Dynamics of a dense gravity current flowing over a corrugation, *Ocean Modelling* **38**, 71–84.

Jackson, L., R. Hallberg, and S. Legg, 2008: A parameterization of shear-driven turbulence for ocean climate models. *J. Phys. Oceanogr.* **38**, 1033–1053.

Jones, H., and J. Marshall, 1993: Convection with rotation in a neutral ocean: a study of open-ocean deep convection. *J. Phys. Oceanogr.* **23**, 1009–1039.

Julien, K., S. Legg, J. McWillaims, and J. Werne, 1996a: Rapidly rotating turbulent Rayleigh-Benard convection. *J. Fluid Mech.* **322**, 243–273.

Julien, K., S. Legg, J. McWilliams, and J. Werne, 1996b: Penetrative convection in rapidly rotating flows: Preliminary results from numerical simulation. *Dyn. Atmos. Oceans* **24**, 237–249.

Julien, K., S. Legg, J. McWilliams, and J. Werne, 1999: Plumes in rotating convection. Part I. Ensemble statistics and dynamical balances. *J. Fluid Mech.* **391**, 151–187.

Kase, R. H., J. B. Girton, and T. B. Sanford, 2003: Structure and variability of the Denmark Strait overflow: Model and observations. *J. Geophys. Res.* **108**, doi:10.1029/2002JC001548.

Lane-Serff, G., and P. Baines, 1998: Eddy formation by dense flows on slopes in a rotating fluid. *J. Fluid Mech.* **363**, 229–252.

Large, W. G., J. C. McWilliams, and S. C. Doney, 1994: Oceanic vertical mixing—A review and a model with a nonlocal boundary-layer parameterization. *Rev. Geophys.* **32**, 363–403.

Legg, S., and J. Marshall, 1993: A heton model of the spreading phase of open-ocean deep convection. *J. Phys. Oceanogr.* **23**, 1040–1056.

Legg, S., H. Jones, and M. Visbeck, 1996: A heton perspective of baroclinic eddy transfer in localized open ocean convection. *J. Phys. Oceanogr.* **26**, 2251–2266.

Legg, S., J. McWilliams, and J. Gao, 1998: Localization of deep ocean convection by a mesoscale eddy. *J. Phys. Oceanogr.* **28**, 944–970.

Legg, S., and J. McWilliams, 2001: Convective modifications of a geostrophic eddy field. *J. Phys. Oceanogr.* **31**, 874–891.

Legg, S., R. W. Hallberg, and J. B. Girton, 2006: Comparison of entrainment in overflows simulated by *z*-coordinate, isopycnal and nonhydrostatic models. *Ocean Modelling* **11**, 69–97.

Legg, S., B. Briegleb, Y. Chang, E. P. Chassignet, G. Danabasoglu, T. Ezer, A. L. Gordon, S. Griffies, R. Hallberg, L. Jackson, W. Large, T. M. Ozgokmen, H. Peters, J. Price, U. Riemenschneider, W. Wu, X. Xu, and J. Yang, 2009: Improving oceanic overflow representation in climate models: The gravity current entrainment climate process team. *Bull. Am. Met. Soc.* **90**, 657–670.

Levy, M. A., and J. S. Fernando, 2002: Turbulent thermal convection in a rotating statified fluid. *J. Fluid Mech.* **467**, 19–40.

Macrander, A., U. Send, H. Valdimarsson, S. Jonsson, and R. H. Kase, 2005: Interannual changes in the overflow from the Nordic Seas into the Atlantic Ocean through Denmark Strait. *Geophys. Res. Lett.* **32**, doi:10.1029/2004GL021463.

Macrander, A., R. H. Kase, U. Send, H. Valdimarsson, and S. Jonsson, 2007: Spatial and temporal structure of the Denmark Strait Overflow revealed by acoustic observations. *Ocean Dynamics* **57**, 75–89.

Marshall, J., and F. Schott, 1999: Open-ocean convection: Observations, theory, and models. *Rev. Geophys.* **37**, 1–64.

Matt, S., and W. E. Johns, 2007: Transport and entrainment in the Red Sea outflow plume. *J. Phys. Oceanogr.* **37**, 819–836.

Mauritzen, C., Y. Morel, and J. Paillet, 2001: On the influence of Mediterranean water on the central waters of the North Atlantic Ocean. *Deep-Sea Res. I.* **48**, 347–381.

Mauritzen, C., J. Price, T. Sanford, and D. Torres, 2005: Circulation and mixing in the Faroese channels. *Deep-Sea Res.* **52**, 883–913.

Maxworthy, T., and S. Narimousa, 1994: Unsteady deep convection in a homogeneous rotating fluid. *J. Phys. Oceanogr.* **24**, 865-887.

Ozgokmen, T. M., and E. P. Chassignet, 2002: Dynamics of two-dimensional turbulent bottom gravity currents. *J. Phys. Oceanogr.* **32**, 1460–1478.

Ozgokmen, T. M., P. F. Fischer, J. Q. Duan, and T. Iliescu, 2004: Three-dimensional turbulent bottom density currents from a high-order nonhydrostatic spectral element model. *J. Phys. Oceanogr.* **34**, 2006–2026.

Ozgokmen, T. M., P. F. Fischer, and W. E. Johns, 2006: Product water mass formation by turbulent dense currents from a high-order nonhydrostatic spectral element model. *Ocean Modelling* **12**, 237–267.

Ozgokmen, T. M., and P. F. Fischer, 2008: On the role of bottom roughness in overflows. *Ocean Modelling* **20**, 336–361.

Padman, L., S. L. Howard, A. H. Orsi, and R. D. Muench, 2009: Tides of the northwestern Ross Sea and teir impact on dense outflows of Antarctic Bottom water. *Deep-Sea Res II* **56**, 818–834.

Paluszkiewisz, T., and R. D. Romea, 1997: A one-dimensional model for the parameterization of deep convection in the ocean. *Dyn. Atmos. Oceans* **26**, 95–130.

Pawlak, G., and L. Armi, 2000: Mixing and entrainment in developing stratified currents. *J. Fluid Mech.* **424**, 45–73.

Peters, H., W. E. Johns, A. S. Bower, and D. M. Fratantoni, 2005a: Mixing and entrainment in the Red Sea outflow plume. Part I: Plume structure. *J. Phys. Oceanogr.* **35**, 569–583.

Peters, H., and W. E. Johns, 2005b: Mixing and entrainment in the Red Sea outflow plume. Part II: Turbulence characteristics. *J. Phys. Oceanogr.* **35**, 584–600.

Pickart, R. S., D. J. Torres, and R. A. Clarke, 2002: Hydrography of the Labrador Sea during active convection. *J. Phys. Oceanogr.* **32**, 428–457.

Pratt, L. J., U. Riemenschneider, and K. R. Helfrich, 2007: A transverse hydraulic jump in a model of the Faroe Bank Channel outflow. *Ocean Modelling* **19**, 1–9.

Price, J. F., M. O. Baringer, R. G. Lueck, G. C. Johnson, I. Ambar, G. Parrilla, A. Cantos, M. A. Kennelly, and T. B. Sanford, 1993: Mediterranean outflow mixing and dynamics. *Science* **259**, 1277–1282.

Price, J. F., and M. O. Baringer, 1994: Outflows and deep-water production by marginal seas. *Prog. Oceanogr.* **33**, 161–200.

Riemenschneider, U., and S. Legg, 2007: Regional simulations of the Faroe Bank Channel overflow in a level model. *Ocean Modelling* **17**, 93–122.

Serra, N., and I. Ambar, 2002: Eddy generation in the Mediterranean undercurrent. *Deep-Sea Res.* **49**, 4225–4243.

Spall, M. A., and J. F. Price, 1998: Mesoscale variability in Denmark Strait: The PV outflow hypothesis. *J. Phys. Oceanogr.* **28**, 1598–1623.

St Laurent, L. C., and A. M. Thurnherr, 2007: Intense mixing of lower thermocline water on the crest of the Mid-Atlantic ridge. *Nature* **448**, doi:10.1038/nature06043.

Straneo, F., M. Kawase, and S. C. Riser, 2002: Idealized models of slantwise convection in a baroclinic flow. *J. Phys. Oceanogr.* **32**, 558–572.

Thomas, L. N., A. Tandon, and A. Mahadevan, 2008: Submesoscale processes and dynamics. In: M. W. Hecht and H. Hasumi (eds.), *Ocean Modeling in an Eddying Regime*, AGU Geophysical Monograph 177, pp.17–38. American Geophysical Union, Washington, DC.

Turner, J. S. 1986: Turbulent entrainment: The development of the entrainment assumption and its application to geophysical flows. *J. Fluid Mech.* **173**, 431–471.

Umlauf, L., L. Arneborg, R. Hofmeister, and H. Burchard, 2010: Entrainment in shallow rotating graviy currents: A modeling study. *J. Phys. Oceanogr.* **40**, 1819–1834.

Visbeck, M., J. Marshall, and J. Jones, 1996: Dynamics of isolated convective regions in the ocean. *J. Phys. Oceanogr.* **26**, 1721–1734.

Visbeck, M., and A. M. Thurnherr, 2009: High-resolution velocity and hydrographic observations of the Drygalski Trough gravity plume. *Deep-Sea Res. II* **56**, 835–842.

Voet, G., and D. Quadfasel, 2010: Entrainment in the Denmark Strait overflow plume by meso-scale eddies. *Ocean Sci.* **6**, 301–310.

Wang, D., 2006: Effects of the earth's rotation on convection: Turbulent statistics, scaling laws and Lagrangian diffusion. *Dyn. Atmos. Oceans* **41**, 103–120.

Wells, M., and P. Nadarajah, 2009: The intrusion depth of density currents flowing into stratified water bodies. *J. Phys. Oceanogr.* **39**, 1935–1947.

Wesson, J. C., and M. C. Gregg, 1994: Mixing at Camarinal Sill in the Strait of Gibraltar. *J. Phys. Oceanogr.* **99**, 9847–9878.

Whitehead, J. A., 1998: Topographic control of oceanic flows in deep passages and straits. *Rev. Geophys.* **36**, 423–440.
Xu, X., Y. S. Chang, H. Peters, T. M. Ozgokmen, and E. P. Chassignet, 2006: Parameterization of gravity current entrainment for ocean circulation models using a high-order 3D nonhydrostatic spectral element model. *Ocean Modelling* **14**, 19–44.

## Appendix: Notation

| Variable | Description |
|---|---|
| $\rho$ | density |
| $\rho_0$ | reference density |
| $\rho'$ | density anomaly |
| $g$ | gravitational acceleration |
| $g'$ | buoyancy anomaly, or reduced gravity, $g' = -g\rho'/\rho_0$ |
| $b$ | buoyancy, $b = -g\rho/\rho_0$ |
| $f$ | Coriolis frequency |
| $N$ | buoyancy frequency, $N^2 = -g/\rho_0 d\rho/dz = db/dz$ |
| $x, y$ | horizontal coordinates (e.g., downstream and cross-stream coordinates for overflow) |
| $z$ | vertical coordinate |
| $U$ | horizontal flow speed (e.g., downstream velocity of overflow) |
| $W$ | vertical velocity |
| $t$ | time |
| $T$ | temperature |
| $q$ | tracer (e.g., temperature or salinity) |
| $\nu$ | viscosity |
| $\kappa$ | diffusivity |
| $h$ | thickness (for plume) or mixed layer depth (for convection) |
| $H$ | total depth of fluid |
| $\theta$ | slope angle |
| $L$ | width of overflow |
| $L_\rho$ | deformation radius, $L_\rho = \sqrt{g'h}/f$ |
| $\delta$ | Ekman thickness, $\delta = \sqrt{2\nu/f}$ |
| $Fr$ | Froude number, $Fr = U/\sqrt{g'h}$ |
| $Ri$ | Richardson number, $Ri = N^2/(dU/dz)^2$ |
| $Ek$ | Ekman number, $Ek = \delta/h$ |
| $Ri_b$ | bulk Richardson number, $Ri = 1/Fr^2$ |
| $Ro$ | Rossby number, $Ro = U/(fL)$, or $Ro = W/(fh)$ for convection |
| $w_e$ | vertical entrainment velocity (e.g., into overflow plume) |
| $E$ | nondimensional entrainment coefficient for overflow, $E = w_e/U$ |
| $V$ | volume (e.g., of thermal) |

| | |
|---|---|
| $S$ | surface area (e.g., of thermal) |
| $V_e$ | entrainment velocity (in all directions, e.g., for thermal/convective plume) |
| $\alpha$ | entrainment coefficient for convective thermal/plume, $\alpha = V_e/W$ |
| $r$ | radius (e.g., of thermal/plume or of convecting region) |
| $A$ | cross-sectional area (e.g., of convective plume) |
| $l$ | circumference (e.g., of convective plume) |
| $F_0$ | buoyancy flux into isolated convective plume |
| $B_0$ | buoyancy flux per unit area |
| $h_{rot}$ | rotational length scale for convection, when $Ro = 1$ |
| $U_{rim}$ | geostrophic rim current |
| $Q$ | potential vorticity |
| $\zeta$ | relative vorticity |

# 6

# An Ocean Climate Modeling Perspective on Buoyancy-Driven Flows

WILLIAM G. LARGE

Ocean models are based on the conservation of vector momentum, and scalar tracers, which give time evolution (prognostic) equations for the velocity field, the active temperature and salinity tracers, and all passive tracers of interest, such as oxygen, carbon dioxide, and nutrients. In contrast, ocean density, $\rho$, is diagnostic and not necessarily conserved. It is computed from the active tracers using an equation of state (EoS). Griffies (2002) and works cited therein discuss the approximations leading to the primitive equations that are typically solved by climate models, and common ways that the global ocean is discretized. Various methods of integrating the equations are also presented, along with their advantages and disadvantages. A summary, including the equations themselves, is provided in Treguier et al. (Chapter 7, this volume). The existence of such excellent references means that this chapter can focus on specific illustrative examples of the workings of buoyancy in particular ocean models, without being comprehensive, or repeating the background.

The great challenge of climate modeling is the roughly 10-decade-wide range of potentially important interacting scales; from the more than $10^7$ m global scale to the less than $10^{-2}$ m viscous scale. Present coupled climate calculations are reaching down to the $10^4$ m horizontal and 10 m vertical scales in the ocean, but still many subgrid scale (SGS) processes and interactions rely on parameterizations. In principle, model fidelity should benefit from increased resolution. Indeed, improvements can be dramatic. For example, the order meter resolution of Large Eddy Simulation (LES) is designed to resolve the most energetic motions, leaving only the less energetic and better understood scales to be parameterized (Deardorff 1970). As reviewed by Wyngaard and Moeng (1993), LES has been very successful in simulating three-dimensional turbulence in planetary boundary layers. At about 62 m resolution, a new class of dynamical motions emerges in hurricane simulations (Rotunno et al., 2009). Also, observed small-scale air-sea interaction is reproduced when an atmospheric model is coupled to a high-resolution, eddy-resolving ocean model, but not when the ocean resolution is lower (Bryan et al. 2010).

In practice, climate models are not configured with the highest possible resolutions, and coupling to eddy-resolving ocean models (Treguier et al., Chapter 7, this volume) has been rare. The reason is that in addition to demands for increased resolution to represent small-scale processes and scale interactions, other science drivers are competing for computational resources. Studies of natural variability and predictability require large ensembles. Data assimilation capabilities are needed for state estimation and initializing predictions. Simulations of past large climate changes (e.g., glacial to interglacial transitions) integrate over thousands of years. There are requirements for more functionality to study potential climate feedbacks and interactions with tropospheric chemistry, the stratosphere, ocean ecosystems, land vegetation, and human activities. Therefore, the demand for computational resources will continue to outstrip availability, so that climate modeling involves compromise between competing demands.

Through most of the 1990s, the most important criteria for evaluating climate models was physical realism in the dynamical solutions of the ocean and atmosphere, but new factors now need to be considered, too. For example, additional functionality, such as the Carbon Cycle, and Ozone Chemistry, places a higher priority on maintaining realistic, monotonic, positive definite tracer concentrations. The advent of petascale computing utilizing hundreds of thousands of processors will place a premium on codes that perform efficiently out to order 100,000 processors. Efficiency here is relative, typically being less than 10% of theoretical peak. Already, in some instances, computing resources are not being allocated to codes that do not scale well enough, regardless of the realism of the dynamics and the tracer fields. Therefore, climate modeling is becoming more of a close collaboration between scientists and software engineers.

The particular focus of the following is buoyancy in climate models. Buoyancy, $B = g\,(1 - \rho/\rho_o)$, is just a way of expressing density relative to a reference density, $\rho_o$, as an acceleration, where g is gravitational acceleration. The accelerations due to pressure gradients that appear in the horizontal momentum equations are just vertical integrals of horizontal buoyancy gradients. The ocean circulation related to these gradients is referred to as the Thermo-Haline Circulation (THC) in recognition of the role of both active tracers. The ocean interior is in approximate steady state, with pressure gradients balanced by Coriolis accelerations. Flow along isopycnals dominates the interior THC because it does not affect this geostrophic balance.

The atmospheric analogues of the THC, such as the Hadley circulation (Malardel, Chapter 8, this volume), are more familiar and better understood. In particular, the thermodynamic work that drives them is akin to the four-phase Carnot cycle. First, solar radiation passing through the atmosphere heats air at high surface pressure. Second, the resulting buoyant convection lifts the heated air to lower pressure where there can be subsequent lateral transport with perhaps some pressure change. Third, the transported air is cooled, primarily through the emission of long-wave radiation.

Fourth, the resulting subsidence of the denser air returns it to high surface pressure where there can be lateral transport to close the cycle. How, or even if, such a cycle operates in the ocean is a very long-standing problem in ocean circulation theory and observations. Therefore, it is not known how well ocean models represent the closed THC. The most suspect aspects are the analogues of phase 1 (heating at depth) and phase 2 (return to low pressure nearer the ocean surface), so these are not considered further. Neither is the phase 2 near-surface (low-pressure) wind-driven transport. However, there are observations and theories for the THC analogues to phases 3 and 4, including lateral transport at depth, where buoyancy plays a fundamental role, so aspects of how these are represented in some climate models are the subjects of what follows. See Griffies (2002) and works cited therein for other aspects and different climate models.

Section 6.1 outlines the treatment of buoyancy in a particular class of Ocean General Circulation Model (OGCM), and discusses surface boundary conditions, or fluxes, and coupling to the atmosphere. Surface cooling and freshwater loss at high latitudes decreases surface buoyancy and is the THC's phase 3 analogue. This buoyancy loss drives turbulent ocean convection, which is considered in Section 6.2 ("Convective Boundary Layers"), following an overview of planetary boundary layers in general, and the Ocean Boundary Layer (OBL) in particular. Penetrative convection and the subsequent interior lateral transports are the subject of Section 6.3 ("Ventilation in Ocean Models"), which covers only some modeling aspects of only two modes of ventilation and the associated sinking. The first mode is shallow thermocline ventilation, including the effects of penetrative convection. The second is deep ocean ventilation, where the focus is the Atlantic Meridional Overturning Circulation (AMOC). Key buoyancy-driven components of the AMOC are overflows from the Mediterranean Sea, the Nordic (Greenland, Iceland, and Norwegian) Seas, and the Antarctic (Weddell and Ross) Seas. Overflows are not resolved in climate models, but Section 6.4 ("Parameterized Overflows") presents how they have been parameterized in an OGCM with some success. Ventilation, as defined by turbulent mixing to depth and interior circulation involving sinking deeper, can be regarded as the THC analogue to phase 4. The importance of these processes to climate arises from the uptake into the deep ocean of heat, of anthropogenic atmospheric constituents like $CO_2$ and CFCs, and of biologically critical nutrients and oxygen.

## 6.1 Buoyancy in Ocean Climate Models

Before tackling the complexity of a global climate model, it is advisable to become familiar with greatly simplified models that can be fully understood. Such models can exhibit extremely complicated behavior that can provide insights into how the climate system may work. Therefore, Section 6.1.1 (Reduced Complexity Models) examines an early example. Section 6.1.2 then introduces OGCMs, as models of very many boxes, with complicated, physically based rules governing the exchanges between boxes. Although they strive for physical realism, OGCMs are subject to numerical constraints and artifacts, so some that affect buoyancy are noted in Section 6.1.3. The major external sources and sinks of ocean buoyancy are the surface heat and freshwater fluxes that combine to give a surface buoyancy flux, so these are the topics of Section 6.1.4 ("Surface Forcing"). In climate models, the surface forcing comes from exchanges with an atmospheric model and some issues are presented in Section 6.1.5 ("Coupling").

### *6.1.1 Reduced Complexity (Box) Models*

The fundamental role of buoyancy in ocean climate has been recognized from the very beginnings of climate modeling and is the driving force behind the earliest examples, including the simplified Stommel (1961) two-box model shown in Figure 6.1. It attempts to capture the basic features of a hydrological cycle of strength $F > 0$. Evaporation ($-F$) from the ocean (left box, subscript 2) decreases buoyancy by increasing salinity. In nature, about 90% of the evaporation falls directly on the ocean, and the remainder returns as continental runoff. Both paths increase ocean buoyancy by decreasing salinity, but in general not where the evaporation occurred, so the horizontal buoyancy gradients generate a circulation. This situation is captured in the two-box model by combining both paths to force ($+F$) the right box (subscript 1) and decrease its salinity and hence buoyancy. The hydrological cycle characterized by the

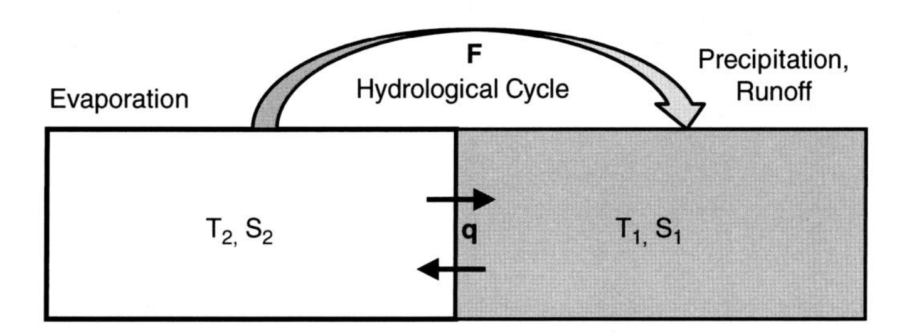


$$\dot{S}_1 = -F + |q|(S_2 - S_1)$$
$$\dot{S}_2 = +F - |q|(S_2 - S_1)$$
Salinity Budget

$$q = -\frac{k}{\rho_0}(\rho_2 - \rho_1)$$
Momentum Balance

$$\rho_2 = \rho_0(1 - \alpha T + \beta S)$$
Equation of State

Figure 6.1. The Stommel (1961) two-box model, showing the hydrological cycle forcing and the four equations (time evolution of salinity in each box, momentum balance, and linear equation of state) in four unknowns ($S_1$, $S_2$, $q$, and $\rho$), where implicit heat fluxes maintain a constant temperature difference, $(T_2 - T_1) > 0$.

forcing, $F$, drives an exchange, $q$, between the boxes, whose strength depends on their normalized density difference with $k$ the constant of proportionality. The simplicity of this model includes implicit heat fluxes that eliminate temperature evolution from the problem, as well as a linear EoS about a reference density $\rho_o$, where the coefficients of thermal expansion, $\alpha$, and of haline contraction, $\beta$, are constant. Nevertheless, it represents the basic global hydrological processes and its analytic equilibrium solutions exhibit intriguing behavior.

Examination of the two-box model solutions reveals that multiple stable equilibrium states are possible for the same forcing, provided it is not too strong relative to the temperature difference $(T_2 - T_1)$. In this regime, this positive temperature difference can keep $\rho_2 < \rho_1$ at equilibrium. In contrast to this thermal mode, the equilibrium salinity difference can be large enough for a haline mode $(\rho_2 > \rho_1)$. However, with stronger forcing there is only the single haline mode. This behavior allows the possibility of hysteresis, which has received further study in a number of extensions, including the six-box model (Welander 1986) and meridional plane models (Marotzke et al. 1988). More recently, models of intermediate complexity (e.g., Rahmstorf et al. 2005) have been developed as a means of increasing realism while maintaining understanding, as well as the computational efficiencies necessary for extensive examination of such possibilities. Other examples of reduced complexity models of buoyancy in the ocean are given by Pedlosky (Chapter 2, this volume) and Spall (Chapter 3, this volume).

### 6.1.2 Ocean General Circulation Models for Climate

An OGCM attempts to include full complexity and feedbacks, which for the climate problem includes the spun-up global circulation, so for the purposes of this chapter an OGCM must be global and integrated for centuries or even millenia. A comprehensive reference for much of what follows is Griffies (2002). Modern OGCMs use a full nonlinear EoS, that includes pressure effects usually through input of the depth. They have a typical resolution of order 1 degree in latitude and longitude and 50 levels in the vertical, and can be regarded as 10 million-box models. Exchanges between boxes are governed by a discretization of the primitive equations given in Treguier et al. (Chapter 7, this volume). Principal differences from Navier-Stokes are the hydrostatic assumption and the traditional approximation's neglect of the horizontal Coriolis terms. The popularity of this order of resolution is determined by computational limits of present-day computers after accounting for the other considerations noted in the Introduction (e.g., integration length, ensemble number, and functionality). Even though the primitive equations can reproduce the most energetic scales of motion in the ocean, the mesoscale, these motions are not resolved by the coarse resolution of such climate OGCMs. However, there has been some success in parameterizing some mesoscale processes following Gent and McWilliams (1990). Recently, computational resources have become available for short, one-off integrations of the physical climate system

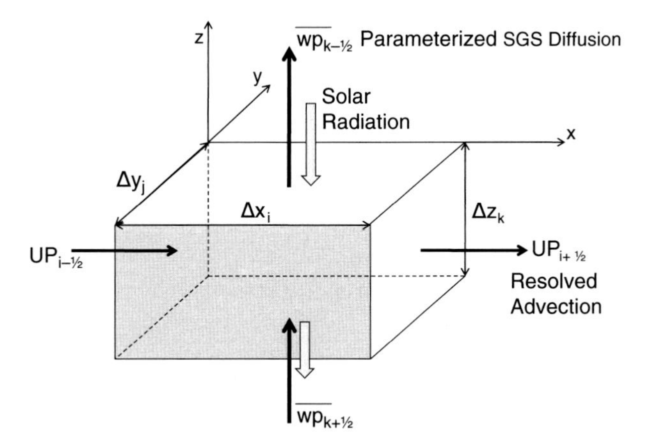


Figure 6.2. Schematic of the OGCM grid-box indexed $(i, j)$ in the horizontal and $k$, increasing with depth. Both parameterized SGS diffusion and resolved advection and appear on all six faces, but are shown only for the $x$-direction and $y$-direction, respectively, for an arbitrary tracer, $P(i, j, k)$. Solar radiation enters at the top and exits out the bottom.

only, at about 0.1 degree ocean resolution, where the mesoscale is mostly resolved. The use of such eddy-resolving, 500 million-box models will continue to increase with computational power, subject to competing demands.

All OGCMs discretize a "continuous" world in space and time by means of a model grid box and a time-stepping scheme, respectively, but these can differ a great deal across models. Nevertheless, there is always inherent discretization error, as well as numerical constraints that compromise a model solution. A particularly simple example of an OGCM grid box is shown in Figure 6.2, for a general prognostic model variable, $P$, defined at the box center. Recall there are also diagnostic quantities like density and buoyancy. The nonlinear terms are partitioned into resolved advection and parameterized SGS diffusion. Conservation is guaranteed by the flux form of the advection, where anything advected through any face leaves one box and enters its neighbor. Using the $x$-direction as an example (the other directions are similar), the tracer tendency is just $(UP_{i+1/2} - UP_{i-1/2})/\Delta x$. Similarly, using the $z$-direction as the example, the tracer tendency due to parameterized (denoted by lowercase) $p$ and vertical velocity $w$ correlations is $(\overline{wp}_{k+1/2} - \overline{wp}_{k-1/2})/\Delta z$, and conservation is assured, because the diffusive flux out the top of box $(i, j, k)$ comes in the bottom of box $(i, j, k-1)$. Solar radiation is a component when $P$ is heat. It enters through the top, a portion is absorbed in the box increasing its temperature, and the remainder exits through the bottom. However, in climate models, all the radiation entering the bottom box must be absorbed, so that none leaves the system. Otherwise, there is systematic heat loss that leads to temperature drifts. The physical argument is that radiation absorbed by the bottom diffuses back into the ocean above.

Typically, tracers are defined at the center of each grid box and velocities along the box boundaries. Tracer values can be interpreted as the mean over the entire box, or as middepth approximations, or in other ways depending on the model discretization. Consistent interpretation could be important for detailed comparison with observations. In the case of temperature observations, they need to be converted to a potential temperature that is consistent with the heat conserved in the model. Clearly, the temperature of uppermost boxes, $T_1$, is not the SST (Sea Surface Temperature), even though it is often taken to be so. Similarly, the velocity, $\mathbf{U}_1$, is not the surface velocity.

### 6.1.3 Numerical Constraints and Artifacts

Perhaps the most limiting and familiar constraint is the CFL (Courant-Frederichs-Levy) condition which requires that the time ($\Delta x/U$) for information to pass through a grid box must be less than the time step ($\Delta t$). Thus, increasing resolution carries the additional computational burden of a proportionally smaller time step. There is also a diffusive analogue that says that the time to diffuse through a grid box must be less than $\Delta t$. In practice, the vertical diffusion can be the most constraining. The corresponding diffusive time scale is ($\Delta x^2/ K_v$), where $K_v$ is the vertical diffusivity. It can be much shorter than $\Delta t$, so implicit integration techniques that are unconditionally stable are often used in models that allow large diffusion. It must be stressed, however, that numerical stability does not necessarily translate to more accuracy than explicit integration with sufficiently short time steps.

Particularly troublesome numerical artifacts are local extrema in tracer fields. They arise from the fact that different wavelengths, and hence Fourier components, propagate through a numerical grid at different speeds. As wavelengths representing a smooth field disperse, the highest wavenumbers can produce local extrema. Although there is overall conservation, with overshoots balancing undershoots, there can be negative tracer concentrations and below-freezing temperatures. The latter can neither be allowed to produce sea-ice nor be used in an EoS. One family of solutions to the problem is to diffuse the extrema in some way, but natural features are then diffused, too. Another is to apply local limiters (e.g., no freezing temperatures, negative concentrations), but these are inherently nonconservative, and hence problematic for climate models. In the end, compromises need to be made.

### 6.1.4 Surface Forcing

OGCMs are primarily forced by the surface fluxes of the two vector components of horizontal momentum, $\boldsymbol{\tau}_o$, heat, $Q_o$, and freshwater, $F_o$. Over the ice-free ocean these are just the air-sea fluxes ($\boldsymbol{\tau}_{as}$, $Q_{as}$, and $F_{as}$) plus continental runoff, $F_R$. In high latitudes, ice-ocean fluxes ($\boldsymbol{\tau}_{io}$, $Q_{io}$, and $F_{io}$) contribute in proportion to the fractional ice coverage, $f_{ice}$, and the air-sea fluxes in proportion to the fraction of open water leads, $f_{\text{ocn}} = (1 - f_{\text{ice}})$. If continental runoff is in the form of ice, it has a latent heat of fusion. The air-sea heat flux, has contributions from solar radiation, long-wave radiation, evaporation, sensible heat flux, and snowmelt. The air-sea freshwater flux is a balance between precipitation gain and evaporation loss. The ocean affects these fluxes almost entirely through SST, where a 1°C increase in SST decreases the net heat flux by 30–70 W/m$^2$, all other things remaining constant. To be more precise, in the ocean there is a near surface "skin" temperature, $T_{\text{skin}}$, at which long-wave radiation, $LW\uparrow$, is emitted, as well as shallow "bulk" temperature, $T_{\text{bulk}}$, which is used in parameterizations of evaporation and sensible heat flux. In practice, neither of these surface temperatures is given by an ocean climate model, so SST $= T_1$, is often used for both. These fluxes only depend weakly on the surface current, which is usually taken as $\mathbf{U}_1$.

The ocean surface buoyancy flux, $B_0$, and in particular its spatial distribution, is the fundamental driver of buoyancy flows in ocean models. The sign and magnitude of buoyancy are arbitrary, depending on $\rho_o$ because it is buoyancy differences and gradients that matter. The full EoS, $\rho(T, S, d)$, where pressure effects are input as depth, $d$, is shown for the ocean surface ($d = 0$) in Figure 6.3 over the range of temperature and salinity where it is most nonlinear. From its freshwater at 0°C value of 1.000 g/cm$^3$ = 1,000 kg/m$^3$ to the highest values of Figure 6.3, ocean density only varies in the third significant figure. To more conveniently convey this small variation, ocean density is traditionally given in sigma units, where $\sigma = \rho - 1{,}000$ kg/m$^3$ and the units of kg/m$^3$ are implicit. Thus, points $A$ and $B$ lie on the $\sigma = 27.5$ density contour. The lines of constant density are not straight as for a linear EoS (e.g.,

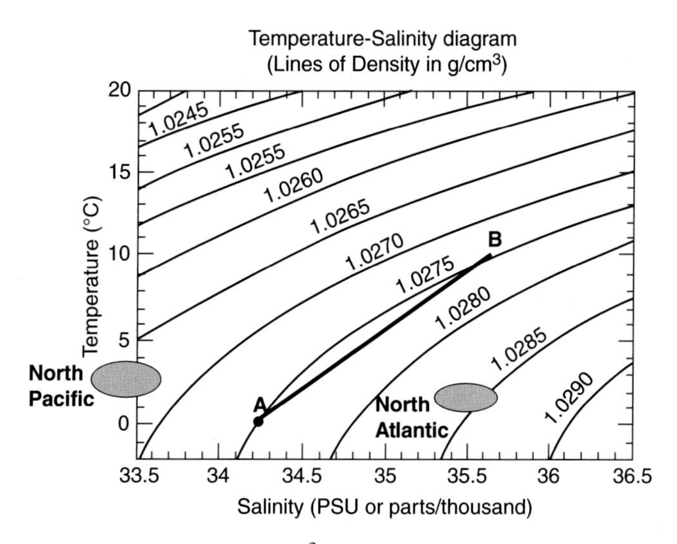


Figure 6.3. Density of seawater (g/cm$^3$) at the surface as a function of temperature and salinity from a full equation of state. Shaded areas represent the waters found in the North Atlantic and North Pacific.

Figure 6.1), and the curvature increases with decreasing temperature. Specifically, the value of the thermal expansion coefficient, $\alpha$, is about three times smaller at point $A$ (0.00005°C$^{-1}$) than at point $B$ (0.00016°C$^{-1}$), and twice as big (0.00032°C$^{-1}$) at 28°C. The haline contraction coefficient, $\beta$, is about 0.00075 psu$^{-1}$ everywhere. To illustrate that density, and hence buoyancy, is not conserved, consider mixing equal parts of equal density water from points $A$ and $B$. The product water would lie at the midpoint of the straight line between $A$ and $B$, and have a higher density. This process is called cabbelling. The pressure/depth effects can be important and are the reason why Antarctic Bottom Water (AABW) at the bottom of Atlantic is more dense than the overlying North Atlantic Deep Water (NADW) from the Nordic and Arctic Seas. At the surface, NADW would be more dense than AABW.

The surface buoyancy flux can be defined as $B_0 = g\ (\alpha\ (\rho_o C_p)^{-1} Q_o + \beta\ S_o F_o\ \rho_o^{-1})$, where $C_p$ is the ocean's specific heat and $S_o$ a reference salinity. As such, all components of both the heat and freshwater fluxes make contributions. Numerically, the effects on salinity and buoyancy of the freshwater flux term are equivalent to the addition or removal of salt through a virtual salt flux, $F_{vs} = -F_o S_o$, with units of kg salt/m$^2$/s. Evaporation, for example, is a negative buoyancy flux because the large magnitude of the latent heat of vaporization means the removal of heat dominates the loss of freshwater. For the same reason, evaporation is much more effective at cooling the ocean than at increasing its salinity. For example, the evaporation necessary to change salinity by 1 psu, would cool the ocean by about 17°C, but this cooling would tend to shut off the evaporation. Therefore, significant evaporation must be accompanied by a compensating large solar radiation, which is a positive buoyancy flux that often counters the buoyancy effects of evaporation during the day. At any time, precipitation can be a locally significant counter, too. At night, however, the ocean is generally convective (losing buoyancy through its surface), except if there is a great deal of rainfall.

In nature, a surface freshwater flux changes the ocean volume. The numerics needed for an ocean model to respond this way are not straightforward, and the changes are minute fractions of the ocean volume. Therefore, some OGCMs assume a constant ocean volume, by arguing that the effects are small compared to those of other model simplifications and discretization. To accommodate this assumption, the freshwater flux is transformed by the constant volume models into a virtual salt flux that appears in the surface boundary condition of the salinity conservation equation. A conservation problem arises in climate models if the reference salinity, $S_o$, is taken to be the local salinity, as physics says it should be. It is illustrated by the following thought problem. First add a volume, $V$, of freshwater to an ocean of salinity $S_1$. It effectively displaces a volume $V$ of salinity $S_o$, decreasing the salinity to $S_2$. Then remove the volume, $V$, of freshwater, which is effectively replaced by a volume $V$ of salinity, say $S_0$. Conservation demands that the final salinity $S_3$ be equal to $S_1$, which is true for $S_o = S_0$, but not if local salinities are used; $S_o = S_1$ and $S_0 = S_2$. Therefore, a constant reference salinity, $S_o$, must be used in climate models forced with virtual salt fluxes. However, the problem then is that rainfall onto freshwater produces a negative salinity, which would not occur if local salinities could be used. As with below-freezing temperatures, negative salinities should not be used in a nonlinear EoS.

What happens in an OGCM when surface cooling makes the ocean temperature fall below freezing? First, the ocean is heated to the freezing point, with the added benefit of avoiding any conflict with the EoS. The exact heat needed to do so appears as the latent heat of fusion of newly formed frazil ice. Frazil ice forms in the water column and is distinct from basal ice that forms at the base of existing ice. Thus, frazil ice becomes an ice-ocean freshwater flux, $F_{io} < 0$, that increases the ocean salinity due to brine rejection given by the difference between the ocean salinity and the salinity of frazil sea-ice, $S_{\text{ice}} \approx 4$ g salt/kg. Melting of sea-ice decreases the ocean salinity, but by less than freshwater because the salt in sea-ice is returned back to the ocean, but generally at a different location. A troublesome complication to avoid is having numerical undershoots below freezing lead to ice formation. One helpful remedy is to form ice only in the uppermost layer of the ocean when it falls below freezing.

### 6.1.5 Coupling

The atmosphere, ocean, sea-ice, and land model components of a climate model are generally connected by some sort of coupling code that exchanges information between components. It can also synchronize the timing of the components, compute some fluxes, and ensure conservation by making sure that a flux from one component is passed exactly to the other components. It is not possible to have the ocean and atmosphere, for example, exactly synchronized, but they must be made conservative. To illustrate using the coupling strategy sketched in Figure 6.4, integration of the atmosphere over a number of time steps can pass its state information, $S_A$, needed for flux calculations, as well as internal atmospheric fluxes, $F_A$, such as solar radiation, to the coupler code as it proceeds. However, the atmospheric forcing includes fluxes, $F_{\text{SST}}$, that depend on the sea surface temperature, SST, from the OGCM and this is only available from the last time the ocean sent its state information ($T_1 =$ SST and $\mathbf{U}_1 =$ surface velocity) to the coupler. Thus, the atmosphere "sees" an SST, through $F_{\text{SST}}$, given by the ocean that "saw" a previous atmosphere, so the integrations are fundamentally asynchronous.

In the example shown in Figure 6.4, though not in all variants, the ocean coupling frequency is only once per 1 day and the upper grid box is about 10 m thick. The coupling can be more frequent, but the increased computational cost may not be acceptable, depending on the time stepping scheme. Most ocean climate models now have a free surface that can move up and down in response to ocean dynamics, but in some models the upper layer advection numerics assume the upper grid box to

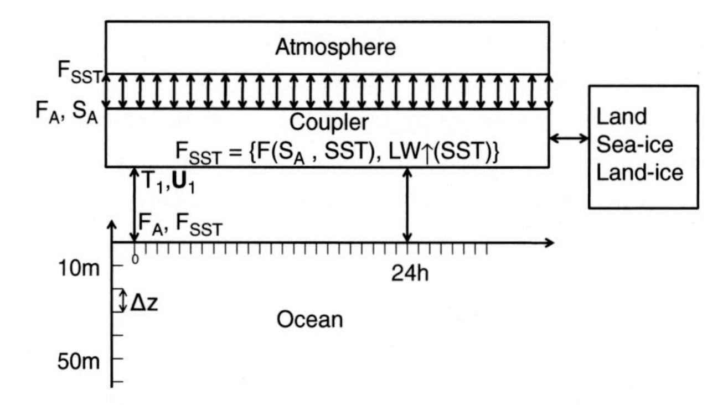


Figure 6.4. Schematic of atmosphere-ocean exchange through a coupler of atmospheric state variables, $S_A$, and radiative fluxes, $F_A$, of upper layer ocean temperature and velocity ($T_1$, $\mathbf{U}_1$), and of fluxes that depend on SST, $F_{\mathrm{SST}}$, that are conservatively passed to the ocean and atmosphere.

be much thicker than these order meter displacements, and hence much too thick to produce observed magnitudes of diurnal surface warming.

The fluxes, $F_A$ and $F_{\mathrm{SST}}$, passed to the ocean could be the time series, or an average of the fluxes used by the atmosphere the previous day, so there is conservation. The time series option gives high-frequency forcing of the ocean and in particular a diurnal cycle of solar flux, so that at night there is always a negative (cooling) heat flux and associated convective (negative) buoyancy flux, except in rare cases of very high precipitation. This physics can also be achieved with daily averages by distributing the daily solar radiation only over the daytime hours for a given latitude and time of year and peaking at local noon. Not surprisingly, the greatest impact of such a diurnal solar cycle is at the equator. But the magnitude of the effect on mean SST (temperature of the $\sim$10 m thick uppermost grid box) is much greater in a fully coupled climate model than the expected order 0.01°C increase. Instead, Figure 6.5 shows an improvement of about a 1°C increase for this "diurnal" case compared to a "1-day" coupling without it. What happens is that the small rectification of SST is seen by the atmosphere, which sets up a positive feedback with wind and clouds and SST that grows until the 1°C warming is reached. This is only one example of the exciting and unexpected behavior of fully coupled climate models. Figure 6.5 also shows that there is great improvement relative to the observations, but little more is gained for the trouble and cost of coupling every 3 hr, or even every 1 hr, as the curves for these cases are nearly indistinguishable. The implication is that the remaining bias is due to other factors, such as vertical mixing in both the ocean and atmosphere components.

### *6.1.6 Concluding remarks on Section 6.1*

A recommendation from Section 6.1 is to become familiar with simplified climate models before becoming too involved in the development and analysis of the OGCMs

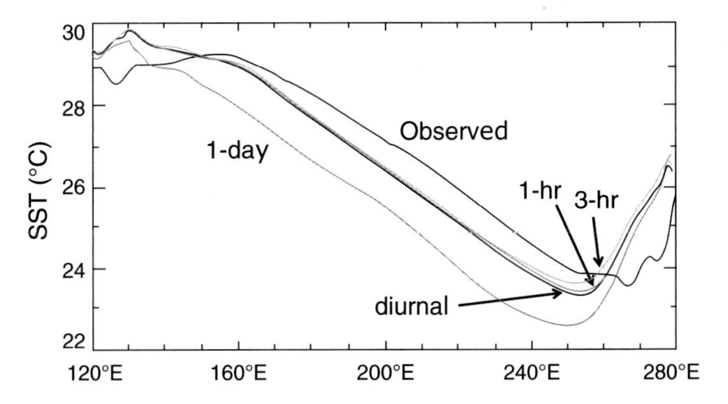


Figure 6.5. Mean SST across the equatorial Pacific from observations compared to a CCM with coupling to the atmosphere every 1 hr, every 3 hr, and once per day without (1 day) and with (diurnal) a diurnal solar cycle. Adapted from Danabasoglu et al. (2006).

introduced in Section 6.1.2. Even very much simplified ocean models can exhibit fascinating behavior that may or may not be relevant to present climate. For example, is the forcing, $F$, of Figure 6.1 a representation of freshwater forcing in the Nordic Seas? Such forcing has been very strong in the past when glacial melt water was rapidly discharged from lakes into the North Atlantic, and it may become so again in the future. However, there are other components of surface forcing (Section 6.1.4), and the density increases that are the THC's analogue of phase 3 discussed in the preceding, depend on freshwater loss (e.g., ice formation), not just cooling. A negative buoyancy forcing drives the convective boundary layers of Section 6.2. Also, the transport, $q$, might be a proxy for the AMOC, an important component of deep ocean ventilation (Section 6.3) and the prime motivation behind the parameterized overflows of Section 6.4. Are all aspects of the AMOC modeled well enough for it to behave like nature in the past, present, and future?

A lesson from Section 6.1 is to be cautious and realistic about OGCMs. Computing resources are finite, and there are conflicting views about how they should be allocated (Introduction). The hope that model resolution will eventually become high enough to satisfy requirements for one's particular problem may never be realized in a career, but other avenues may be open. For example, very high vertical resolution near the surface is necessary to explicitly compute $T_{\text{skin}}$ and $T_{\text{bulk}}$, but there may be adequate, more feasible alternative methods (e.g., Zeng and Beljaars 2005). Certainly the diurnal distribution of a daily averaged solar radiation is a successful (Figure 6.5) approach to satisfying constraints on coupling frequency (Figure 6.4). Numerical modeling is fraught with constraints (Section 6.1.3), which need to be appreciated in order to avoid misinterpretations and to deal with frustrations. Sometimes novel approaches can be employed. For example, the parameterizations of Section 6.4, appear to overcome many of the difficulties with representing deep ocean overflows in climate OGCMs.

## 6.2 Convective Boundary Layers

A planetary boundary layer is that portion of a geophysical fluid that is directly influenced (forced) by fluxes across the boundary, and a geophysical fluid feels the Earth's rotation. Examples are the atmospheric boundary layer at the bottom of the atmosphere and the benthic boundary layer at the bottom of the ocean. The latter is stress driven, whereas the lower atmosphere is often strongly convective following daytime heating of the underlying land or ocean. In contrast, the surface Ocean Boundary Layer (OBL) is almost always convective at night. Boundary layers are characterized by high Reynolds number and dominated by fully developed three-dimensional turbulence, in contrast to the subordinate role of turbulence in the geostrophic interior. Boundary layer turbulence is relatively well understood thanks to semiempirical similarity theory (Section 6.2.2), which can be applied to the OBL (Section 6.2.1) of climate models, with Large et al. (1994) a general reference for all of Section 6.2.

The OBL provides the connection between the ocean interior and the atmosphere, but the stratification below the OBL can be an effective barrier to communication, and hence to ocean ventilation (Section 6.3). Turbulent transport in the OBL can produce deep winter mixed layers, take surface fluid to considerable depths itself (deep ocean convection; Legg, Chapter 5, this volume), and penetrate into the stratified geostrophic interior. This penetration into and through the stratification barrier is most effective when the OBL is convective, as defined by surface buoyancy loss (Section 6.1.4). Therefore, the processes of penetrative convection in ocean climate models are the focus of Section 6.2.3.

### *6.2.1 The Ocean Boundary Layer*

OGCM boundary layer schemes, sometimes called mixed-layer models, must fit within the numerical constraints. Since the models conserve momentum, kinetic energy is often diagnostic, and potential energy is given by the buoyancy. Ideally, schemes should recognize the distinctive nature of boundary layer turbulence and should include a diurnal cycle of solar heating (Figure 6.5). They have been as simple as the fixed 50 m upper layer of Pacanowski and Philander (1981) but can much more complicated, high-order turbulence closures (Mellor and Yamada 1982). Large Eddy Simulation (LES), although very high fidelity, is too computer intensive to even consider in an OGCM, but it is very useful for ocean process studies and verification (Large and Gent 1999).

A schematic of the layered structure of the upper ocean is shown in Figure 6.6. Moving from top to bottom, there are three distinct regimes within the OBL, with distance, $d$, from the surface the important length scale. In the upper few centimeters, there is a viscous skin layer where $d$ is small enough to keep the Reynolds number ($Re = dU/\upsilon$, where $\upsilon$ is molecular viscosity and $U$ is the speed of a characteristic

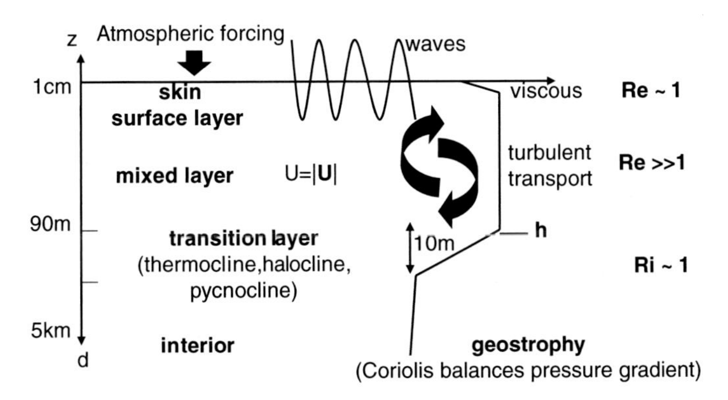


Figure 6.6. Schematic showing the density profile of an ocean boundary layer of depth $h$. Layers are a consequence of the changing balance of momentum terms as distance, $d$, from the surface increases. These balances are characterized by the Reynolds number, *Re*, and the Richardson number, *Ri*, defined in the text.

horizontal velocity, **U**) of order 1. As $d$ increases, the Reynolds number increases until there is a well "mixed" layer of fully developed three-dimension turbulence. The turbulence and mean structure (logarithmic profiles) of roughly the upper 10% of the boundary layer is relatively well studied and understood, as discussed in Section 6.2.2. Therefore, this ocean surface layer should be relatively well represented in OGCMs, though OBL modifications may be called for once there is a consensus view on surface waves effects that are not already accounted in the surface forcing.

Although wind mixing is often sufficient to maintain boundary layers depths of order 100 m, the surface buoyancy flux is responsible for the very large range. At one extreme, depths of only a few meters are found under conditions of strong stabilizing forcing (solar heating, rainfall) and weak winds that are commonly found in the tropics. At the other extreme, boundary layer depths of more than 1,000 m are found in few select regions (e.g., Labrador Sea) during late winter convection forced by strong surface cooling.

Below the mixed layer, there is a transition layer of large vertical gradients of temperature (thermocline), salinity (halocline), and buoyancy (pycnocline) that separates the OBL from the geostrophic ocean interior. The upper reaches of this gradient layer can become turbulent when the boundary penetrates into the stratification of the pycnocline ($h >$ mixed layer depth) but stays for too short of time to completely mix. Otherwise, turbulence in this layer is governed by the local Richardson number, *Ri* $= N^2/|\partial_z \mathbf{U}|^2$, where the square buoyancy frequency, $N$, equals the vertical buoyancy gradient ($N^2 = \partial_z B$). For *Ri* greater than about 1, there is insufficient kinetic energy (scales with $|\partial_z \mathbf{U}|^2$) for a turbulent overturning to supply the needed increase in potential energy (scales with $N^2$). Empirically, for *Ri* less than about 0.25, there is sufficient energy to maintain a fully turbulent fluid.

### 6.2.2 Similarity Theory

According to the semiempirical Monin-Obukov similarity theory (Monin and Yaglom 1971), in the surface layer not too close to the surface roughness elements themselves, the only important turbulence parameters under neutral forcing ($B_0 = 0$) are distance, $d$, from the boundary and the surface kinematic fluxes. These fluxes define the turbulent velocity scale, $u*$, and the turbulence scales, $x*$, for any scalar, including temperature, $t*$, and salinity, $s*$:

$$|\tau_o|/\rho = u^* u^* = u^{*2}$$
$$Q_e/(\rho_o \zeta_p) = u^* t^*$$
$$F_o S_o/\rho_o = u^* s^*$$

In neutral forcing, dimensional analysis says that nondimensional groups formed by these parameters must equal a constant. Empirically, the group, $x* / (d\, \partial_z X)$, equals the von Karman constant, $\kappa = 0.4$. Integration of this relation gives the familiar "law of wall," logarithmic vertical profiles in the surface layer, $X(d) = (x*/\ \kappa)\ \ln(d/z_x)$, where the so-called roughness lengths $z_x$ are the constants of integration.

The surface buoyancy flux gives the kinematic surface buoyancy flux, $< wb >_o = -B_0$ and is the rate of potential energy loss (or gain if negative), while the mechanical production of turbulent kinetic energy is $u^{*2}\ \partial_z U$, which from the earlier empiricism scales as $u^{*3}/(\kappa\ d)$. Therefore, the depth at which these two quantities match is a fundamental turbulent length scale governing the role of the buoyancy flux. It is known as the Monin-Obukov depth, $L = u^{*3}/(\kappa\ B_0)$, and it means there is now one more parameter than dimension, so dimensional analysis then says that nondimensional groups should be universal functions, $\phi_x(d/L) = (\kappa d/x^*)\ \partial_z X$. These nondimensional profiles have been empirically determined from observations in the atmosphere's surface layer and in the absence of evidence to the contrary (e.g., an additional role of surface waves beyond the surfaces fluxes), they are assumed to be the same in the ocean. They equal 1 in neutral ($d/L = B_0 = 0$), are less than 1 in convective ($d/L < 0$; $B_0 < 0$), and are greater than 1 in stable ($d/L > 0$; $B_0 > 0$) conditions.

If the scalar diffusivity of a tracer, $K_x$, is defined as usual by the expression $u*\ x* = -K_x\ \partial_z X$, then in the surface layer similarity theory says

$$K_x \rightarrow -u^* x^* / \partial_z X = \kappa d u^* / \phi_x(d/L) \rightarrow \kappa u^* d \text{ in neutral conditions.}$$

The eddy diffusivity can be interpreted as an eddy size proportional to $d$ times a velocity, $u^*/\phi_x(d/L)$. This turbulent velocity scale is just $u*$ in neutral conditions and smaller in stabilizing conditions. In a convectively driven boundary layer, $B_0 < 0$, it is larger, but a different definition of diffusivity is needed, because the turbulent fluxes can be nonlocal, as illustrated in Figure 6.7 (right profile).

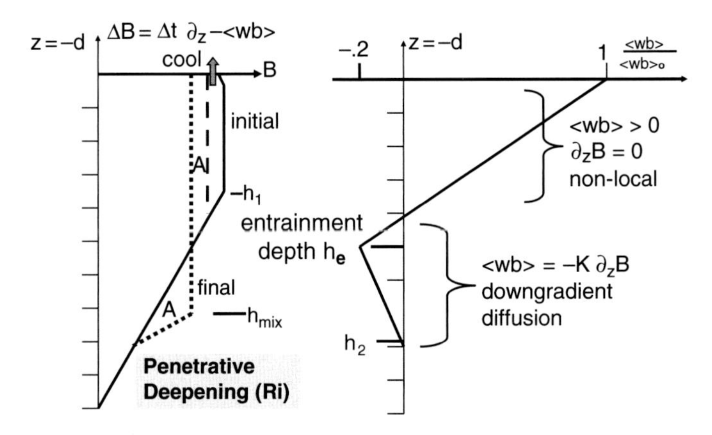


Figure 6.7. Schematic of the penetrative convective deepening of an OBL. (*Left*) The initial buoyancy (solid), final nonpenetrative (dashed), and final penetrative (dotted). (*Right*) The buoyancy flux of penetrative convection.

### *6.2.3 Penetrative Convection and Spice Injection*

Deardorff (1972) uses the temperature variance equation to derive the general form for the buoyancy flux under convective forcing, $<wb> = K_b(\partial_z B + \gamma)$, with $\gamma = C^* <wb>_o\ (hu^*/\phi_x)^{-1}$ and $C^*$ an empirical constant. This means that there can be a flux of buoyancy even where there is no local buoyancy gradient. In the convective limit, $u^* \to 0$, turbulent fluxes in a boundary layer of depth, $h$, should scale with the convective velocity $w^* = (-B_0 h)^{1/3}$, which implies the form $\varphi_x(d/L) = (1 - c\ (d/L)\ )^{-1/3}$, for c another empirical constant. The turbulent velocity scale then becomes $(u^{*3} + c\,\kappa\ w^{*3})^{1/3}$, and there is a smooth transition from wind driven ($w^* = 0$) to convective boundary layers ($u^* = 0$; $B_0 < 0$). Figure 6.7 illustrates another characteristic of purely convective boundary layers that models ought to represent. The boundary layer deepens because $<wb>$ decreases with depth until it changes sign and becomes equal to about $-0.2\ <wb>_o$ at the entrainment depth, $h_e$ (Ball 1960; Lenschow 1974; Lenschow et al. 1980).

In Figure 6.9, buoyancy loss (e.g., cooling) through the surface results in reduced boundary layer buoyancy according to $-\Delta \mathrm{B} = \Delta \mathrm{t}\ \partial_z <wb>$. Conservation would be satisfied with the dashed, nonpenetrative profile. However, convective boundary layers do penetrate into stratification, as represented by the dotted profile on the left and the buoyancy flux on the right. Conservation demands that the two areas denoted $A$ be equal. Below the final mixed-layer depth, $h_{\mathrm{mix}}$, there is a gradient layer where the buoyancy has increased. The isopycnals (surfaces of constant density) represented by the range of buoyancies in this layer can all be seen to have deepened from the initial (solid) profile down to the dotted, but none outcrop, a term used to describe an isopycnal in direct contact with the surface.

As illustrated by Figure 6.3, ocean density decreases with temperature and increases with salinity. Over large regions of the subtropical gyres, in and downstream of the evaporation zones in particular, a salty surface lies over a fresh interior. The stability of the water column is maintained by a positive temperature gradient, unless winter cooling (recall 90% of the buoyancy loss from evaporation is due to cooling) removes sufficient mixed-layer buoyancy to generate penetrative convection. In an OGCM, the buoyancy is seen to behave similar to that shown in Figure 6.7, but the presence of a destabilizing salinity gradient means that in the gradient layer the increase in buoyancy is due more to the mixing of heat from above than fresh water from below. Therefore, the deepening isopycnals are more saline than before the penetrative convection. Since salinity variability on an isopycnal is known as spice, this process is like a spice injection into isopycnals lodged within the pycnocline. The oceanographic parameter used to describe density compensating temperature and salinity structures is the Turner angle, *Tu*. It is most clearly defined by means of Figure 6.8. The regime of interest for the penetrative convection example is for $45° < Tu < 90°$, where the water column is stable because the destabilizing salinity gradient is more than compensated by the stabilizing temperature gradient.

The differential mixing of temperature and salinity in the gradient layer of Figure 6.7 also occurs above $h_{\text{mix}}$, where there are temperature and salinity gradients that compensate in density. A search of the hydrographic database for observations of a well-mixed layer in density, but with significant density-compensating temperature and salinity gradients (spice) in the lower portion yields the rare example from the Southeast Pacific in 1967 shown in Figure 6.9. The temperature and salinity profiles are well mixed to about 100 m, but deeper they nearly exactly compensate in density, so the Turner angle is nearly 90° and the density is very well mixed to 139 m. The Turner angle is nearly 90° also between about 160 and 170 m. The latter is likely a remnant of earlier penetrative convection farther upstream (south) when the $\sigma = 25.8$ isopycnal

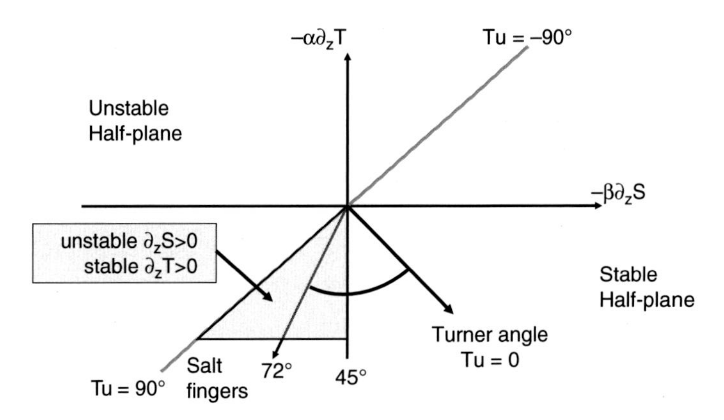


Figure 6.8. Turner angle defined as a function of the stabilizing effect of the salinity gradient ($-\beta\,\partial_z S$), and the destabilizing effect of the temperature gradient ($-\alpha\partial_z T$).

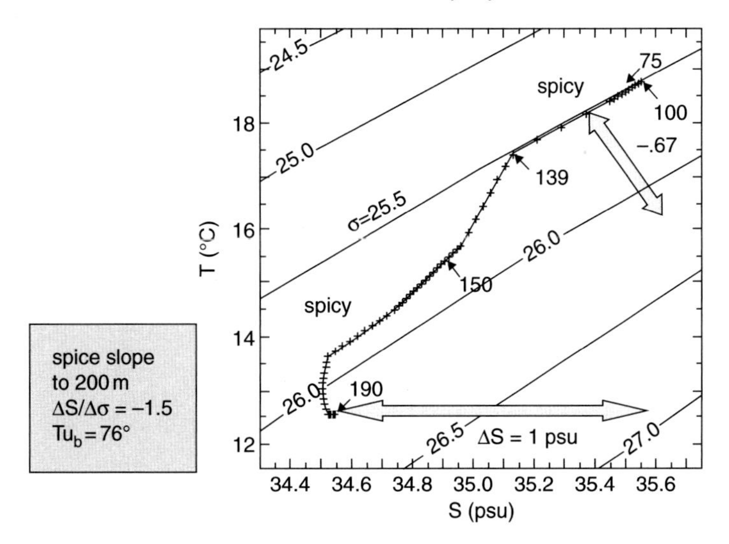


Figure 6.9. Temperature versus salinity from the surface to 200 m depth at 16°S, 88°W in September of 1967. There is maximum spice ($Tu \approx 90°$) in the water column between 100 and 139 m, and also from 160 to 170 m (Yeager and Large 2004).

was in the gradient layer. Over the upper 200 m, the density increases by only 0.67 $\sigma$-units, whereas salinity alone would lead to about a 3.5 $\sigma$-unit decrease, with the difference due to the stabilizing temperature. The overall density compensation from the surface to 200 m is described by a bulk Turner angle, based on the temperature and salinity differences over the upper 200 m. In Figure 6.9, $Tu_b$ is equal to about 76° making the whole water column favorable to salt fingering. This buoyancy process is driven by the higher molecular diffusivity of heat compared to salt and generates turbulent mixing despite a stable buoyancy gradient (Schmitt 1988).

### *6.2.4 Concluding Remarks on Section 6.2*

If positive surface buoyancy forcing (Section 6.1.4) was dominant everywhere, then a likely equilibrium ocean would have a very buoyant OBL above a very strong pycnocline (Section 6.2.1). With the interior isolated from the OBL and atmosphere there would be no ocean ventilation (Section 6.3), and no reason to be concerned about the overflow parameterizations of Section 6.4. In nature, however, the buoyancy forcing is convective (negative) almost always at night, and extremely convective at high latitudes due to strong winter cooling and brine rejection during sea-ice formation. The key message from Section 6.2 is that convective boundary layers can penetrate into the pycnocline and even through the pycnocline, and play a role in shallow (Section 6.3) and deep (Sections 6.3 and 6.4) ventilation. Furthermore, Section 6.2 summarizes the semiempirical Monin-Obukhov similarity theory (Section 6.2.2), which is the foundation for our understanding of convective boundary layers. This physics can be incorporated into ocean climate models (e.g., Large et al. 1994).

Section 6.2 also illustrates some of the characteristics of convective ocean boundary layers (Figure 6.7), introduces the Turner angle as a useful diagnostic of spice injection during penetrative convection (Figure 6.8), and presents observations that suggest that ocean behaves in a similar manner (Figure 6.9). It presents an essentially one-dimensional (vertical) view and sets the stage for the three-dimensional perspective of Section 6.3, where the spice injection into subducted isopycals due to penetrative convection modifies our conception of shallow thermocline ventilation in ocean climate models (Section 6.3.3). A bulk Turner angle is used as a diagnostic for further verification against observations that the process is at work in nature, too.

## 6.3 Ventilation in Ocean Models

Under present climate conditions the entire ocean is ventilated; that is, surface water is mixed down throughout the OBL and circulated to virtually everywhere in the ocean, including the bottom of the deepest basins. Recall, ventilation, as defined by turbulent mixing to depth and interior circulation involving sinking deeper can be regarded as the THC analogue to phase 4. How do OGCMs ventilate? Section 6.3 begins by treating two modeling techniques for quantifying ventilation: ideal age (Section 6.3.1) and transit time distributions (Section 6.3.2). It then moves on to specific examples of ventilation. However, only the two such circulations deemed most important to climate modeling will be considered. Section 6.3.3 examines the shallow ventilation as intermediate water of deep winter mixed layers formed in the subtropics flows along isopycnals to the equator within the pycnocline. The second (Section 6.3.4) is the spread and sinking (Spall, Chapter 3, this volume; Legg, Chapter 5, this volume) of North Atlantic Deep Water (NADW), where the Atlantic Meridional Overturning Circulation (AMOC) is used as a measure of the rate of NADW ventilation. The deep ocean convection in the Nordic Seas and Labrador Sea relevant to NADW is considered in general in Section 6.2 and specifically in Chapter 5 (Legg, this volume), so will not be considered further. However, before spreading as NADW, source water from the Nordic Seas must pass through the shallow ridges between Europe and Greenland and these overflows are the subject of Section 6.4.

### *6.3.1 Ideal Age*

How long ocean water has effectively been away from the surface is its age. Observational estimates of age have been derived from measurement of the relative concentrations of tracers. For example, if surface water (Age $=$ 0) is assumed to be saturated with radioactive tritium, $^3$H, and deplete in its daughter helium, $^3$He, then the known decay rate allows an age to be computed from measured $^3$H and $^3$He concentrations. However, the calculation is nonlinear, and mixing produces a young bias.

A climate model alternative, not observable and not subject to mixing biases, is the ideal age tracer, $A$. It is computed from a straightforward advection-diffusion equation:

$$\partial_t A = -U \cdot \nabla A + \nabla \kappa \nabla A + 1.$$

The first term on the right-hand-side describes ageing due to transport by the resolved flow, $\mathbf{U}$. The second term is a parameterization of mixing as down-gradient diffusion with a possibly spatially dependent coefficient, $\kappa$. The third term is a unity source term that would age a stationary water parcel with no mixing by 1 $\Delta t$, every time step. A simple boundary condition sets $A = 0$ in all the uppermost grid levels in a model, and turbulent mixing keeps $A$ near 0 throughout the OBL. Therefore, $A$ is a quantitative measure of ocean ventilation because it is the effective time since a water parcel was last exposed to atmosphere.

Ideal age is a very useful tracer for quantifying differences between different model configurations. For example, Figure 6.10 shows the volume of Nordic Seas water with an age less than 10 yr, as an indicator of Nordic Seas ventilation. The nominal horizontal resolutions are 5 degrees (low), 4 degrees (medium), and 2 degrees (high) in the atmosphere model and 3 degrees (low) and 1 degree (medium) in the ocean. The three resolution combinations are high atmosphere, medium ocean (HM), medium in both components (MM), and low in both (LL). The steady-state values for present-day conditions are about 0.41, 0.33, and 0.10 million km$^3$, respectively. Also shown for each, are two transient simulations of the response to global warming scenarios. Ventilation in the MM configuration responds faster and stays lower than in HM. In

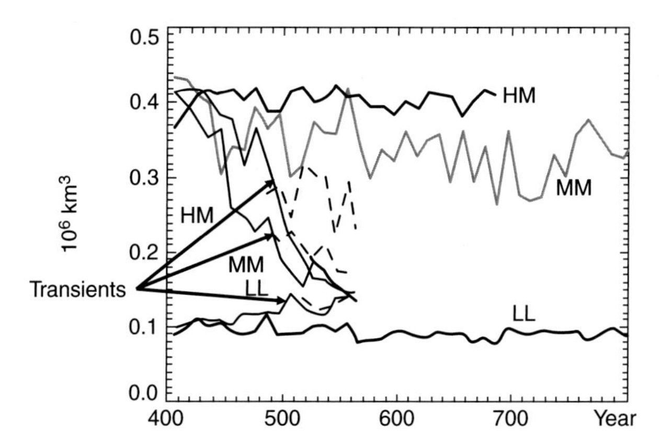


Figure 6.10. Volume of Nordic Seas water with Age <10 yr over time from different CCM component resolutions given in the text. Shown for each are the present-day (thick solid traces) simulation, and two transient responses to increasing $CO_2$ scenarios (thin solid and dashed traces). Adapted from Bryan et al. (2006). ©American Meteorological Society. Reprinted with permission.

contrast, the LL case appears to have weak Nordic Seas ventilation even in the present day and shows little response to warming. These results suggest that LL resolutions may not be adequate for studies involving ventilation of the Nordic Seas.

### 6.3.2 Transit Time Distributions

Ventilation is a nonlinear combination of the effects of both advection and diffusion, such that the concentration of a tracer $C(x,t)$, of a water parcel at a point, $x$, at time $t$, is the result of multiple pathways, each originating at a different surface point of a region, $\Omega$, and taking a different time to reach $x$. For a statistical steady-state ocean with no interior sources or sinks of the tracer, and a time varying surface concentration, $Co(t)$, that is uniform over the domain, $\Omega$, the Transit Time Distribution (TTD) function, $G(x, t, \Omega)$, is defined by the convolution integral

$$C(x,t) = \int_0^\infty d\xi \; Co(t-\xi)\; G(x,\xi,\Omega).$$

The TTD is computed as the solution to a model's transport equations forced by a delta-function impulse surface concentration, $Co(t = 0)$, applied uniformly over $\Omega$. In practice, the surface tracer concentrations are applied for 1 yr to account for the seasonal cycle (e.g., the seasonal thermocline and deep winter mixed layers). These solutions take thousands of years to equilibrate, so TTDs are computationally expensive. In the case of an impulse surface concentration and pure advection taking, say 34 yr to reach $x$, $G(x, t)$ is just a delta function at $t = 34$ yr, as depicted in Figure 6.11. The other examples in Figure 6.11 include diffusion, which has the effect of spreading the distribution and moving the peak to a shorter time. These TTDs quantify the rapid ventilation of the Labrador Sea down to 1500 m by winter deep convection almost every year, the fast ventilation of the shallow (500 m) subtropics in 5 to 10 years, and the still longer and more spread ventilation of the western boundary at 1,500 m. Examples from the deep ocean would stretch much longer in time. The longest is about 5,000 years in the North Pacific below the pycnocline where ventilated water originates far away in the North Atlantic and off Antarctica.

It is possible to divide the ocean surface into a number of domains, $\Omega$, and apply an impulse of a different independent tracer for each domain. A single, albeit expensive, model integration can then be used to compute TTDs for each domain, and a sum over these domains gives the global domain answer. For example, Peacock and Maltrud (2006) have 11 domains. They spun up an OGCM for about 300 yr and then applied a different tracer source function uniformly over each region for 1 yr only. The sum of the 11 individual TTDs gives the $\Omega =$ global TTD, but their regional results are of particular interest. For example, at a 183 m location in the Indian Ocean there is rapid 10 yr, mostly nonlocal ventilation from the South Indian domain. Similarly, an eastern tropical North Pacific site is ventilated at 94 m in 14 yr, mostly from the

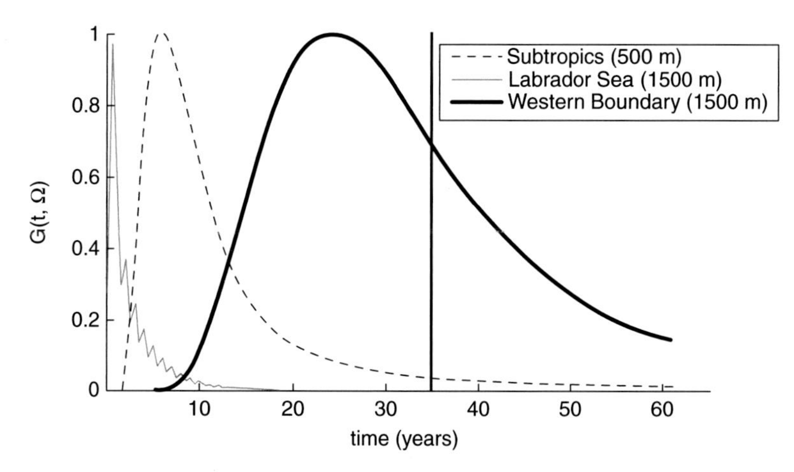


Figure 6.11. Normalized TTD for $\Omega =$ the global ocean, at three locations, given by the time series of OGCM concentration at each, following an "impulse" boundary condition. Adapted with permission from Khatiwala et al. (2001).

southern Pacific. The ventilation of the South Atlantic at 2,014 m peaks at 65–70 yr, the time for water to arrive from the subpolar North Atlantic. None of these results is unexpected, but the TTDs do provide a quantitative measure of ventilation due to both advection and diffusion. Clearly, with more regions, the results become more quantitative and informative, but the computational and data processing costs escalate.

There are a number of practical applications of TTDs. Integration of a TTD gives the cumulative probability function, which converges to 1 as time approaches ideal age. Therefore, these integrals provide an indication of when ideal age becomes meaningful. Solving the previous convolution integral is an economical way of estimating the uptake of anthropogenic gases, provided the time history of source concentration, $Co(t)$, is known, and the TTDs have been computed. TTDs are a quantitative diagnostic for model intercomparisons, especially of ventilation. Similarly, they are a quantitative measure of the advection and diffusive effects of resolved versus parameterized mesoscale eddies as well as of other resolution effects. For example, Figure 6.10 implies that TTDs from the Nordic Seas would peak at shorter times for the HM configuration than for LL. Finally, TTDs can be used in simple experiments to test hypotheses.

### *6.3.3 Shallow Ventilation*

Shallow ventilation of the ocean's thermocline is thought to be relatively well observed and understood, so climate models should strive to represent the processes faithfully. Using the South Pacific as an example, the zonally averaged conception is illustrated in Figure 6.12. During southern winter deep mixed layers expose dense isopycnals to the surface, where surface fluxes in the outcrop region set the temperature and salinity.

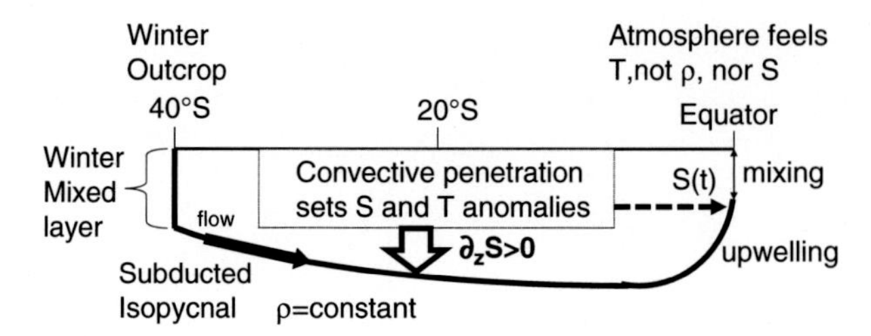


Figure 6.12. Schematic of shallow ventilation of the pycnocline by water formed in a winter mixed layer around 40° S flowing along an isopycnal toward the equator. Spice injection can occur around 20° S where the salinity gradient is strongly destabilizing, producing anomalies that are seen at the equator much later.

Interannual variability of the outcrop, the mixed layer depth, and the fluxes produces interannual variability on the isopycnal, including spice. Water parcels on isopycnals lodged in the permanent pycnocline are isolated from the surface by a seasonal summer pycnocline. The process is known as subduction, and flow toward the equator is along subducted isopycnals. Perhaps years later the parcel approaches the equator, where wind-driven upwelling and mixing bring it back in contact with the surface, so that the parcel temperature is a contributor to the SST seen by the atmosphere, and hence the surface fluxes of heat and evaporation. However, the atmosphere feels neither the density nor salinity directly, so the influence of a water parcel on a given isopycnal depends on its spice, as set back at its winter outcrop.

At least one particular OGCM differs from this simple concept in one critical aspect. As shown in Figure 6..13, the monthly variance of salinity anomalies is a maximum, not over the winter outcrop region, but farther toward the equator where the $\sigma = 25.5$ isopyncal remains subducted throughout the year. The implication is that the locations of these maxima are where the spice anomalies are set for this isopycnal. This signal is strongest on $\sigma = 25.5$ in the southeast Pacific (Figure 6.13), but on other isopycnals in other basins. In accord with Figure 6.12, the variance map of Figure 6.13 suggests propagation to the Equator, but via a pathway across the Pacific, in both hemispheres (dotted trajectories). From Section 6.2, recall that the large variance occurs in the same regions of high surface salinity overlying fresh Intermediate water where penetrative convection can mix salt into the pycnocline. This is the additional process depicted in Figure 6.12 as setting $S$ and $T$ anomalies on the isopycnal around 20°S.

A signature of the spice injection process in the OGCM at 23°S, 100°W in the southeast Pacific, is shown in Figure 6.14, by mid-winter (August) to end of winter (October) increase in salinity on isopycnals found below the mixed layer. Unlike the previous simple concept of thermocline ventilation, these subducted isopycnals are not isolated from the atmosphere. Injections evident in Figure 6.14 well below the mixed layer (e.g., 1985) likely occurred farther upstream (south) in the model, which would be in accord with the observed deeper spicy layer in Figure 6.9. In Figure 6.13, the $\sigma = 25.5$ isopycnal is shown to outcrop farther south (hatched), where local air-sea

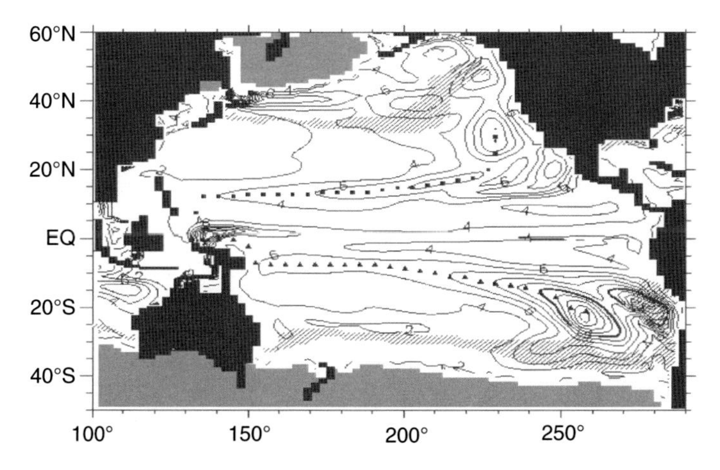


Figure 6.13. Contours at 0.02 psu intervals of rms monthly salinity on $\sigma = 25.5$, from 40 yr of an OGCM, showing source regions of spice variability on the isopycnal. The envelope of winter outcrops is hatched. From Yeager and Large (2004). ©American Meteorological Society. Reprinted with permission.

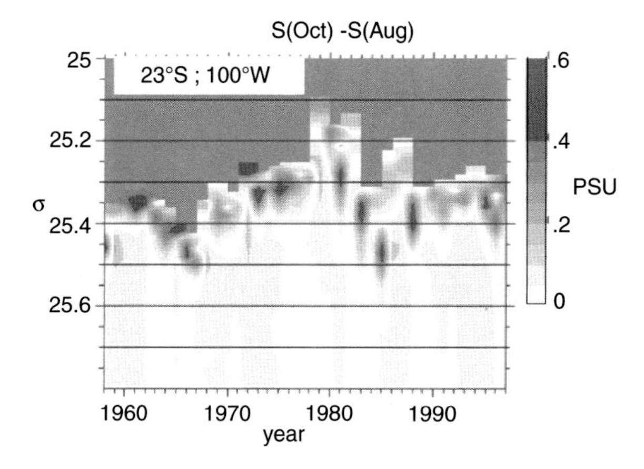


Figure 6.14. Annual time series from an OGCM at 23°S, 100°W, showing the mid-winter (August) to end of winter (October) increase in salinity on isopycnals. The mixed layer density is the maximum density of the solid gray shading, and lower densities do not exist Adapted from Yeager and Large (2004). ©American Meteorological Society. Reprinted with permission.

flux anomalies would not be expected to be related to the spice injections of Figure 6.14. This disconnect may explain the Kessler (1999) finding of an observed general 0.3 psu rise in $S$ on $\sigma = 24.5$ over the 1990s south of the equator at 165°E. In this study "an attempt was made to trace the source of these changes to changes in the surface fluxes at the outcrop, where $E$-$P$ variations suggested salinity anomalies of the same magnitude. However, the implied salinity changes were of the wrong sign."

However, this is only indirect evidence that shallow ventilation in the OGCM is anything like nature. The possibility of searching for direct evidence on a global scale has recently become possible with the advent of ARGO profiling floats (Roemmich et al., 2004). These floats are parked at a nominal depth of 1,000 m where their trajectories provide a measure of flow at this depth. Every 10 days, or so, the float generates buoyancy and rises to the surface measuring temperature, salinity, and hence density through the water column. At the surface, the data are transmitted to shore via satellite, and the location is determined. The float is then made to lose buoyancy and sink back to its parking depth.

What is needed to compare OGCM behavior with nature globally is a simple diagnostic of penetrative convection and spice injection on subducted isopycnals, which can be computed from ARGO data. One such signature of the process is a distinct increase in the bulk Turner angle (Figure 6.9) computed from gradients from the surface to below the gradient layer (250 m). The OGCM results in Figure 6.15 show this angle increasing from about 65° to more than 75°from summer (January through March) to late winter (July through August) in the southeast Pacific and South Atlantic. An even greater change is seen in the southeast Indian Ocean. In the Northern Hemisphere, similar changes are seen from summer (July through September) to winter (January through March) in the northeast Pacific and eastern North Atlantic. These seasonal bulk Turner angles from the OGCM (right panels) are indeed similar to those from ARGO (left panels) in both magnitude and pattern. The areas of high (>75°) Turner angles in winter match very well, except for the northwest Atlantic where the path of the Gulf Stream in OGCM is well known to be too zonal and too far south. In these regions of agreement, angles in the summer, and, therefore, the increase in winter, are very similar, too. Agreement (Figure 6.15) between the model and ARGO data in the seasonal distributions and changes in $Tu_b$ is strong evidence that shallow ventilation via penetrative convection occurs in nature, too, and verification of the model results such as Figures 6.13 and 6.14.

### *6.3.4 NADW and the AMOC*

Why are North Atlantic Deep Water (NADW) and the Atlantic Meridional Overturning Circulation (AMOC) chosen to illustrate deep ventilation? It is because surface water sinks to form deep water in the far North Atlantic and Nordic Seas, and the overflows parameterized in Section 6.4 are sources of NADW, with the AMOC strength a quantitative measure of deep Atlantic ventilation. There is also an Antarctic Bottom Water (AABW) component of the AMOC, which appears as northward flow near the bottom of the Atlantic. In the North Pacific, Figure 6.3 shows that surface water is so fresh that even if it were cooled to the freezing point, its density would still be significantly more buoyant than the saltier NADW. The North Atlantic also produces the most deep

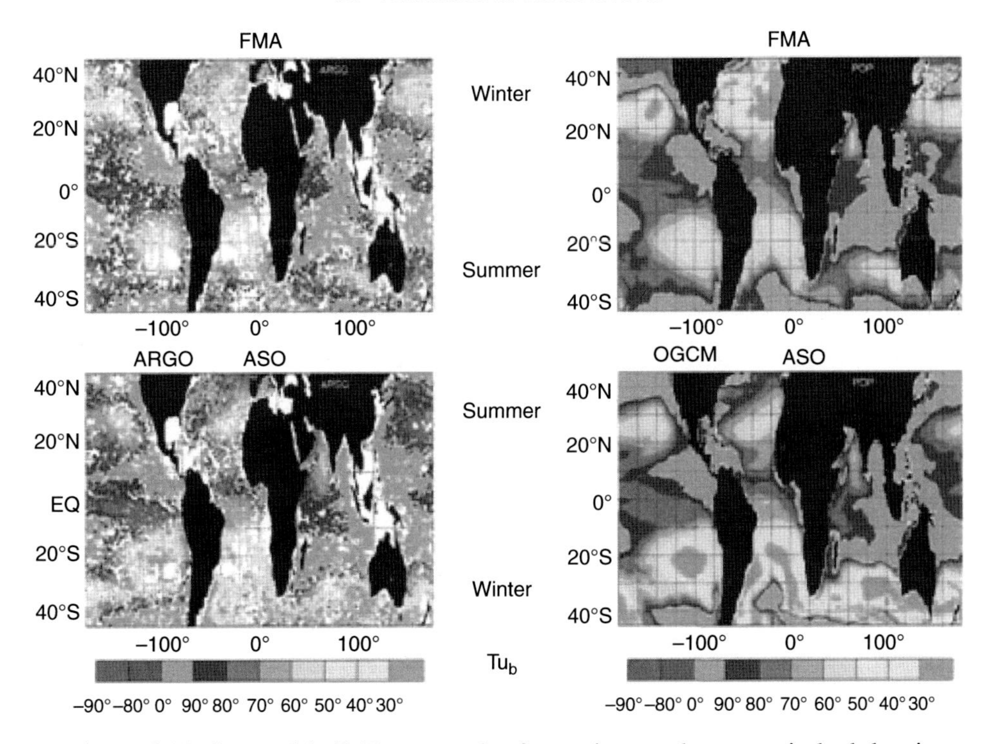


Figure 6.15. Seasonal bulk Turner angles from winter and summer in both hemispheres from an OGCM (*right*) and from individual ARGO floats (*left*). Adapted from Yeager and Large (2007). For a color version of this figure please see the color plate section.

water, and NADW can found at depth in all the ocean basins. This vigorous ventilation means that the North Atlantic is a key region for the uptake of atmospheric gases. Perhaps the most climatically important of these gases is $CO_2$. The column inventory of observed anthropogenic $CO_2$ of Sabine et al. (2004) reveals that most is found the North Atlantic where there are regions with more than 80 moles-$CO_2/m^2$. Over much of the North Pacific, there is only about 20 moles-$CO_2/m^2$. Antarctic ventilation would be second choice because of the large area of values over 30 moles-$CO_2/m^2$. Another very useful anthropogenic gas to include in ocean models is CFC-11. It was virtually nonexistent before the 1930s, and the quantity manufactured and emitted each year is well known. Following the Montreal Protocol its use declined, so there is very distinctive time history of its source function (ocean surface flux). Models employing this function have been successful at reproducing the observed depletion and apparent recovery of ozone, especially in the stratosphere over Antarctica.

A useful and common metric for the strength of North Atlantic ventilation and the AMOC is the local maximum in the meridional stream function, which should be about 18 Sv (Cunningham et al. 2007; see Section 6.4). NADW is the major contributor to the southward transport, with the northward return flow mostly in the upper ocean, including the near surface wind-driven circulation and some from the buoyancy driven AABW at depth. This metric is believed to be a reliable indicator of the ocean's capacity to uptake heat as well as anthropogenic gases, and therefore, something climate models strive to get right. Unfortunately this has proven to be a difficult task because it is very sensitive to both the ocean and atmospheric models, and in particular their resolutions. Using the same three combinations (LL, MM, HM) as Figure 6.10 of high, medium, and low atmospheric resolution, and medium, and low ocean, Figure 6.16 shows the time history of AMOC transport under present climate conditions as the top curves in the three left panels of Figure 6.16. Recall, that a high-resolution, eddy-resolving ocean is too computationally demanding for coupled climate experiments of such lengths (1,000 yr). The AMOC is seen to strengthen with resolution and is relatively steady at about 15, 19, and 21 Sv in LL, MM, and HM, respectively. The average decrease of the respective three member ensemble in AMOC in response to warming is 3.3, 4.6, and 5.4 Sv per century. Therefore, resolution differences are a likely contributor to the more than 15-Sv range in the responses to climate change forcing in Intergovernmental Panel on Climate Change (IPCC)-coupled models.

The structure of the AMOC after stabilizing at the end of the warming is shown in the right-hand panels of Figure 6.16. Noted, are the strengths of 10.6, 13.2, and 15.1 for the LL, MM, and HM resolutions, respectively. A notable difference in the low ocean resolution case, LL, is less sinking around 60°N. However, there are notable similarities. In all three cases the depth of zero transport is between 2,500 and 3,000

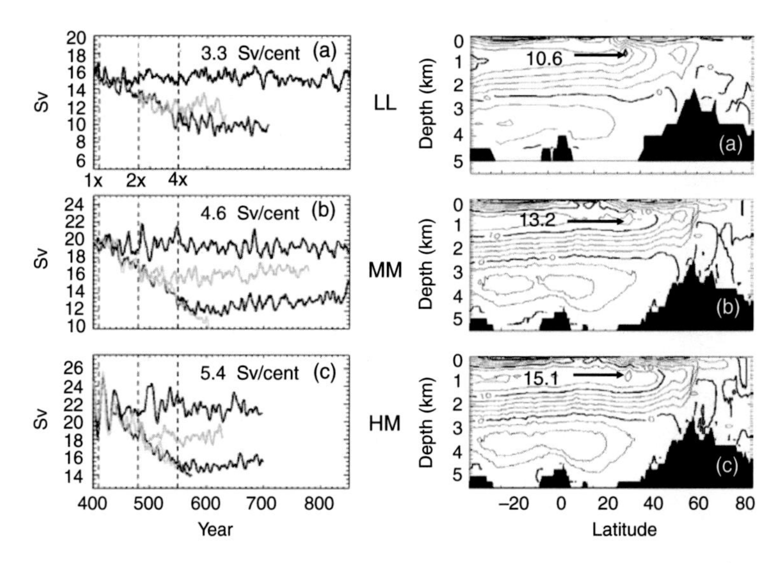


Figure 6.16. The AMOC response to global warming from the same CCM integrations as Figure 6.10. (*Left*) The controls (upper traces) and the transients (lower traces) as well as the overall decline given in Sv per century. The dotted lines are at the number of times $CO_2$ increased. (*Right*) The final streamfunctions. Adapted from Bryan et al. (2006), ©American Meteorological Society. Reprinted with permission.

m, and the southward flow only penetrates to about 3,500 m. These are long-standing shallow biases that do not seem to be related to resolution. A related bias appears to be the northward flow of mostly AABW near the bottom of the model. A possible solution to these problems is presented in Section 6.4.

### *6.3.5 Concluding Remarks on Section 6.3*

In modern climate conditions the entire ocean is ventilated, with the longest time about 800 yr for the deep North Pacific. Many processes are believed to be involved, including high-latitude buoyancy loss (Section 6.1), penetrative convection both into and through the pycnocline (Section 6.2), the overflow of dense water from marginal seas into the deep ocean interior (Section 6.4) and the spread of intermediate and deep water throughout the ocean basins (Sections 6.3 and 6.4). With such complexity, the assumption that all the necessary ingredients will always be present to ventilate the ocean may not be correct. Indeed, the Paleo Climate record suggests that this has not always been true, and there are hypothesis that ventilation will change in a warmer future climate. At times, the Black Sea has been isolated from the surface long enough to lose all its oxygen and become anoxic. Could such conditions become more frequent and widespread? How would altered ventilation affect ocean heat transport, as well as ocean storage and uptake of heat and atmospheric greenhouse gases, GHG, such as $CO_2$ and methane? These and related questions are prime motivations behind including the ocean in climate models, which have become a unique tool for studying past, present and future climates (Section 6.3.4). They can represent potentially important feedbacks. Positive feedbacks include enhanced warming due to reduced ocean heat uptake as the ventilation weakens (Section 6.3.4).

With this growing and important role comes the responsibility of evaluating how faithfully climate models represent nature. Section 6.3.3 presents compelling evidence (Figure 6.15) that this is the case for penetrative convection, provided the atmospheric forcing is accurate. However, there are biases in the AMOC solutions shown in Figure 6.16, which lowers the confidence in the transient results and motivates improving the representation of overflows in Section 6.4.

## 6.4 Parameterized Overflows

An overview of the CLIVAR Climate Process Team (CPT) on Gravity Current Entrainment is provided in Legg et al. (2009). Included is a summary of observational estimates of various quantities related to overflows and their ranges and sources. Some of these numbers are quoted here. The overview also describes the two approaches taken by the CPT. The first is to explicitly resolve overflow dynamics in process models, but with convergence not yet reached at $\Delta x = \Delta y = 2$ km, there are no immediate prospects for doing so in a climate OGCM. Some results are provided in Legg

(Chapter 5, this volume), where there are further details of ocean overflows at numerous known locations. The alternative is to parameterize important overflows. Section 6.4 highlights the outcome of this approach, including buoyancy driving (Section 6.4.1), a Parameterized Mediterranean Overflow (PMO) at Gibraltar (Section 6.4.2), and Nordic Sea Overflow Parameterizations (NSOP) through both Denmark Strait and Faroe Bank Channel (Section 6.4.3). Section 6.4.4 then continues the evaluation begun in Section 6.3 of Climate Model representation of various aspects North Atlantic ventilation and the AMOC. The references for everything in Section 6.4 plus additional details are Wu et al. (2007) for the PMO, and Danabasoglu et al. (2011) for the NSOP.

### 6.4.1 Characteristics of Buoyancy-Driven Overflows

In the common situation where a source region (e.g., marginal sea) of dense water is separated from the ocean interior by shallow topography, the dense water fills the source basin until it finds the deepest escape route to the open ocean. Given the nature of ocean bottom topography, this route is usually through a confined channel of sill depth, $d_s$, and restricted width, $W_s$, such that the flow is hydraulically controlled. These and other characteristic features of overflows are sketched in Figure 6.19: The density of the source water exceeds that of the interior; the outflow through the sill accelerates a distance, $X_{\mathrm{ssb}}$, down the slope, $\alpha$, of the continental shelf to the shelf break; vigorous entrainment near the shelf break at depth, $h_e$, dilutes the source water; and the resulting product water flows down the continental slope either to the depth where its density matches the interior density or to the bottom. Overflows are important contributors to the deep water masses of the world's oceans. The Mediterranean Overflow through the Straits of Gibraltar is the source of the warm salty Mediterranean salt tongue that spreads over much of the North Atlantic at about 1,100 m. The Nordic Overflows through Denmark Strait (DS) between Greenland and Iceland and through the Faroe Bank Channel (FBC) between Iceland and Scotland are major sources of North Atlantic Deep Water. Antarctic Bottom Water owes its existence to the Antarctic Overflows from the Ross and Weddell seas.

At climate model resolutions, both the sill dimensions and the scales of the down slope flows are subgridscale. A common practice has been to ignore the problem and simply carve out overly wide passages and let the model equations generate whatever flows they can (e.g., Figure 6.16). The problems with this method are well documented and depend on the vertical coordinate. For example, with coordinates that discretize the continental shelf and slope as staircase topography, the overflow becomes a series of lateral advection followed by vertical convection processes over each step. Such a sequence is known to entrain too much and produce too buoyant and shallow product water (Chapter 7, Treguier et al., this volume). In contrast, layer coordinate models tend not to entrain enough and need some sort of parameterization to prevent overflows from descending too deep.

Given that computational resources will not soon be available to resolve the order 1 km overflow processes, the attractive alternative examined here is to parameterize the important overflows as single entities, and not to try and explicitly resolve any of the overflow processes. For each parameterized overflow, all the SGS sill and shelf topography needs to be specified as constants (e.g., $d_s$, $W_s$, $X_{\mathrm{ssb}}$, $\alpha$, $h_e$). The inputs from the climate model are just the properties of the source water upstream of the sill and of the interior ocean at all depths on the downstream side. As in nature, parameterized overflows are driven by the excess density of the source water relative to the interior, which can be expressed as a reduced gravity, $g_S'$. Therefore, they rely on other aspects of the climate model, especially the ocean surface buoyancy fluxes over the source region, to give the mass transport of source water, $M_S$. For overflows such as the Mediterranean at Gibraltar, the return inflow, $M_I$, is restricted to be collocated with the overflow of source water, so the computation is given by (A) in Figure 6.17, with $\mathrm{M_I} = \mathrm{M_S}$, neglecting about 0.05 $\mathrm{M_S}$ of net input to the Mediterranean needed to balance excess evaporation over precipitation plus runoff. When there is no such restriction (e.g. Denmark Strait and Faroe Bank Channel) $\mathrm{M_S}$ is given by (B) (Figure 6.17), and the inflow returns wherever the equations of the OGCM dictate. In both cases, $M_S$ depends on the thickness of the source water, $h_S$, which in (A) is given by topography, $h_S = d_S/2$. In (B) it is either specified from observations, or diagnosed from the Source water properties from the OGCM.

The entrainment is determined by $M_S$ and a reduced gravity $g_E'$ based on the difference in density between the source water and interior at the depth of the shelf

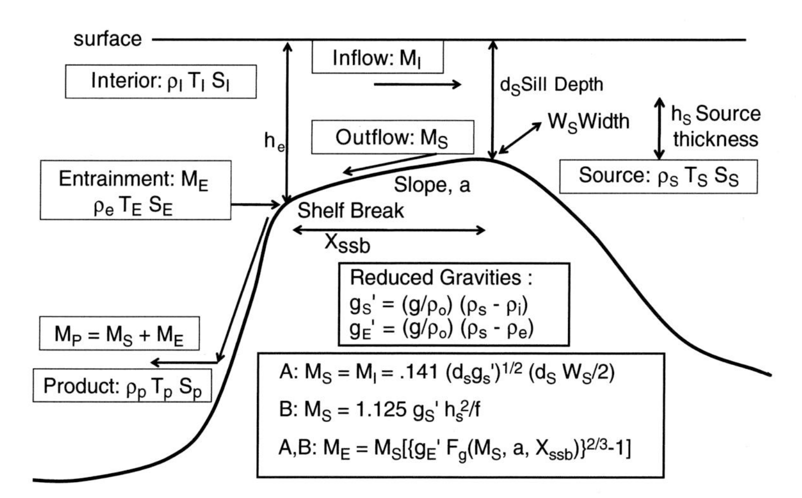


Figure 6.17. Sketch of the topography controls and buoyancy driving by reduced gravities of dense source water overflow through a channel into the ocean interior with (A) and without (B) return inflow confined to the channel.

break. The dependence of the entrainment on the topography is also strong and comes through a geostrophic Froude number (Legg, this volume), $F_g$, which in Figure 6.17 depends also on $M_S$ (Wu et al. 2007).

The product water transport is just $M_P = M_S + M_E$, and an entrainment parameter, $M_E / M_P$, is used to determine the product transport of all tracers, including temperature, salinity, and ideal age. However, determining the depth and location where the product water is injected into the interior is more complicated. It depends first on buoyancy, because the injection is at the depth where the product water first becomes buoyant relative to the interior. In general, an $i, j$ grid location needs to be specified for where to inject the product water at each depth below the entrainment depth. The path of these locations is the implicit overflow pathway down the continental slope. In practice, the path need not go beyond the steep topography of the continental slope because flow over the flatter abyssal topography is not a particular problem in ocean climate models.

### *6.4.2 A Parameterized Mediterranean Overflow*

The numerical implementation of a Parameterized Mediterranean Overflow is greatly simplified by the restriction that both the inflow and outflow are through the Straits of Gibraltar (Wu et al. 2007). A further simplifying assumption is that the inflow and outflow always balance, so that the barotropic (depth averaged) flow through the strait is always zero. This condition is satisfied by construction in models with virtual salt fluxes, by closing the strait and implementing the numerics to allow nonzero flow through side boundaries. However, with a closed strait, a natural solution is for the product water to fill the water column just on the interior side and supply both the inflow and entrainment such that there is conservation with $M_P = M_I + M_E$. In practice, there is only a tendency toward this solution because of the horizontal density gradients, but it can cause the product water depth to drift to shallower depths. There are two parts to the numerical solution. First, the side boundary conditions are applied across a side of at least three grid boxes, regardless of resolution. Second, the interior, entrainment, and product properties (temperature, salinity, and density) are taken from horizontal averages over a large domain of at least nine grid boxes, again regardless of resolution. Numerical problems are ameliorated by similar boundary conditions and averaging applied on the source side, too.

The final implementation step is diagnostic tuning. A number of the specified topographic and flow parameters (e.g., topography and bottom drag) have a large range of acceptable values. In practice, they are tuned within these ranges to produce diagnostic transports of $M_I$, $M_S$, $M_E$, and $M_P$ within their uncertainties (Table 6.1), given observed ocean temperature and salinity. Despite the large observed range, this exercise becomes the first step in evaluating the PMO and its success is clear from Table 6.1. The next step is the performance of an uncoupled OGCM forced with observed

Table 6.1. *Comparison of Mediterranean Overflow transports (Sv) for each of the three stages of verification; Diagnostic (PMO), uncoupled OGCM*, and fully Coupled Climate Model (CCM*)*[a]

| | Observed | Diagnostic PMO | Uncoupled OGCM* | Coupled CCM* |
|---|---|---|---|---|
| $M_I$ | $0.93 \pm 0.27$ | 0.82 | 0.74 | 0.84 |
| $M_S$ | $0.7 - 1.3$ | 0.82 | 0.74 | 0.84 |
| $M_E$ | $0.9 - 2.4$ | 1.75 | 0.96 | 1.03 |
| $M_P$ | $1.9 - 3.7$ | 2.57 | 1.70 | 1.87 |

[a]Asterisks denote a parameterized overflow.

surface boundary conditions. Here the performance of the ocean model becomes a factor, but the forcing is as realistic as possible. From Table 6.1, both the source and entrainment are on the lower limit of the observed range, and combined they give too little product water. The final evaluation step is an assessment of the climate impacts in a fully coupled climate model (CCM). Hereafter, configurations with parameterized overflows will be denoted with an asterisk, for example OGCM* when uncoupled and CCM* when coupled. The corresponding configurations with no such parameterizations are OGCM and CCM, respectively. In CCM*, the fidelity of the atmospheric model becomes another factor, and parameterizations cannot be expected to work as intended if the coupled inputs are not reasonable. Nevertheless, its performance is somewhat improved over OGCM*, such that the product water transport increases to the lower limit of the observations. Both the OGCM* and CCM* configurations converge to steady solutions within about 100 yr. This is an essential requirement because climate applications generally have a low tolerance of drifts. Considering that the OGCM* has the low (L) 3-degree ocean resolution and the CCM* is the LL configuration, the results do support using the PMO in climate experiments.

Further evaluation of PMO in the far-field North Atlantic is shown in Figure 6.18 Specifically, both OGCM* and CCM* configurations develop a very representative Mediterranean salt tongue at about 1,100 m depth, that compares favorably to high-resolution results (Treguier et al., Chapter 7, this volume). The climatology of observations from the World Ocean Atlas show the salinity of the core falling about 1 psu, to about 35.2 psu, from Gibraltar to 40°W. The fall in CCM* is about the same, but to a more salty 35.6 psu. In OGCM* there is more of a fall (1.2 psu), but to a more agreeable 35.4 psu compared to the observed 35.2 psu. With no PMO and a blocked Gibraltar Strait there is no evidence of a salt tongue and the salinity at 1,100 m is about 35.0 psu everywhere in the Figure 6.18 domain, except off Newfoundland. However, a PMO may not be needed. By chance, the entrainment with an open Gibraltar, even when it is carved out to be much too wide, can produce a salinity structure similar

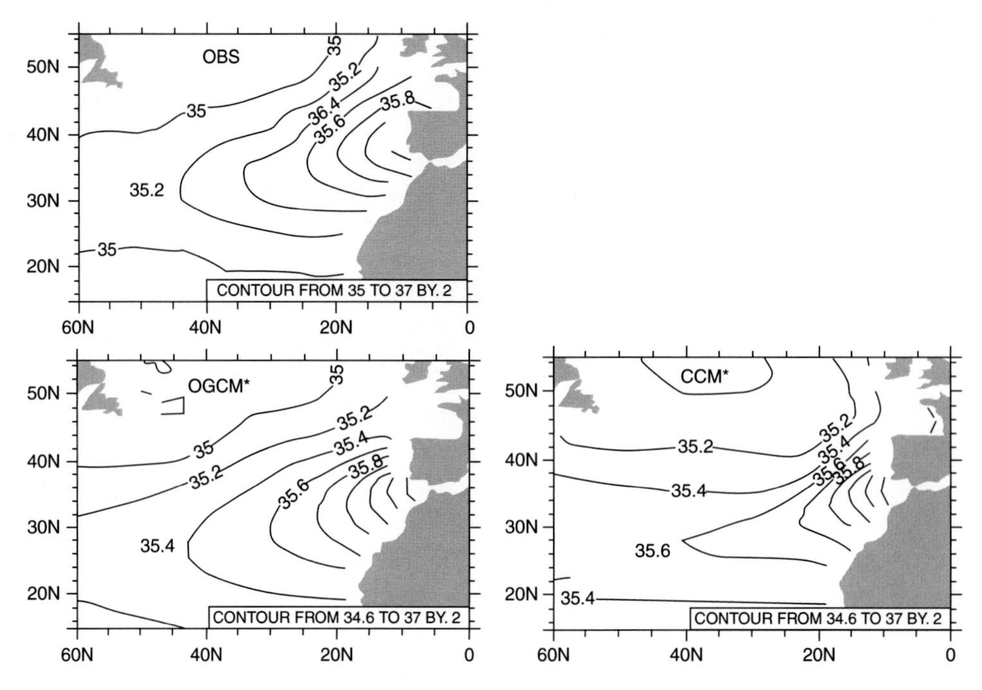


Figure 6.18. North Atlantic climatological salinity (contoured at 0.2 psu) at 1,100 m depth from observations (OBS), compared to an uncoupled OGCM* and a coupled climate model (CCM*). The asterisks denote a PMO in both. Adapted with permission from Wu et al. (2007).

to Figure 6.18, provided the neighboring Atlantic topography falls off abruptly like a cliff all in one grid spacing. However, if the ocean model topography is made to fall as a staircase, then as noted previously, there is excessive entrainment, so that a too weak tongue forms at too shallow a depth.

### *6.4.3 Nordic Sea Overflows (Denmark Strait; Faroe Bank Channel)*

Parameterizations of the Nordic Sea overflows are more challenging because the ocean model's general circulation determines the return flow, and hence plays an important role in maintaining stable solutions that are reasonable. Observations suggest that return flow into the Nordic Seas is mostly to the east of Iceland, some through the Bering Strait, and little or none to the west of Iceland, or through the Canadian Archipelago. The success of any Nordic Sea overflow, parameterized or resolved, depends on the capability of the climate model to develop such a return flow. Antarctic overflows are similar, but there are fewer possible pathways for the return flow.

Figure 6.19 is a schematic of an implementation of a Nordic Sea Overflow Parameterization. It corresponds to Case B of Figure 6.17. As with the Mediterranean, the key is to remove the source water from the prognostic OGCM, but now the topography is

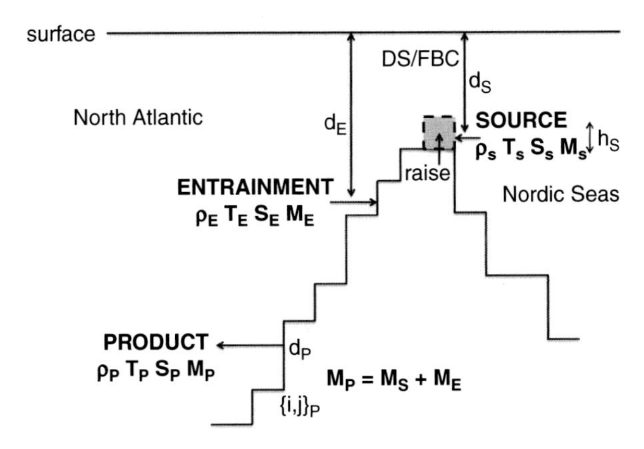


Figure 6.19. Schematic of a parameterized Nordic Seas overflow, showing the raised topography where the source boundary conditions are applied.

only raised to the grid level (dashed box) containing the sill depth. The source flow becomes a boundary condition on the Nordic Sea side of this raised cell. Again the specified topography gives necessary parameters and the location and depth of the entrainment, but not $h_S$. In the following first implementations for present day $\pm$ 100 yr climates, $h_S$ is specified from observations. In general, the topography of the continental slope is such that the location, $\{i, j\}_P$, of product water injection moves farther downstream as its depth, $d_P$, increases. The choice of these points is somewhat subjective, but beyond the steep continental slope, it is sufficient to just inject the product water as a dense bottom flow and let the OGCM take over where its topography is much less steep. The injection or removal of mass (source, entrainment, and product) through side boundary conditions without compensating local return flows greatly complicates integration of the depth-averaged (barotropic) flow. These numerics are beyond the scope of this work, but a numerical solution is detailed in Danabasoglu et al. (2010).

The evaluation of the NSOP follows the same procedure as the preceding PMO. Following the tuning exercise, observed ocean temperatures and salinities produced acceptable transports for overflows at both Denmark Strait (DS) and Faroe Bank Channel (FBC). In the subsequent numerical experiments evaluation includes comparisons against observations both near the overflow and far downstream in the Atlantic Ocean. The latter are measures of deep ventilation and are shown in Figures 6.20 and 6.21. First, uncoupled OGCMs were run with observed atmospheric forcing both without (denoted OGCM) and with (OGCM*) the NSOP. The only difference in the ocean model is the raised grid box in the latter at both DS and FBC. Second, a fully coupled climate model was run in a control case without the NSOP in its ocean component (denoted CCM), and then in a case with the NSOP (CCM*). All four cases were run for 170 yr, by which time the mass transports, and associated temperatures and salinities became stable.

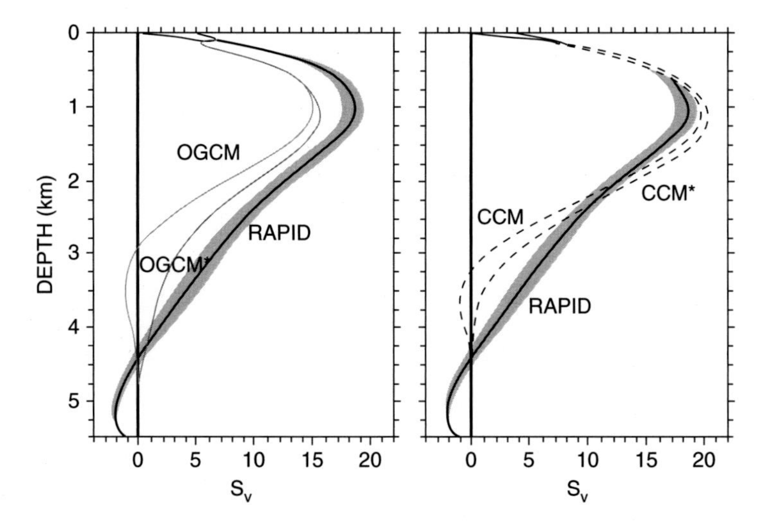


Figure 6.20. Depth integral of the zonal averaged AMOC Transport (Sv) at 26.5°N, as observed by the RAPID array and from an uncoupled ocean model with (OGCM*) and without (OGCM) the NSOP, and from a coupled climate model with (CCM*) and without (CCM) the NSOP. Adapted from Danabasoglu et al. (2010).

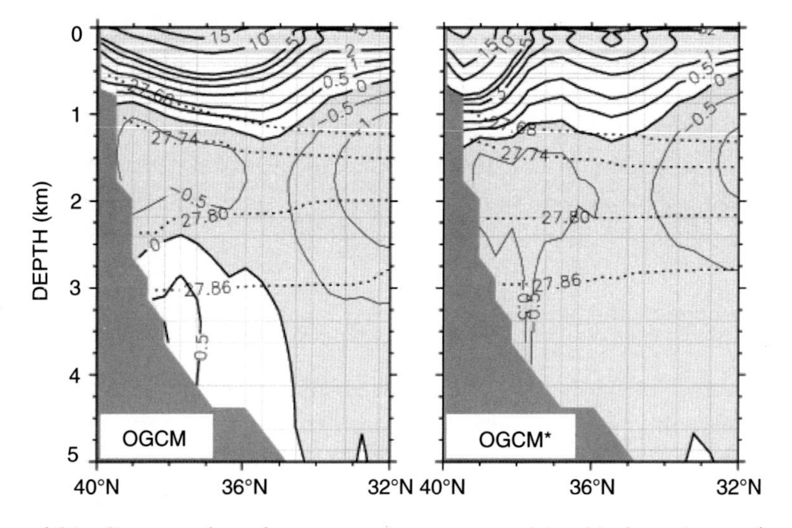


Figure 6.21. Cross-section of mean zonal current speed (cm/s) along the continental slope across 69°W from uncoupled models with (OGCM*) and without (OGCM) the NSOP. Dotted contours are isopycnals. Adapted from Danabasoglu et al. (2010).

### *6.4.4 Comparison with Observations of Ventilation*

A longstanding bias in OGCMs has been the depth of the deep western boundary current in the North Atlantic. Among the problems this bias creates are the depth and extent of NADW, and the depth structure of the AMOC. These have recently been quantified by a new time series of observations from the RAPID array (Cunningham et al. 2007). Figure 6.20 shows RAPID results as a depth profile at 26.5°N of the integrated northward transport (in Sverdrups) from the surface to each depth. Depths of increasing (decreasing) transport indicate northward (southward) zonal averaged flow. In RAPID, there is net northward transport down to about 4,400 m, but in OGCM and CCM the zero crossing is much shallower at about 2,800 m and 3,200 m, respectively. Interestingly, from very near the surface to below 3,500 m, the CCM is more like RAPID than is OGCM, suggesting that errors in the coupled forcing are tending to compensate in this measure for some model problems, such as the Gulf Stream path. The success of the NSOP in deepening the AMOC is evident in both uncoupled OGCM* and coupled CCM*, whose zero crossings are at about 4,500 m and 4,200 m, respectively, and hence in very good agreement with RAPID. However, neither shows the northward bottom transport of AABW, perhaps because the Antarctic Overflows are not yet parameterized in these cases. Again, the coupled CCM* appears to be in better overall agreement with RAPID than OGCM*, but not at all depths. Below about 3,000 m, for example, OGCM* is more like RAPID.

The main reason for the NSOP impacts on the AMOC is the dramatic effect on deep western boundary currents shown in Figure 6.21. In OGCM*, the currents along the continental slope across 69°W, including the DWBC, are negative (to the west and to the equator) at all depths below about 1,300 m. In OGCM, however, flow to the equator is confined between about 800 and 2,500 m. Deeper there is a spurious poleward current (not shaded) that results in the peculiar AMOC transport curve below about 2,800 m in Figure 6.20. This strange deep countercurrent can be traced into the Labrador Sea in OGCM and is responsible for much of the bias found in the deep transport along the western boundary of the North Atlantic. An inference is that such a countercurrent would exist in nature if the Nordic Sea overflows overly entrained, as in OGCM, weakened or shallowed. Therefore, it may have existed in climates with less ventilation either due a warmer surface Atlantic, or to fresher Nordic Seas. Do they form in such climate experiments even with NSOP?

A more rigorous test of model solutions in the western North Atlantic is the volume transport of density classes. Observations for densities greater than $\sigma = 27.8$ across 44°W, 49.3°W, and 69°W are, respectively 13.3 Sv (Dickson and Brown 1994), 14.7 Sv (Fischer et al. 2004), and 12.5 Sv (Joyce et al. 2005). In OGCM the transports are all much smaller (5.3, 3.5, and 0.2 Sv, respectively), with the previous deep countercurrent a major contributor to the problem. They become significantly greater and improved in OGCM* at 10.7, 9.3, and 2.0 Sv, respectively. Similarly, the observed 26-Sv transport of densities greater than $\sigma = 27.4$ across 44.3°W (Fischer et al. 2004) is very well represented by OGCM* (26.7 Sv), but too weak in OGCM (17.3 Sv).

The NSOP in OGCM* takes dense Nordic Sea overflow water deeper into the North Atlantic because there is less entrainment of lighter, warm, salty water during its descent. Therefore, North Atlantic deep water in OGCM* should be colder and fresher than OGCM, and this signal should spread over much of the Atlantic. Not only do the model results display this signal, but it represents an almost universal improvement relative to observations. For example, at 2,649 m the mean temperature bias in Atlantic Ocean temperature is improved from 0.45°C in OGCM to −0.04°C in OGCM*, whereas the rms temperature difference from observations is reduced from 0.5°C to only 0.13°C. Locally, downstream of the Nordic overflows the NSOP reduces the mean temperature bias at 2,649 m by about 0.7°C almost everywhere north of 40°N. In this region, the salinity bias in OGCM is typically reduced from more than 0.07 to less than 0.03 Psu. Although smaller in magnitude, similar temperature and salinity bias reductions are found at this depth in the South Atlantic, too.

One of the chief motivations behind the NSOP was to increase confidence in the uptake of atmospheric tracers, including anthropogenic greenhouse gases, into the oceans of climate models. The uptake into the deep ocean depends on deep ventilation making ideal age a convenient model proxy. More, younger water at deeper depths is indicative of more ventilation and a capacity for greater uptake of all tracers into the deep ocean and away from possible exchange with the atmosphere. Figure 6.22 shows that indeed there is much greater ventilation with NSOP than without, in both uncoupled (OGCM* – OCN) and fully coupled models (CCSM* – CCSM).

An interpretation of Figure 6.22 is that there is greater entrainment above about 2,000 m without the NSOP, so the water column downstream of the Nordic Sea overflows around 50°N is more ventilated in OGCM and CCM. This effect is seen as OCN* and CCSM* water in this region with an ideal age up to more than 20 yr older. The NSOP delivers its young product water at about 2,000 m depth, and the much

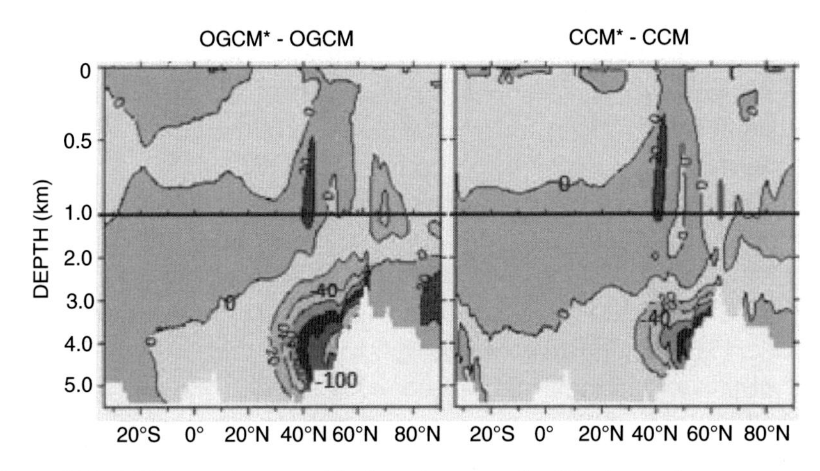


Figure 6.22. Difference between ideal age (contoured at 20 yr), zonal averaged across the North Atlantic: (a) uncoupled OGCM* with NSOP minus OGCM without; (b) coupled climate CCM* with NSOP minus CCM without. Adapted from Danabasoglu et al. (2010).

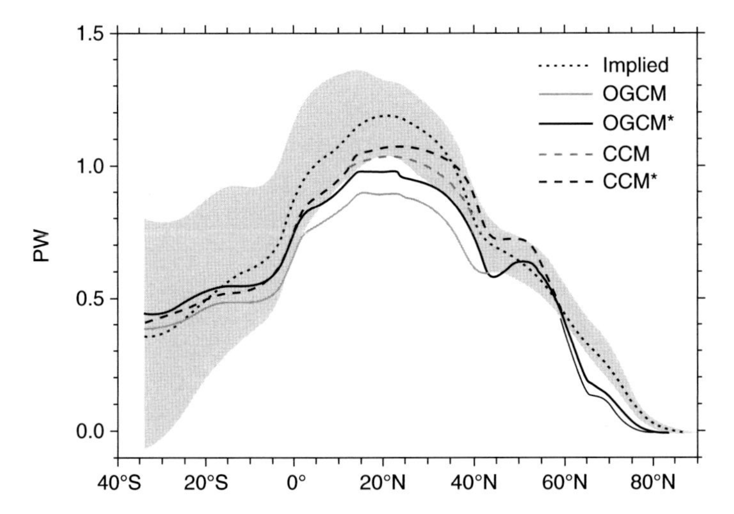


Figure 6.23. North Atlantic heat transport (PW) from uncoupled (OGCM and OGCM*) and coupled (CCM and CCM*) models, and as implied by an air-sea flux climatology (Large and Yeager 2008), whose range from 26 annual means is shaded. Adapted from Danabasoglu et al. (2010).

younger water found deeper between 3,000 and 5,000 m in both OGCM* and CCM* indicates that the ocean model is capable of moving overflow water to these depths, provided excessive entrainment is avoided.

Perhaps the most appreciated role of the ocean in the world's climate its transport of heat from equatorial regions toward the poles. Most of this capacity is due to the temperature difference between the cold buoyancy driven NADW in the DWBC, and the return flow of the warm near surface wind driven currents such as the Gulf Stream. Both coupled and uncoupled ocean models have had problems with too little northward heat transport. Figure 6.23 shows that this is still true of OGCM and CCM, but that the NSOP improves the situation in both OGCM* and CCM*.

### 6.4.5 Concluding Remarks on Section 6.4

The successes of Section 6.4.4 provide an overarching theme for all parts of this chapter. They depend on the fidelity of the forcing and the numerics (Section 6.1), of the modeled OBL and its penetrative convection (Section 6.2), and of its deep ventilation (Section 6.3), as well as of the NSOP (Section 6.4). They lower the priority of allocating sufficient computational resources to resolve overflows, thereby allowing other demands noted initially to be satisfied. However, the overflow parameterizations are site specific, requiring topographic information as input. A desirable development would be to diagnose, $h_S$, from the OGCM source properties, but a nontrivial effort would be needed to find a suitable algorithm. It would act to prevent an OGCM that produces excess source water from spilling over the raised topography (Figure 6.19), by increasing $M_S$ in response to the greater source water volume and thickness. Also the parameterizations could then be applied with more confidence to the future and past climates provided the topography remains fixed. With this proviso, should an OGCM produce new overflows in different regimes or climates, a parameterization could be developed for the site using known present topography. The expectation would be an improved model, but verification would be difficult if not impossible. Clearly overflow parameterizations are not an option, for example, for studies of deep Paleo climate with a different distribution of continents and unknown ocean bottom topography at order 1 km scales.

## Acknowledgment

The generosity of remarkable colleagues Frank Bryan, Synte Peacock, Gokhan Danabsoglu, Kieth Lindsay, Steve Yeager, Bruce Breigleb, and Markus Jochum is gratefully acknowledged for their time, thoughts, encouragement, and material.

## References

Ball, F. K., 1960: Control of inversion height by surface heating. *Q. J. R. Meteorol. Soc.* **86**, 483–494.

Bryan, F. O., G. Danabasoglu, N. Nakashiki, N. Yoshida, Y. Kim, D. H. Tsutsuij, and S. C. Doney, 2005: Response of North Atlantic thermohaline circulation and ventilation to increasing carbon dioxide in CCSM3. *J. Climate* **19**, 2382–2397.

Bryan, F. O., R. Thomas, J. Dennis, D. B. Chelton, N. G. Loeb, and J. L. McClean, 2010: Frontal Scale air–sea interaction in high-resolution coupled climate models. *J. Climate* **23**, 6277–6291, doi: 10.1175/2010 JCLI3665.1.

Cunningham, S. A., and Coauthors, 2007: Temporal variability of the Atlantic meridional overturning circulation at 26.5°N. *Science* **317**, 935–938, doi:10.1126/science.1141304.

Danabasoglu, G., W. G. Large, J. J. Tribbia, P. R. Gent, B. P. Briegleb, and J. C. McWilliams, 2006: Diurnal coupling in the tropical oceans of CCSM3. *J. Climate*, **19**, 2347–2365.

Danabasoglu, G., W. G. Large, and B. P. Briegleb, in press: Climate impacts of parameterized Nordic Sea overflows. *J. Geophys. Res. Oceans* **115**, C11005, doi:10.1029/2010JC006243.

Deardorff, J. W., 1970: A numerical study of three-dimensional channel flow at large Reynolds number. *J. Fluid Mech.* **41**, 453–480.

Deardorff, J. W., 1972: Theoretical expression for the countergradient vertical heat flux. *J. Geophys. Res.* **77**, 5900–5904.

Dickson, R. R., and J. Brown, 1994: The production of North Atlantic Deep Water: Sources, rates, and pathways. *J. Geophys. Res.* **99**, 12319–12341.

Fischer, J., F. A. Schott, and M. Dengler, 2004: Boundary circulation at the exit of the Labrador Sea. *J. Phys. Oceanogr.* **34**, 1548–1570.

Gent, P. R., and J. C. McWilliams, 1990: Isopycnal mixing in ocean circulation models. *J. Phys. Oceanogr.* **20**, 150–155.

Griffies, S. M., 2002: *Fundamentals of Ocean Climate Models.* Princeton University Press, Princeton, NJ.

Joyce, T. M., J. Dunworth-Baker, R. S. Pickart, D. Torres, and S. Waterman, 2005: On the Deep Western Boundary Current south of Cape Cod. *Deep-Sea Res. II* **52**, 615–625.

Kessler, W., 1999: Interannual variability of the subsurface high salinity tongue south of the equator at 165°E. *J. Phys. Oceanogr.* **29**, 2038–2049.

Khatiwala S., M. Visbeck, and P. Schlosser, 2001: Age tracers in an ocean GCM. *Deep-Sea Res. I* **48**, 1423–1441.

Large, W. G., J. C. McWilliams, and S. C. Doney, 1994: Oceanic vertical mixing: A review and a model with a nonlocal boundary layer parameterization. *Rev. Geophys.* **32**, 363–403.

Large, W. G., and P. R. Gent, 1999: Validation of vertical mixing in an equatorial ocean model using large eddy simulations and observations. *J. Phys. Oceanogr.* **29**, 449–464.

Large, W. G. and S. G. Yeager, 2008: The Global Climatology of an Interannually Varying Air-Sea Flux Data Set. *Climate Dyn.* **33**, 341–364, doi: 10.1007/s00382-008-0441-3.

Legg, S., B. Briegleb, Y. Chang, E. P. Chassignet, G. Danabasoglu, T. Ezer, A. L. Gordon, S. Griffies, R. Hallberg, L. Jackson, W. Large, T. M. Ozgokmen, H. Peters, J. Price, U. Riemenschneider, W. Wu, X. Xu, and J. Yang, 2009: Improving oceanic overflow representation in climate models: The Gravity Current Entrainment Climate Process Team. *Bull. Am. Meteorol. Soc.* **90**, 657–670.

Lenschow, D. H., 1974: Model of the height variation of the turbulence kinetic energy budget in the unstable planetary boundary layer. *J. Atmos. Sci.* **31**, 465–474.

Lenschow, D. H., J. C. Wyngaard, and W. T. Pennell, 1980: Mean-Field and second-moments budgets in a baroclinic, convective boundary layer. *J. Atmos. Sci.* **37**, 1313–1326.

Marotzke, J., P. Welander, and J. Willebrand, 1988: Instability and multiple steady states in a meridional-plane modle of the thermohaline circulation. *Tellus* **40A**, 162–172.

Mellor, G. L., and T. Yamada, 1982: Development of a turbulence closure model for geophysical fluid problems. *Rev. Geophys.* **20**, 851–875.

Monin, A. S., and A. M. Yaglom, 1971. *Statistical Fluid Mechanics*, vol. 1. MIT Press, Cambridge MA.

Pacanowski R. C., and S. G. H. Philander, 1981: Parameterization of vertical mixing in numerical models of the tropical ocean. *J. Phys. Oceanogr.* **11**, 1443–1451.

Peacock, S., and M. Maltrud, 2006: Transit-time distributions in a global ocean model. *J. Phys. Oceanogr.* **36**, 474–495.

Rahmstorf, S., M. Crucifix, A. Ganopolski, H. Goosse, I. V. Kamenkovich, R. Knutti, G. Lohmann, R. Marsh, L. A. Mysak, Z. Wang, and A. J. Weaver, 2005: Thermohaline circulation hysteresis: A model intercomparison. *Geophys. Res. Lett.* **32**, L23605, doi: 10.1029/2005GL023655.

Roemmich, D., S. Riser, R. Davis, and Y. Desaubies, 2004: Autonomous profiling floats: Workhorse for broadscale ocean observations. *Mar. Techn. Soc. J.* **38**(1), 31–39.

Rotunno, R., Y. Chen, W. Wang, C. A. Davis, J. Dudhia, and G. J. Holland, 2009: Large-eddy simulation of an idealized tropical cyclone. *Bull. Am. Meteorol. Soc.*, doi: 10.1175/2009BAMS2884.1.

Sabine, C. L., R. A. Feely, and R. Wanninkhof (2008): The global ocean carbon cycle. In: D. H. Levinson and J. H. Lawrimore (eds.), *State of the Climate in 2007. Bull. Am. Meteorol. Soc.* **89**(7), S52–S56.

Schmitt, R. W., 1988: Mixing in a thermohaline staircase. In: J. C. J. Nihoul and B. M. Jamart (eds.), *Small Scale Turbulence and Mixing in the Ocean*, pp. 435–452. Elsevier, New York.

Stommel, H., 1961: Thermohaline convection with two stable regimes of flow. *Tellus* **13**, 224–230.

Welander, P., 1986: Thermohaline effects on the ocean circulation and related simple models. In: D. L. T. Anderson and J. Willebrand (eds.), *Large-Scale Transport Processes in the Oceans and Atmospheres*, pp. 163–200. NATO ASI Series, Reidel.

Wu, W., G. Danabasoglu, and W. G. Large, 2007: On the effects of parameterized Mediterranean overflow on North Atlantic ocean circulation and climate. *Ocean Modelling* **19**, 31–52.

Wyngaard, J. C., and C.-H. Moeng, 1993: Large eddy simulation in geophysical turbulence parameterization: An overview. In: B. Galper in (eds.), *Large Eddy Simulation of Complex Engineering and Geophysical Flows*, pp. 349–368. Cambridge Univ. Press, New York.

Yeager, S. and W. G. Large, 2004: Later winter generation of spiciness on subducted isopycnals. *Phys. Oceanogr.* **34**, 1528–1547.

Yeager, S. and W. G. Large, 2007: Observational evidence of winter spice injection. *J. Phys. Oceanogr.* **37**, 2895–2919.

Zeng, X., and A. Beljaars, 2005: A prognostic scheme of sea surface skin temperature for modeling and data assimilation. *Geophys. Res. Lett.*, **32**, doi:10.1029/2005GL023030.

# 7

# Buoyancy-Driven Currents in Eddying Ocean Models

ANNE MARIE TREGUIER, BRUNO FERRON,
AND RAPHAEL DUSSIN

## 7.1 Introduction

### *7.1.1 Dynamics of Water Mass Formation and Spreading*

Small-scale buoyancy-driven flows, such as the overflows from marginal seas, are the main process by which the distinct water masses of the deep ocean are formed. For example, the flow of Antarctic Bottom Water (AABW) from the continental shelf down to the bottom of the Weddell Sea influences water mass properties all the way to the North Atlantic Ocean. The large range of spatial scales and mechanisms involved in the formation and spreading of these water masses poses a formidable challenge to numerical models. Legg (Chapter 5, this volume) reviews the main dense overflows of the world ocean. The width of an overflow is set either by the width of the strait or channel through which it flows (in the case of the Red Sea overflow, for example) or by the Rossby radius of deformation, which is the main dynamic scale for stratified rotating fluids. For an overflow of thickness $h$, with density anomaly $\delta\rho$ relative to the density $\rho$ of the surrounding fluid, the reduced gravity $g'$ is defined as $g\delta\rho/\rho$ ($g$ being the acceleration of gravity) and the Rossby radius $L_r$ is defined as $L_r = (g'h)^{1/2}/f$ with $f$ being the Coriolis parameter. $L_r$ decreases with latitude and its magnitude is only a few kilometers in the Nordic Seas. The dynamics of the plumes of dense water and the amount of entrainment that takes place as they descend along topographic slopes set the properties of the newly formed water masses (e.g., Chapter 5 by Legg). However, understanding the ocean thermohaline circulation requires addressing not only the formation of deepwater masses, but also their spreading and decay (by mixing with less-dense water masses). This mixing often occurs at specific locations, such as fracture zones or deep valleys, where deep overflows are observed (e.g., Ferron et al. 1998). Because of the range of scales involved (from the global scale down to a few kilometers), the formation, spreading, and decay of deep water masses is represented very crudely in climate models that have typical grid spacings of ~100 km (Large, Chapter 6, this volume).

One key element that is missing in climate models is the spreading of water masses by mesoscale eddies. Oceanographers define "mesoscale eddies" as features that have scales of 10–200 km that arise from baroclinic instabilities and whose variability corresponds to a peak in the kinetic energy spectrum. Eddies have long been ignored in the traditional large-scale view of the spreading of dense waters, stemming from the original study of Stommel and Arons (1960). In that model, dense waters flow into ocean basins as deep western boundary currents and are consumed by slow, quasi-uniform mixing in the interior. However, observations have shown that deep western boundary currents are unstable and generate eddies. One example is the breakup of the deep western boundary current in the Atlantic at 8° S into large eddies of 60-km radius (Dengler et al. 2004). Another example is the complex pathway of North Atlantic Deep Water (NADW) between the Labrador Sea and the tropical Atlantic. The region off Newfoundland is a region of high eddy activity, due to instabilities of the North Atlantic Drift (Figure 7.1). The eddies act to divert the flow of NADW into the interior, as shown by the measurements of Lagrangian floats (Fischer and Schott 2002). The role of eddies is further illustrated by Getztlaff et al. (2006) who use a high-resolution

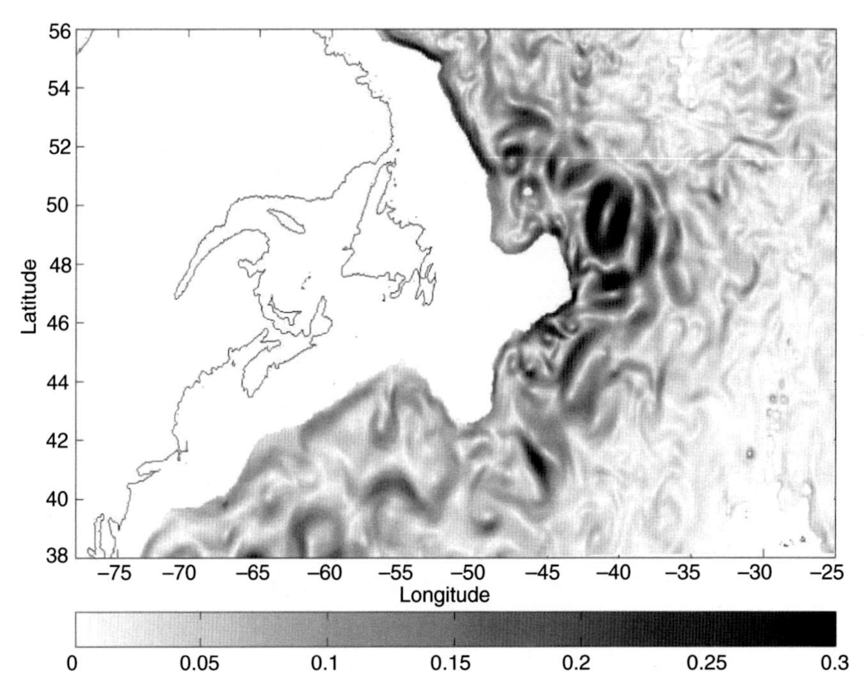


Figure 7.1. Velocity magnitude (m/s) at 1,940 m depth in the North Atlantic west of Newfoundland in the NATL12 model (see Table 7.1 for the characteristics of models used in this chapter). This is a 5-day average of model results chosen at random (21–25 April 1989). This figure shows the turbulent character of the flow and the absence of a laminar, continuous western boundary current to carry the North Atlantic Deep Water from the Labrador Sea to the tropics.

(1/12°) numerical model to simulate float trajectories starting in the NADW layer in the Labrador Current. Only one out of six modeled trajectories eventually reaches the deep western boundary current at 32° N, and these model floats reveal complex pathways, wandering as far as the Mid-Atlantic Ridge before eventually rejoining the American continental slope.

To study the role of buoyancy-driven overflows in the formation and spreading of global water masses, we need numerical models that can simulate both the overflows themselves (the formation process) and the mesoscale eddies (the main spreading process) at the scale of an ocean basin. This chapter is an overview of these numerical models, loosely called "eddy permitting" (for typical mesh sizes of 20–30 km) or "eddy resolving" (for mesh sizes smaller than 10 km). Baroclinic instability is the most important source of eddy activity in the ocean, so the most energetic eddies have diameters of a few deformation radii. In the open ocean, the Rossby radius of deformation varies from a few kilometers in the subpolar regions to 50 km in the tropics.

### *7.1.2 Representing Eddies in Numerical Models: A Historical Perspective*

Numerical models began to explicitly represent mesoscale eddies at the basin scale or global scale in the 1990s, thanks to the continual increase in computing resources. These high-resolution models encompassing large domains represented a considerable computational challenge and were often developed as community modeling efforts. Examples of these early eddy-permitting models are the Community Model Experiment (CME) 1/3° model of the North Atlantic (Bryan et al. 1995), the Fine Resolution Antarctic Model (FRAM) at 1/2° in the Southern Ocean (the FRAM group 1991), and the global 1/4° Semtner and Chervin and 0.28° POP models (McClean et al. 1997; Maltrud et al. 1998). These efforts brought a wealth of scientific results and soon were followed by basin-scale eddy-resolving models. At a winter school in Les Houches (Chassignet and Verron 1998), McWilliams (1998) showed the first pictures of POP10, a new North Atlantic model at 1/10°. For the first time, a model covering the Mediterranean and the North Atlantic represented the eddying structure of the high salinity tongue of Mediterranean water at 1,100 m in the northeastern Atlantic (Figure 17 in McWilliams 1998, qualitatively similar to our Figure 7.2). These developments could make one optimistic that eddy-resolving models would soon become adequate tools for studying the major outflows from marginal seas into the global ocean. Early examples of eddy-resolving models are the above-mentioned North Atlantic 1/10° POP (McWilliams 1998; Smith et al. 2000), the North Atlantic 1/12° MICOM (Paiva et al. 1999), and the global 1/10° POP (Maltrud and McClean 2005).

However, when the results from these high-resolution models were compared with observations, many shortcomings were discovered. Global and basin-scale models are used to study the ocean circulation, the water masses, and their variability on

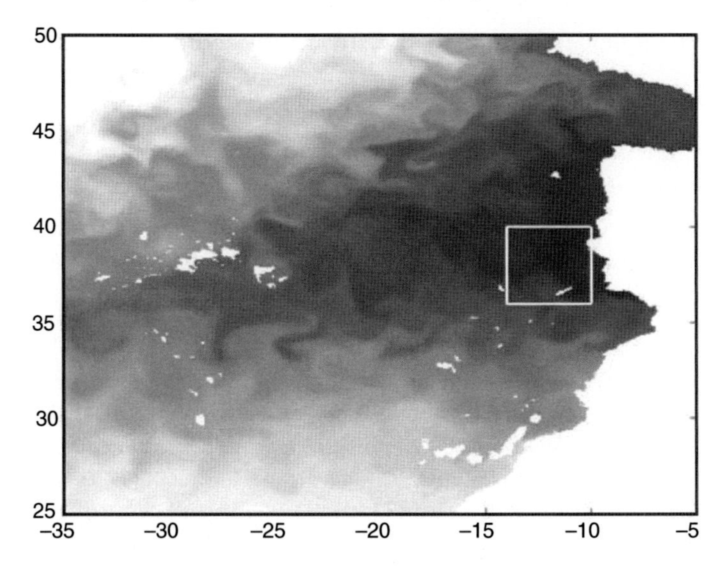


Figure 7.2. Instantaneous map of salinity at 1040 m in the NATL12 model, showing the high-salinity tongue of Mediterranean origin. The region outlined is used to calculate the profiles of Figure 7.3.

time scales of years to decades or centuries. Thus, the water masses resulting from density-driven overflows must be assessed not only at the formation site (usually a region of high entrainment where the overflow accelerates downstream of a sill), but also farther away in the open ocean. An example of such a model-data comparison is given in Figure 7.3. In a technical report (Treguier et al. 2002), profiles of temperature and salinity averaged in a box (outlined in Figure 7.2) were compared for the Levitus climatology and three eddying models covering the whole North Atlantic basin: POP10 (Smith et al. 2000), MERCATOR-PAM (a subsequent version of this model is documented in Drillet et al. 2005), and MICOM (Papadakis et al. 2003). Characteristics of these models are given in Table 7.1 (see next section). The POP10 model produced Mediterranean water that was too warm by 2°, too salty by almost 0.5 psu, and too shallow, with a maximum anomaly located near 900 m instead of at 1,100–1,200 m as in the observations. The salinity was better in the MERCATOR-PAM model, but the overflow was still too warm and even shallower (located at 800 m depth). The MICOM model had larger differences with the observations, even though the entrainment scheme had been tuned by Papadakis et al. (2003) before it was applied to the basin-scale model. These models were initialized by an observed climatology of temperature and salinity and forced by similar atmospheric data, reanalyses, or operational analyses from centers such as NCEP (National Center for Environmental Prediction) or ECMWF (European Centre for Medium-Range Weather Forecasts). Sensitivity experiments carried out by the different groups suggest that the differences in water-mass properties shown in Figure 7.3 occur rapidly during the model spin-up (within a few months or a few years) and cannot be attributed to different

Table 7.1. *Characteristics of the Main Numerical Model Solutions Discussed in This Chapter*[a]

| Model, Code | Horizontal Grid | Vertical Grid | Parameterizations (lateral/*vertical*) | Reference |
|---|---|---|---|---|
| POP10, LANL POP | B grid 1/10° Mercator | $z$-coordinate, 40 levels | $z$-biharmonic *Ri* | Figure 7.2 and 7.3 Smith et al. (2000) |
| MICOM | C grid 1/12° Mercator | Density ($\sigma_0$), 20 layers | Iso-Laplacian *Ri* | Figure 7.3 Papadakis et al. (2003) |
| PAM-05, NEMO code | Rotated C grid, 1/15°–1/6° | $z$-coordinate, 42 levels | $z$-biharmonic *TKE* | Figure 7.3 Drillet et al. (2005) |
| "LEVEL," GFDL MOM | B grid 1/3° Mercator | z-coordinate, 40 levels | $z$-biharmonic *Laplacian, 1/N* | DYNAMO project Willebrand et al. (2001) |
| "SIGMA," SPEM | C grid 1/3° Mercator | $\sigma$ coordinate, 20 levels | $z$-biharmonic *Laplacian, 1/N* | DYNAMO project Willebrand et al. (2001) |
| "ISO," MICOM | C grid 1/3° Mercator | Density ($\sigma_0$), 19 layers | Iso-Laplacian (variable) *Laplacian, 1/N* | DYNAMO project Willebrand et al. (2001) |
| Romanche, OPA/NEMO | C grid, 10 × 5 km | $z$-coordinate, 31 levels | $z$-biharmonic *TKE* | Figures 7.8 and 7.10 Ferron et al. (2000) |
| NATL12, NEMO | C grid 1/12° Mercator | $z$-coordinate, 64 levels | Iso-Laplacian (T,S) $z$-biharmonic $(u,v)$ *TKE* | Figures 7.1, 7.2, 7.3, 7.13, and 7.14 Dussin et al. (2010) |
| HYCOM | C grid 1/12° Mercator | Density ($\sigma_2$), 32 layers | Laplacian + biharmonic (variable) *Ri* (Xu et al. 2006) | Figure 7.14 Xu et al. (2007) |

[a]Documention and up-to-date versions of the model codes are found on the models websites: MOM on www.gfdl.noaa.gov/ocean-model; HYCOM (successor of MICOM) on www.hycom.org; POP on www.cesm.ucar.edu/models/ccsm4.0/pop; NEMO on www.nemo-ocean.eu; ROMS (successor of SPEM) on www.myroms.org. B and C grids are sketched in Figure 7.4. A "Mercator" grid is isotropic, with a latitudinal grid spacing varying like the cosine of latitude. Regarding lateral parameterizations, "$z$-biharmonic" is used for a horizontal biharmonic mixing and "iso-Laplacian" for a Laplacian rotated along isopycnals. The Laplacian and biharmonic coefficient for viscosity in MICOM/HYCOM is variable (Smagorinsly formulation or grid-size dependent). For vertical parameterizations, "Ri" means a simple closure dependent on Richardson number, "$1/N$" means a background vertical diffusion inversely proportional to the stratification, and TKE is a more complete second-order closure with Turbulent Kinetic Energy equation (see each model reference for details).

buoyancy forcings over the Mediterranean Sea (which are important over longer time scales).

The failure of these models was, of course, primarily because their spatial resolution was still much too coarse to resolve the dynamics of the Mediterranean overflow (Legg, Chapter 5, this volume): The interactions between the hydraulically controlled

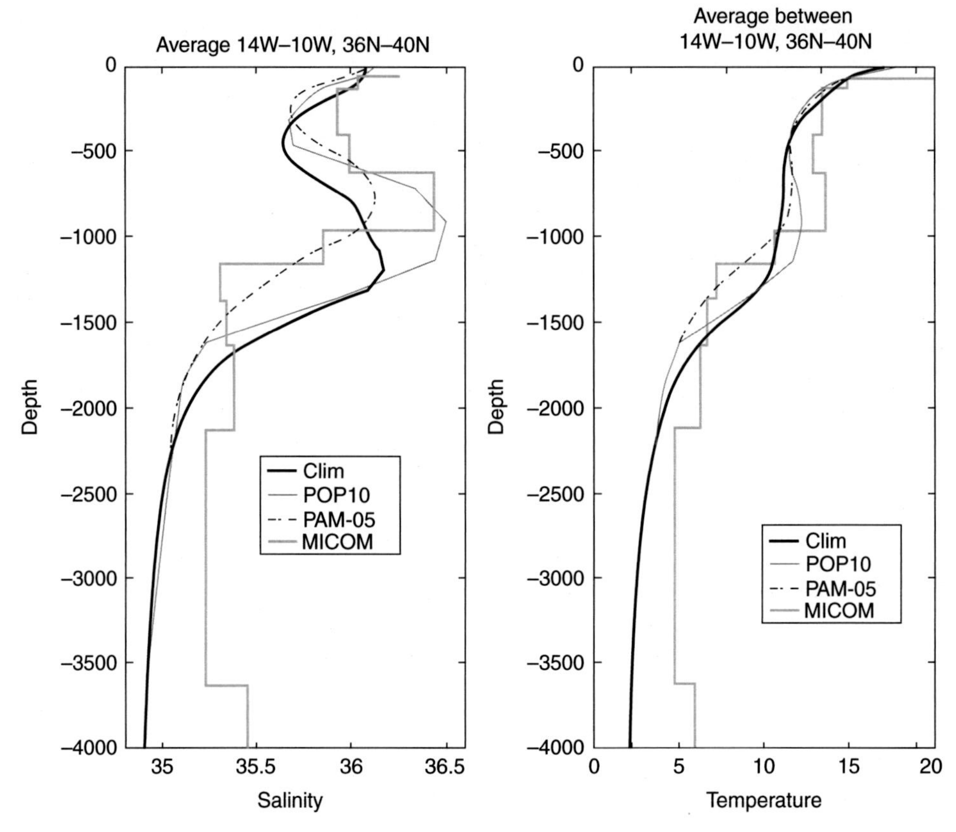


Figure 7.3. Salinity and temperature profiles in a 4° × 4° box off Portugal (see Figure 7.2). A climatology (thick lines, Reynaud et al. 1998) is compared with three high-resolution models: POP10 (Smith et al. 2000); PAM-05 (MERCATOR, Drillet et al. 2005) and MICOM (similar to Papadakis et al. 2003, but with a slightly different resolution in density space, and in a wider domain, covering the whole North Atlantic). Model results are averaged over one year, after several years of spin-up. The MICOM solution is shown by constant values over the average thickness of each density layer, emphasizing the relatively low resolution of this model configuration in the deep layers.

outflow and the tides in the Strait of Gibraltar, the acceleration of the gravity current down the slope, the shear-driven entrainment at the top of the overflow plume, or the preferred pathways through narrow canyons. The eddying models of today (meshes of a few kilometers, vertical layer thicknesses of 50–200 m) are in an intermediate range where overflows dynamics are neither completely absent nor represented satisfactorily. The most obvious way to improve the models is to increase spatial resolution, as shown by the encouraging results of high-resolution models targeted at one specific overflow (Riemenschneider and Legg 2007; Peters et al. 2005; Sannino et al. 2009). However, increases in spatial resolution on basin and global scales come slowly and painstakingly. The evolution of numerical models toward higher resolution occurs on decadal time scales. More than 10 years elapsed between the first "eddy-permitting" basin-scale models (50-km mesh, Semtner and Chervin 1992; 30-km mesh, Maltrud et al. 1998) and the first "eddy-resolving" global models (4- to 12-km mesh, Maltrud and McClean 2005). It is thus useful to review the representation of overflows in the eddying models that are available today. All of these models have important uses for operational oceanography (Drillet et al. 2005; Chassignet 2011). Some are being run in ocean-atmosphere coupled mode and they will probably be routinely used as components of coupled earth system models 10 to 20 years from now.

The aim of this chapter is to discuss model results from the perspective of the link between the overflows and the basin-scale flow (exemplified by Figure 7.2) through comparison with observations (such as in Figure 7.3). In the models considered here, the representation of overflows and the resulting water masses are dependent on numerical schemes and subgrid scale parameterizations. Section 7.2 provides a general description of ocean models; a historical perspective of the links between numerics and parameterizations is presented in Section 7.3. Section 7.4 shows how these links apply to one specific example (the flow of Antarctic bottom water through the Romanche fracture zone). In Section 7.5, we give the reader a glimpse of the complex dynamics involved in shaping the large-scale salinity anomaly originating from the Mediterranean outflow (Figure 7.2); we consider two recent models to measure the progress accomplished during the last 10 years (since the comparison illustrated in Figure 7.3).

## 7.2 Characteristics of Numerical Models of the Ocean

For completeness, we present the characteristics of the numerical models used in this chapter; the reader familiar with ocean models may skip this section. Most ocean models are based on the "primitive equations." For oceanographers, this means the Navier-Stockes fluid dynamics equations with hydrostatic, Boussinesq, and traditional approximations. The latter assumes a thin fluid relative to the radius of the sphere and neglects the horizontal Coriolis terms. The conservation of momentum and mass, written in Cartesian coordinates for simplicity, are

$$\frac{\partial u}{\partial t} - fv + Lu = -\frac{1}{\rho_0}\frac{\partial P}{\partial x} + F_l u + F_v u,$$

$$\frac{\partial v}{\partial t} - fu + Lv = -\frac{1}{\rho_0}\frac{\partial P}{\partial y} + F_l v + F_v v,$$

$$\frac{\partial P}{\partial z} = -\rho g,$$

$$\frac{\partial u}{\partial x} + \frac{\partial v}{\partial y} + \frac{\partial w}{\partial z} = 0,$$

with $u, v, w$ the components of velocity, $P$ the pressure, $\rho$ the density, $g$ the acceleration of gravity, $\rho_0$ a constant reference density, and $f$ the Coriolis parameter.

$L$ represents the nonlinear advection and $F$ the lateral and vertical mixing processes. As a result of both Boussinesq and hydrostatic approximation, the equation for vertical momentum reduces to the hydrostatic equilibrium and the mass equation becomes diagnostic (as in the case of an incompressible fluid). This means that vertical acceleration of the fluid due to gravity is not explicitly represented and that static instability and convective plumes (see Chapter 5 by Legg, this volume) must be parameterized. The thermodynamics equations are written in terms of conservation of potential temperature $\theta$ and salinity $S$, completed by an equation of state to calculate density (note that the effect of pressure is important for deep water masses and cannot be ignored for the evaluation of pressure gradients when modelling the global ocean).

$$\frac{\partial \theta}{\partial t} + L\theta = F_l\theta + F_v\theta,$$

$$\frac{\partial S}{\partial t} + LS = F_l S + F_v S,$$

$$\rho = \rho(S, \theta, P).$$

The books by Haidvogel and Beckmann (1999) and Griffies (2004) are excellent introductions to ocean modeling. Griffies et al. (2000) also provide a useful description and assessment of ocean models used as part of the coupled climate modeling platforms. Parameterization and eddy dynamics are discussed in the proceedings of the 1998 Les Houches summer school (Chassignet and Verron 1998) as well as in a recent AGU monograph (Hecht and Hasumi 2008).

The majority of ocean models use structured grids and classical finite difference techniques with numerical schemes that are typically of second order. Basin-scale and global models take great care in ensuring conservation of mass, momentum, heat, and salt so that the schemes are often equivalent to a "finite volume" formulation. Horizontal grids are generally curvilinear orthogonal grids but models differ in their placement of variables on the grid (Griffies et al. 2000): In the B grid the two components of the horizontal velocity are collocated at the corner of cells, whereas in the C grid the velocities are perpendicular to the cell boundaries. The B grid best represents the geostrophic flow and the Coriolis force because the acceleration and Coriolis force are collocated. The C grid has better properties for inertiogravity waves and for the continuity equation. The choice of grid has important implications for the representation of overflows because the B grid does not allow mass to flow through a channel unless it is at least two grid points wide, while one grid point is enough to allow throughflow with the C grid (Figure 7.4).

The choice of model grid also influences the lateral boundary conditions. A "no slip" boundary condition is natural on a B grid (all components of the velocity are zero at the coast), whereas "free slip" boundary conditions are often used on C grids (only the normal component of the velocity is set to zero, and the vorticity is zero at the coast).

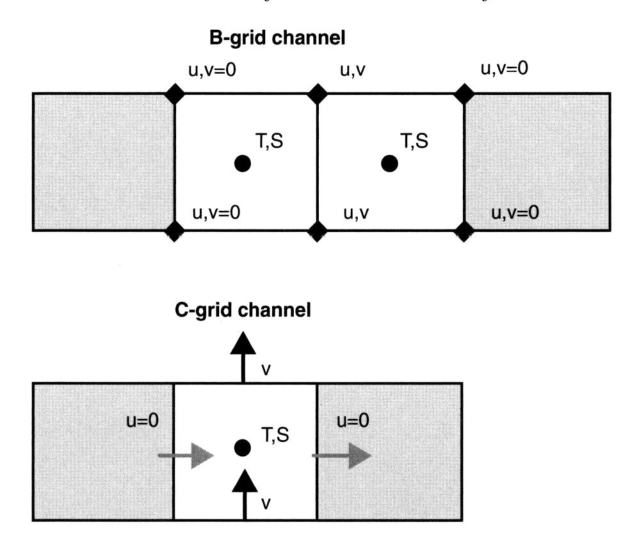


Figure 7.4. Sketch of the narrowest channel allowing a throughflow on a model grid (view is from above; ocean grid cells are white and land cells are gray). The location of variables ($T$, $S$: temperature and salinity; $u$, $v$: horizontal velocities) differs between the B and C grids. On a B grid (*top*), two grid points are necessary for throughflow in a channel, whereas on the C grid (*bottom*) one grid point is sufficient.

Ocean models use different vertical coordinates, which result in different representations of the bathymetry. All these vertical coordinates are "vertical" (i.e., directed along the direction of gravity) to avoid serious numerical errors in the hydrostatic equilibrium. Figure 7.5 illustrates two classes of models: the "$z$-coordinate" models with horizontal levels and the terrain-following coordinates. The first class ("$z$") was used in the first primitive equation ocean models; it was found to be very robust, and most ocean components of coupled Earth system models currently use this coordinate. Among the terrain-following coordinates, the simplest is the standard "sigma" coordinate defined as $\sigma = z/h$, h being the ocean depth. However, the use of a more arbitrary function $s(z, h)$ (sometimes called $s$-coordinate) allows freedom in the definition of layer thicknesses (Figure 7.5). A third class of models (isopycnic models, not illustrated in Figure 7.5) use potential density as a vertical coordinate. Advantages and disavantages of these models are discussed extensively by Griffies et al. (2000) and Griffies (2004) and are only summarized here. Regarding deep overflows that tend to flow down the bathymetry, terrain-following coordinates should work well since vertical resolution is high near the bottom. The main disavantage of these models is the numerical error in calculating the horizontal pressure gradient and the numerical diffusivity due to advection along the sloping coordinates. Isopycnic models provide interesting solutions because dense water can spread at the bottom without excessive mixing. Indeed, the isopycnic model is the only one in which an overflow can maintain its density without any numerical entrainment or mixing (entrainment must be parameterized). $Z$-coordinate models seem less realistic because they represent the

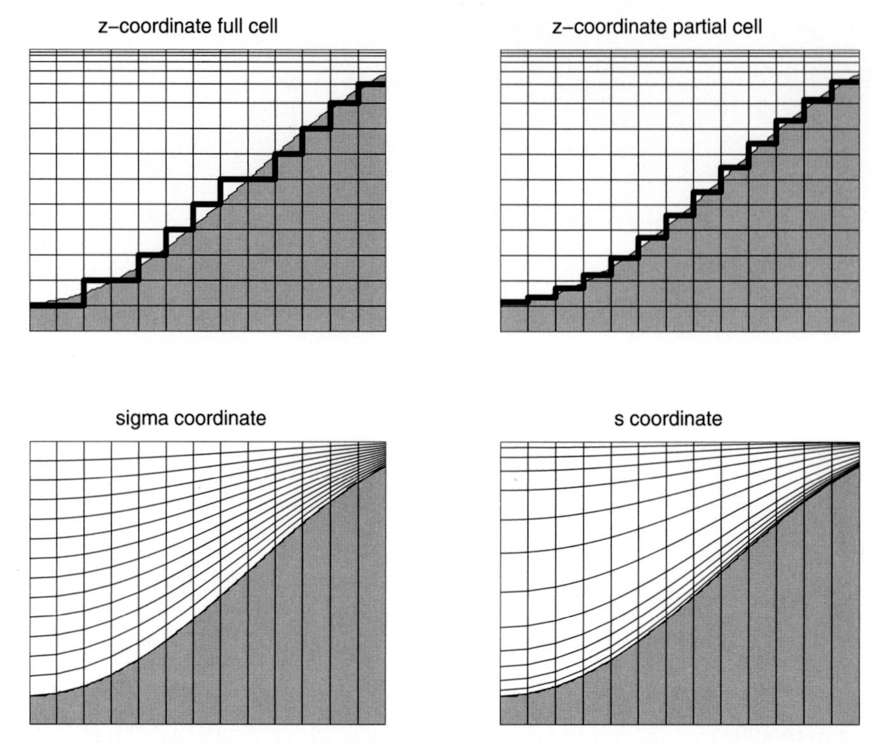


Figure 7.5. Vertical coordinates and representation of the bathymetry (arbitrary vertical section from the top to the bottom of the ocean). The true ocean bathymetry is represented in gray. With $z$-coordinates (*top left*), the bathymetry is represented by staircases (thick black line). The "partial cell" representation allows the bottom cell to have a variable thickness to follow the bathymetry accurately (*top right*). The standard sigma coordinates follow the bathymetry (*bottom left*). The use of a coordinate with stretching dependent on depth (sometimes called $s$-coordinate) allows the levels to remain quasi-horizontal near the top of the ocean and/or to increase resolution in the bottom boundary layer (*bottom right*).

ocean bottom as a series of steps (Figure 7.5), which results in the application of spurious lateral boundary conditions on the bathymetry (in reality, the ocean has no sidewalls, only a bottom). This is the reason $z$-coordinate models are very sensitive to the lateral boundary condition, "free slip" or "no slip" (Penduff et al. 2007). The distinction between these classes of models becomes blurred with the introduction of "hybrid" coordinate models, which can use different coordinates in different parts of the domain (Chassignet et al. 2006; 2009).

In z-coordinate models at coarse resolution (20–200 km), the dense overflows are not well represented. This has been attributed to the "staircase" representation of bathymetry. Each time dense water flows off a step, it mixes with less dense water below it through convective mixing. It is easy to see from Figure 7.5 that in low-resolution $z$-coordinate models, dense water cascading from step to step occupies a volume that is too large compared to a well-resolved bottom layer in $s$-coordinates,

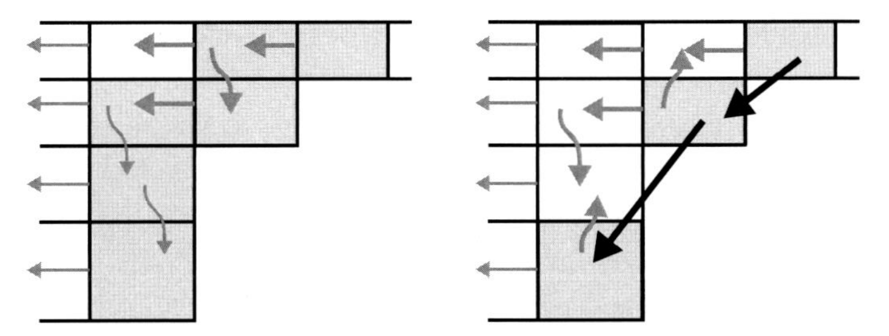

Figure 7.6. Schematic representation of downslope flow in a $z$-coordinate model (after Beckmann and Döscher 1997). (*Left*) Without BBL parameterization and (*right*) with BBL parameterization. Horizontal gray arrows represent advection and vertical arrows advection and/or convective mixing (when dense water occurs above lighter water). The BBL scheme allows direct fluxes of tracers between bottom cells (black arrows). As a result, the volume of water affected by the dense overflow (gray area) is reduced, and mixing occurs over a smaller surface.

and therefore may be subject to too much lateral and vertical mixing. The comparison between $\sigma$-coordinates and the $z$ staircases led Beckmann and Döscher (1997) to propose a bottom boundary layer (BBL) parameterization. Their first BBL parameterization consisted in providing direct advective and diffusive fluxes between two adjacent bottom grid boxes whenever the densest bottom water occurred over shallower topography, resulting in reduced entrainment and mixing (Figure 7.6). Killworth and Edwards (1999) proposed a BBL parameterization valid for momentum as well as tracers. However, in that case, the momentum equation suffers from the same pressure gradient errors as a $\sigma$-coordinate model does. Campin and Goosse (1999) presented a variant of the BBL parameterization in which the equilibration depth of the dense water is calculated. (The paramerized fluxes are not necessarily between bottom cells; instead, dense water can flow from a bottom cell down to an intermediate depth in the water column.) These BBL parameterizations have given mixed results; they do not seem efficient enough in low- resolution climate models, which has motivated the development of a new "streamtube" parameterization in the NCAR POP model (Large, Chapter 6, this volume).

Parameterizations are needed not only for the bottom boundary layer, but also for other subgrid scale processes, such as interior lateral viscosity and diffusivity (e.g., the $F_l$ and $F_v$ operators in the primitive equations), and these parameterizations differ between models (Table 7.1). Most of the parameterizations are heuristic because the dynamics responsible for mixing (between scales of a few kilometers down to the dissipation scale) are numerous and not well observed (e.g., generation of filaments by eddies, breaking of internal waves, tidal mixing, inertial instability, Langmuir cells, surface swell, double diffusion). Lateral diffusion is usually represented by a Laplacian operator (classical Fickian diffusion). Rotating the operator along the direction of isopycnal surfaces is advised since eddies tend to mix tracers along isopycnals at the meso and basin scales. In eddying models, a biharmonic operator is often preferred for viscosity because this operator acts selectively on small spatial scales and helps maintain a larger level of eddy kinetic energy for a given model resolution. Regarding the vertical direction, the simplest parameterization for the turbulent vertical flux of a tracer $T$ as a function of the vertical gradient of the mean tracer is

$$w'T' = -\kappa \frac{\partial T}{\partial z}$$

The coefficient $\kappa$ varies across many orders of magnitude between the surface mixed layer where diffusivities are of order $10^{-2}$ m s$^{-2}$ or larger, and the stratified interior where diffusivities are very small, of order $10^{-5}$ m s$^{-2}$. $\kappa$ is sometimes defined as a function of the Richardson number $Ri = N^2/U_z^2$, where $N$ is the Vaisala frequency and $U_z$ the vertical gradient of horizontal velocity. Most recent ocean models use a turbulent closure for vertical diffusivity and viscosity. Usually, these assume that diffusivity (or viscosity) is the product of the rms velocity (square root of the turbulent kinetic energy) and a mixing length. The equation for turbulent kinetic energy is generally prognostic, and there is an additional equation (prognostic or diagnostic) for either the mixing length or for the dissipation (see Griffies 2004 or Large et al. 1994 for reviews of vertical mixing parameterizations). Both turbulent closures and Richardson-dependent mixing allow the model to represent (albeit crudely) the entrainment of water into a dense overflow through Kelvin-Hemholtz instabilities (Legg, Chapter 5, this volume).

We acknowledge that only a limited set of ocean models, summarized in Table 7.1, is considered in this chapter. We do not describe results from more innovative models using finite elements or adaptive grids because these models have not yet been applied extensively at the basin or global scale. Useful references on these promising techniques can be found in special issues of *Ocean Modelling*, one example is the proceedings of the workshop "Unstructured Mesh Numerical Modelling of Coastal, Shelf and Ocean Flows" (Ham et al. 2009).

## 7.3 Interplay of Numerics and Parameterizations

It is important to realize that the parameterization of subgrid scale effects and the numerical schemes of a given ocean model determine the solution as much as do the boundary conditions of the problem (e.g., atmospheric forcing) or the physical parameters (Treguier 2006). The vertical coordinate has often been singled out as the most influential numerical factor for ocean model solutions at the basin scale, and this has motivated international model intercomparisons between the different vertical coordinates. We consider, as an example, the DYNAMO intercomparison of three models of the North Atlantic circulation at 1/3° resolution. The models (Table 7.1) are the $z$-coordinate MOM (GFDL), the isopycnic model MICOM, and the sigma-coordinate SPEM (ancestor of the ROMS model). Beckmann (1998) showed that these three models had very different bottom currents. In their synthesis of the DYNAMO project, Willebrand et al. (2001) stated: "Of fundamental importance is the conclusion that the large-scale thermohaline circulation is strongly influenced by rather localised processes; in particular by the overflow crossing the Greenland-Scotland region, their water mass properties, and the details of mixing within a few hundred kilometers south of the sills." None of the DYNAMO models was found satisfactory in representing the Nordic seas overflows. It is important to realize that for each model, the deficiencies were not due to a single numerical choice but rather to an interplay between different numerical schemes and parameterizations.

The SIGMA model was supposed to be ideally suited to represent bottom-trapped currents (the bottom being a surface of constant $\sigma$, Figure 7.5). However, the dense water was found to mix too much downstream of the sills; the overflow lost its properties too quickly as compared with the observations, due to excessive diapycnal mixing. Two hypotheses, one related to parameterizations and one to numerics, can explain this. Willebrand et al. (2001) suggested that the horizontal biharmonic mixing, acting in the regions of strong isopycnal slopes downstream of the sills, was responsible for this diapycnal mixing. Another possible shortcoming of the SIGMA model was the large change of the thickness of the bottom layer (from 50 m at the sill of Denmark Strait to 470 m in 4,000 m depth), causing numerical entrainment. Some terrain-following coordinate model solutions have been more successful in representing density-driven overflows by concentrating the vertical resolution in the bottom boundary layer. An example is Jungclaus and Mellor (2000): Their limited-area model represented the Mediterranean outflow quite realistically in the Gulf of Cadiz. It had 40 layers (twice as much as SIGMA), and 9 of these layers were distributed logarithmically near the bottom. So far no terrain-following coordinate model has been implemented with such a high resolution near the bottom at the basin scale, and we are not aware of a North Atlantic model configuration that would represent the overflows from marginal seas significantly better than the SIGMA model in Willebrand et al. (2001). Although terrain-following models can be tuned for regional domains and integration times of a few years to decades, for longer times and at large spatial scales, a proper representation of mixing (parameterized and/or numerical) remains an important issue. One example is the recent realization that advection along terrain-following coordinates can generate a large spurious mixing (Marchesiello et al. 2009).

In theory, isopycnic models (or hybrid-coordinate models such as HYCOM, which follow isopycnal surfaces in the interior but allow other coordinates as well) provide a "clean" framework in which it is possible to separate numerical artifacts from physical parameterizations. In such models, with no parameterization of entrainment and with low "background" diapycnal mixing, overflow waters sink to the bottom whenever the source water is dense enough (which is the case for the Mediterranean Sea or the Nordic Seas). This actually happened in the DYNAMO model. To represent entrainment driven by vertical current shear in isopycnic models, a parameterization was first proposed by Hallberg (2000) and implemented in Papadakis et al. (2003). Recent progress is reported in Legg et al. (2009) and in Chapter 5 (Legg, this volume). The definition of an extended Turner parameterization (Chang et al. 2005) and the addition of the Hallberg Bottom Boundary Layer (Legg et al. 2006) result in an improved representation of an idealized overflow, by comparison with very high-resolution nonhydrostatic models. On the other hand, comparison with observations still shows discrepancies (see Section 7.5) and, not surprisingly, recent studies show that parameterized entrainment is sensitive to numerical choices (e.g., advection schemes or spatial resolution) as well as other parameterizations (e.g., viscosity). This is in agreement with the findings presented in Figure 7.3, where it was surmised that the Mediterranean water was not represented correctly in the MICOM simulation, partly because the entrainment parameterizations had been tuned in a regional version of the model (Papadakis et al. 2003) that had more density layers. Parameterizations of entrainment will give different results, dependent on the resolution in density space as well as on the horizontal resolution. Legg et al. (2006) discuss in more detail this interplay between numerics and parameterizations.

Returning to the DYNAMO project, let us consider now the $z$-coordinate model. It had too much diapycnal mixing of dense water downstream of the overflows, resulting in a North Atlantic meridional overturning circulation that was too shallow (see also Dengg et al. 1999). This was interpreted by Willebrand et al. (2001) as an interaction between the step-wise topography (numerics) and vertical convection and horizontal diffusion (parameterizations). The excessive diapycnal mixing of dense waters was perceived as a major problem because it was already a prominent problem in low-resolution climate models, and the DYNAMO project had made clear that the issue was not solved at 20-km spatial resolution. This motivated the development of the BBL parameterization by Beckmann and Döscher (1997) and indeed the first encouraging results were reported by Dengg et al. (1999) in a 1/3° model similar to the LEVEL model of DYNAMO. However, more than a decade later, it is clear that BBL parameterizations have not brought a general, massive improvement in $z$-coordinate models. We have tested different BBL parameterizations in model configurations based on the NEMO code at 1/4° or 1/12° resolutions. We have noted that the results are very dependent on the advection scheme; Beckmann and Döscher (1997) used a diffusive upstream advection scheme, but less diffusive schemes give different results. Moreover, the existing BBL parameterizations take into account the bottom layer only (a "numerical" thickness scale, not a physical thickness). This is clearly a problem in the case of a partial cell representation of bottom topography. With partial cells (Figure 7.5), the thickness of the bottom layer is adjusted at each grid point to fit the bathymetry and therefore can vary dramatically from one grid point to the next. To be compatible with partial cell bathymetry, BBL parameterizations would need to be based on a physical thickness scale and encompass an arbitrary number of model layers above the bottom.

To conclude this section, let us point out that the results of the DYNAMO project go beyond the numerical issues we have discussed here; the paper by Willebrand et al. (2001) is part of a special issue devoted to the project. For example, it was the DYNAMO intercomparison that brought attention to the large-scale influence of the Mediterranean overflow, not only on the temperature and salinity of the North Atlantic (Figure 7.2), but also on the dynamics. Jia (2000; Jia et al. 2007) demonstrated that increasing the entrainment in the Mediterranean overflow plume increases the strength of the Azores Current (a quasi-zonal eastward current found across the eastern Atlantic basin between 32°N and 35°N). This happens because entrainment causes downwelling in the Gulf of Cadiz, introducing an extra forcing for the large-scale vorticity balance of the North Atlantic. The response is nonlocal because of the westward propagation of Rossby waves (Özgökmen et al. 2001), a feature explained in detail by Pedlosky (Chapter 2, this volume).

## 7.4 Modeling Deep Flow Through the Romanche Fracture Zone

More specific examples are useful to illustrate successes and difficulties in modeling buoyancy-driven flows. In this section, we consider the flow of Antarctic bottom water (AABW) through the Romanche fracture zone (Figure 7.7). The Romanche Fracture Zone (FZ) is an equatorial passage in the Mid-Atlantic Ridge with a sill depth of

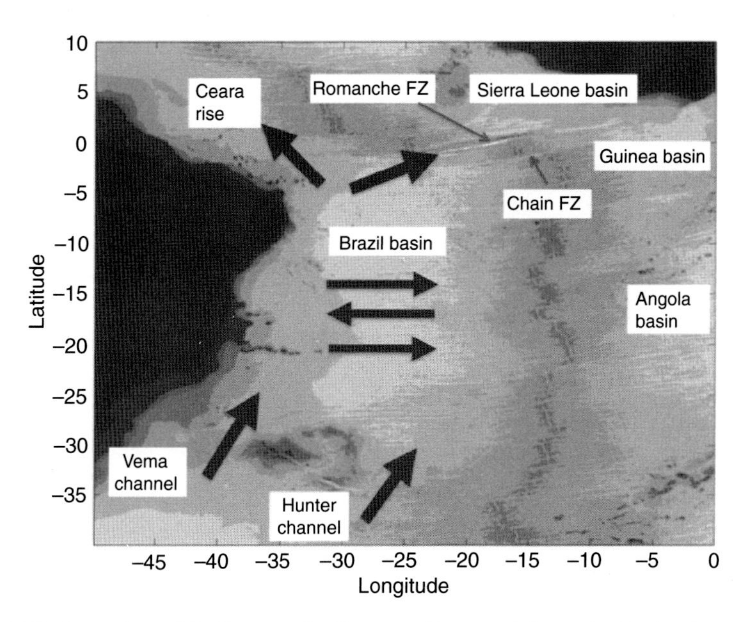


Figure 7.7. Location of the Romanche Fracture Zone and the major basins it connects. The path of Antarctic Bottom Water is represented as black arrows. AABW flows into the Brazil Basin through two channels and exits at Ceara rise and the Romanche and Chain Fracture Zones. The flow of deepwater masses in the Brazil Basin is complex and characterized by alternating zonal jets (Treguier et al. 2003).

4,350 m, allowing an eastward flow of AABW from the Brazil Basin to the Sierra Leone and Guinea abyssal plains. This example is chosen to point out the importance of understanding not only the formation of the dense water masses through overflows (from the continental shelves of the Weddell and Ross seas, in the case of AABW), but also their decay. Most of the mixing of AABW occurs in specific "hot spots" such as Romanche and Chain. Although these fracture zones represent 0.4% of the area covered by the Guinea and Sierra Leone abyssal plains, half the mixing across the 1.4° isotherm occurs there (Ferron et al. 1998). The swift flow of AABW through the Romanche (15–25 cm/s), driven by the pressure gradient between adjacent basins, has been measured by a current meter array during two years (Mercier and Speer 1998). The flow is subject to hydraulic control by a sill and a narrow. Hydraulic control occurs when the Froude number reaches unity: $Fr = U/(g'h)^{1/2} = 1$ (see Linden, Chapter 1, or Legg, Chapter 5, this volume). Vertical mixing in the overflow plume downstream of the control point at the sill has been evaluated by microstructure profilers (Polzin et al. 1996) as well as by indirect calculations (Ferron et al. 1998). It is illustrated by the large vertical excursions of isotherms in Figure 7.8 (top panel).

We have built a numerical model of this flow based on the OPA primitive equation code (now part of the NEMO system, Madec 2008). A closed rectangular basin ($1,430 \times 830 \times 5,500$ m depth) was separated into two basins by a ridge (culminating at 3,000 m depth). In the western basin, water masses were restored to a profile representative of the Brazil Basin. The Romanche Fracture Zone (RFZ) was represented by a 20-km-wide gap in the ridge, with a realistic bathymetry of the main sills and narrows. The model resolution was 10 km in the zonal direction and 5 km in the meridional direction (Table 7.1). The vertical mesh size was 20 m near the interface of Antarctic bottom water and increased upward and downward. To follow the direction of the fracture zone, the $x$-axis of the model was rotated by 10° from the equator. The initial pressure gradient forced a flow of AABW through the fracture zone, with a recirculation above 3,500 m depth. Because of the proximity to the equator, the flow is close to a nonrotating regime. Figure 7.8 shows that the model reproduces the observed isopycnal slopes characteristic of the hydraulic jump downstream of the main sill. The transport of Antarctic bottom water (Ferron et al. 2000), as well as its intense mixing downstream, is well represented. Antarctic bottom water enters the RFZ with temperatures between 0.7° C and 0.9° C, but the bottom temperature east of the RFZ is higher than 1.5° C. This good agreement of modeled and observed entrainment was obtained by using the standard scheme for vertical mixing in the OPA model, the TKE second-order closure (Blanke and Delecluse, 1993), based on the equation of turbulent kinetic energy.

We have compared different schemes for vertical mixing (Ferron et al. 2004). One sensitivity experiment uses a vertical diffusivity dependent on the Richardson number ($Ri$), with the same dependency as in the "interior mixing" part of the KPP (K-profile Parameterization) of Large et al. (1994). The $Ri$-dependent mixing reproduces the

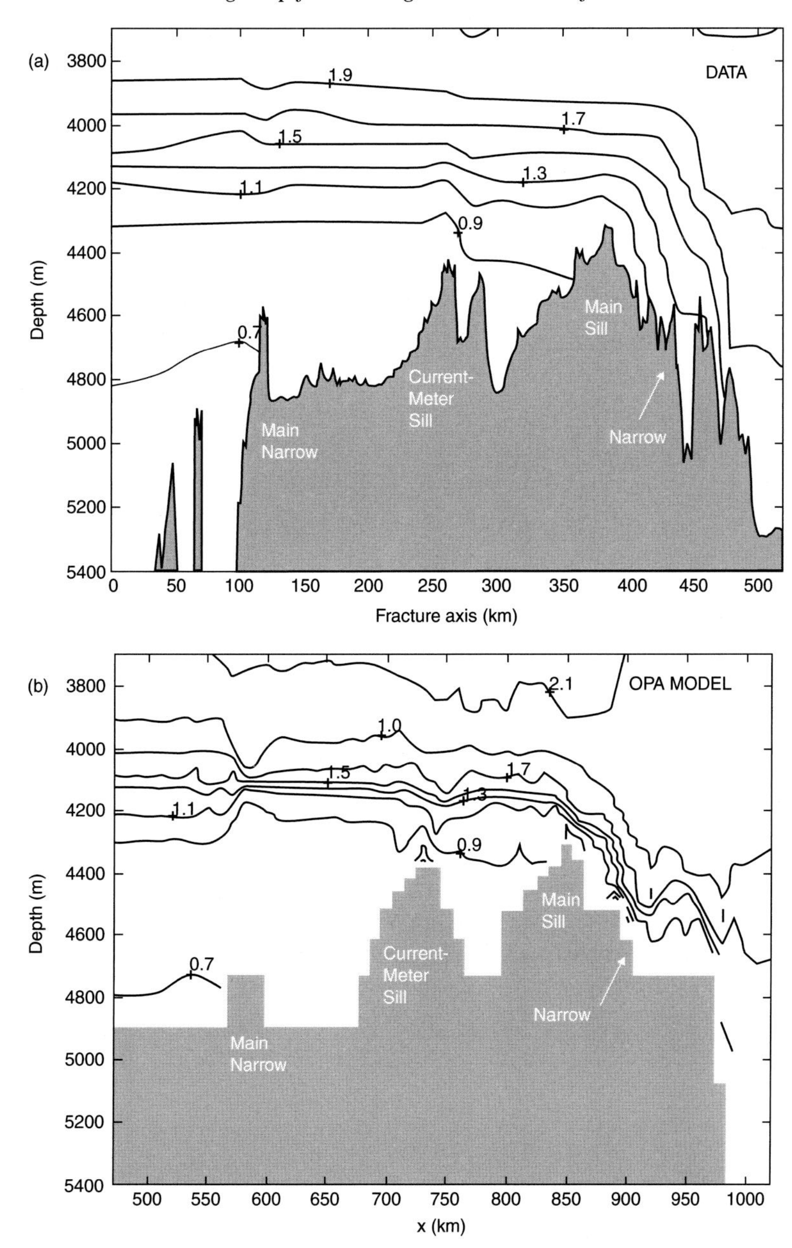


Figure 7.8. Observed potential temperature section along the RFZ axis (*top*); modeled potential temperature along the RFZ axis (*bottom*). Reproduced from Ferron et al. (2000).

observed temperature profile at the exit of the RFZ, provided the critical Richardson number is increased from $Ri = 0.7$ (as assumed in Large et al. 1994) to $Ri = 1$. Note that the critical Richardson number for the onset of Kelvin Hemholtz instabilities is well defined ($Ri = 0.25$), but in numerical models we have to let it depend on vertical resolution according to how well the vertical current shears are resolved. In our model, both the TKE turbulent closure and the simpler Richardson-dependent formulation of vertical mixing give satisfactory results. In another sensitivity experiment, the vertical mixing coefficient is a constant weak background (excepted for convective adjustment in case of hydrostatic instability). In that case, the temperature of the bottom water exiting the RFZ is too low by 0.2°C and the overflow is too dense. This is contrary to the commonly held view that $z$-coordinate models with staircase topography always produce too much mixing of overflows (Legg et al. 2009). Ferron et al. (2004) clarified the issue by considering the balance of a temperature in a column just downstream of the sill (Figure 7.9). If the temperatures in the upstream and downstream reservoir are fixed, the temperature $\theta_m$ of the column reaches equilibrium when horizontal and vertical diffusion are large enough to balance advection. In the RFZ model, advection is dominant and $\theta_m$ can equilibrate at a low temperature. In the absence of a turbulent closure to represent the enhanced vertical mixing due to shear-driven instability of the overflow, the modeled temperature is lower than observed. This would not be the case in a model with coarser resolution. In that case, horizontal diffusion would balance advection for a higher temperature (a temperature closer to the value in the downstream reservoir).

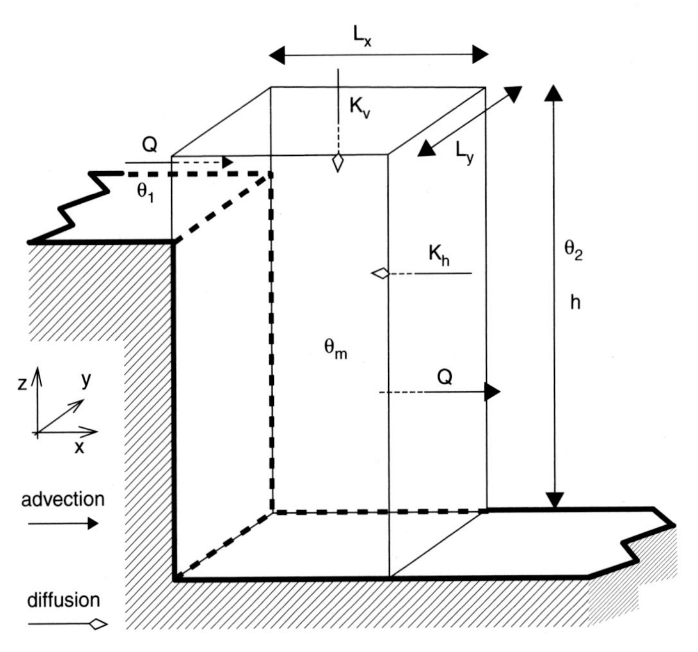


Figure 7.9. Schematic diagram for convection downstream of a sill in a $z$-coordinate model. Cold water of temperature $\theta_1$ is advected with a transport $Q$ into the downstream valley grid cell whose volume is lx ly.h. Initially the temperature of the grid cell $\theta_m$ is equal to the downstream reservoir temperature $\theta_2$. The vertical diffusion $K_v$ and the horizontal diffusion $K_h$ must balance the advection to get a steady $\theta_m$ (from Ferron et al. 2004).

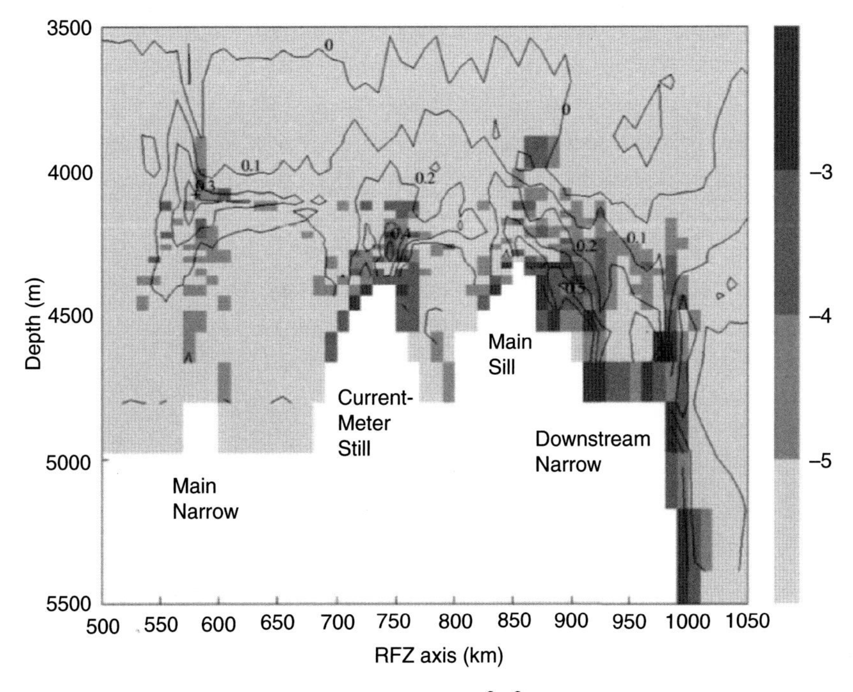


Figure 7.10. Decimal logarithm of the TKE ($m^2.s^2$) along the RFZ-axis (grayscale) and alongstream velocity contours (m $s^{-1}$). From Ferron et al. (2004).

Ferron et al. (2004) compared the turbulent kinetic energy and mixing lengths predicted by the TKE model to direct observations of turbulent mixing. The locations of enhanced turbulent kinetic energy (Figure 7.10) correspond to the topographical constraints on the flow (sills and narrows). The most intense mixing was observed in the region downstream of the main sill (Polzin et al. 1996; Ferron et al. 1998). An examination of the different terms of the TKE balance show that TKE is generated by the vertical shear in the model and that the turbulent length scale is in agreement with the observations. This work suggests that a turbulent closure such as the TKE parameterization, originally designed for the surface mixed layer, may have some skill for representing mixing of deep overflows. However, the idealized model of Ferron et al. (2004) does not allow a quantitative validation because of its simplified geometry and forcing. For example, the vertical temperature gradient is stronger in the model than in the observations (Figure 7.8).

This attempt to model the Romanche Fracture Zone with a $z$-coordinate model and its TKE parameterization of vertical mixing showed that given high enough horizontal and vertical resolution, $z$-coordinate primitive equation models with standard turbulent energy closures should provide good representations of overflows. Winton et al. (1998) considered dense water sinking in an idealized setting with a $z$-coordinate model at various resolution and concluded that $z$-coordinate models can converge to the same solution as an isopycnic model when the horizontal and vertical resolution are fine enough to resolve the bottom slope as well as the thickness of the bottom layer. This is the case in the model of Ferron et al. (2004), and it explains why this $z$-coordinate model requires a parameterization of shear-driven vertical mixing (here, the TKE scheme), just as an isopycnic model does. We emphasize, however, that spatial resolution requirements are dependent on the overflow considered. For the Romanche Fracture Zone a satisfactory solution was achieved with moderate horizontal resolution ($10 \times 5$-km grid), perhaps due to the proximity of the equator. In the sensitivity studies of Legg et al. (2006) the dependency on resolution seems lower in the nonrotating case, compared with the rotating cases.

## 7.5 Modeling the Spreading of Mediterranean Water in the Atlantic

The outflow of Mediterranean water into the Atlantic Ocean creates a large-scale salinity anomaly that covers most of the northeastern Atlantic (Figure 7.2). Water of Mediterranean origin impacts the transport of warm and saline water toward the subpolar gyre and the Nordic Seas, although the pathways and time scales of these influences are not well known (references are found in Jia et al. 2007). It is thus important for climate models to represent correctly the lateral and vertical extent of this salinity anomaly. The failure of early high-resolution models to do so (Figure 7.3) is not surprising, considering the complexity of the dynamics that create this large-scale anomaly. Following the outflow away from Gibraltar, three regimes must be modeled. The first two, the initial descent and the Mediterranean undercurrent, are sketched on a map in Figure 7.11; the third regime is the eddy spreading of the salt tongue represented in Figure 7.2. In this section, we review the models used in Figure 7.3 and the attempts that have been made during the last 10 years to improve their representation of the Mediterranean water in these three regimes.

### *7.5.1 The initial descent*

The outflow first follows a channel that is an extension of the Strait of Gibraltar (Price et al. 1993). Reaching the edge of the continental slope, the outflow descends more steeply, accelerates, and undergoes a geostrophic adjustment process, veering to the right because of the Coriolis force. Most of the entrainment occurs in that part of the outflow, and bottom friction plays a key part (Johnson et al. 1994). The problems found in the $z$-coordinate models of Figure 7.3 were first attributed to a misrepresentation of this initial descent. The large temperature and salinity anomaly of the POP model of Smith et al. (2000) was perhaps partly due to artifacts of the centered advection scheme. When a current advects anomalies across a large gradient, which is the case in the initial descent, a centered advection scheme creates false extrema (a salinity of 41 was found at one grid point). Almost all the eddying models have

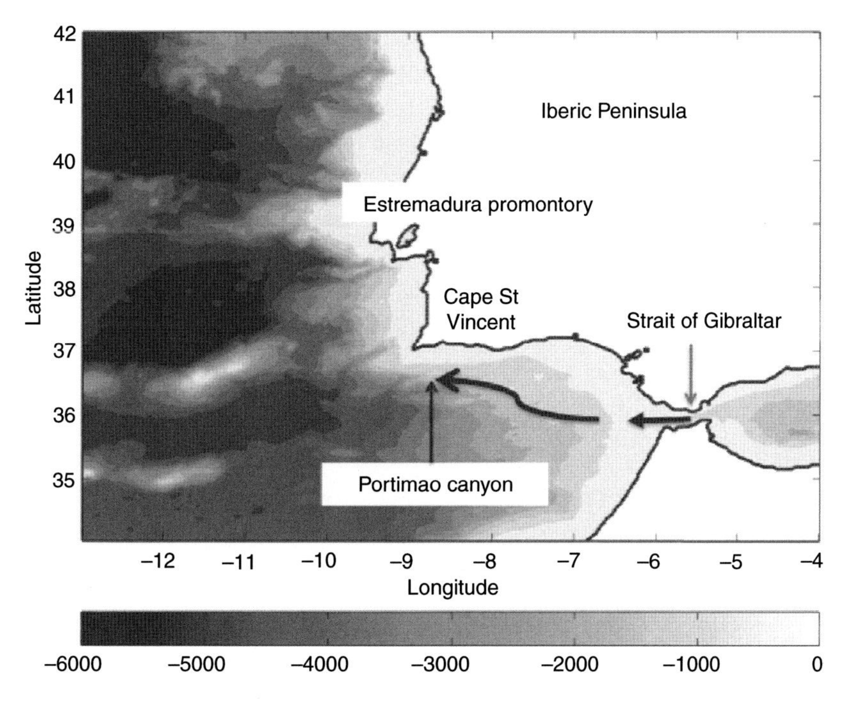


Figure 7.11. Bathymetric map of the Mediterranean outflow region. The two thick black lines with arrows represent the initial descent of the plume and the Mediterranean undercurrent, respectively.

now abandoned second-order centered advection schemes for temperature and salinity and use positive-definite numerical schemes or higher-order schemes to avoid these problems. The numerical experiments run with the POP model subsequent to the Smith et al. (2000) pioneering paper used an updated code and model grid (the same grid as the global model of Maltrud and McClean 2005). The salinity and temperature anomalies were reduced compared with the POP profiles in Figure 7.3, but maxima were located too high in the water column (thus resembling the Mercator solution in Figure 7.3). M. Maltrud and R. Smith (personal communication) have performed a number of sensitivity experiments but have failed to identify the source of the difference between the model solutions of Smith et al. (2000) and Maltrud and McClean (2005). The behavior of the Mercator model in Figure 7.3, with an overflow that is not dense enough because of its high temperature, is the most common behavior among z-coordinate models (similar to the OCCAM model of Jia et al. 2007). We have performed a large number of tests to try to represent better the final density of the plume in that model (Treguier et al. 2002). No significant improvement was found with different tracer advection schemes, vertical mixing schemes, or lateral diffusion (horizontal or isopycnal). A bottom boundary layer parameterization following Beckmann and Döscher (1997) allowed a slightly better descent of the outflow. Finally, bottom friction was found to be the most influential parameter. We tested a very high drag coefficient

($Cd = 4 \times 10^{-3}$) following a similar study by Stratford and Haines (2000) in the Mediterranean Sea. However, even with this high bottom drag, the solution, although improved, was not satisfactory. In the end, an artificial relaxation to climatological temperature and salinity had to be added in the Gulf of Cadiz in the operational Mercator model (Drillet et al. 2005). A relatively weak relaxation allowed the model to represent the dynamics and instabilities of the upper core of the outflow while maintaining acceptable water mass properties at the large scale.

### *7.5.2 The Mediterranean undercurrent*

After the initial descent, the Mediterranean water splits into two cores, probably because of the steep canyon that allows part of the outflow to descend while the upper core continues along the continental slope (Johnson et al. 2002). This splitting was successfully reproduced by Jungclaus and Mellor (2000) in the terrain-following coordinate POM model, with very high vertical resolution in the bottom boundary layer. At 8°W, the outflow is considered to be gravitationally stable and consists of two cores located near 850 and 1,100 m. The two cores appear on a hydrographic section in Figure 7.12 (SEMANE cruise, Carton et al. 2010). In Figure 7.13, we compare these

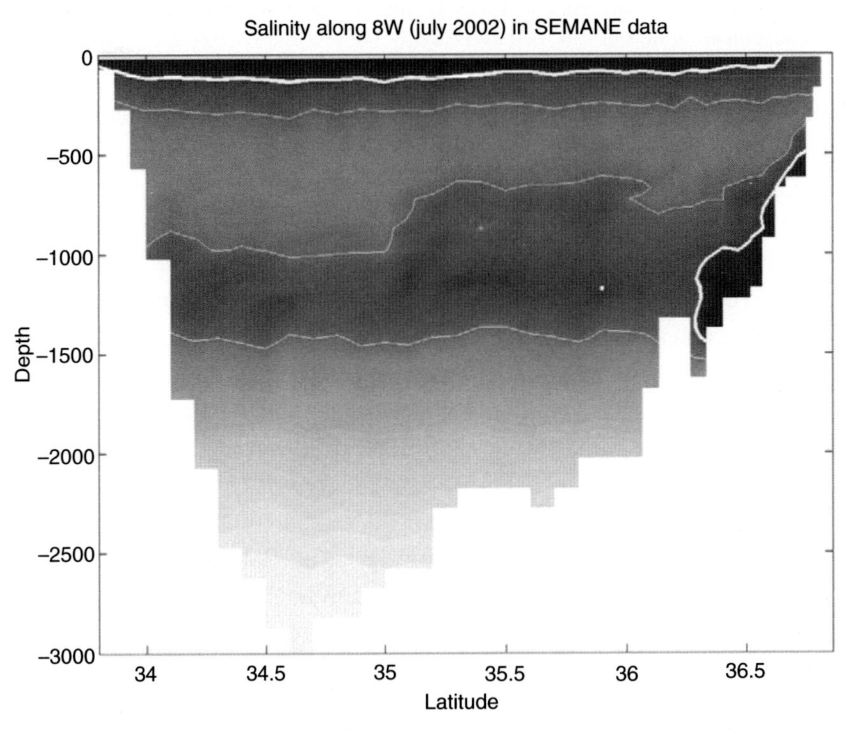


Figure 7.12. Salinity along a section at 8°W (SEMANE cruise). The light white contour is salinity 35.8 and the thick white contour 36.2.

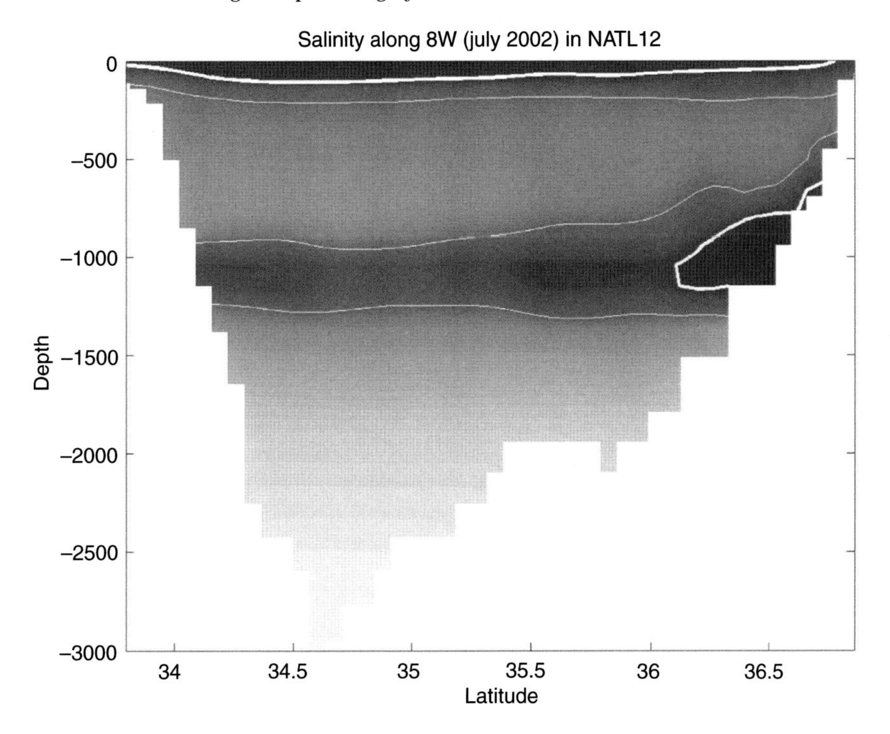


Figure 7.13. Salinity along the same section for the NATL12 model. The light white contour is salinity 35.8 and the thick white contour 36.2.

data with the results of a recent $z$-coordinate model. This model (NATL12) is based on the NEMO numerical code (Madec 2008) and is developed as part of the DRAKKAR project (Barnier et al. 2006). Its horizontal mesh is extracted from a global tripolar grid with resolution 1/12° at the equator; there are 64 levels in the vertical (thicknesses range from 6 m near the surface to 200 m at depth). With standard parameterizations, the first version of this model produced an overflow that was too high in the water column (a little better than the Mercator model in Figure 7.3, but still unsatisfactory). Following our extensive and unsuccessful set of tests, we tried a simple ad hoc parameterization: a local modification of the topography. A few grid points were deepened downstream of Gibraltar to allow the outflow to sink more rapidly. This may be seen as a crude parameterization of the unresolved canyons that cut the continental slope. With this modified bathymetry, we have been able to obtain a quite reasonable solution without any relaxation to climatology in the Gulf of Cadiz. There is even a hint of the two cores of high salinity at about the right depth (Figure 7.13). This solution was obtained with a mean exchange of 0.59 Sv at the Strait of Gibraltar (computed over the 1990–2006 period, after 10 years of spin-up from 1980 to 1989). A higher exchange rate resulted in an increase of the salinity in the salt tongue, relative to the observations.

In the real ocean, the spreading of the Mediterranean water is determined by the undercurrent transport and properties, but also by its instabilities, which create detached blobs of Mediterranean water called meddies (references are found in Bower et al. 1997 or Drillet et al. 2005). Meddies probably account for most of the transport of salt out of the undercurrent. Float data such as that gathered by Bower et al. (1997) or Paillet et al. (2002) show that there are preferred sites for their generation, due to topographic accidents along the Iberian Peninsula (Figure 7.11): Portimao Canyon in the Gulf of Cadiz, Cape St. Vincent, the Estremadura Promontory, and Cape Ortegal to the north. The instability process is a complex one, dependent on the horizontal and vertical velocity shears (mixed barotropic/baroclinic instability) as well as on the topographic slope and its curvature. At model resolutions of a few kilometers, we expect meddy formation to be very dependent on numerics (horizontal and vertical resolution, advection schemes) as well as on the parameterizations. Papadakis et al. (2003) found that the level of eddy activity in their experiments was dependent on the amount of entrainment and thus on the depth of the equilibrated undercurrent (the higher the undercurrent in the water column, the more unstable). Similarly, the overflow plume was too stable in the simulations of Xu et al. (2007) using HYCOM: The number of meddies at a given time and their rate of formation were lower than those estimated from observations. With the NATL12 model, we have found another example of sensitivity to parameterizations. In addition to the modification of bathymetry described previously, efforts to improve the initial descent of the plume in the model include increasing the bottom drag coefficient over a wide area, almost all the way to Cape St. Vincent. We have found that the eddy kinetic energy was much reduced between 8° W and Cape St. Vincent compared with another model experiment without this enhanced bottom friction (not shown).

### 7.5.3 The Mediterranean Salt Tongue

The overall shape of the salinity anomaly in the North Atlantic (Figure 7.2) depends on the dynamics of the undercurrent and its instabilities: When these are not represented correctly, the salt tongue may take an unrealistic shape. For example, in the CLIPPER ATL6 model (Treguier et al. 2005) the Mediterranean outflow was represented by an open boundary in the Gulf of Cadiz, based on data from the SEMANE cruise (Carton et al. 2010). In this way, the depth of the outflow and its properties were correct, but the undercurrent did not have the right dynamics. Part of the undercurrent continued as a westward jet, leaving the boundary at Cape St. Vincent instead of turning northward following the continental slope of the Iberian Peninsula. This resulted in a salt tongue that extended too far to the west and not enough to the north. In the MICOM model used in Figure 7.3, the undercurrent turned north but shed comparatively too many northern meddies and not enough southern meddies, resulting in a salinity anomaly extending too far west at the latitude of Cape Ortegal (the northwest corner of the

Iberian Peninsula). Ultimately, the shape of the salt tongue and its variability also depend on the large-scale circulation, as shown by Bozec et al. (2011) in a recent study using the HYCOM model at moderate resolution (1/3°).

To conclude this section, let us consider in Figure 7.14 profiles similar to those in Figure 7.3 but for two recent 1/12° models, NATL12 (Dussin and Treguier 2010) and HYCOM (Xu et al. 2010). The HYCOM model shows a large improvement compared with MICOM (Figure 7.3) because of a better vertical resolution, the use of potential density referenced to 2,000 m (instead of the surface), the increase of vertical resolution in the surface mixed layer allowed by the hybrid coordinate system, and an improved parameterization of entrainment. However the Mediterranean water is still too deep. The NATL12 model fits the data much better, which is not surprising since the bathymetry and parameterizations in this model have been adjusted with the purpose of improving the spreading of Mediterranean water, whereas the HYCOM solution has not been tuned. The large intermodel difference in Figure 7.14 reminds us that at these spatial resolutions (a few kilometers), parameterizations such as the one used in HYCOM cannot be universal and work for all overflows. The amount of entrainment is not the same for a current flowing on a smooth continental slope and

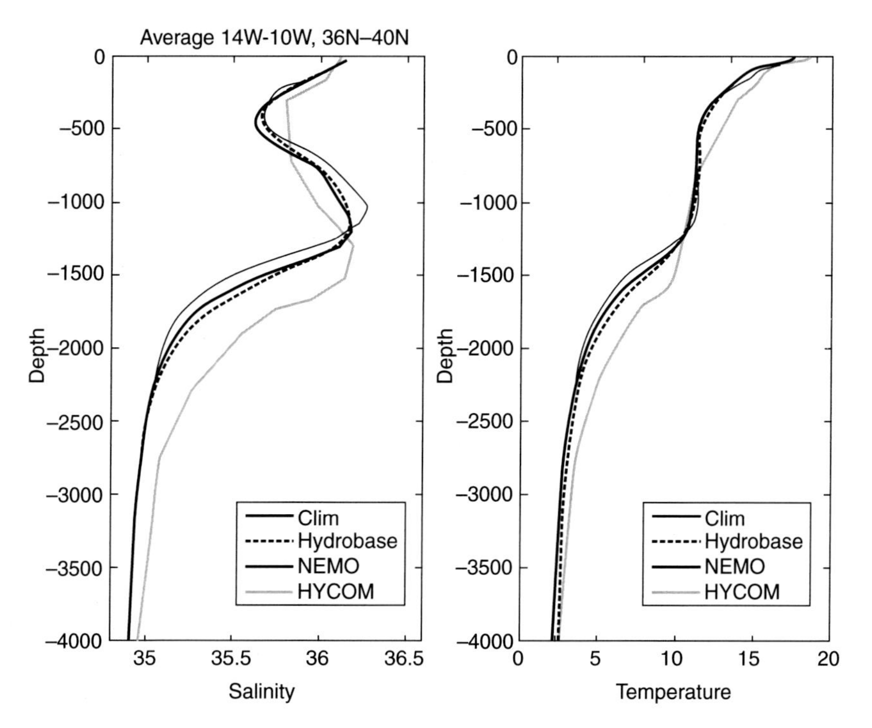


Figure 7.14. Salinity and temperature profiles in a 4° × 4° box off Portugal (see Figure 7.1). Two climatologies are represented: Reynaud et al. (1998) by a thick continuous line and Hydrobase (Curry 1996) by a thick dashed line. Model results are 5-year averages of the NEMO NATL12 configuration (Dussin and Treguier 2010) and HYCOM (Xu et al. 2010).

a current plunging into a canyon. Either the bathymetry or the parameterization of entrainment has to be tuned for each overflow independently to represent the effect of the unresolved bathymetric features that control the dynamics of the descending plume.

## 7.6 Conclusion

In this short presentation of how buoyancy-driven currents are (mis)represented in eddying models, we have contrasted a relatively easy exercise (representing the non-rotating, constrained flow of Antarctic bottom water in the Romanche Fracture Zone) with a difficult exercise (representing the flow of Mediterranean water in the Atlantic at all scales: the initial descent, the undercurrent, and the high salinity tongue). In both cases, $z$-coordinate models can give acceptable performances with adequate tuning; hybrid coordinate models also require tuning to reach a similar level of accuracy. These model solutions at resolutions of a few kilometers are not fully satisfactory to represent buoyancy-driven overflows in the world ocean. The necessity for tuning numerics and parameterizations is due in large part to the influence of unresolved bathymetric features.

Advantages and disadvantages of different numerical models and parameterizations have been mentioned in this chapter, but it was not our objective to provide an extensive review. With the growing number of models, coordinate systems, numerical schemes, and parameterizations, the task requires an entire textbook, such as the one by Haivodgel and Beckmann (1999) or Griffies (2004). Here we have tried to point out that the performance of one given model does not depend on one single characteristic. The choice of coordinates, numerical schemes, and subgrid-scale parameterizations is important, and all interact in a complex (and sometimes unexpected) fashion. Will this problem remain in the future? We may hope that increasing computing power will allow better spatial resolution and reduce the burden placed on subgrid-scale parameterizations. Another opportunity for progress is the development of more physically based parameterizations. The bottom boundary layer parameterization, for example, cannot be adequate unless the thickness of the boundary layer is controlled by physical parameters and not by the vertical resolution of the grid.

Models must be validated against observations, and this is extremely challenging in the case of buoyancy-driven overflows because observations of the deep ocean are scarce and buoyancy-driven flow is highly variable in time. Models are often compared with gridded climatology products, such as the World Ocean Atlas, which are also used to initialize model experiments, but dense water plumes and boundary currents such as the Mediterranean undercurrent are not represented in these climatologies. The Hydrobase climatology (Curry 1996) provides an interesting analysis of bottom properties, but its 1° grid is still too coarse for model validation. There is a critical need for easily available observations at a higher spatial resolution to validate and improve the representation of buoyancy-driven flows in eddying models.

## List of Acronyms

AABW, AntArctic Bottom Water
BBL, Bottom Boundary Layer
DYNAMO, DYnamics of North Atlantic MOdels
ECMWF, European Centre for Medium-Range Weather Forecasts
FRAM, Fine Resolution Antarctic Model
GFDL, Geophysical Fluid Dynamics Laboratory
HYCOM, HYbrid Coordinate Ocean Model
LANL, Los Alamos National Laboratory
MERCATOR-PAM, Mercator-Ocean Prototype Atlantic Mediterranean Sea
MICOM, Miami Isopycnal Coordinate Ocean Model
MOM, Modular Ocean Model
NADW, North Atlantic Deep Water
NCAR, National Center for Atmospheric Research
NCEP, National Center for Environmental Prediction
NEMO, Nucleus for European Modelling of the Ocean
OCCAM, Ocean Circulation and Climate Advanced Modelling
POM, Princeton Ocean Model
POP, Parallel Ocean Program
RFZ, Romanche Fracture Zone
ROMS, Regional Ocean Modelling System
SPEM, Sigma-coordinate Primitive Equation Model
TKE, Turbulent Kinetic Energy

## References

Barnier, B., G. Madec, T. Penduff, J. M. Molines, A. M. Treguier, J. Le Sommer, A. Beckmann, A. Biastoch, C. Böning, J. Dengg, C. Derval, E. Durand, S. Gulev, E. Remy, C. Talandier, S. Theetten, M. Maltrud, J. McClean, and B. De Cuevas, 2006: Impact of partial steps and momentum advection schemes in a global ocean circulation model at eddy permitting resolution. *Ocean Dynamics*, 56 (5–6), 543–567, doi: 10.1007/s10236-006-0082-1.

Beckmann, A. and R. Dorscher, 1997: A method for improved representation of dense water spreading over topography in geopotential-coordinate models. *J. Phys. Oceanogr.* **27**, 581–591.

Beckmann, A., 1998: Represetnation of bottom boundary layer processes in numerical ocean circulation models. In: E. P. Chassignet and J. Verron (eds.), *Ocean Modeling and Parameterization*, Kluwer Academic Publishers. pp. 135–154.

Blanke, B., and P. Delecluse, 1993: Variability of the tropical Atlantic ocean simulated by a general circulation model with two different mixed-layer physics. *J. Phys. Oceanogr.* **23**, 1363–1388.

Bower, A. S., L. Armi, and I. Ambar, 1997: Lagrangian observations of Meddy Formation during a Mediterranean undercurrent seeding experiment. *J. Phys. Oceanogr.* **27**, 2545–2575.

Bozec, A., E. P. Chassignet, M. S. Lozier, and G. R. Halliwell, 2011: On the variability of the Mediterranean outflow water in the Atlantic Ocean. Part I. Source of the Mediterranean outflow water variability. *J. Geophys. Res.*, in press.

Bryan, F. O., C. W., Böning, Holland, W.R., 1995: On the mid-latitude circulation in a high resolution model of the North Atlantic. *J. Phys. Ocean.* **25**, 289–305.

Campin, J. M., and H. Goosse, 1999: Parameterization of density-driven downsloping flow for a coarse-resolution ocean model in *z*-coordinate. *Tellus* **51**, 412–430.

Carton, X., N. Daniault, J. Alves, L. Cherubin, and I. Ambar, 2010: Meddy dynamics and interaction with neighboring eddies southwest of Portugal: Observations and modeling. *J. Geophys. Res.* **115**, C06017, doi: 10.1029/2009JC005646.

Chang, Y. S., X. Xu, T. M. Ozgokmen, E. P. Chassignet, H. Peters, and P. F. Fischer, 2005: Comparison of gravity current mixing parameterizations and calibration using a high resolution 3D nonhydrostatic spectral element model. *Ocean Modelling* **10**, 342–368.

Chassignet, E. P., and J. Verron (eds.), 1998: *Ocean Modeling and Parameterization.* Kluwer Academic Publishers.

Chassignet, E. P., H. E. Hurlburt, O. M. Smedstad, G. R. Halliwell, A. J. Wallcraft, E. J. Metzger, B. O. Blanton, C. Lozano, D. B. Rao, P. J. Hogan, and A. Srinivasan, 2006: Generalized vertical coordinates for eddy-resolving global and coastal ocean forecasts. *Oceanography* **19**(1), 20–31.

Chassignet, E. P., H. E. Hurlburt, E. J. Metzger, O. M. Smedstad, J. Cummings, G. R. Halliwell, R. Bleck, R. Baraille, A. J. Wallcraft, C. Lozano, H. L. Tolman, A. Srinivasan, S. Hankin, P. Cornillon, R. Weisberg, A. Barth, R. He, F. Werner, and J. Wilkin, 2009: U.S. GODAE: Global Ocean Prediction with the HYbrid Coordinate Ocean Model (HYCOM). *Oceanography* **22**(2), 64–75.

Chassignet, E. P., 2011: Isopycnic and hybrid ocean modeling in the context of GODAE. In: A. Schiller and G. Brassington (eds.), *Operational Oceanography in the 21st Century,* Springer.

Curry, R. G., 1996: HydroBase: A database of hydrographic station and tools for climatologic analysis. WHOI Technical Report 96-01.

Dengg, J., C. Böning, U. Ernst, R. Redler, and A. Beckmann, 1999: Effects of an improved model representation of overflow water on the subpolar North Atlantic. *Int. WOCE Newsletter* **37**, 10–15.

Dengler, M., F. A. Schott, C. Eden, P. Brandt, J. Fischer and R. J. Zantopp, 2004: Break-up of the Atlantic deep western boundary current into eddies at 8°S. *Nature*, **432**, 1018–1020, doi:10.1038/nature03134.

Drillet, Y., R. Bourdallé-Badie, L. Siefridt, and C. Le Provost, 2005: Meddies in the Mercator North Atlantic and Mediterranean Sea eddy-resolving model. *J. Geophys. Res.* **110**, C03016, doi:10.1029/2003JC002170.

Dussin, R., and A. M. Treguier, 2010: Evaluation of the NATL12-BRD81 simulation. LPO report 10-03.

Ferron, B. , H. Mercier, K. G. Speer, A. E. Gargett and K. L. Polzin, 1998: Mixing in the Romanche Fracture Zone. *J. Phys. Oceanogr.* **28**, 1929–1945.

Ferron, B., H. Mercier, and A. M. Treguier, 2000: Modelisation of the flow of bottom water through the Romanche Fracture Zone with a primitive Equation model. Part 1: Dynamics. *J. Mar Res.* **58**, 837–862.

Ferron, B., H. Mercier, and A. M. Treguier, 2004: Modelisation of the bottom water flow through the Romanche Fracture Zone with a primitive Equation model. Part 2: Comparison of vertical mixing parameterizations with observations. *Ocean Modelling* **6**, 177–190.

Fischer, J., and F. A. Schott (2002), Labrador Sea Water tracked by profiling floats—From the boundary current into the open North Atlantic, *J. Phys. Oceanogr.* **32**, 573–584.

The FRAM group, 1991: Initial results from a fine resolution model of the Southern Ocean. *Eos Trans*. AGU **72**, 174–175.

Getzlaff, K., C. W. Böning, and J. Dengg, 2006: Lagrangian perspectives of deep water export from the subpolar North Atlantic. Geophys. Res. Lett. **33**, L21S08, doi:10.1029/2006GL026470.

Griffies, S. M., C. Böning, F. O. Bryan, E. P. Chassignet, R. Gerdes, H. Hasumi, A. Hirst, A.-M. Treguier, and D. Webb (2000): Developments in ocean climate modelling, *Ocean Modelling*. **2**, 123–192.

Griffies, S. M. (2004): *Fundamentals of Ocean Climate Models*. Princeton University Press, Princeton, NJ.

Haidvogel, D. B., and A. Beckmann, 1999: *Numerical Ocean Circulation Modeling*. Imperial College Press, London.

Hallberg, R. W., 2000: Time integration of diapycnal diffusion and Richardson number-dependent mixing in isypycnal coordinate ocean models. *Mon. Weather Rev.* **128**, 1402–1419.

Ham, D. A., C. C. Pain, E. Hanert, J. Pietrzak, and J. Schroter, 2009: Special Issue: The sixth international workshop on unstructured mesh numerical modelling of coastal, shelf and ocean flows. Imperial College London, September 19–21, 2007, *Ocean Modelling* **28**, (1–3) The Sixth International Workshop on Unstructured Mesh Numerical Modelling of Coastal, Shelf and Ocean Flows, doi: 10.1016/j.ocemod.2009.02.005.

Hecht, M., and H. Hasumi (eds.), 2008: *Ocean Modelling in the Eddying regime*, Geophysical Monograph Series, vol. 177. American Geophysical Union, Washington, DC.

Jia, Y., 2000: Formation of an Azores Current due to Mediterranean overflow in a modeling study of the North Atlantic. *J. Phys. Oceanogr.* **30**, 2342–2358.

Jia, Y., A. C. Coward, B.A. de Cuevas, D. Webb, and S. S. Drijfhout, 2007: A model analysis of the behavior of the Mediterranean water in the North Atlantic. *J. Phys. Oceanogr.* **37**, 764–786.

Johnson, G. C., T. B. Sanford, and M. O'Neil Baringer, 1994: Stress on the Mediterranean Outflow plume: Part I. Veolcity and water property measurements. *J. Phys. Oceanogr.* **24**, 2072–2083.

Johnson J., I. Ambar, N. Serra, and I. Stevens, 2002: Comparative studies of spreading of Mediterranean water through the Gulf of Cadiz. *Deep-Sea Res.* **49**, 4179–4193.

Jungclaus, J. H., and G. Mellor, 2000: A three-dimensional model study of the Mediterranean outflow. *J. Mar. Sys.* **24**, 41–66.

Killworth, P., and N. Edwards, 1999: A turbulent bottom boundary layer code for use a numerical ocean models, *J. Phys. Oceanogr.* **29**, 1221–1238.

Large, W. G., J. C. McWilliams, and S. C. Doney, 1994: Oceanic vertical mixing: A review and a model with a nonlocal boundary layer parameterization. *Rev. Geophys.* **32**, 363–403.

Legg, S., R. W. Hallberg, and J. B. Girton, 2006: Comparison of entrainment in overflows simulated by *z*-coordinate, isopycnal and non-hydrostatic models. *Ocean Modelling* **11**(1–2), doi:10.1016/j.ocemod.2004.11.006.

Legg, S., et al. 2009: Improving oceanic overflow representation in climate models: The Gravity Current Entrainment Climate Process Team. *BAMS* **90**, 657–670, doi: 10.1175/2008BAMS2667.1.

Madec, G., 2008: NEMO ocean engine, Note du Pole de modélisation, Institut Pierre-Simon Laplace (IPSL), France, No **27**, ISSN 1288–1619.

Maltrud, M. E., R. D. Smith., A. J. Semtner, and R. C. Malone, 1998: Global eddy-resolving ocean simulations driven by 1985–1995 atmospheric winds. *J. Geophys. Res.* **103**, C13, 30825–30853.

Maltrud M. E., and McClean J. L., 2005: An eddy resolving global 1/10° ocean simulation. *Ocean Modelling* **8**, 31–54.

Marchesiello, P., L. Debreu, and X. Couvelard, 2009: Spurious diapycnal mixing in terrain-following coordinate models: The problem and a solution. *Ocean Modelling* **26**, 156–169.

McClean, J. L., A. J. Semtner, and V. Zlotnicki, 1997: Comparisons of mesoscale variability in the Semtner-Chervin quarter-degree model, the Los Alamos sixth-degree model, and TOPEX/POSEIDON Data. *J. Geophy. Res.* **102**(C11), 25203–25226.

McWilliams, J. C., 1998: Oceanic general circulation models. In: E. Chassignet and J. Verron (eds.), *Ocean Modelling and Parameterizations*. NATO Science Series, Kluwer.

Mercier, H., and K. G. Speer. 1998. Transport of bottom water in the Romanche Fracture Zone and the Chain Fracture Zone. *J. Phys. Oceanogr.* **28**, 779–790.

Özgökmen, T. M., E. P. Chassignet, and C. G. H. Rooth, 2001: On the connection between the Mediterranean outflow and the Azores Current. *J. Phys. Oceanogr.* **31**, 461–480.

Paillet, J., B. Le Cann, X Carton, Y. Morel, and A. Serpette, 2002: Dynamics and evolution of a northern meddy. *J. Phys. Oceanogr.* **32**, 55–79.

Paiva, A. M., J. T. Hargrove, E. P. Chassignet, and R. Bleck, 1999: Turbulent behavior of a fine mesh (1/12 degree) numerical simulation of the North Atlantic. *J. Mar. Sys.* **21**, 307–320.

Papadakis, M. P, E. P. Chassignet, and R. W. Hallberg, 2003: Numerical simulations of the Mediterranean outflow: Impact of the entrainment parameterization in an isopycnic coordinate ocean model. *Ocean Modelling* **5**, 325–356.

Penduff, T., J. Le Sommer, B. Barnier, A. M. Treguier, J. Molines, and G. Madec, 2007: Influence of numerical schemes on current-topography interactions in 1/4° global ocean simulations. *Ocean Science* **3**, 509–524.

Peters, H., W. E. Johns, A. S. Bower, and D. M. Fratantoni, 2005: Mixing and entrainment in the Red Sea outflow plume. Part 1: plume structure. *J. Phys. Oeanogr.* **35**, 569–583.

Polzin, K. L., K. G. Speer, J. M. Toole, and R. W. Schmitt, 1996: Intense mixing of Antarctic Bottom Water in the equatorial Atlantic Ocean. *Nature* **380**, 54–57.

Price, J. F., M. O'Neil Baringer, R. G. Lueck, G. C. Johnson, I. Ambar, G. Parilla, A. Cantos, M. A. Kennelly, and T. B. Sanford, 1993: Mediterranean outflow mixing and dynamics. *Science* **259**, 1277–1282.

Reynaud T., P. Legrand, H. Mercier, and B. Barnier (1998): A new analysis of hydrographic data in the Atlantic and its application to an inverse modelling study. International WOCE Newsletter, Number 32, 29–31.

Riemenschneider U., and S. Legg, 2007: Regional simulations of the Faroe Bank Channel overflow in a level model. *Ocean Modelling* **17**, 93–122.

Sannino, G., M. Hermann, A. Carillo, V. Rupolo, V. Rugiero, V. Artale, and P. Heimbach, 2009: An eddy-permitting model of the Mediterranean Sea with a two-way grid refinement at the Strait of Gibraltar. *Ocean Modelling* **30**, 56–72.

Semtner, A. J., and R. M. Chervin, 1992: Ocean general circulation from a global eddy-resolving model. *J. Geophys. Res.* **97**, 5493–5550.

Smith, R. D., M. E. Maltrud, F. O. Bryan, and M. W. Hecht, 2000: Numerical simulation of the North Atlantic Ocean at 1/10°. *J. Phys. Oceanogr.* **30**, 1532–1561.

Stommel, H., and A. Arons, 1960: On the abyssal circulation of the World Ocean. Part I: Stationary planetary flow patterns on a sphere. *Deep-Sea Res.* **8**, 140–154.

Stratford, K., and K Haines, 2000: Frictional sinking of the dense water overflow in a *z*-coordinate OGCM of the Mediterranean Sea, *Geophys. Res. Lett.* **27**, 3973–3976.

Treguier, A. M., C. Talandier, and S. Theetten, 2002: Modelling Mediterranean water in the North East Atlantic. LPO internal report LPO-02-14, 16pp.

Treguier, A. M., N. G. Hogg, M. Maltrud, K. Speer, and V. Thierry, 2003: Origin of deep zonal flows in the Brazil Basin. *J. Phys. Oceanogr.* **33**, 580–599.

Treguier, A. M., S. Theetten, E. P. Chassignet, T. Penduff, R. Smith, L. Talley, J. O. Beisman, and C. Boening, 2005: Salinity distribution and circulation of the North Atlantic subpolar gyre in high resolution models. *J. Phys. Oceanogr.* **35**, 757–774.

Treguier, A. M., 2006: Models of ocean: which ocean? In: E. P. Chassignet and Verron (eds.), *Ocean Weather Forecasting: An Integrated View of Oceanography.* Dortrecht, Springer.

Willebrand, J., B. Barnier, C. Boening, C. Dieterich, P. Hermann, P. D. Killworth, C. Le Provost, Y. Jia, J.M. Molines, and A. L. New, 2001: Circulation characteristics in three eddy-permitting models of the North Atlantic, *Prog. Oceanogr.* **48**, 123–161.

Winton, M., R. Hallberg, and A. Gnanadesikan, 1998: Simulation of density-driven frictional downslope flow in $z$-coordinate ocean models. *J. Phys. Oceanogr.* **28**, 2163–2174.

Xu, X., Y. S. Chang, H. Peters, T. M. Ozgokmen, and E. P. Chassignet, 2006: Parameterization of gravity current entrainment for ocean circulation models using a high order 3D nonhydrostatic spectral element model. *Ocean Modelling* **14**, 19–44.

Xu, X., E. P. Chassignet, J. F. Price, T. M. Ozgokmen, and H. Peters, 2007: A regional modeling study of the entraining Mediterranean outflow. *J. Geophys. Res.* **1112**, C12005.

Xu, X., W. J. Schmitz Jr., H. E. Hurlburt, P. J. Hogan, and E. P. Chassignet, 2010: Transport of Nordic Seas overflow water into and within the Irminger Sea: An eddy-resolving simulation and observations, *J. Geophys. Res.* **115**, C12048, doi:10.1029/2010JC006351.

# 8

# Atmospheric Buoyancy-Driven Flows

SYLVIE MALARDEL

## 8.1 Introduction

### *8.1.1 The Atmosphere*

The atmosphere, like the ocean, is a stratified fluid highly influenced by the rotation of the Earth.

But, unlike the ocean, the atmosphere is a mixture of gases known as *air*.

The composition of the gas layer around the Earth has evolved very slowly since the time of its formation. Thanks to the appearance of life about 3.5 billion years ago, the main constituents of the air are now nitrogen ($N_2$, about 78%) and oxygen ($O_2$, about 21%). Other minor constituents are argon (1%), ozone, carbon dioxide, and water vapor.

The air near the surface is about 1,000 times lighter than the water in the ocean. It is also much more compressible. The mean state of the atmosphere is stably stratified and is in hydrostatic equilibrium (Figure 8.1). The first 10–15 km of the atmosphere, known as the troposphere, contain nearly 90% of the atmospheric mass. This is the layer where the weather occurs. The bottom of the troposphere, the boundary layer (BL), is directly influenced by the surface (land or ocean). Its mean depth is about 1 km, but depth can reduce to a few tens of meters on a cold winter day and expand to several kilometers on a warm, turbulent, summer day. The layer above the troposphere, called the stratosphere, is stably stratified. The stratosphere is the layer where the ozone chemistry protects the air and the surface below from incoming ultraviolet. The tropopause is the interface between the troposphere and the stratosphere. It behaves as a lid for the troposphere because the vertical motions are quickly dampened in the stable stratosphere above.

For a given spatial scale, atmospheric motions are generally faster than oceanic currents, so the atmospheric time scales are shorter; however, energy levels are comparable because of the density difference.

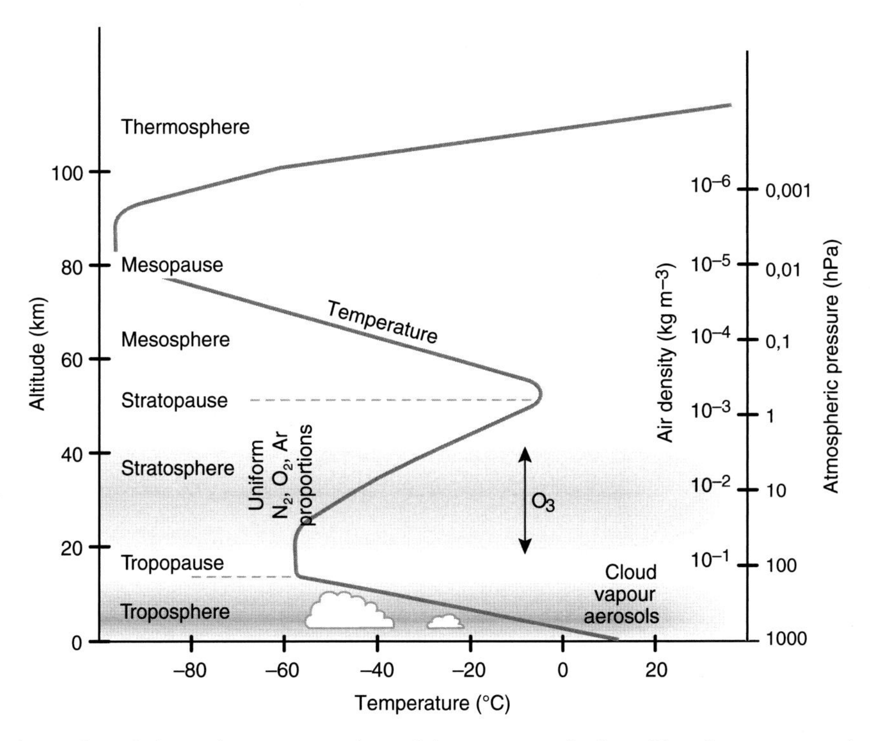


Figure 8.1. Schematic representation of the mean vertical profile of temperature in the atmosphere with some indication of the vertical variation of density and pressure (right axis). From McIlveen (1992).

### *8.1.2 The Weather and the Climate*

The climate system's main source of energy is the Sun. The atmosphere absorbs less than 20% of the incoming radiation and about 30% is reflected back, leaving 50% to be absorbed by the surface (35%, ocean; 15%, land). The atmosphere absorbs most of the infrared (IR) radiation emitted by the surface. The atmospheric fluid is then mainly heated from below, and, as in a saucepan of boiling water, the heat source near the surface strongly influences buoyancy-driven flows like atmospheric convection.

Larger-scale circulations are mainly driven by the differential heating in latitudes (Figure 8.2).

The meridional mixing is done both by the oceans and by the atmosphere (Figure 8.3); the ocean does most of the mixing in the tropics, and the atmosphere does most of the mixing in the midlatitudes and the polar regions. In the vertical, the mean radiative cooling at the top of the atmosphere is balanced by convective motions. The water phase transitions play an important role in the vertical redistribution of energy, especially in the tropics.

In the tropics, most of the mixing is done by the zonally averaged flow. The low-level branches of the Hadley cells concentrate sensible and latent heat near the equator,

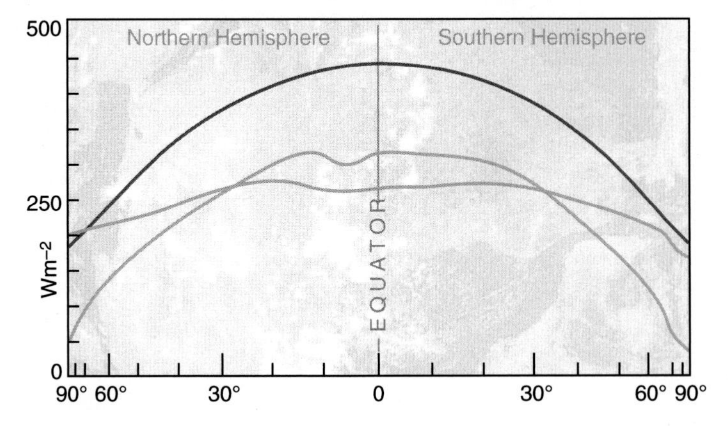


Figure 8.2. Zonally averaged radiative budget of the system Earth/Atmosphere. The red curve is the mean solar input at the top of the atmosphere. The blue curve shows the mean radiation absorbed by the system (the difference between the red and the blue curve is due to the reflection of the solar radiation). The green curve shows what is emitted back into space. The radiative budget between the blue and the green curve shows excess in the tropical regions and deficits in the midlatitudes and polar regions. From Gill (1982). For a color version of this figure please see the color plate section.

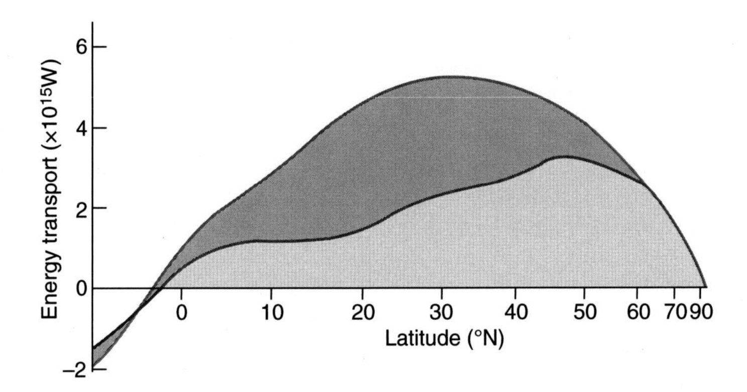


Figure 8.3. Equator/pole energy transport by the geophysical fluids in the Northern Hemisphere. The surface under the red curve is proportional to the total transport of energy necessary to balance the meridional differential heating. The green shaded area is the part transported by the oceans, and the blue shaded area is the part transported by the atmosphere. From Gill (1982). For a color version of this figure please see the color plate section.

a region that already has a positive energy budget. In the Intertropical Convergence Zone (ITCZ) atmospheric convection transports energy upward and converts it into potential energy. Finally, the upper branches of the Hadley cells export potential energy toward the higher latitudes. The final budget of the Hadley cells transport is a transfer of energy from the equator to the poles, but the net efficiency of these circulations that may be seen as "heat engines" is low.

In the midlatitudes, the meridional mixing is mainly due to the stationary (barotropic Rossby waves) and transient eddies (mainly baroclinic waves).

The spectra of atmospheric circulations involved in the climatic equilibrium are very broad. The buoyancy is an important controlling factor in most of atmospheric circulation even if, for the large scales, the work of gravity toward a minimal potential energy state is made difficult because of the rotation of the Earth.

As in the ocean, the Rossby radius of deformation $R_d$ gives an order of magnitude of the scale separation between the processes mainly driven by the buoyancy and the processes mainly controlled by the rotation. In the atmosphere, $R_d$ is of the order of 1,000 km. Circulations larger than $R_d$ are in quasi-geostrophic balance. The corresponding state of the atmosphere still contains available potential energy that cannot be extracted by direct overturning cells. More complex structures such as Rossby waves or baroclinic waves are important for the mixing where the rotation is dominant.

The circulations that are a direct response to a buoyancy forcing are usually circulations that have a scale smaller than the Rossby radius of deformation. At this scale, the buoyancy-driven flows are usually direct circulations, with warm air parcels ascending vertically and compensating flows of cooler air subsiding. Horizontally, the wind goes "directly" from high pressure to low pressure.

Thermal/density gradients are also driving circulations larger than the Rossby radius of deformation. But these circulations are generally in hydrostatic balance and not far from the geostrophic balance. The vertical component of the motion in these circulations is not directly forced by the buoyancy force, and the horizontal component of the motion is almost parallel to the isobars. These circulations also convert potential energy into kinetic energy, but this conversion involves much more complex processes than the ones in a direct circulation.

As in the ocean, local buoyancy-driven flows like atmospheric convection trigger larger-scale waves that then play an important role in the redistribution of energy around the planet. It is especially true along the equatorial wave guide (Gill 1982). The interaction between atmospheric deep convection and large-scale waves is one key feature of the intraseasonal variability of the tropical atmosphere (Madden-Julian Oscillation, El Niño-Southern Oscillation).

### *8.1.3 Buoyancy in a Perfect Gas*

In a compressible fluid like a gas, the buoyancy is usually not expressed in terms of density, which is not a conserved quantity, but in terms of *potential temperature*. In the atmosphere, the buoyancy also depends on the moisture.

### 8.1.3.1 Dry Case

To a good level of approximation, air can be considered a perfect gas. The equation of state is then the perfect gas law

$$p = \rho R_a T,$$

where $p$ is the pressure of the gas, $T$ is the temperature, $\rho$ is the density, and $R_a = R^*/M_a = 287\ \mathrm{J\,kg^{-1}\,K^{-1}}$ the gas constant for dry air ($R^* = 8.314\ \mathrm{J\,mol^{-1}\,K^{-1}}$ is the universal gas constant and $M_a \simeq 29$ g the molar mass of dry air).

The first principle of thermodynamics (Bohren and Albrecht 1998) applied to a perfect gas can be written as an equation for the enthalpy of the gas

$$\frac{Dh}{Dt} = -\frac{1}{\rho}\frac{Dp}{Dt} + \dot{Q}, \tag{8.1}$$

where the enthalpy $h = c_p T$, $c_p = 1004.5\ \mathrm{J\,kg^{-1}\,K^{-1}}$ is the specific heat at constant pressure, and $\dot{Q}$ is the heating rate. In an adiabatic process ($\dot{Q} = 0$), the evolution of the temperature is associated with a reversible process between the enthalpy of the gas and the pressure work associated with the volume changes. In the atmosphere, an adiabatic cooling is then mainly a consequence of a vertical ascent.

It is possible to define a thermodynamics variable that is conservative with respect to the work of the pressure force in an adiabatic evolution. A combination of the left-hand side and the first term on the right-hand side of equation (8.1) (for $\dot{Q} = 0$) gives

$$\frac{D\left(\ln\left(\frac{T}{p^{R/c_p}}\right)\right)}{Dt} = 0.$$

The ratio $T/p^{R/c_p}$ is then conserved in an adiabatic evolution.

For a pressure $p_o = 1{,}000$ hPa, this ratio is $\theta/p_o^{R/c_p}$, and $\theta$ is called the potential temperature. The potential temperature of a parcel of temperature $T$ and pressure $p$ is then

$$\theta = T\left(\frac{p_o}{p}\right)^{R/c_p}.$$

$\theta$ is also the temperature that an air parcel would have when adiabatically moved to a pressure level of 1000 hPa.

$\theta$ is conserved in an adiabatic (dry) atmospheric evolution. It evolves because of the heat exchange with the environment if the evolution is diabatic.

A classical simplified approach to studying the vertical motion of an air parcel driven by the buoyancy force is to consider this air parcel as "isolated" from its environment like a bubble or a buoy.

The equation for the vertical velocity of the parcel $w_p$ is given by the vertical momentum equation applied to the parcel

$$\frac{Dw_p}{Dt} = -g - \frac{1}{\rho_p}\frac{\partial p_p}{\partial z}.$$

A usual hypothesis is to consider that the pressure of the parcel $p_p$ instantaneously adjusts to the pressure of the hydrostatic environment $p_e$ when the parcel is moving vertically. The air parcel is also supposed to follow an adiabatic evolution.

In this context, the vertical acceleration of the parcel is directly related to the buoyancy, which can be written as a difference of density, temperature, or potential temperature between the parcel and the environment

$$\frac{Dw_p}{Dt} = -g\frac{\rho_p - \rho_e}{\rho_p} = -g\frac{T_e - T_p}{T_e} = -g\frac{\theta_e - \theta_p}{\theta_e} = \mathcal{B}. \tag{8.2}$$

Here $\mathcal{B}$ is the equivalent for the atmosphere of the reduced gravity $g'$ defined in the introduction to this book.

A first-order development of equation (8.2), for a small vertical displacement around the equilibrium level of the parcel shows that

- The environment is stable with respect to a small vertical displacement if $\partial\theta_e/\partial z > 0$.
- The environment is unstable if $\partial\theta_e/\partial z < 0$.
- The environment is neutral if $\partial\theta_e/\partial z = 0$.

If the air is stable, parcels will oscillate around their equilibrium level with the Brunt-Vaïsälä frequency $N_e = \sqrt{(g/\theta_e)(\partial\theta_e/\partial z)}$.

If the air is unstable, parcels will be accelerated upward or downward until they reach a level of neutral buoyancy (or the surface).

The mean state of the atmosphere is stably stratified (Figure 8.4) as $\theta$ increases with the vertical.

#### *8.1.3.2 Buoyancy in Moist Air*

The molar mass of water vapor (18 g) is lighter than the molar mass of air (about 29 g).

One important consequence of this difference is that the more moist the air, the more buoyant it is.

To compare a moist air parcel of density $\rho_h$ and molar mass $M_h$ with a (differently) moist environment, meteorologists use the virtual temperature $T_v$, which is the temperature that a parcel of dry air of molar mass $M_a$ would have if it was with the same pressure $p$ and density $\rho_h$ as the moist parcel

$$p = \rho_h R_h T = \rho_h R_d T_v.$$

As $R_d = R^*/M_d < R_h = R^*/M_h$, $T_v > T$.

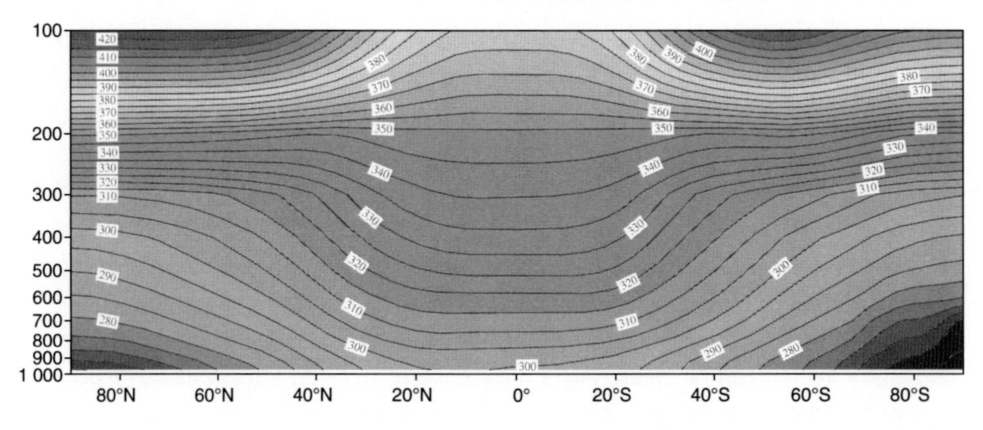


Figure 8.4. Annually and zonally averaged potential temperature (K) between 1,000 and 100 hPa computed from the 44 years of ERA40 reanalysis. From Malardel (2009).

$T_v$ is a practical way to express the gain in buoyancy due to the presence of water vapor. The difference between $T$ and $T_v$ may be of the order of a few degrees (air at 20°C with a moisture content of 12 g/kg, $T_v - T = 2$°C).

The vertical momentum equation for a parcel of moist air is then written as

$$\frac{Dw_p}{Dt} = \mathcal{B} = -g\frac{\theta_{ve} - \theta_{vp}}{\theta_{ve}},$$

where

$$\theta_v = T_v \left(\frac{p_o}{p}\right)^{R/c_p}.$$

#### 8.1.3.3 Buoyancy in Saturated Air

If a parcel of moist air reaches a level where the temperature is low enough for condensation to start, the parcel is warmed by latent heat release. If the ascending motion of the parcel continues, the vertical rate of decrease of the temperature of the parcel is smaller (about 4 K/km instead of 9.8 K/km), and the chance for the parcel to be more buoyant than its environment is much greater. Most of atmospheric convection is actually possible because of the condensation of water vapor.

The positive effect of the condensation on the buoyancy is generally attenuated by the effect of the loading of the air parcels by the rain and ice precipitation. In big thunderstorm clouds, this effect can reduce significantly the strength of the ascents.

## 8.2 Circulations

In this section, we will discuss three atmospheric circulations. The first, the atmospheric front, is often presented as a gravity current. We will try to show that the dynamics of cold and warm fronts result from a different mechanism for which the Coriolis force cannot be neglected.

We will then explain the basic principles of atmospheric deep convection. In particular, we will introduce the role of wind shear and downdrafts in the organization of convective systems with longer life cycles.

In the last part of this section, we will list different types of direct cell circulations whose upward or downward branches are directly driven by the buoyancy force.

### 8.2.1 Atmospheric Frontal Systems

Extratropical cyclones and atmospheric cold and warm fronts are often presented as an undulation of a preexisting limit between two air masses of different temperatures and densities. In the "Norwegian theory" this preexisting limit, called the polar front (Bjerknes and Solberg, 1922), is seen as one result of the general circulation. Extratropical cyclone and anticyclones are seen as an undulation of this thermal limit, and the vertical motion and the clouds along the cold and the warm fronts are explained as consequences of the "fight" between the cold and the warm air, the warm air being lifted above the cold mass as in a gravity current.

This description of the midlatitude frontal systems was exceptionally realistic compared to the very limited observation system available in 1920. However, we know now that there is no evidence of the existence of the polar front as it was described by the "Norwegian theory."

The theory now commonly accepted to explain the development of extratropical cyclones and frontal systems is based on baroclinic instability (Gill 1982; Pedlosky 1987). The fronts are a secondary product of the baroclinic waves. The temperature gradients are amplified by frontogenesis in a positive feedback process controlled by the Earth's rotation (Hoskins and Bretherton 1972; Hoskins 1975).

#### *8.2.1.1 The Baroclinic Zone*

The conservation of the angular momentum of the air parcels in the upper branch of the Hadley cells results in the formation of a strong westerly jet around 30° (James 1994). This jet (or the vertical wind shear associated with the jet) is in thermal wind balance with the meridional gradient of temperature. The stronger the jet is, the stronger the thermal gradient (the baroclinicity) is. As the scale of the jet is large compared with the Rossby radius of deformation, the mean state of the atmosphere in the midlatitudes is characterized by a large reservoir of mean available potential energy (Gill 1982), which cannot be released by mean direct circulations because of the Earth's rotation.

This baroclinic zone in the midlatitudes is the reservoir of energy for the baroclinic waves and the corresponding extratropical cyclones, anticyclones, and frontal systems. These circulations may then be considered as driven by a horizontal gradient of temperature or density. But the motion of the air in these circulations in quasi- (or semi-)

geostrophic balance cannot be explained by the intuitive conceptual model of a direct cell.

#### *8.2.1.2 Baroclinic Development*

An analysis of the structure of developing extratropical cyclones and anticyclones shows that the main amplitude of the perturbation of the mean baroclinic zone is stronger near the surface and at the level of the tropopause. Such a structure is in agreement with the circulations obtained by the inversion of the quasi-geostrophic potential vorticity (QGPV) in a layer (the troposphere) of quasi-uniform QGPV.

The circulations inside a uniform QGPV layer are completely determined by the structure of the thermal field at its upper and lower boundaries (the surface and the tropopause).

A positive anomaly of temperature at the tropopause is associated with an anti-cyclonic circulation (high pressure anomaly, negative vorticity in the Northern Hemisphere). Near the surface, it is the opposite; a warm anomaly corresponds to a cyclonic circulation (low pressure anomaly, positive vorticity in the Northern Hemisphere).

Such perturbations near the boundaries propagate with a mechanism similar to that of barotropic Rossby waves, westward at the tropopause but eastward near the surface (Figure 8.5).

The interaction between a cyclonic anomaly and the baroclinic zone generates vertical velocity (this is not true if the cyclonic anomaly is in a barotropic environment), with ascending motion upstream and subsiding motion downstream of the anomaly (Figure 8.6).

When a cyclonic tropopause anomaly is situated upstream with respect to a cyclonic low-level anomaly, the vertical velocity generated by each anomaly amplifies the vorticity of the other anomaly by vortex stretching (Figure 8.6). The tropopause and the surface waves are then in a phase of baroclinic development. The two waves may remain locked for a few days. During this phase, the amplitude of the waves increases, the system propagates westward (Figure 8.7) and the baroclinic perturbation transports sensible and latent heat from the tropics to the poles.

But slowly the upper anomaly overtakes the surface anomaly, and a baroclinic decay (vorticity shrinking) starts.

This approach summarizes the main principles of baroclinic wave development in the midlatitudes (other, more academic approaches can be found in Gill 1982 and Holton 1992). However, several other factors can influence the real development of these synoptic extratropical systems: large-scale weather regimes, interaction with the entrance/exit of a jet streak, upstream/downstream developments, surface fluxes, large-scale cloud formation, convection, and so on.

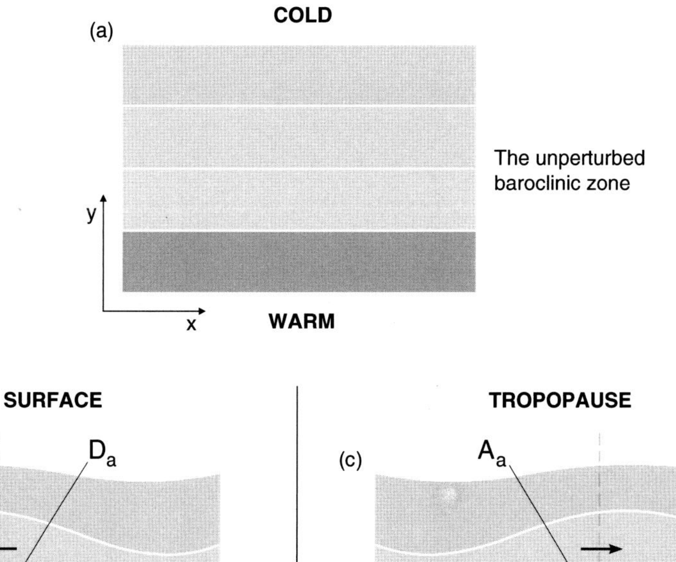


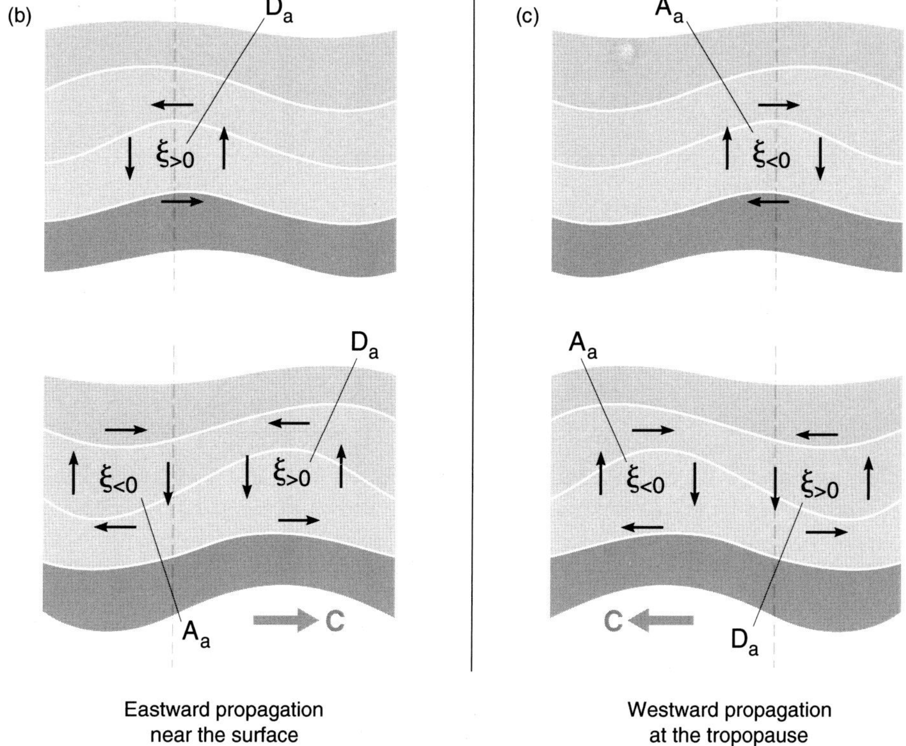


Figure 8.5. Horizontal propagation of a temperature and vorticity perturbations of an initial baroclinic zone. Near the surface, a warm anomaly is associated with a cyclonic perturbation (positive vorticity in the Northern Hemisphere). The circulation will propagate the positive anomaly eastward (the cyclonic wind advects warm air from the south to the north on the east side of the perturbation). Near the tropopause, a warm anomaly is associated with an anticyclonic circulation, which will propagate the warm undulation to the west. From Malardel (2009).

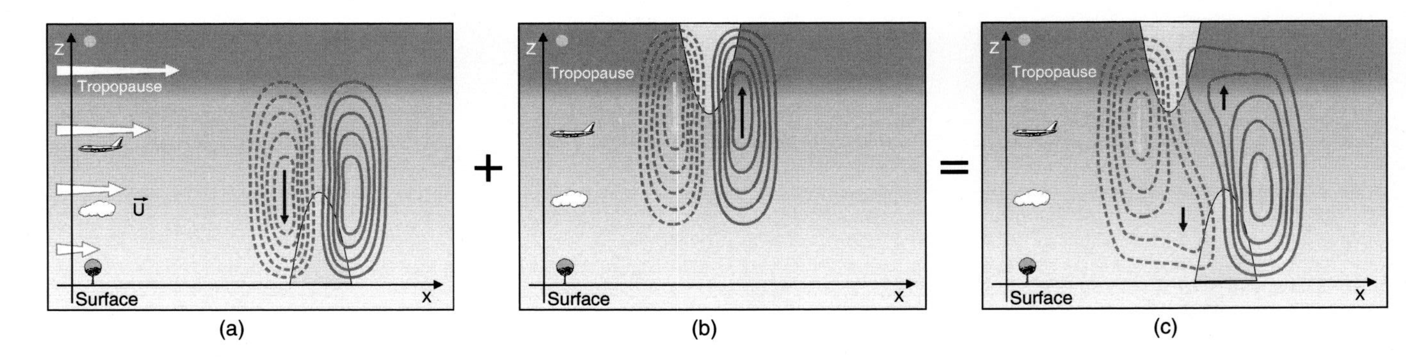


Figure 8.6. Schematic vertical cross-section of the vertical velocity generated by cyclonic circulations in a baroclinic zone (zonal wind shear), (a) near the surface, (b) near the tropopause, (c) superposition of (a) and (b). From Malardel (2009).

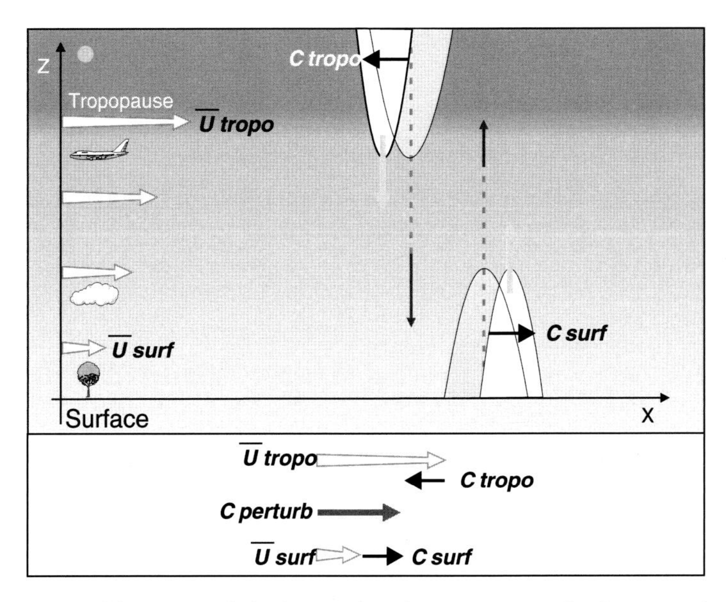


Figure 8.7. Positive baroclinic interaction between a cyclonic anomaly at the tropopause and a cyclonic anomaly at the surface. The difference of zonal advection of the system near the surface and at the tropopause due to the vertical wind shear is compensated by the propagation of the anomalies in the opposite direction. The full system propagates westward (with a characteristic speed of about 10 m/s). From Malardel (2009). For a color version of this figure please see the color plate section.

### *8.2.1.3 Frontogenesis*

During the life cycle of the baroclinic waves, the quasi-geostrophic cyclonic anomalies amplify the deformation of the baroclinic zone mainly to the southwest and to the northeast of the cyclone (Figure 8.8) because of a mechanism called *frontogenesis*.

In both zones of frontogenetic deformation, a convergent ageostrophic circulation (still controlled by the rotation) appears near the surface and at the tropopause (Figure 8.9).

Because of positive feedback between the ageostrophic motion and the reinforcement of the thermal gradient, regions with very high temperature gradients form near the surface and at the tropopause. They are the cold and warm fronts associated with the extratropical cyclones.

Even if they finally look like an interface between two air masses of different temperatures, these types of fronts are not gravity currents.

The main wind and displacement of the air parcels is along the front and not in the cross-front direction as in a classical gravity current. Parcels are slowly ascending from the tropical warm and moist region to the higher latitudes or slowly subsiding from the high latitudes to the tropics in the "cold" part of the wave. Stratiform clouds are formed during this slow ascending motion, which is globally parallel to the fronts.

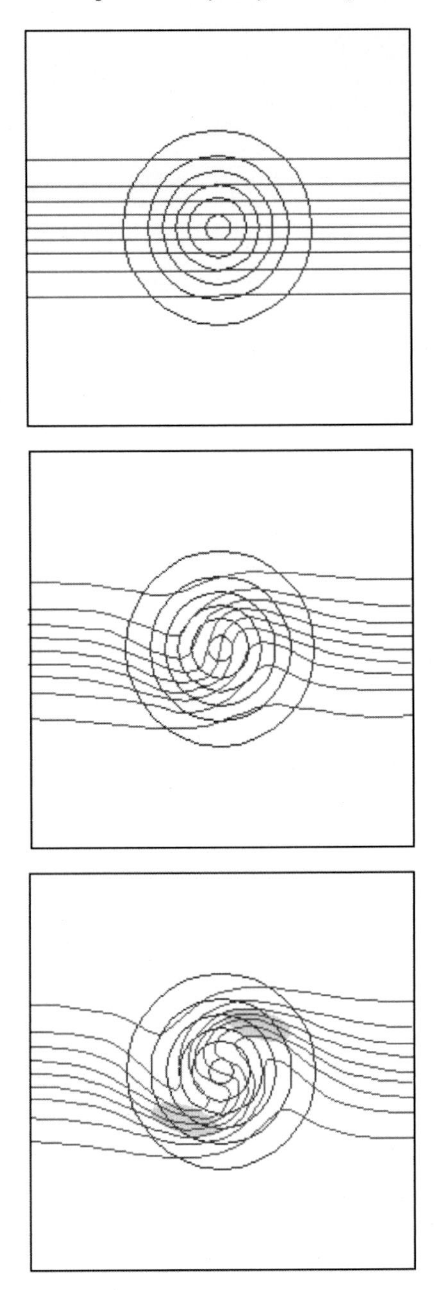

Figure 8.8. Deformation of a baroclinic zone by a cyclonic vortex. The original thermal gradient of the baroclinic zone is reinforced both to the southwest and to the northwest of the cyclone. From Malardel (2009).

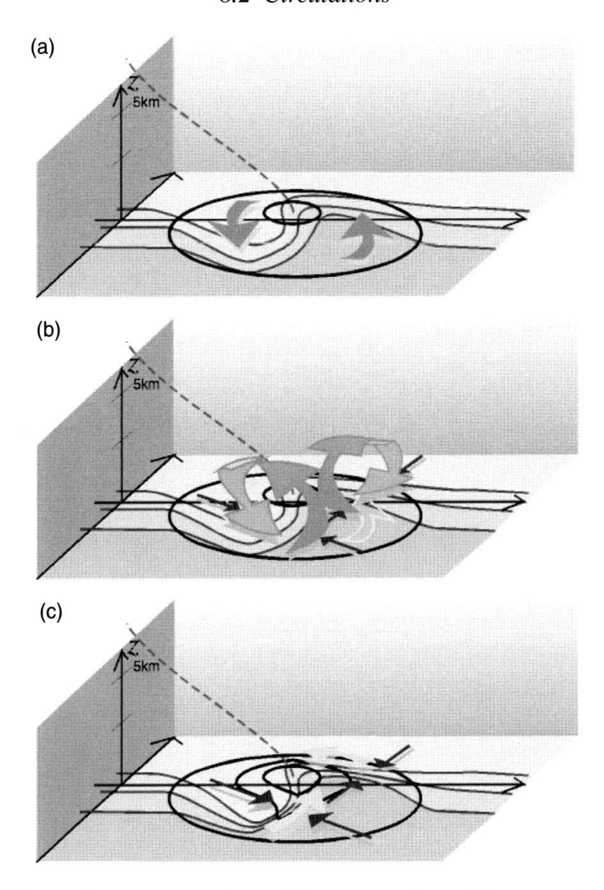


Figure 8.9. Schematic representation of the retroaction of the ageostrophic circulation on a thermal gradient in both a cold and a warm front. (*top*) The deformation of the meridional thermal gradient of the baroclinic zone starts with the advection by the quasi-geostrophic circulation around a growing cyclonic vortex (baroclinic development), (*middle*) a vertical and cross-frontal ageostrophic circulation is necessary to maintain the thermal wind balance, (*bottom*) the convergence of the ageostrophic wind reinforces the thermal gradient. However, a stronger convergent ageostrophic circulation is necessary to maintain the thermal wind balance, which reinforces the gradient. Thermal quasi-discontinuities form to the southwest and northeast of the initial cyclone: the cold and warm fronts. From Malardel (2009). For a color version of this figure please see the color plate section.

Secondary smaller-scale convective updrafts that are more directly driven by buoyancy may be embedded in the stratiform clouds. But the mesoscale quasi-vertical circulations associated with these clouds are only superimposed to the main, larger-scale frontal circulation.

### 8.2.2 Atmospheric Convection

Dry and adiabatic convective plumes are present mainly during sunny days in the boundary layer (BL). Their time scale is of the order of minutes.

These plumes are very turbulent, and most of the transport associated with these convective ascents can be described as isotropic turbulent mixing. More organized dry thermals also exist, but their vertical extension is usually limited to the top of the BL.

Deeper convection always involves condensation.

#### *8.2.2.1 Convective Inhibition and Convective Available Potential Energy*

Most convective plumes start from the lower levels of the atmosphere. Quite often, the lowest levels are stable with respect to dry adiabatic motions. But the convective inhibition, which is the energy necessary to lift the air parcels until they reach a level where they become positively buoyant, that is, warmer than their environment (CIN, Figure 8.10), can be compensated for by a source of ascending motion like turbulence, dry thermals, orographic forcing, or large-scale ascent.

Once the level of free convection is reached, the CAPE is converted into vertical motion until the level of neutral buoyancy is reached. Air parcels overshoot this level thanks to inertia, but the tropopause is an efficient lid, and most of the air is spread under the tropopause in a stratiform icy cloud called an anvil.

The CAPE is the vertical integral of the buoyancy $\mathcal{B}$ (equation (8.2)) between the level of free convection and the level of neutral buoyancy

$$\mathrm{CAPE} = \int_{\mathrm{LFC}}^{\mathrm{LNB}} \mathcal{B} dz.$$

We obtain an estimation of the maximal vertical velocity of the air parcel if we suppose that all of the CAPE is converted into kinetic energy

$$w_{\mathrm{max}} = \sqrt{2 \int_{\mathrm{LFC}}^{\mathrm{LNB}} \mathcal{B} dz.}$$

But $w_{\mathrm{max}}$ overestimates the real vertical velocities in the updraft, which are reduced by non-hydrostatic pressure effects, entrainment of non-buoyant air, precipitation, and turbulence.

In the no-wind, or no-wind-shear cases, the convective ascents are quickly killed by the downdraft generated by the evaporation of the precipitations under the cloud (Figure 8.11). The characteristic time scale of an ordinary precipitating cumulus is 15/30 minutes.

#### *8.2.2.2 Downdrafts and Cold Density Currents*

The melting of snow, graupel, and hail and the evaporation of rain cools the air underneath the cloud. This air becomes negatively buoyant and sinks. When the downdraft reaches the ground, the air spreads along the surface and forms cold gravity currents.

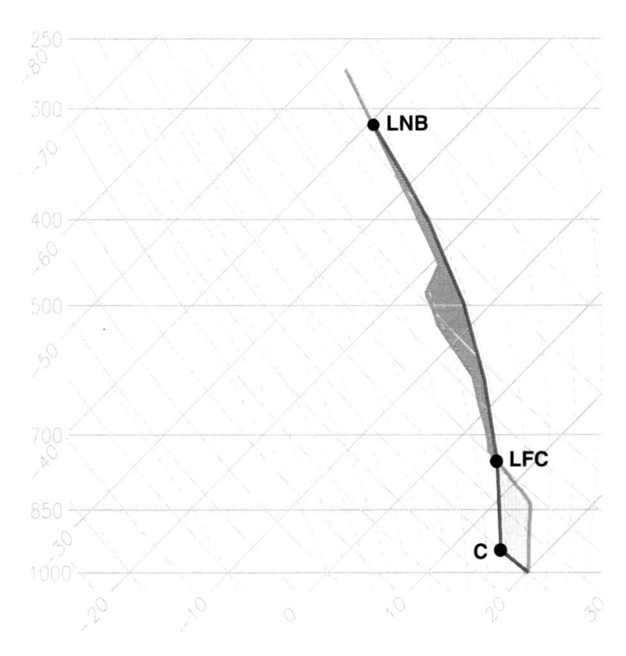


Figure 8.10. Representation of convective instability on a $p$-$T$ diagram (horizontal brown lines are isobars; diagonal brown lines are isotherms). A parcel of air is lifted from an initial level of 1,000 hPa (red line). Until its level of condensation (C), the parcel follows a dry adiabat (solid green curves). Above the level of condensation, the parcel follows a saturated adiabat (dashed green curves). The temperature of the parcel decreases more slowly then because of latent heat release. Below the level of free convection (LFC), the parcel is cooler than the environment (blue line). The surface between the temperature profile of the parcel (red line) and the temperature profile of the environment (blue line) is proportional to the convective inhibition (yellow shading). Above the LFC, the saturated adiabat is warmer than the environment; the parcel is buoyant and accelerated upward until it reaches its level of neutral buoyancy (LNB). The surface between the temperature profile of the parcel and the temperature profile of the environment (orange shading) is proportional to the convective available potential energy (CAPE). From Malardel (2009). For a color version of this figure please see the color plate section.

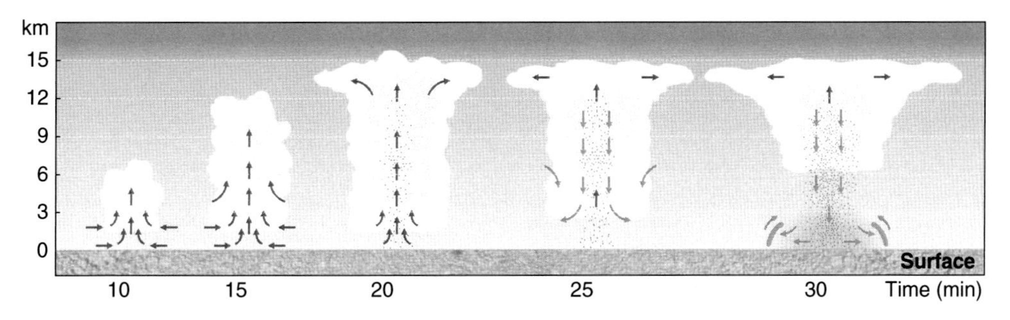


Figure 8.11. Schematic representation of the life cycle of an ordinary convective cell. From COMET (1996).

The horizontal velocity (wind) of the gravity current under the cloud (gust front) may be estimated by (Houze 1993)

$$\frac{1}{2}U_{CD}^2 = \frac{(\rho_{CD} - \rho_e)}{\rho_e}gh.$$

The gravity currents may also trigger new convective cells if the wind shear of the environment is favorable.

#### *8.2.2.3 Organization of Convection*

The CAPE is a necessary ingredient for the formation of deep convective clouds. But other parameters such as the wind shear (rotational or not) of the environment, the role of microphysics (in particular the ice phase), and interaction with the surface (buoyancy fluxes) also influence the real development of convection in the atmosphere.

The wind shear is a key factor in the organization of longer-lived convective systems. Unidirectional low-level wind shear is necessary for the development of multicell systems (Figure 8.12). With deeper rotational shear, the convective circulations are associated with mesoscale vortex, and very active systems called supercells may develop (Figure 8.13). The supercells are often responsible for the formation of hail and sometimes for tornadoes.

Ordinary cells, multicells, and supercells may be elementary cells of much bigger convective systems of more than 100 km in one of their horizontal directions. The mesoscale circulation characteristic of these mesoscale convective systems is usually

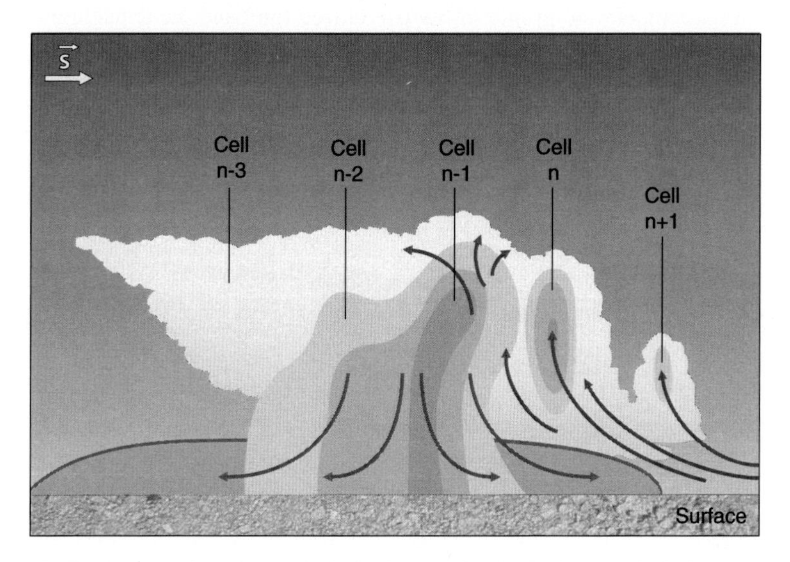


Figure 8.12. Schematic representation of a multicell. Cell $n-1$ is fully developed. Cells $n$ and $n+1$ are growing because of the interaction between the gravity current generated by older cells and the vertical wind shear of the environment. Cells $n-2$ and $n-3$ are dissipating. From COMET (1996).

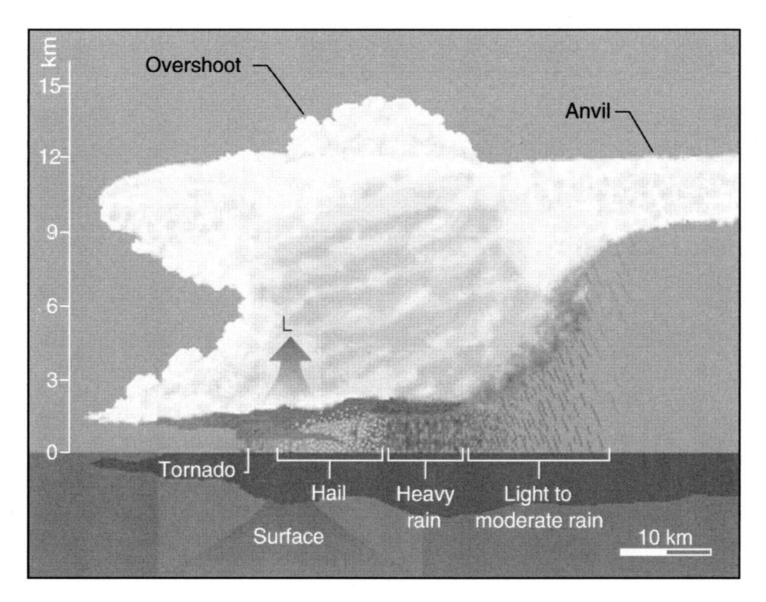


Figure 8.13. Conceptual model of a supercellular thunderstorm. From COMET (1996). For a color version of this figure please see the color plate section.

influenced by the Coriolis force, but the time and space scales of these circulations are not large enough for the circulation to be controlled by the quasi-geostrophic balance.

One of the most studied mesoscale convective systems is the squall line. Squall lines often start as a multicell system, with new cells developing in the upstream side of the wind shear. Older cells are slowly integrated into the anvil cloud and a large stratiform system forms behind the active convective cells. In the mature system, a mesoscale circulation enters the system from the back of the squall line and slowly descends toward the density currents in the front of the system (Figure 8.14). An ascending flow is forced by the gravity current at the front of the squall line. This ascending flow is associated with active convective cells and continues with a smoother slope in the stratiform part of the system, where it entrains ice particles collected in the convective cells. The sublimation of these precipitating ice particles amplifies the negative buoyancy of the circulation, which feeds the system from the back.

### *8.2.3 Direct Cells*

Direct overturning circulations can be large-scale circulations in the tropical regions, where the Coriolis force is not too strong (Hadley Cells, Monsoon). In the midlatitudes, direct cells are mesoscale circulations driven by strong but short-time-scale thermal contrasts near the surface usually associated with the daily cycle. At higher latitudes, the intense cooling along an ice slope generates katabatic wind in the boundary layer.

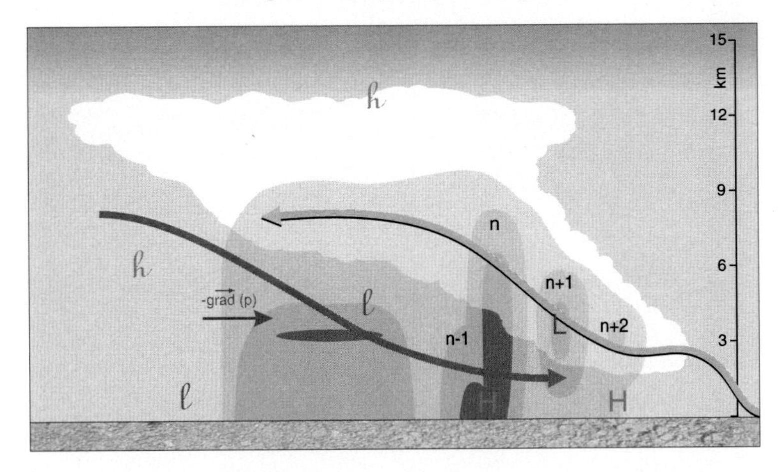


Figure 8.14. Conceptual model of a squall line. Cross-section perpendicular to the squall line. From Weisman and Przybylinski (1999). For a color version of this figure please see the color plate section.

On the Antarctic continent, these winds, which are very persistent and strong, interact with the larger-scale, quasi-geostrophic polar vortex (Egger 1985; 1994).

### *8.2.3.1 Land/Sea Breeze*

During a sunny day, the solar radiation is absorbed by a thin layer of land along the coast and by a much deeper layer of water in the ocean. The coastal land then becomes much warmer than the nearby sea. The air above the land is heated from the bottom, and a convergent/divergent circulation is triggered as the thickness of the layer above the surface increases (Figure 8.15). A direct cell organizes, which has convective ascending motions and cumulus formation over the land and a compensating subsiding branch over the sea. During the day, a refreshing breeze blows near the surface from the sea to the land. If the breeze remains for several hours, it is affected by the Earth's rotation, and the wind acquires a component parallel to the coast.

### *8.2.3.2 Mountain Breeze*

The differential heating between the surface along a slope and the air at the same level creates direct circulations perpendicular to the slopes. We usually distinguish between the valley breeze that blows in the main direction of a valley and the slope breeze generated on the steeper slopes on each side of the valley. This type of breeze is characterized by katabatic winds (subsiding gravity currents) along the slope during the night and anabatic winds (ascending buoyant currents) along the slope during the day (Figure 8.16). During the day the valley breeze and the slope breeze may interact and more complex helicoidal circulations may be observed in the valley.

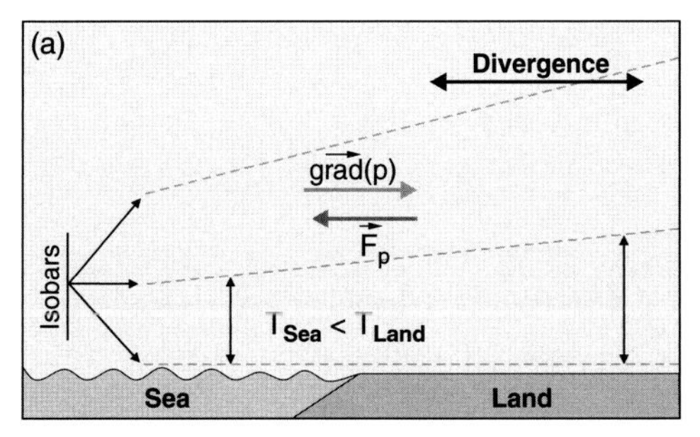


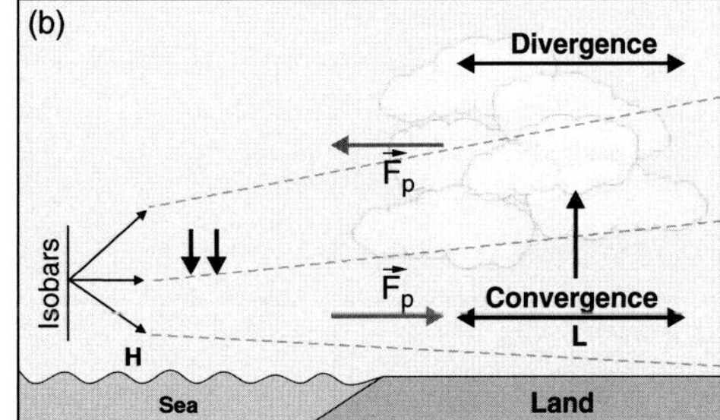


Figure 8.15. Initiation (a) and installation (b) of the land/sea breeze circulation during the day. As the air above the land surface becomes warmer than the air above the sea, the thickness between isobars above the land surface deepens. The pressure gradient force $\vec{F}_p$ is pointing toward the sea at the top of the boundary layer. Mass is then moved from the land to the sea at the top of the BL. The surface (hydrostatic) pressure in the air columns above land decreases. The pressure gradient force near the surface points toward the land, and it accelerates the air from the sea toward the land in a low-level wind called a *sea breeze*. From Malardel (2009). For a color version of this figure please see the color plate section.

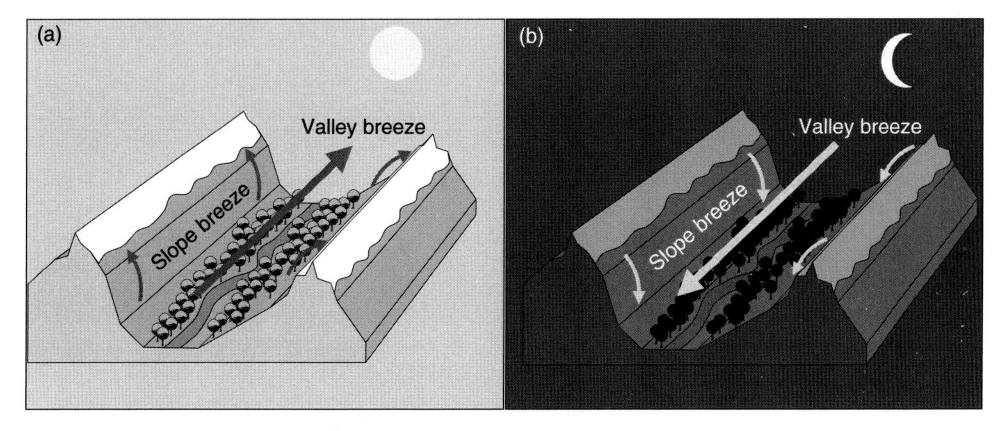


Figure 8.16. Valley and slope breezes: (a) anabatic circulations during the day and (b) katabatic circulations during the night. From Malardel (2009). For a color version of this figure please see the color plate section.

## 8.3 Simulations

### *8.3.1 Overview of Atmospheric Simulations*

A summary of the different types of atmospheric simulations is given in Figure 8.17. Global operational models and climate models still have resolutions that resolve processes in the hydrostatic range. Use of non-hydrostatic (NH) "convection-permitting" models, with resolutions of 1–3 km, as operational numerical weather prediction (NWP) tools over limited areas (LAM) is just beginning. Higher-resolution models with resolutions ranging from less than 1 km (cloud-resolving models, or CRM) to

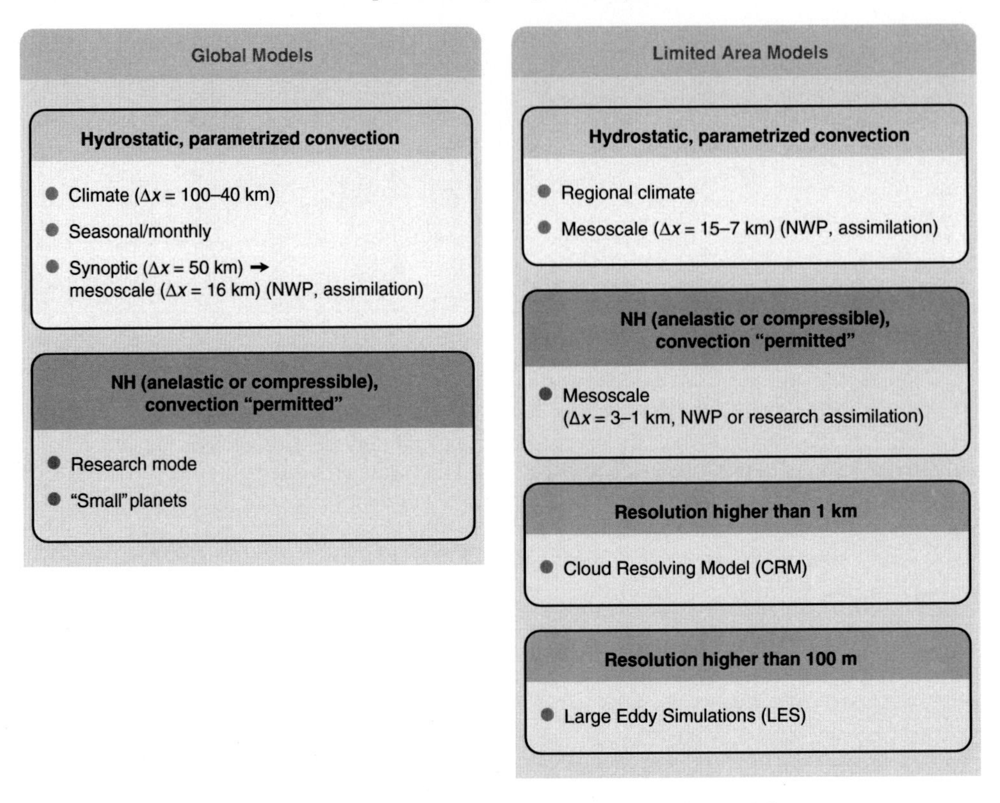


Figure 8.17. The different types of atmospheric models.

10 m (large eddy simulation, or LES, models) are used for process studies and for the development and the validation of the physical parameterizations of the larger-scale models.

When the hydrostatic approximation is released and the vertical velocity becomes a prognostic equation in the model, the acoustic modes become a solution of the system of equations. These waves may be filtered by the anelastic approximation or treated with some degree of implicitness in the time scheme.

Most large-scale hydrostatic models are written with a hybrid coordinate ($\sigma$ coordinate near the surface and pressure coordinate above). Semi-Lagrangian schemes associated with a semi-implicit treatment of the gravity waves allow the longest time steps. Higher-resolution models often use a $z$-type coordinate and explicit flux schemes on C-grid to resolve smaller-scale processes for which the local conservative properties of the scheme become more important.

Global hydrostatic models are used for longer-range forecasts but are also used as lateral boundary conditions for the higher-resolution, limited-area models.

The predictability of the atmospheric flows decreases with the scale of the processes. Small errors in the initial conditions and in the model may have a very fast, nonlinear growth rate and give a very different forecast. A probabilistic approach usually replaces the deterministic forecast for the medium ranges of NWP (4–10 days) and for the seasonal forecasts. A full ensemble of forecasts starting with a slightly different initial condition or slightly different models is run and this ensemble of forecasts is then interpreted with statistical methods. Currently, even if the predictability of individual convective event is low, the cost of convection-permitting models is still too high to run ensembles of high-resolution simulations.

The assimilation of observations is a very important step of NWP. Observations from the synoptic surface network, radio-soundings, and satellite and radar observations are used to create the initial conditions of the forecast. For a synoptic forecast, every 6 hr the assimilation cycle produces an assimilation that "captures" structures that are still mainly controlled by the geostrophic balance. At mesoscale, the assimilation cycle is more rapid, and the analysis is less constrained by the rotation of the Earth. The assimilation of new types of observations like Doppler radar winds and temperature or moisture measurements from dense GPS networks are important to improve the mesoscale structures in the initial condition, especially in the boundary layer. A correct representation of the stratification and the moisture near the surface is a necessary condition for realistic simulations of the mesoscale buoyancy waves and instabilities.

### 8.3.2 Modeling Buoyancy-Driven Flows

The simulation of convective updrafts and downdrafts is possible with a non-hydrostatic model and a parameterization of the cloud microphysics with a good representation of the ice phases (ice crystal, snow, graupel, and hail) and a resolution of at least 3 km (Figure 8.18).

A non-hydrostatic dynamics is also necessary for the simulation of short, orographic-trapped gravity waves (Figure 8.19). In the larger-scale models, the drag associated with the unresolved orographic gravity waves has to be parameterized.

Mesoscale buoyancy-driven flows like land/sea breezes and hydrostatic orographic waves are simulated by a hydrostatic model with resolutions between 7 and 15 km. The shallow or deep convection that is often associated with these processes is then parameterized.

But for large-scale models with a resolution lower than 15 km, most of the mesoscale buoyancy-driven flows are filtered or have to be (at least partially) parameterized.

The turbulent, subgrid-scale mixing has to be parameterized in any case, but the structure of the most energetic subgrid vortices depends on the resolution. For resolution lower than about 1 km, the vertical transport by the turbulent processes is dominant, and a parameterization of the vertical diffusion of momentum and of sensible and latent heat is sufficient. For higher resolution, a 3-D turbulent scheme is necessary. In the simplest schemes, the turbulent exchange coefficients $K$ are a function of the grid-scale features of the flow only (1 order closure). However,

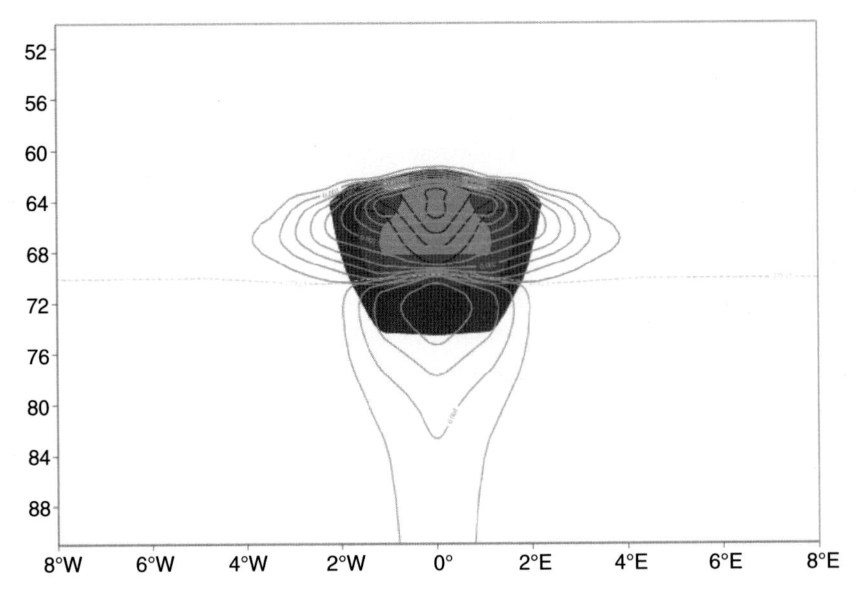


Figure 8.18. Academic simulation of a resolved tropical cumulonimbus on a small planet (radius of the Earth divided by 100) with the non-hydrostatic version of the model of the European Centre for Medium Range Weather Forecast. Vertical cross-sections through the cloud include cloud liquid water content (shading, kg/kg), cloud ice water content (green contours, kg/kg), and rain content (blue contours, kg/kg). The yellow line is the isotherm 0°C.

operational mesoscale models now often use schemes with a prognostic turbulent kinetic energy used for the computation of the $K$ coefficient (1.5 order closure). Higher-order closures exist in research models.

A parameterization of the convection is used in all hydrostatic models. In the range between 2 and 7 km, the problem of the parameterization of the atmospheric convection is not very well posed. The separation between the scale of the convective processes and the size of the grid is not possible anymore, but the resolution is not high enough to resolve the convective processes. This "gray" zone is currently avoided by most atmospheric models.

The interaction between the atmosphere and the surface below are treated by the surface scheme. The surface scheme is often a group of several parameterizations or simple models for different types of surfaces: ocean (from very simple parameterizations to full ocean models, but also 1-D surface layers), lake, nature (vegetation), and town. The surface at the bottom of each column of the model corresponds either to one single type of surface or to several tiles of different types of surface (e.g., a coastal grid box with 30% of sea and 70% of soil/vegetation).

The atmosphere sends to the surface the state of the atmosphere at the first model level, radiation fluxes (from the radiation scheme), and precipitations (from the microphysics).

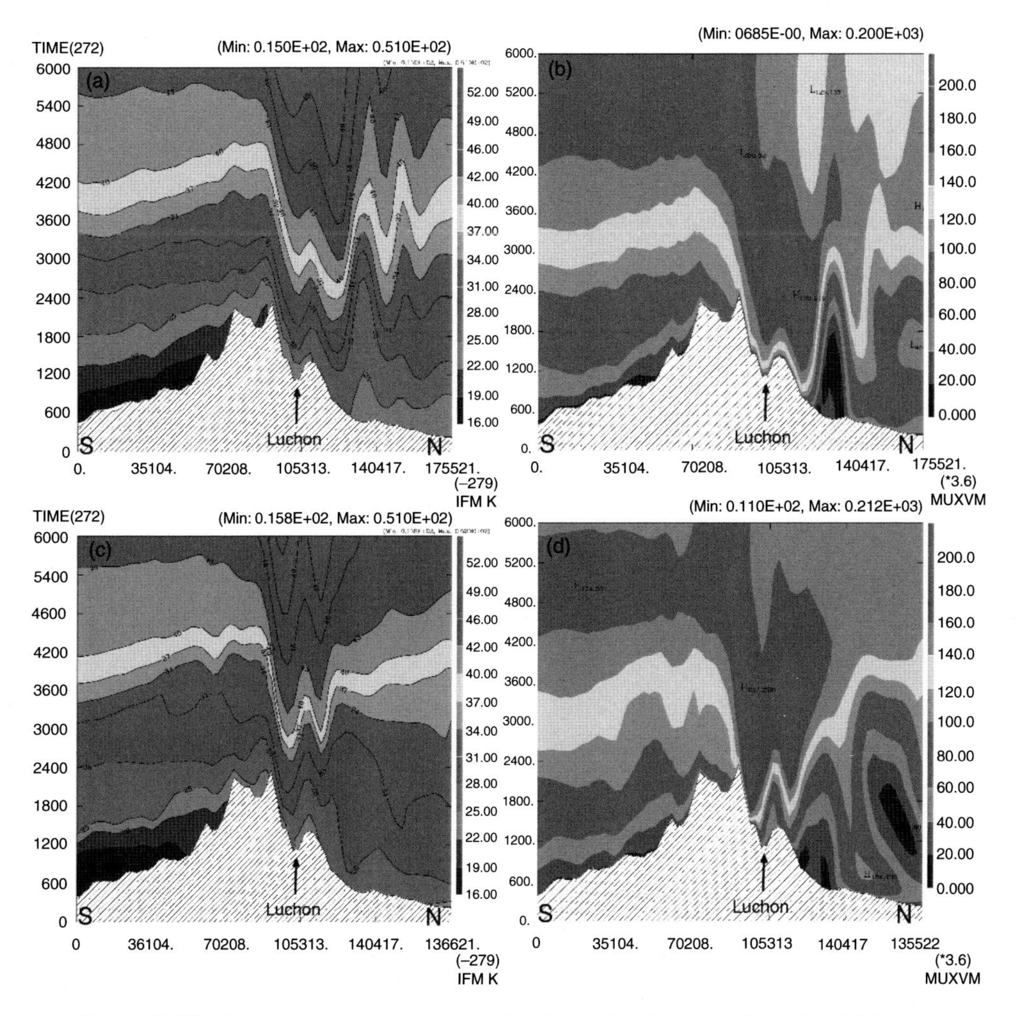


Figure 8.19. Orographic gravity waves simulated by the operational LAM Arome (Meteo-France) on 27 February 2010 on the north side of the Pyrénées. Potential temperature along a south/north cross-section of the Pyrénées: (a) non-hydrostatic simulation and (c) hydrostatic simulation. The trapped lee waves (shorter wave lengths) are filtered by the hydrostatic version. The stronger surface winds of the NH version are much more realistic. Wind intensity: (b) non-hydrostatic simulation and (d) hydrostatic simulation. For a color version of this figure please see the color plate section.

The surface sends back to the atmosphere momentum, heat and moisture fluxes (bottom boundary condition for the vertical diffusion scheme, sometimes also for the convection scheme), a "surface temperature," and an albedo (for the radiation scheme). If a grid box is covered by several tiles, the fluxes for each type of surface are aggregated before being sent to the atmospheric model.

For large-scale models, stratiform clouds and precipitation are parameterized by a very simple scheme. In most models, the cloud (nonprecipitating) condensates are prognostic (known from one time step to the next and advected by the advection scheme), but the precipitating condensates are often supposed to reach the ground in less than a time step (no airborne precipitation). In higher-resolution models, a more sophisticated microphysics of the cloud and precipitations is needed. The simpler ones have three or four reservoirs of hydrometeors (cloud droplets, rain, snow, sometimes graupel), which are described only by their mean mass content relative to the total mass in the grid box. More sophisticated schemes may have more classes of hydrometeors and take into account prognostically or diagnostically the number of drops or crystals in the grid box.

For high-resolution models, the feedback between the cloud microphysics and the dynamics (in particular the vertical advection linked with the buoyancy) is the driving process of the "resolved" convection and of the cold gravity current below the convective clouds.

Global convection-permitting simulations are not yet available for operational weather forecasting, but most weather centers are actively researching this domain.

An interesting alternative used to study the behavior of the global model is to reduce the radius of the planet. This type of experimentation, which is convection permitting and is available at a reasonable cost, is used to prepare and validate the global model of the European Centre for Medium Range Forecast (ECMWF) to the resolutions of the future.

On the small planet, dry and moist bubbles triggered by a low-level warming near the equator in an initial no-wind environment rise until reaching their level of neutral buoyancy. When a condensation scheme is activated, the updraft reaches the tropopause and a large cloud develops (Figure 8.18).

Even with a relatively simple microphysics of clouds and precipitation, the model reproduces the feedback between the vertical acceleration and the warming caused by condensation.

Such academic experiments are also used to study the interaction between atmospheric convection and large-scale waves. A better understanding of these processes is important to improve the parameterization of convection in the larger-scale models, which still have difficulty with correctly forecasting complex processes like the Madden-Julian Oscillation (MJO).

## References

Bjerknes, J., and H. Solberg, 1922: Life cycle of cyclones and the polar front theory of atmospheric circulation. *Geofys. Publ.* **3**, 1–18.

Bohren, C. F., and B. A. Albrecht, 1998: *Atmospheric Thermodynamics*. Oxford University Press, Oxford, New York.

Comet, F. M. L. (1996). Anticipating convective storm structure and evolution. CD-ROM.

Egger, J., 1985: Slope winds and the axisymmetric circulation over antartctica. *J. Atmos. Sci.* **42**, 1859–1867.
Egger, J., 1994: Antarctic slope winds and the polar stratospheric vortex. *J. Atmos. and Terrest. Phys.* **56**, 1067–1072.
Gill, A. E., 1982: *Atmosphere–Ocean Dynamics*, vol. 30. Academic Press, New York, London.
Holton, J. R., 1992: *Dynamic Meteorology*, 3rd ed., Academic Press, New York, London.
Hoskins, B. J., 1975: The geostrophic momentum approximation and the semigeostrophic equations. *J. Atmos. Sci.* **32**, 233–242.
Hoskins, B. J., and F. P. Bretherton, 1972: Atmospheric frontogenesis models: Mathematical formulation and solutions. *J. Atmos. Sci.* **29**, 11–37.
Houze, R. A., 1993: *Cloud Dynamics*. Academic Press, San Diego.
James, I. N., 1994: *Introduction to Circulating Atmospheres*. Cambridge University Press, Cambridge.
Malardel, S., 2009: *Fondamentaux de Météorologie*. Cépadues Editions, Toulouse.
McIlveen, R., 1992: *Fundamentals of Weather and Climate*, 2nd ed. Chapman & Hall, Oxford.
Pedlosky, J., 1987: *Geophysical Fluid Dynamics*. Springer-Verlag, Berlin.
Weisman, M., and Przybylinski, R., 1997–1999: Mesoscale convective systems: Squall lines and bow echoes. Web site. E. A. O. COMET Program.

# 9

# Volcanic Flows

ANDY WOODS

## 9.1 Introduction

Volcanic systems are controlled by a wide range of fluid mechanical processes, including the subsurface migration and ensuing accumulation of molten rock in crustal reservoirs, know as magma chambers, and the subsequent explosive eruption of ash and transport high into the atmosphere (Sparks et al. 1997). The philosophy of this chapter is to develop a series of simplified physical models of the flow processes in order to gain insights about the dominant controls on the processes; many of the models have been developed based on geological field evidence and have been tested with laboratory experiments. The aim is to build understanding rather than to simulate the processes, which are extremely complex and often for which there is insufficient data for a complete characterization of the physical and chemical state of the system. We provide a brief introduction to the overall range of processes which occur, and then immerse ourselves in some of the fascinating buoyancy-driven flow processes. For a geological introduction to some of the processes and context, textbooks such as McBirney (1985) include a comprehensive description of many of the geological observations and processes.

In magma reservoirs, which may lie 5–10 km below the surface, magma is exposed to the relatively cold surrounding crust, and this may lead to cooling and crystallization of the magma, as well as melting of the surrounding crustal rock, often called country rock. In addition, new magma may be supplied to the magma reservoir leading to pressurization of the system. Typically the new magma (we refer to this as basaltic or basic unevolved magma) will be hotter and of greater density than the evolved, and hence compositionally distinct, magma (we refer to this as silicic or evolved magma) already in the chamber. As a result of these different processes, magma reservoirs may develop two or more layers of molten rock, typically of different composition and temperature, leading to a density stratified system (Figure 9.1).

The controls on the fate of the magma in a magma reservoir are complex: If the system becomes sufficiently pressurized, magma may erupt at the surface, but if the magma cools and crystallizes rapidly, it may solidify in the crust. In many cases, the magma contains dissolved volatile species, especially water and $CO_2$, and the solubility of these gases increases with pressure. Owing to the decrease in pressure as the magma rises through the crust, it eventually reaches a point at which the gas comes out of solution and gas bubbles form. The gas bubbles lead to a rapid decrease in the bulk density of the mixture and also increase the compressibility of the magma (Bower and Woods 1996). The decrease in density tends to increase the pressure, while the increase in compressibility tends to increase the volume that is erupted for a given decrease in pressure; hence, these effects are both key in controlling eruptions. Also, the presence of the bubbles and the associated density changes can lead to magma mixing. A second process that is key for the evolution of the magma is the cooling and crystallization that occurs in the crust, since the magma may be several hundred degrees centigrade hotter than the surrounding rock. Many of the mineral phases that crystallize from the magma during this cooling tend to be relatively anhydrous compared to the magma, and so the crystallization leads to a progressive concentration of the dissolved gases in the remaining molten magma. This in itself can lead to the remaining melt becoming supersaturated in volatiles and the ensuing formation of gas bubbles (Tait et al. 1989).

As indicated in this short summary, there are many complex processes in operation in the subsurface volcanic systems. In this chapter, we focus on some of the processes associated with eruption triggering through recharge of magma. We also explore the ensuing cooling and crystallization of the magma including a discussion of the potential for mixing of the different magma layers during eruption. The discussion is based around a simplified picture of a magma reservoir as shown in Figure 9.1.

In the second part of the chapter, we review some of the processes responsible for the injection of ash high into the atmosphere or for the formation of dense ash flows

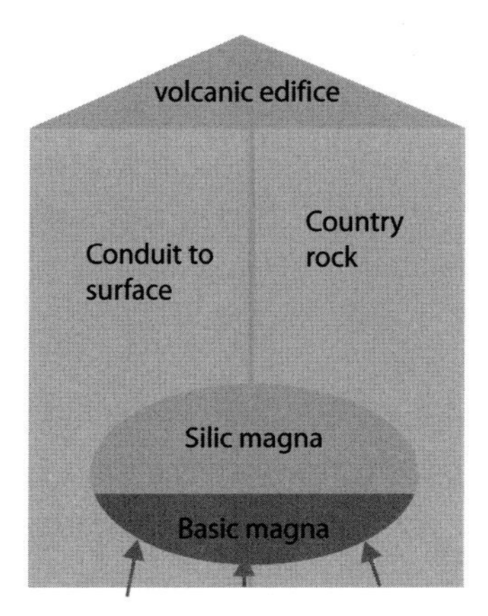


Figure 9.1. Cartoon of a magma reservoir-volcano system showing a two-layer magma chamber, with cold, less-dense, evolved magma (silicic magma) overlying hotter, denser unevolved magma (basic magma).

that travel tens of kilometers along the ground (Sparks et al. 1997). We also describe how other natural explosive processes may be modeled within a similar framework, including hydrothermal plumes and $CO_2$-driven lake eruptions, and we briefly discuss the complexities of submarine volcanic eruptions.

## 9.2 Magma Injection and Eruption Triggering

The injection of magma into a crustal magma reservoir can increase the pressure in the reservoir and eventually lead to fracture of the crust and formation of a pathway, known as a dike or conduit, to the surface along which magma may flow and erupt. The rate of pressurization of the magma reservoir as new magma is injected into the system depends on the compressibility of the magma and the surrounding crust which, up to the point of fracture, may be described as an elastic material (e.g., Touloukian et al. 1981). Since the magma typically contains both exsolved volatiles and liquid melt, we can describe the density of the magma in terms of the gas mass fraction $n$, the gas density $P/RT$, where $P$ is pressure, $R$ the gas constant for the gas, which we take as water vapour for simplicity, with $R = 460$ J/kg/$C$, and $T$ is the temperature, measured in Kelvin. Also, the density of the liquid magma is $\rho_m$ and this leads to the expression for the bulk density in terms of the mass fraction of the gas and the solid material present in the flow

$$\rho = \left[\frac{nRT}{P} + \frac{1-n}{\rho_m}\right]. \tag{9.2.1}$$

To calculate the exsolved gas mass fraction, we use Henry's law (e.g., Tait et al. 1989), which relates the mass of dissolved gas to the pressure. As the pressure increases, more of the gas is able to dissolve into the melt, and so the mass fraction of free (exsolved) gas decreases until it reaches a critical pressure when it falls to zero. If the variable $n$ is used to denote the gas mass fraction that has been exsolved and is in the gas state, and if the total mass fraction of gas in the mixture is denoted $n_o$, then the mass fraction of gas dissolved in the melt is given to leading order by the Henry's law relation

$$n = n_o - sp^{1/2}, \tag{9.2.2}$$

where $s$ is the solubility coefficient. With these relations, we can calculate the compressibility of the bulk magma, which is typically much greater than the surrounding rock if there is exsolved gas present. We can also calculate the increase in pressure of the magma chamber as a function of the addition of new magma. Indeed, if the volume of the magma chamber increases by an amount $dV$, from the original volume $V$, then the increase in pressure associated with the deformation of the surrounding crust is given in terms of $dV$, $V$, and the bulk modulus of the crust, $\beta$, according to the relation (cf. Tait et al. 1989)

$$dp = \beta \frac{dV}{V}, \tag{9.2.3}$$

assuming that the crust is an isotropic elastic material and that the increase in volume is small compared to the original volume, so we can linearize the relation between the change in pressure and the change in volume.

Figure 9.2 illustrates the change in density of magmas with total gas mass fraction 0.03 and 0.04, and also the increase in pressure as a function of the fractional increase in volume of a magma chamber for both volatile unsaturated and saturated magmas.

This model illustrates the control of the magma compressibility on the volume which may be injected into a chamber prior to triggering an eruption. On eruption, the pressure in the chamber falls, and so a similar result applies, in that with compressible volatile saturated magma, the volume of the chamber that erupts may be as much as 5–10% of the volume of the chamber (red line, Figure 9.2b), whereas with a volatile unsaturated magma, the volume that may erupt is only a fraction 0.1–1% of the volume of the chamber (blue line, Figure 9.2b). Further details of this control of eruption volume on magma compressibility is given in Bower and Woods (1997) and Woods and Huppert (2003).

A key element from this analysis is that if only a small fraction of the magma in a crustal chamber actually erupts before the pressure returns to the original crustal pressures, then a large volume of magma will remain in the crust. This volume is typically surrounded by cold country rock, and so there will be substantial heat transfer between the rock and the magma. Since the magma may be of temperature 800–1,200°C, while the country rock may have a substantially lower temperature of order 500–600°C (dependent on the depth), then some of the crust may melt in addition to cooling the magma, especially in the early stages of the process. We can model this cooling and melting in a simplified manner by using a parameterized model for the heat transfer associated with high Rayleigh number convection (Turner 1979). In essence, with high Rayleigh number convection, the heat flux is controlled by the release of plumes from the heated or cooled boundaries of the system, and these boundary layers are much shallower than the depth of the whole layer. In this limit, the convective heat flux, $F$, is independent of the depth of the layer, and so takes the form

$$F = 0.1 Ra^{1/3} \frac{\rho C_p k \Delta T}{H} = \gamma \Delta T^{4/3}, \tag{9.2.4}$$

with

$$Ra = \frac{g \Delta\rho H^3}{\kappa \mu}. \tag{9.2.5}$$

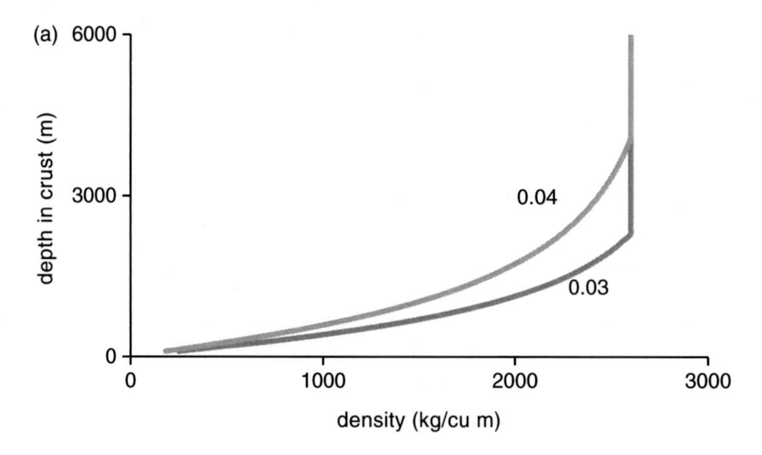


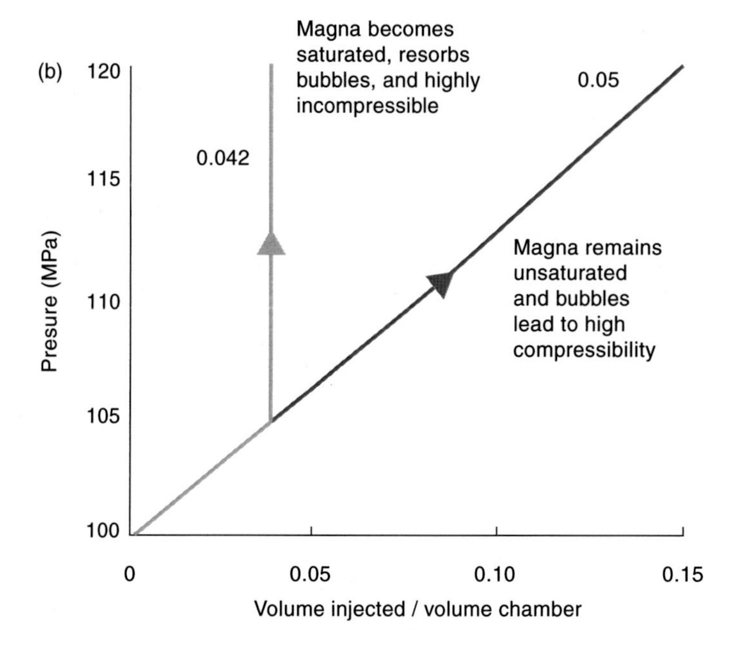


Figure 9.2. (a) Density of magma as a function of depth in the crust, and hence pressure, for magma with gas mass fraction 0.03 and 0.04, and (b) illustration of the increase in pressure of the magma reservoir as a function of the fractional increase in volume associated with injection of new magma into the magma reservoir, with the blue line showing the increase in pressure in a magma, which has no exsolved gas (total gas mass fraction 0.042) for pressures in excess of 105 MPa, where the blue line becomes nearly vertical, and the red line showing the increase pressure in a magma which still contains exsolved gas and hence is much more compressible (total gas mass fraction 0.05). For a color version of this figure please see the color plate section.

Here $g$ is the acceleration of gravity, $H$ is the depth of the convecting layer, $\kappa$ is the thermal diffusivity, $\mu$ is the viscosity, and the constant $\gamma$, defined by (9.2.4) and (9.2.5) along with a relation $\Delta\rho = \beta\Delta T$, where $\beta$ is an effective thermal expansion coefficient, is introduced for convenience. Note that this model for the heat flux is based on dimensional arguments and the assumption that the heat flux is independent of the depth of the layer. There have been a series of experiments, including the work of Niemala et al. (2000), which suggests that at very high Rayleigh numbers, the exponent in the relation between the heat flux and the Rayleigh number is smaller than 1/3 (Equation 9.2.4), and has exponent of about 0.31. However, for the purposes of the present work we use the classical model (9.2.4).

In this model, the density contrast across the heated or cooled boundary $\Delta\rho$ is related to the temperature contrast across the boundary $\Delta T$. However, the origin of the density contrast may arise from the difference in the concentration of crystals, which are typically relatively dense, or bubbles, which may be buoyant, in the cold (hot) fluid in the boundary layer compared to the hot (cold) fluid in the interior of the melt. As a result, there is no simple formula for $\beta$. Further details of models that quantify the buoyancy as a function of the crystal content or the bubble content are given in Cardoso and Woods (1999).

Using this model for the heat flux, we can write a simplified one-dimensional model for the melting of the roof and the cooling of the hot magma, accounting for the partial crystallization of the hot magma, and the fact that the roof material is less dense and so forms a new layer above the original magma. Furthermore, we note that the roof material may detach and enter the upper layer of melt as a partially crystalline material with sufficient melt so as to be mobile, and so we need to account for the melting of the crystals in this upper layer as it is heated by the hot underlying layer of magma. The model takes the form of three heat transfer equations, which are shown here for interest. Namely equations for (i) the lower layer, whose temperature evolves through heat transfer to the upper layer and through crystallization in this layer (Equation 9.2.6); (ii) the roof melt layer, which is heated from below, and also cooled through heat transfer to the overlying rock, as well as evolving through crystallization (Equation 9.2.7); and (iii) the melting of the roof which results from the heat transfer from the upper layer of melt (Equation 9.2.8).

$$\frac{d}{dt}(\rho C_l T_l h_l) = -\gamma_l (T_l - T_i)^{4/3} + \rho L_l \frac{d}{dt}(x_l h_l), \tag{9.2.6}$$

$$\frac{d}{dt}(\rho C_n T_u h_u) = \gamma_u (T_i - T_u)^{4/3} - \gamma_u (T_u - T_b)^{4/3} + \rho L_u \frac{d h_u x_u}{dt}, \tag{9.2.7}$$

$$\frac{d}{dt}(\rho L_u h_u) = \gamma_u (T_u - T_b)^{4/3}, \tag{9.2.8}$$

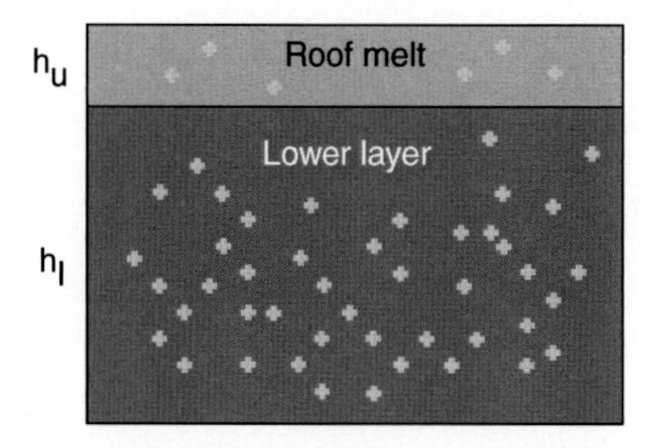


Figure 9.3. Schematic of the two layer chamber system, consisting of an upper layer of roof melt and a lower layer of injected hot magma. The hot magma melts the roof to create a buoyant overlying layer of melt; this layer may be partially crystalline, and some of the crystals may settle into the lower layer. In addition, the lower layer cools and hence crystallizes. The crystals ultimately settle to the base of the system forming a cumulate layer.

where $x$ denotes the crystal content and the subscripts $l$ and $u$ denote the properties of the lower and upper layers of melt. The layer depths are denoted by $h$, the temperatures are denoted by $T$, and the interface between the upper and lower layer of melt has temperature $T_i$, while the melting temperature of the roof is $T_b$ and the latent heat of melting is $L$. The model follows from a classic model of Huppert and Sparks (1988), except for the inclusion of the crystals in the upper melt layer, associated with the slumping of partially molten roof material into the upper layer. Later, we will see that this provides a process for some of the upper layer material to mix into the lower layer through crystal settling. As a result, this may generate mixed magma of different composition and so a hybrid cumulate layer: The crystal settling may occur if the crystals are denser than the melt, as is often the case (cf. Cowan and Woods 2011a, sub-judice). See Figure 9.3.

Owing to the fact that each of the layers is turbulently convecting, there will be some crystal settling, both onto the base of the chamber from the lower layer and from the upper layer into the lower layer.

The crystal settling can be described by the classic sedimentation law of Hazen (1904) for a turbulent suspension of depth $h$

$$\frac{dx}{dt} = -\frac{vx}{h} + \text{Source}, \tag{9.2.9}$$

where $x$ is the crystal concentration, $h$ is the depth of the magma, and $v$ is the settling speed of the crystals, with the term "Source" denoting the rate of production of crystals per unit volume of magma. If the crystal pile (i.e., cumulate layer) at the base of the chamber has a solid fraction $y$, and depth $h_c$, which is related to the packing efficiency of the crystals, then the depth of the lower layer of convecting melt decreases with

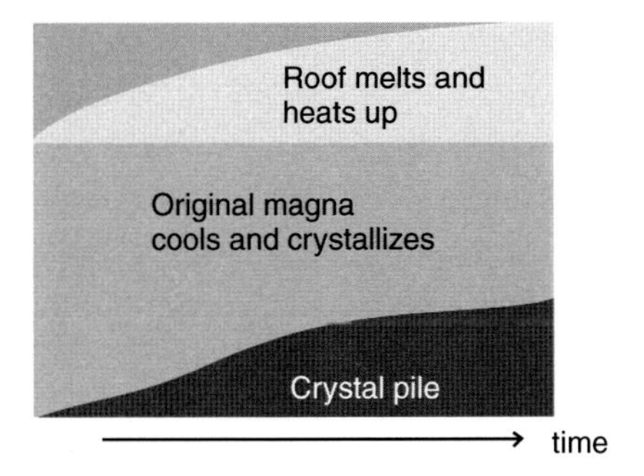


Figure 9.4. Illustration of the evolution of the chamber as a function of time. Initially, the original magma cools and crystallizes as it heats up the roof and a buoyant roof melt layer develops. A cumulate layer then builds up from the floor of the chamber.

settling and accumulation of the cumulate layer according to

$$\frac{dh_l}{dt} = -\frac{h_l}{y}\frac{dx}{dt}. \tag{9.2.10}$$

Solution of this system of equations requires a model for the crystal content as a function of the temperature of the melt, and simple parameterizations of this are given by Huppert and Sparks (1988). The predictions of this model, including the crystal settling, may be expressed in terms of the time evolution of the chamber as shown in Figure 9.4 (see Cowan and Woods 2011a). Note that in the model presented previously, we have assumed that the convection is of high Rayleigh number and turbulent from the onset of the process; in practice as the roof melt layer initially forms, there is a relatively short period during which the roof melt heat transfer is controlled by conduction prior to the onset of turbulent convection (further details of this may be found in Huppert and Sparks (1988)).

Provided that the magma in the chamber remains unsaturated with respect to the gases dissolved in the magma, the system remains stably stratified, since the roof melt is typically composed of a magma with a less-dense composition than the new input of lower-layer magma, which is expected to be hot and dense. As the lower layer crystallizes, the bulk density of the mixture initially increases, since the crystals are denser than the melt. However, as the crystals begin to settle, the lower-layer density may decrease. Meanwhile, the upper layer also becomes progressively hotter, and the crystals in the upper layer either settle out or melt. As a result, this layer typically remains less dense than the lower layer. Later in the chapter, we identify that if the lower layer becomes saturated in dissolved gases, and exsolves bubbles of vapor, then the situation may be different, and overturn and mixing may occur, leading to hybrid magma.

It is of interest to consider the evolution of the crystal pile after it has formed. As the pile forms, it traps melt between the crystals and so gradually builds up a porous bed. Since the magma is continually cooling and crystallizing relatively dense crystals, the remaining melt becomes less dense with time, since the composition tends to dominate temperature in determining the density. As a result, the melt trapped in the crystal pile has an evolving composition and decreasing density with height. Since the melt is stably stratified, one therefore expects the crystal pile to be compositionally zoned since there will be little convective exchange between the overlying body of melt and the interstitial melt in the crystal pile as it solidifies. This is illustrated schematically in Figure 9.5 in which we show the trajectory of the melt along the so-called liquidus, on which the liquid and solid are in equilibrium. As the melt evolves from condition A to B to C, the composition and temperature fall, while the melt crosses successive lines of constant and decreasing density. In turn, this melt is trapped in the pore spaces between the crystals, which constitute the lower crystal pile (cumulate layer), and since the initial trapped melt, A, is of higher composition than B or C, it is of higher density and therefore stably stratified.

At longer times, as the base of the crystal pile begins to cool and the interstitial melt itself solidifies, then some slow mixing may arise in the crystal pile. This occurs as melt at the lowest (point A) evolves in composition through crystallization of

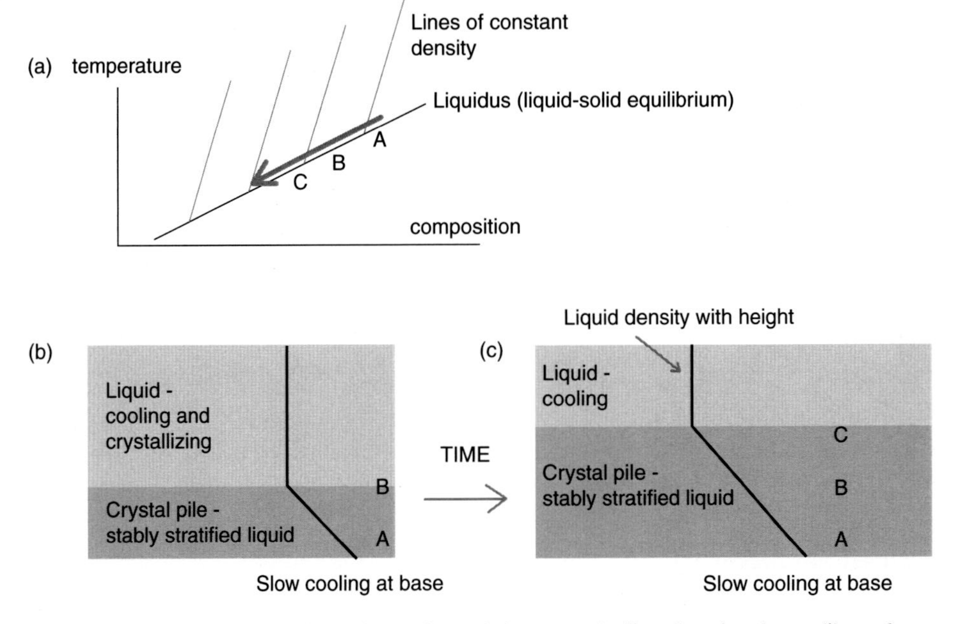


Figure 9.5. Schematic of the formation of the crystal pile, showing how, (i) as the melt evolves along the liquidus on the temperature-composition diagram, from A to B to C, through cooling and crystallization, (ii, iii) the interstitial melt trapped in the crystal pile, becomes stably stratified, with melt of composition A underyling melt B which in turn underlies melt C.

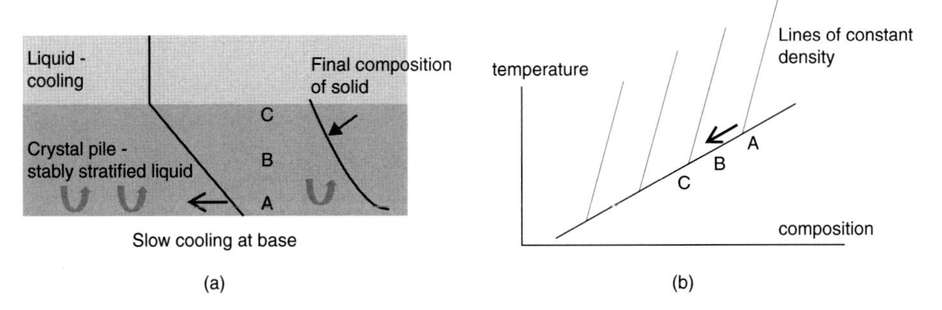


Figure 9.6. Final solidification and onset of weak mixing in the crystal pile as the interstitial melt solidifies. (a) Illustration of the time evolution of the melt density at point A and of the final compositional profile and (b) illustration of the evolution of the melt along the liquidus, as this melt cools and crystallizes.

relatively dense minerals, and the remaining melt at point A becomes less dense than the melt immediately above (point B, Figure 9.6). However, the original trend of the compositional zonation of the interstitial melt remains in the final solidified system since the dense crystals are required to precipitate at the base of the melt to drive convective overturning.

## 9.3 Second Boiling and Eruption Triggers

The preceding discussion has focused on the evolution of a layer of hot molten magma in a reservoir below the surface, exploring the impact of thermal convection on the cooling and crystallization of the magma and also the potential to melt back some of the original crust around the hot molten magma. In the present section, we explore how these processes might lead to the eruption of the magma from the chamber as a result of any pressure buildup associated with the cooling and crystallization.

Indeed, as the melt in the chamber continues to cool, at some stage we expect the melt to become saturated in the dissolved gases (volatiles), and this will lead to formation of gas bubbles and pressurization of the system, possibly leading to eruption. This process is known as second boiling and may be responsible for a continuing sequence of eruptions in a number of volcanoes, following the original explosive eruption associated with the injection of magma into the crust. The potential for a given magma to erupt multiple times arises because, as we illustrated earlier, in a single eruption, typically only a few percent of the total chamber volume erupts.

We first consider a single-layer system and explore the pressure buildup in a closed chamber of volume $V$, temperature $T$, crystal fraction $x$, and gas content $n$. As the magma cools, we can describe the system with a series of simplified model equations that describe the relation between the crystal content and the magma temperature (Equation 9.3.1; Huppert and Sparks, 1988); Henry's law for the exsolution of the gas phase as a function of the magma pressure (Equation 9.3.2: cf. Equation 9.2.2); the equation for the elastic deformation of the magma reservoir (Equation 9.3.3); and the equations for mass conservation (Equation 9.3.4) and the bulk density of the mixture (Equation 9.3.5).

Crystal formation

$$x_b = \frac{(T_o - T)}{200} \tag{9.3.1}$$

Volatile release

$$n = n_o - sp^{1/2}(1 - x) \tag{9.3.2}$$

Rock deformation

$$dp = \beta \frac{dV}{V} \tag{9.3.3}$$

Mass conservation

$$M = \rho V \tag{9.3.4}$$

Density

$$\rho = \left[ \frac{nRT}{p} + \frac{1-n}{\rho_1} \right] \tag{9.3.5}$$

In Equation (9.3.1), $T_o$ represents the initial temperature at which crystallization occurs, and the crystal content increases as the temperature $T$ decreases to smaller values. With this system of equations, we can express the increase in pressure of the magma in terms of the decrease in temperature, leading to predictions of the rate of pressurisation with cooling. Indeed, in Figure 9.7 we illustrate how the pressure in two model magmas varies with crystal content, which is a proxy for the cooling. The magma with a volatile mass fraction of 0.02 remains unsaturated until the crystal content has increased to a mass fraction of about 0.4; before this the pressure decreases owing to the formation of dense crystals and the associated contraction of the magma. In contrast, for the magma which has a volatile mass fraction of 0.04 , the magma is saturated prior to the commencement of crystallization, and as crystallization procedes more bubbles form, building up the pressure to critical conditions for eruption. Further details of this process are given by Tait et al. (1989) and Phillips and Woods (2002).

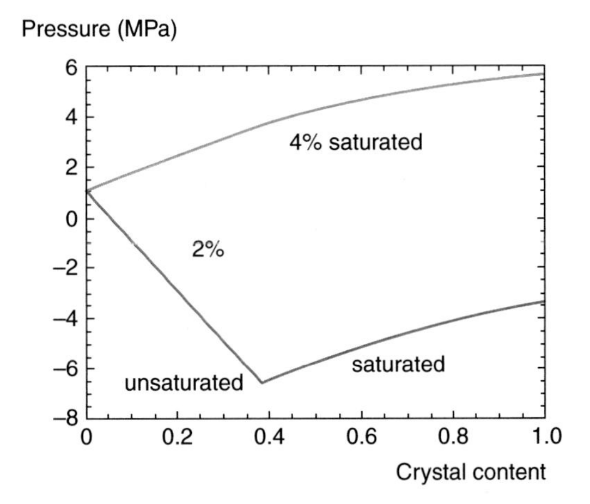


Figure 9.7. Pressure change (Pa) as a function of crystal mass fraction, in a cooling and crystallizing magma. The two illustrative magmas have volatile mass fractions 0.02 and 0.04, and therefore show different pressure responses. The magma with 0.02 mass fraction of gas remains undersaturated until there has been sufficient crystallization, while the more volatile rich magma, with 0.04 mass fraction of gas, is saturated from the onset of crystallization, and hence releases volatiles throughout the evolution of the cooling (after Phillips and Woods 2002).

## 9.4 Magma Mixing

The production of bubbles as magma cools and crystallizes also lowers the density of the magma and hence has the potential to drive some convective mixing or overturn in the magma reservoir if there are two layers of magma present. This situation can arise following recharge of an evolved and relatively cool chamber with new hot but relatively dense magma, or as a result of roof melting as described in Section 9.2.

### *9.4.1 Mixing Prior to Eruption*

If the lower layer becomes saturated in volatiles as it cools, the density will then begin to decrease as gases are released from the melt. Two situations can be envisaged. First, bubbles may rise through the lower layer and accumulate at the interface with the more viscous upper layer of evolved melt; the accumulation may occur owing to the reduction in the rise speed in the upper layer, so that a zone of high concentration develops at the interface. This bubbly layer is less dense than the main body of fluid in the upper layer and so may develop convective plumes rising from the interface. The speed of the convective plumes is given in simple scaling terms by the relation

$$u \sim \frac{2g\phi(\rho_m - \rho_b)h^2}{9\mu}, \tag{9.4.1}$$

where $\phi$ denotes the bubble fraction in the layer accumulating at the interface and $h$ is the thickness of this layer with $\mu$ the viscosity of the upper layer fluid, which controls the rise speed of the plume, and the density difference is between the density of the melt and the bubbles; the factor 2/9 arises from Stokes law, but there may also be a scaling parameter associated with the relation between the layer depth $h$ and the size of the plume/thermals. The layer will tend to grow until the speed of the convective plumes matches the rate of accumulation of the layer so that a quasi-steady balance is established. The layer builds up at a rate

$$\phi \frac{dh}{dt} \sim u_b c \tag{9.4.2}$$

where $u_b$ is the rise speed of the bubbles in the lower layer, $H$ is the depth of the lower layer, and $c$ is the concentration of bubbles in the lower layer.

Matching the two speeds leads to the scaling for the plume size

$$h \sim \left[ \frac{9 \mu u_b c}{2g\left(\rho_m - \rho_g\right)\phi^2 )} \right]^{1/2} . \tag{9.4.3}$$

Inserting typical values for a two layer magmatic system, we can predict the size of incipient plumes that rise into the upper layer, although we note the specific predictions depend on the density of the bubbles (Figure 9.8). However, the prediction of size of the plumes of the lower-layer magma and bubbles that rise into the other layer are consistent with field evidence, which often shows centimeter-size parcels of the lower-layer magma in the upper-layer magma (Blundy and Sparks 1992; Conrad and Kay 1984).

The second mode of mixing arises when the lower layer is of sufficiently high viscosity that the bubbles accumulate in this layer, and there is a large-scale overturn of the magma chamber as a result of the lower layer becoming less dense en masse than the upper layer. The Hazen (1904) law for sedimentation also applies to bubble ascent, and so the bubble concentration in the lower layer, $c$, of depth $H$ is expected to evolve according to the relation

$$\frac{dc}{dt} = Q_b - \frac{u_b c}{H}, \tag{9.4.4}$$

where $Q_b$ is the bubble production rate, given in terms of the cooling and crystallization (Section 3) and the bubble rise speed is $u_b$.

With typical rise speeds and cooling rates (Phillips and Woods 2002), we find that the bubble concentration increases to a steady balance which depends on the rise speed and the production rate (Figure 9.9). If the bulk density of the lower layer, including these bubbles, falls below that of the upper layer, then the magma layers will overturn. We therefore expect that with high cooling rates or with more viscous lower layers of magma, the bubble concentration will reach sufficient values for overturn.

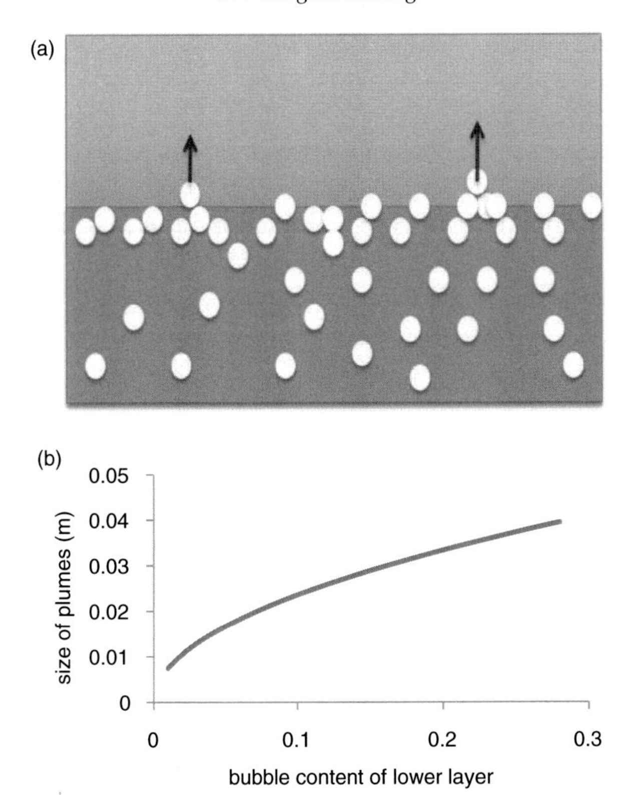


Figure 9.8. (a) Illustration of plumes rising from a bubbly layer at the interface between two layers of magma. (b) A calculation of the typical size of plumes rising from the lower to the upper layer associated with bubble ascent and accumulation at the magma-magma interface, as a function of the bubble content of the lower layer of magma.

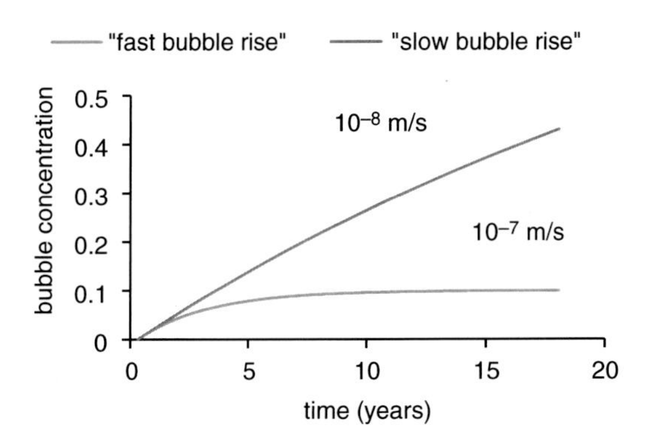


Figure 9.9. Illustration of typical bubble accumulation rates in the lower layer of a magma chamber as a function of the bubble rise speed. The larger bubble rise speed leads to escape of most bubbles from the layer, whereas the smaller rise speed leads to a significant increase in the concentration and associated reduction in density.

In contrast with less-viscous magma, which leads to higher bubble rise speeds, or with slower cooling and crystallization, which leads to slower bubble production, the magmas will not overturn, and instead plumes of bubbly lower-layer magma will migrate into the upper layer. These two modes of mixing have been demonstrated by Woods and Phillips (2002) using a two-layer system of polymer solutions, with a saline and hence dense lower layer and a fresh less-dense upper layer. Bubbles were produced in the lower layer through electrolysis and accumulated in the lower layer with some bubbles reaching the interface with the upper layer. With a low-viscosity lower layer, overturn did not occur, but plumes of bubbles rose through the interface carrying small parcels of the lower layer fluid into the upper layer (Figure 9.10a: orange fluid) whereas with a high-viscosity lower layer, large-scale overturn occurred (Figure 9.10b: blue fluid).

The experimental observations have some features in common with the image of an intermingled sample of two magmas erupted from Tolbachek volcano (Figure 9.10c).

In a magma reservoir, there is a competition between the rate at which the density of the lower layer evolves through bubble production and the rate at which the pressure of the reservoir evolves. As the lower layer cools and crystallizes, it will heat up the upper and colder layer of more evolved magma. As we have seen, the rate of pressurization then depends on the volatile content of the two layers of magma, and with high volatile contents, the pressure will evolve slowly, whereas with smaller volatile contents, the pressure will evolve more rapidly. We therefore expect that high volatile content magmas in the shallow crust are more likely to overturn and mix prior to eruption than magmas in deeper chambers and with lower volatile contents (Phillips and Woods 2002).

### *9.4.2 Mixing During Eruption*

As well as the buoyancy-driven mixing processes that can occur prior to eruption, it is also possible for magma mixing to occur during an eruption, as a result of the decompression of the magma reservoir (Woods and Cowan 2009). Indeed, as an eruption procedes, the reservoir pressure may fall by several megapascals and this allows the bubbles which are already exsolved in the magma to expand, and perhaps some additional bubbles to exsolve from the melt.

We consider the case in which the lower layer of magma is more volatile rich than the upper layer of magma, but, at the original chamber pressure, it is also a little denser owing, for example, to compositional differences. Following the onset of eruption, the chamber then decompresses, and so the lower layer of magma will expand more rapidly than the upper layer. As a result, its density will drop below that of the overlying magma, and the system becomes gravitationally unstable (Figure 9.11). In order for the mixing to occur during the eruption, the time required for the overturn should be smaller than the duration of the eruption. To establish that this may occur we now estimate this time for overturn.

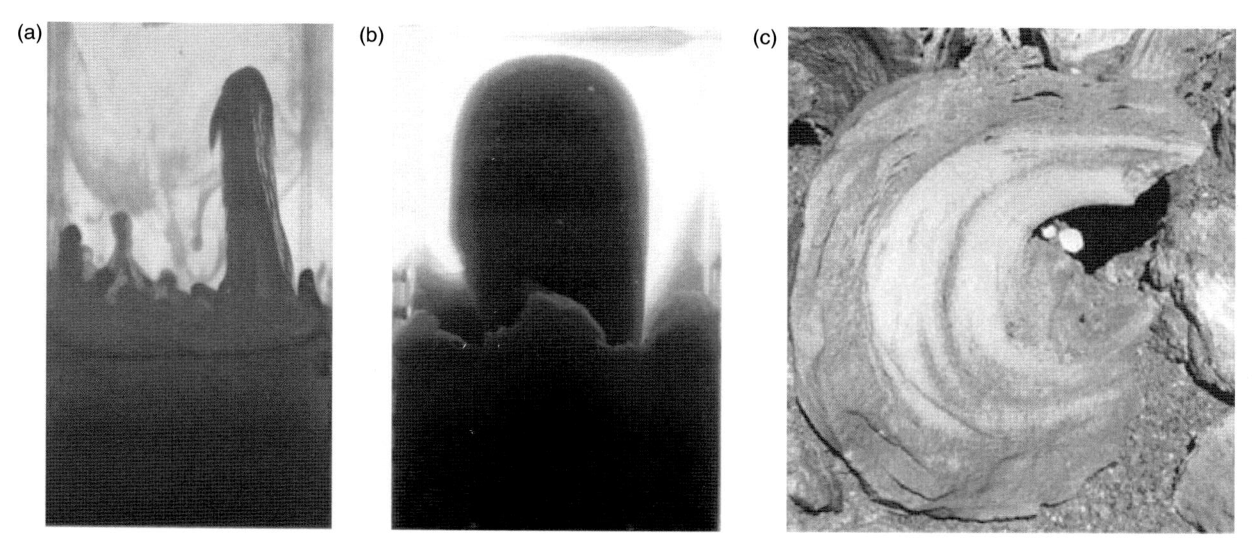


Figure 9.10. (a) The red experiment shows plumes migrating through the interface as a result of bubbles accumulating at the interface. (b) The blue experiment shows large-scale overturn as a result of bubbles in the lower layer leading to a reversal in the sign of the relative density of the two layers (adopted from Woods and Phillips 2002). (c) Illustration of a pumice erupted from Tolbachek in Kamchatka, showing both brown and white pumice intermingled in the clast. For a color version of this figure please see the color plate section.

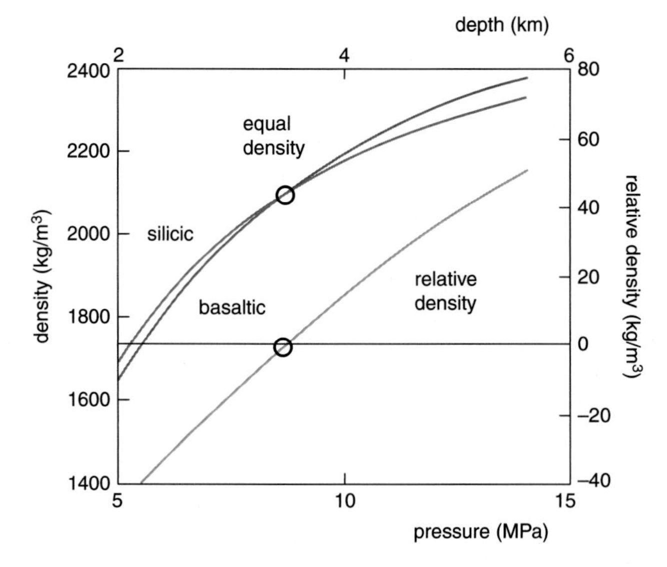


Figure 9.11. Illustration of the variation of the density of magma as a result of the decompression of a magma chamber during eruption, which can lead to overturn of the chamber. The blue and red lines show the bulk density of each of the layers (left-hand side vertical axis), whereas the green line shows the difference between the two densities (right-hand side vertical axis). For a color version of this figure please see the color plate section.

First, we note that the speed of the convective plumes which rise from the lower layer, driven by the density contrast, scales as

$$u \sim \frac{2\Delta\rho g H^2}{9\mu}. \tag{9.4.5}$$

In a typical magma chamber system, this may have values of order 0.01–100.0 m/s, suggesting that overturn and mixing occurs within 10s of minutes (Woods and Cowan 2009). In many large explosive eruptions, the eruption rate is of order $10^8$–$10^9$ kg/s, and the eruption may discharge $10^{12}$–$10^{15}$ kg so that the eruption duration is of order $10^3$–$10^6$s. We deduce that in many cases, there would be sufficient time for the overturn to occur.

Recognition that such overturn can occur may be relevant for interpreting a number of eruptions in which two different magmas erupt in sequence, such as the AD 79 eruption of Mt. Vesuvius, since as well as requiring a change in the magma during the eruption, we note from the earlier discussion that a typical eruption only leads to discharge of a few percent of the total volume of the chamber (Woods and Cowan 2009). The process of overturn through decompression has been demonstrated experimentally using a two-layer system of aqueous solutions, each laden with polymer (cf. Figure 9.10), but in this case the upper layer is decompressed prior to the experiment to remove any dissolved gas. The system is then placed in a vaccuum system,

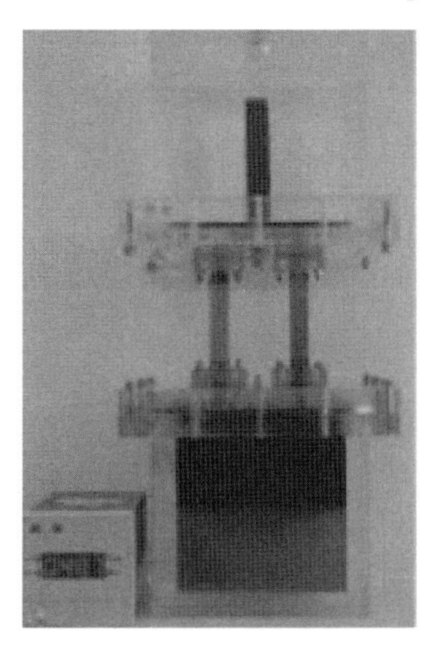

Figure 9.12. Illustration of the overturn of a magma reservoir during eruption, which leads to a change in the erupting material. In this experiment, the conduit on the right is erupting the lower layer initially (red), but following the overturn in the reservoir, the blue volatile poor layer sinks to the base of the chamber and erupts from this vent. Conversely, the conduit on the left issues the shallow fluid from the system, and therefore undergoes a transition from erupting blue to red fluid owing to the overturn. For a color version of this figure please see the color plate section.

and the pressure decreased until the lower layer begins to exsolve volatiles; the density of the lower layer then decreases, and eventually large-scale overturn occurs. A vent originally issuing the upper-layer magma would then begin to erupt the lower-layer magma, or vice versa, if a side vent originally erupts the lower layer (Figure 9.12; from Cowan and Woods 2011b, in review).

## 9.5 Eruption Dynamics

Once a magma chamber has reached sufficient pressure to open a pathway to the surface, the magma begins to rise to the surface initiating an eruption. In highly explosive eruptions, as the magma rises it decompresses and the bubbles expand. Also, further bubbles may be released from the melt owing to the decreasing solubility with pressure. Since the magma is very viscous, on the time scale of the magma ascent from the chamber to the surface, the relative motion of the bubbles and magma is very small. We can therefore assume that they remain locked with the liquid melt, and that the mixture becomes progressively more foamy as it rises to the surface (Sparks 1978). Eventually, as the liquid films around the bubbles become very thin, the viscous stress required to stretch the bubbles exceeds the fracture strength of the magma. At this point, the liquid films fracture so that eventually the gas becomes the continuous phase, suspending fragments of the liquid. The mixture then continues to the surface, expanding and decompressing, until venting into the atmosphere. Since the flow accelerates to very high speed as it decompresses, the flow becomes choked at the vent, reaching the speed of sound of the mixture of about 120–150 m/s, with a pressure in the range of 0.1–5.0 MPa (Woods, 1995). The flow rates at volcanic vents are in the range of $10^5$–$10^6$ $m^3/s$ in large explosive eruptions with the conduits having dimensions in excess of 20–30 m in order to carry the magma to the surface. There are many studies of the eruption process along conduits, but many research problems remain related to the interaction of the conduit walls and the ascending magma (cf. Sparks et al. 1997; Barmin et al. 2002).

### *9.5.1 Eruption Columns*

We now focus on the dynamical processes above the surface, and in particular on the dynamics of eruption columns and ash flows. Reviews (Woods 1995; Woods 2010) have summarized much of the understanding of eruption column dynamics, and so we do not repeat that work here, but expand on some of the processes not considered in such detail in that review.

In order to gain insight into the behaviour of eruption columns, and also the transition in behavior from eruption columns to ash flows, a model for the eruption column is needed. Here, we review the original model of Woods (1988), which describes the key features of the flow. The model is based on an integral description of the plume dynamics, in which the evolution of the quasi-steady, horizontally averaged properties of the plume are followed with height in the atmosphere. This includes the mass flux, $Q$, momentum flux, $M$, and the plume radius, $b$, temperature, $\theta$, gas mass fraction, $n$, and upward speed $u$. The model accounts for the turbulent entrainment of air as the material rises by assuming there is a horizontal inflow to the plume from the surrounding atmosphere, which occurs at a rate $\varepsilon \sim 0.1$ of the mean upward velocity at that height. This is known as the entrainment law of Morton et al. (1956), and has been used extensively to model the dynamics of turbulent buoyant plumes in a variety of environments.

The eruption column model also accounts for momentum conservation, the steady flow energy or enthalpy equation, and an equation for the bulk density, $\rho$, and the conservation of gas through entrainment of air. In addition, the model requires a description of the atmospheric stratification (e.g., Gill 1981). In the model equations that follow, subscript $a$ denotes ambient properties.

Mass conservation

$$\frac{dQ}{dz} = 2\pi \varepsilon u \rho_a b \tag{9.5.1}$$

Momentum conservation

$$\frac{dM}{dz} = g(\rho_a - \rho)b^2 \tag{9.5.2}$$

Enthalpy conservation

$$\frac{d[Q(u^2/2 + gz + C_p\theta)]}{dz} = 2\pi\varepsilon ub(gz + C_p\theta_a) \tag{9.5.3}$$

Density

$$\rho = \left[\frac{nRT}{p} + \frac{1-n}{\rho_m}\right] \tag{9.5.4}$$

Gas conservation

$$\frac{dnQ}{dz} = 2\pi\varepsilon u\rho_a b \tag{9.5.5}$$

Mass flux; momentum flux

$$Q = \rho ub^2;\ M = \rho u^2 b^2 \qquad (9.5.6;\ 9.5.7)$$

where $\varepsilon$ is the entrainment coefficient (Morton et al. 1956), $C$ is the specific heat, and $g$ is the acceleration of gravity. These equations are solved for given source conditions, in a model atmosphere, and the predictions of the model indicate the height of rise of the eruption column.

As the ash and gas mixture rises from the volcanic vent, there is an initial decompression phase over which the flow decelerates to atmospheric pressure and the momentum then carries the flow upward where it mixes with and entrains air. Since the ash has a temperature in the range of 800–1,200°C, then depending on the precise magma which is erupting, it is able to heat up the air. As a result, the density of the mixture progressively decreases from initial values, which may be as high as ten times the air density. During this initial phase of ascent, the relatively dense flow decelerates rapidly. There is a competition between the reduction in the density through entrainment and mixing with ambient air and the reduction in the velocity of the flow. If the density falls below the ambient density prior to the velocity falling to zero, then the flow continues upward to form a buoyant convecting plume. However if the momentum falls to zero while the mixture is still dense, then the flow collapses back to the ground to form a dense ash flow.

The model may be solved to give predictions for the height of the eruption column with flow rate, and also to give predictions of the critical eruption velocity required for the flow to become buoyant and indeed form a buoyant plume.

Figure 9.13 illustrates the critical velocity for a buoyant convective plume to develop. As the mass flux increases, the radius of the flow increases; since the entrainment rate is proportional to the perimeter and the flux is proportional to the cross-sectional area, larger flow rates require greater source velocities. The figure also illustrates the weak dependence on the initial gas content of the erupting magma.

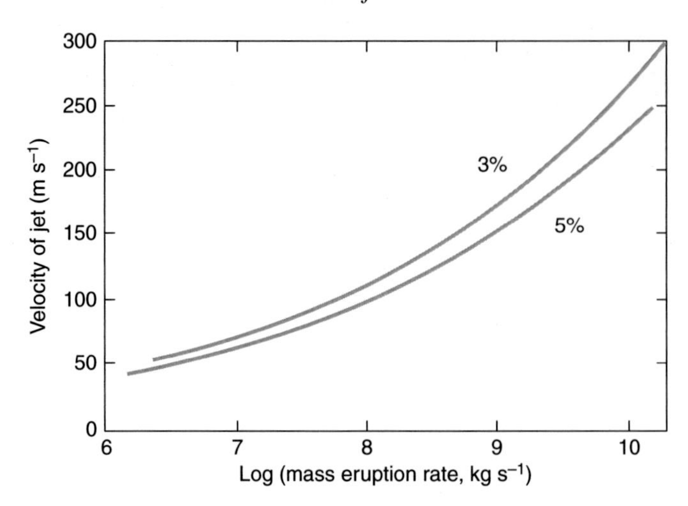


Figure 9.13. Critical velocity to form a buoyant eruption column as a function of the mass flux. The two curves correspond to magma with volatile mass fractions of 0.03 and 0.05.

With more gas, the initial density of the mixture is lower, and hence less entrainment is required to generate a buoyant column. This leads to a small reduction in the critical velocity for an eruption to develop.

Once the material has risen to the neutral buoyancy height above the plume, the ash–air mixture forms a horizontally spreading intrusion. In many cases, the ambient wind field causes the flow to spread downwind from which the ash particles sediment (Figure 9.14), although there have been some examples of radially symmetric spreading in relatively wind-free environments. If we model the flow as being carried downwind a distance $x$ with speed $u$, and spreading laterally under gravity, with width $w(x)$, and thickness $h(x)$, we can track the evolution of the flow and of the ash particles.

If we neglect entrainment or mixing and ambient air into the intrusion, then the volume flux remains constant leading to the relation

$$Q_v = whu. \tag{9.5.8}$$

If we assume the flow spreads under gravity in the cross-flow direction as it moves downwind with the ambient wind speed, then the spreading occurs at a rate that depends on the ambient stratification $N$ according to (Sparks et al. 1997)

$$u\frac{dw}{dx} = \lambda N h \tag{9.5.9}$$

where $\lambda = O(1)$ is an empirical constant. The ash particles fall out of this intrusion at a rate given by Hazen's law, in terms of the fall speed $v_s$, the current depth $h$, the

(a)

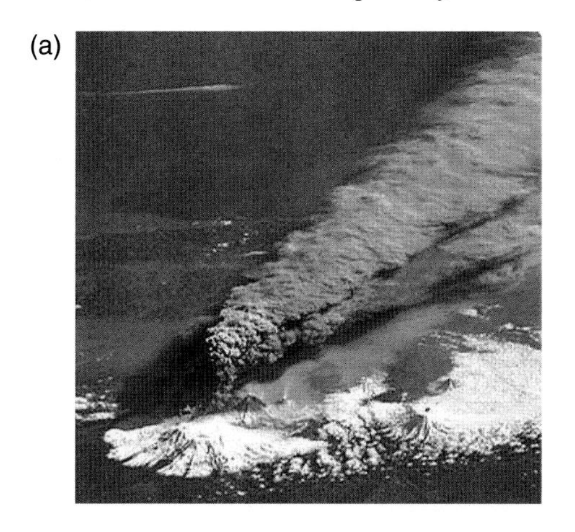

(b)

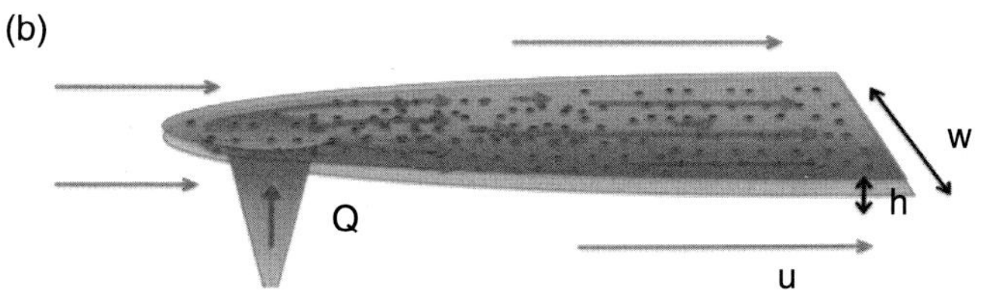


Figure 9.14. (a) Photograph from Mt. Cook, Alaska, and (b) a schematic of the dispersal of ash downwind of the eruption column. For a color version of this figure please see the color plate section.

particle concentration $c$, and the speed of the flow $u$, according to

$$u\frac{dc}{dx} = -\frac{v_s c}{h}. \tag{9.5.10}$$

The system has solutions for the width, thickness, and ash concentration given in terms of the downwind distance as follows (Sparks et al. 1997):

Width

$$w = \left[\frac{2\lambda N Q x}{u^2}\right]^{1/2} \tag{9.5.11}$$

Thickness

$$h = \left[\frac{Q}{2\lambda N x}\right]^{1/2} \tag{9.5.12}$$

Concentration

$$c(x) = c(0)\exp\left[-\frac{2v_s}{3u}\left(\frac{2\lambda N}{Q}\right)^{1/2} x^{3/2}\right], \tag{9.5.13}$$

where we assume there is a dominant ash particle size, although the model can be readily extended to account for the differential settling when there are a range of

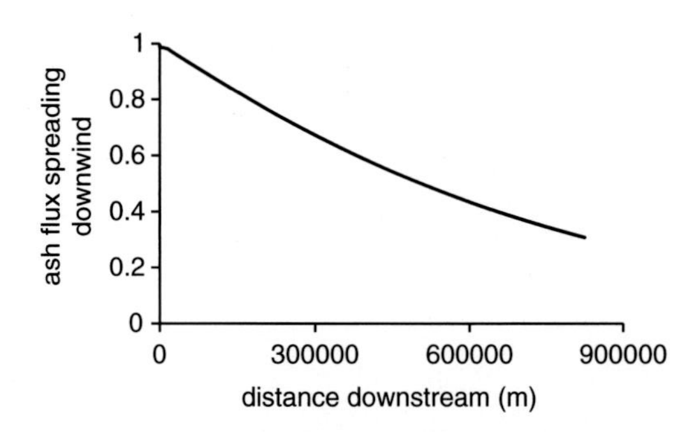


Figure 9.15. Model calculations of ash fallout from the laterally spreading neutral cloud in a typical eruption. The ash cloud propagates 900 km downwind in this calculation prior to the ash concentration falling below 0.2 of the initial value. The fall speed here is taken to be 0.001 m/s.

Figure 9.16. Illustration of a small developing ash flow at the Soufriere Hills Volcano, Montserrat. For a color version of this figure please see the color plate section.

particle sizes. The flux deposited by this laterally spreading cloud is given by $Qdc/dx$ (Figure 9.15). Once the material has sedimented from the intruding cloud, it then falls through the atmosphere, being dispersed by the ambient turbulence, as it is carried downwind, until eventually being deposited on the ground.

### 9.5.2 Ash Flows

In the case that the column collapses and an ash flow develops instead of a buoyant plume, the dense hot mixture runs along the ground from the volcano, sedimenting material and entraining air as it advances (Figure 9.16). This leads to a gradual reduction in the density of the flow, and eventually the flow becomes less dense than the ambient air and lifts off the ground. The dynamics of the ash flows are complex owing to the range of grain sizes involved in the flow, and the possible segregation of the fine ash into a dilute cloud and the coarser-grained material into a ground hugging avalanche-type flow. However, it is possible to develop a simplified picture of some of the dynamics in order to provide bounds on the dynamics, by assuming the flow is highly turbulent and well mixed, as may be the case in the most explosive systems.

As for an eruption column, the flow may then be described in terms of equations for mass, momentum, and energy conservation, with the entrainment into the flow now being dependent on the Richardson number of the flow, which is a measure of the potential energy required to mix air into the flow, and the kinetic energy available to do this. In the limit of low *Ri*, there is substantial entrainment, whereas with high *Ri* there is little mixing (Bursik and Woods 1996). This leads to predictions of the run-out distance of the flow, at which point the flow lift off and rise into the atmosphere. If the run-out distances are compared with historic eruptions, we can estimate the eruption rates, within some bounds of uncertainty, based on the range of grain sizes in the flow (Figure 9.17).

One important feature of ash flows is their ability to scale topography; indeed, there are many observations of ash flows traveling over ridges 1–2 km high up to tens of kilometers from the source of the eruption. Ash flows are able to climb such topography owing to the reduction in the density of the flow as they propagate, sediment ash, and entrain air. As a result, the buoyancy force they need to overcome to scale topography far from the vent decreases with distance (Woods et al. 1998).

## 9.6 Related Volcanic Processes

In this section, we extend some of the results of the previous section on turbulent buoyant plumes to discuss models of submarine eruptions, hydrothermal eruptions and also $CO_2$ driven lake eruptions, such as occurred at Lake Nyos, Cameroon.

### *9.6.1 Submarine Eruptions*

In recent years, it has become evident that a number of volcanoes erupt in submarine settings, and this leads to some different physical processes compared to atmospheric ash clouds, although there are very few observations relating to submarine eruption processes. However, there have been some interesting observations of pumice floating on the sea surface near the Kavachi volcano in the South Pacific, and there is geological evidence of submarine deposits in Greece. When a fragmented mixture of ash, pumice, and gas erupts on the seafloor, the pumice may be less dense than the seawater, thereby providing some substantial additional buoyancy to the rising flow. The gas phase in

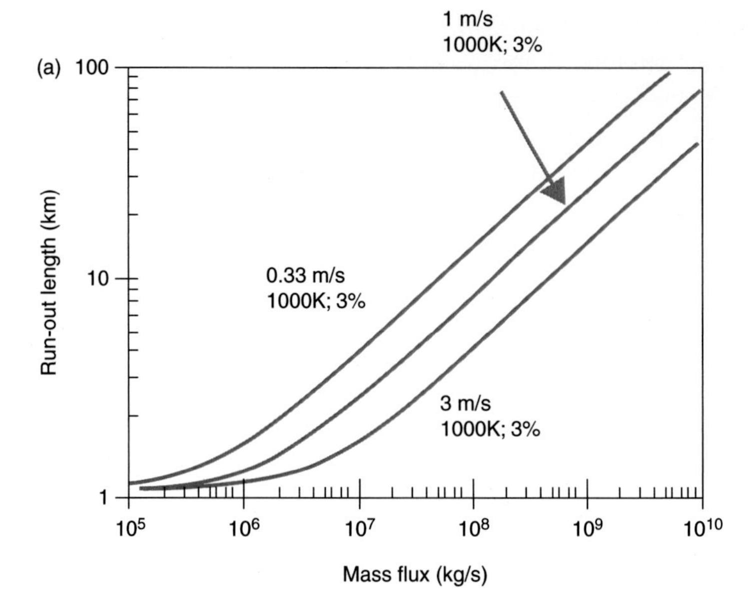


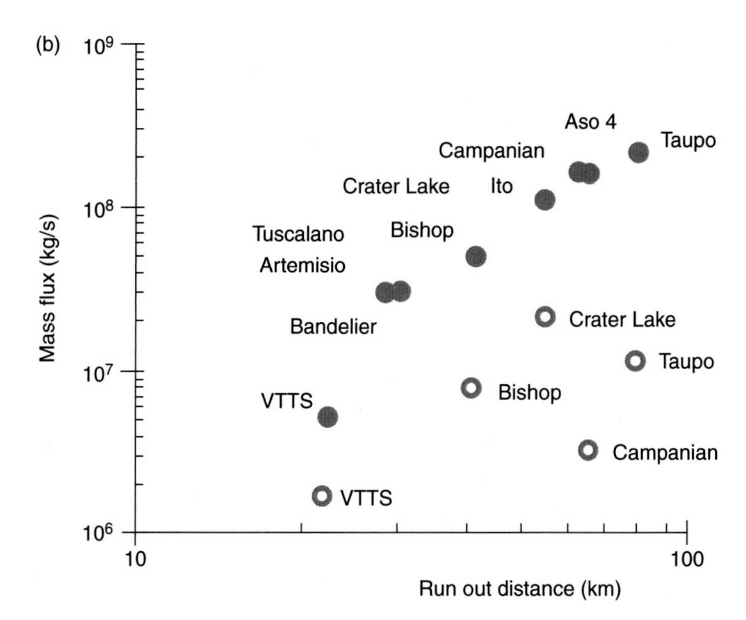


Figure 9.17. (a) Model predictions of the rub out distance as a function of the mass flux, for different eruption temperatures, magmatic gas content and and fall speed of the particles. (b) Geological data, illustrating the mass flux of the largest eruptions in the gelogical record. These calculations follow from the model of Bursik and Woods (1996).

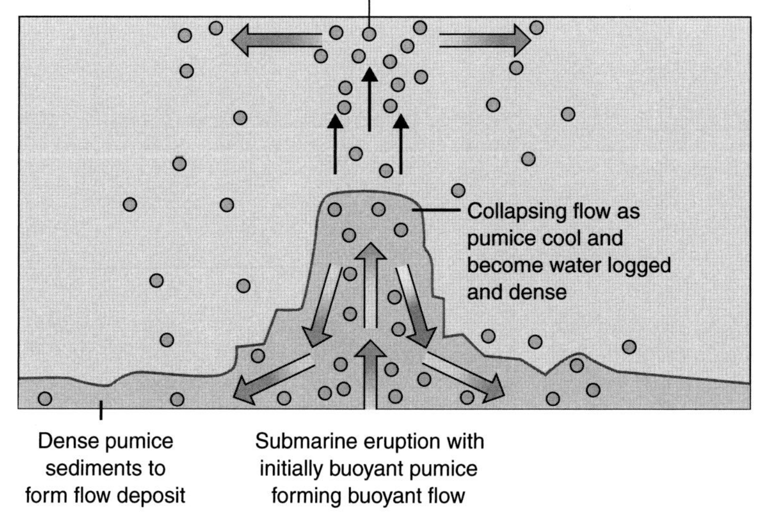


Figure 9.18. Dynamics of a submarine eruption column, in which the pumice particles entrain water, and become dense, thereby leading to collapse back to the seafloor.

this flow rapidly condenses on mixing with cold seawater, leading to a flow of buoyant pumice, ash, and warm water. As this flow rises, the pumice become progressively waterlogged and cool. Their density then increases above that of the seawater, thereby reducing the buoyancy of the overall flow. Eventually the mixture becomes dense relative to the seawater, and the flow collapses, with the mixture falling back to the seafloor to form a warm but dense flow of ash and pumice, which gradually sediments from the flow (Woods 2011, sub-judice).

In order to model the dynamics of such flows, Woods (2011) has developed a model in which the pumice are assumed to cool gradually through contact with the seawater. As a result, the density of the pumice, and hence the overall flow, increases, in accord with some recent experiments (Figure 9.18). This then leads to predictions of the evolution of the bulk density of the flow with height. Initially the flow is very buoyant and develops considerable momentum to carry the mixture upward through the water column. However, as the system cools, and the pumice density increases, the flow becomes negatively buoyant, decelerates, and comes to rest. An illustrative calculation of the evolution of both the bulk density and the pumice density as a function of height is shown in Figure 9.19. It is seen that the variation of the water density with temperature is not significant in the dynamics, compared to the contribution of the pumice since the thermal expansion coefficient of the water is very small. The flow therefore collapses back to the seafloor as pumice becomes dense. As it spreads along the seafloor the coarser pumice particles settle from the flow, and the remaining mixture

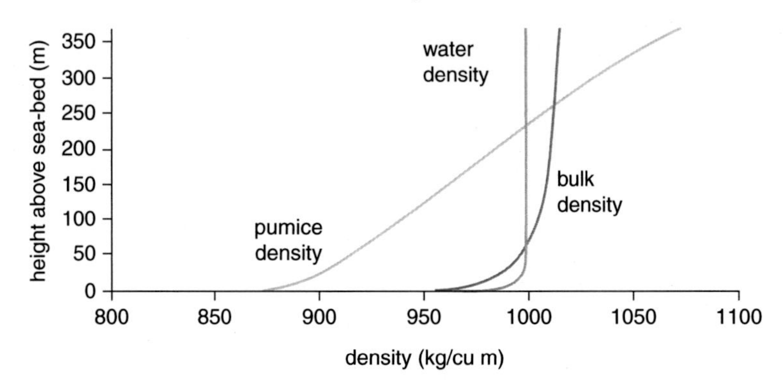


Figure 9.19. Evolution of the density of the flow as it rises from the seafloor, with the pumice cooling and the bulk density gradually increasing to form a relatively dense flow which collapses back to the seafloor. The pumice density increases as it ingests water, and this leads to the increase in the bulk density. The momentum of the originally buoyant jet carries the mixture above the neutral buoyancy height at which the density of the bulk mixture and the seawater match.

of warm water, heated by the erupting solids, and the finer ash particles, lift off from the seafloor to form a warm, fine-grained particle plume.

### *9.6.2 Hydrothermal Eruptions*

Hydrothermal plumes develop at mid-ocean ridge spreading centers, where hot water, laden with minerals, vents from the seabed having been heated up in the intense convective flows of seawater through the newly formed ocean-crust at the mid-ocean ridge spreading centers. The water is typically close to the boiling temperature of the seafloor. As the hot water rises through the cold deep sea water, it precipitates minerals and takes on the characteristic "black smoker" appearance (Figure 9.20), while lower temperature vents produce "white smokers" as anyhydrite precipitates.

The motion of the black smoker plumes through the water column can be modeled using the classical theory of tubulent buoyant plumes (Morton et al. 1956). The height of rise is dominated by the stratification of the ambient and the initial thermal energy of the water erupting through the seafloor. Since the ocean is stratified in both temperature and salinity, as the plume reaches its neutral height, it typically has an anomaly in both salinity and temperature (Figure 9.21).

In order to estimate the height of rise of the plume, $H$, we use the result of Morton et al. (1956), which relates $H$ to the buoyancy flux $B = g'Q$ of the plume, where $g'$ is the buoyancy and $Q$ is the volume flux at the source, and the Brunt Vaiasala frequency of the water column, $N$, according to

$$H = 5B^{1/4}N^{-3/4}. \tag{9.6.1}$$

Figure 9.20. Image of two neighboring black smoker plumes, showing the turbulent nature of the particle laden, hot water. For a color version of this figure please see the color plate section.

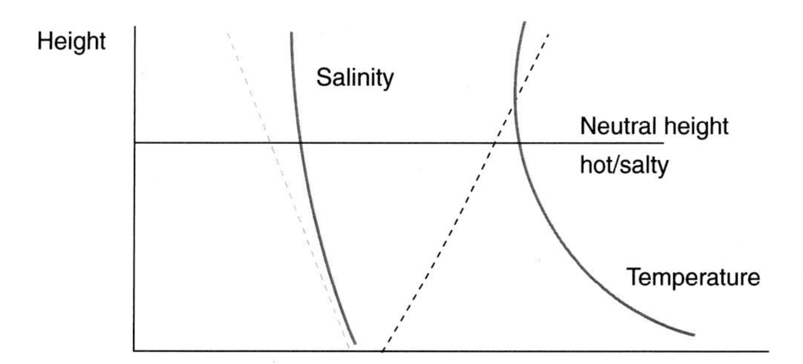


Figure 9.21. Typical (schematic) evolution of the properties in a hydrothermal plume as a function of height above the seafloor. The ambient salinity and temperature are shown with dotted lines and the properties of the plume are shown with solid lines. The fluid starts off hot on the seafloor and entrains ambient water, thereby becoming relatively saline as it rises. The temperature gradually falls, owing to the cold ambient and the mixture reaches a neutral height when relatively saline and hot. The subsequent overshoot can change the temperature relative to the ambient at the top of the plume compared to the value at the neutral height.

Here, since the plume is very hot at the source, and the thermal expansion coefficient of the water is nonlinear at such high temperatures, we introduce a modified relation for the buoyancy flux in terms of the heat flux at the vent

$$B = \frac{Q_H g \alpha}{\rho C} \tag{9.6.2}$$

and the thermal expansion coefficient of water $\alpha$ (Turner and Campbell 1987; Woods and Bush 1999). This predicts heights of rise of order 200 m for the most intense hydrothermal plumes in the deepsea, which is very weakly stratified.

However, since hydrothermal plumes are relatively long-lived phenomena persisting for months or longer, the neutral cloud that spreads radially from the top of the plume may become influenced by the Earth's rotation, which has frequency of rotation $f$. As a result, it may form a geostrophically adjusted lens of fluid with the ratio $NH/fR \sim O(1)$, where $H$ is the thickness of the neutral cloud and $R$ is the radius. This expression may be regarded as the ratio of the Rossby radius of deformation $NH/f$ to the scale of the lens $R$. The eddies do not spread beyond this radius, but, instead, they detach from the top of the plume and are carried off by ambient currents as a series of discrete clouds (Helfrich and Battisti 1991).

### *9.6.3 Lake Nyos Explosion*

We conclude with a brief mention of the eruption of $CO_2$ from Lake Nyos in 1986. During this event, a large mass of $CO_2$ of order 0.1 km$^3$ issued from the lake in a period estimated to be of order 4 hours or less. Little is known about the saturation of $CO_2$ in the lake prior to the event, but monitoring of the lake since the explosion shows it is continually being recharged, and the concentration is increasing toward saturation levels. However, the lake does have a vertical gradient of dissolved $CO_2$, which leads to a very stable density stratification, since the density increases with $CO_2$ content. Since the solubility of the lake to $CO_2$ increases with depth and hence pressure, the water deep in the lake contains more $CO_2$ than the equivalent water near the surface. Hence, for the $CO_2$ to be released, some mechanism is needed to enable lake water to rise upward in the lake and overturn. Various mechanisms for this have been proposed, although only two processes have been quantified with simple models to establish whether they are viable (Woods and Phillips 1999; Mott and Woods 2010). Both are plausible processes, and both rely on the $CO_2$ content of the deep lake water being close to saturation.

First, a turbulent bubble plume may develop if $CO_2$ continually recharges the water at the lake bed. As bubbles rise deep in the lake, from a localized source, they carry water upward in their wake. As a result of the ensuing decompression, the water releases more $CO_2$, and there is nonlinear feedback, leading to a greater flux of bubbles. Eventually, the bubble–water mix begins to move as a bulk convective flow, leading to the development of an intense bubble plume. This can then entrain large quantitites of the deep lake water toward the surface, leading to release of a large quantity of $CO_2$ (Woods and Phillips 1999).

To model the process, Woods and Phillips (1999) proposed a model of a turbulent buoyant plume, accounting for the conservation of mass, momentum, and total flux

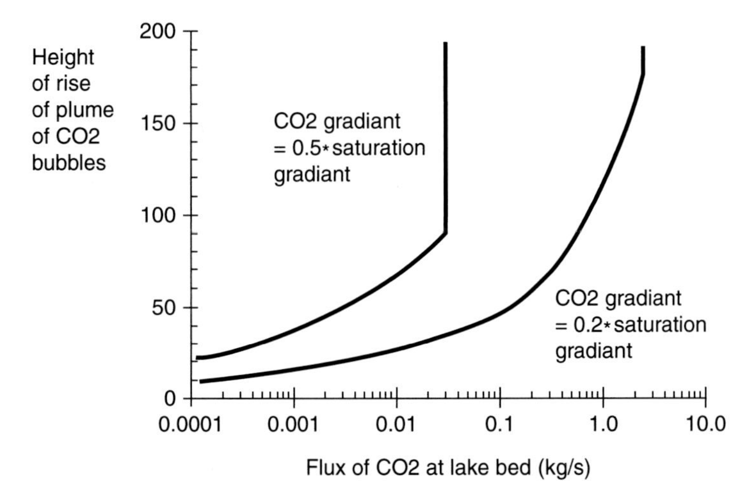


Figure 9.22. Calculation illustrating the height of rise of a bubbly $CO_2$ plume as a function of source $CO_2$ flux, rising through a model lake, 200 m deep, containing a linear gradient of $CO_2$ with height. The numbers on the curves 0.2, 0.5 denote the ratio of the $CO_2$ content in the lake to the saturation $CO_2$ content at that pressure. The greater the mass of $CO_2$ in the lake, the smaller the flux of $CO_2$ at the lake bottom needed for the plume to reach the surface since the plume entrains more $CO_2$ from the low level lake water, and the buoyancy of the released gas dominates the subsequent dynamics.

of $CO_2$ in the plume, together with an equation for the mass fraction of $CO_2$, which is in solution and which is exsolved at each height in the plume. This is evaluated using a Henry's law relation for the solubility of $CO_2$ in water as a function of pressure. The model illustrates the competition between the strong stable stratification of the deep lake water, owing to the vertical gradient of dissolved $CO_2$, and the buoyancy of the plume associated with the presence of buoyant bubbles. Small plumes tend to be arrested by the stable stratification and lead to mixing of the water in the lower part of the water column, while larger plumes are able to reach the surface and release $CO_2$ to the atmosphere. See Figure 9.22.

In the second scenario, a vigorous storm or landslide can generate some deep boundary flows on the lake bed which, owing to roughness on the lake bed, may mix the deep water leading to formation of a supersaturated zone of water just above the lake bed. Gas bubbles may exsolve from the supersaturated water into this layer producing a zone of buoyant fluid. Discrete thermals of water may then detach from this layer, entraining large quantities of water and $CO_2$ as they rise. If the thermals are sufficiently large, they can in fact reach the lake surface, again leading to a large and rapid release of $CO_2$.

Mott and Woods (2010) have modeled this second process with a simple laboratory experiment, using saturated lemonade at the base of a tank, and releasing a current a

Figure 9.23. Experimental model of a $CO_2$ lake eruption triggered by a landslide. Salt solution with salt crystals is released in a tank of water containing a thin layer of lemonade at the base of the tank. On mixing with salt, the lemonade releases bubbles, and a series of convective bubbly thermal clouds rise from the flow of dense saltwater, which continues to run along the base of the tank, as a model avalanche. For a color version of this figure please see the color plate section.

saline water containing salt particles into the lemonade to seed bubble growth. This leads to formation of a series of buoyant thermals, carrying the released gas to the surface (Figure 9.23).

## 9.7 Summary

In this chapter, we have studied a range of processes in explosive volcanic eruptions, focusing on the evolution of the fluids and their density as pressure or temperature changes, or they mix with other fluids. We have identified numerous processes below the surface and above the surface in which the resulting buoyancy contrasts lead to dramatic natural phenomena.

In the subsurface, the large scale of the flows, typically kilometers, have high Rayleigh number and so lead to turbulent convection, even though the motion of individual particles and bubbles is of low Reynolds number. This leads to a fascinating range of problems concerning sedimention and separated flow, heat transfer, and mixing. Above the surface, there are again small-scale flow processes associated with millimeter to micron size ash particles sedimenting from the flows, as well as high-speed, highly turbulent flows over scales of kilometers.

Much of the modeling described in this chapter has been inspired or triggered by geological observations. Some of these relate to highly complex processes, and the models are very simplified. This allows for insights about some of the dominant processes controlling the flows, but in some cases there are likely to be other flow regimes and other important processes that influence the flow dynamics. Continued interaction of geological observations and modeling will lead to new insights and explanations for many other complex processes.

This is a rich topic for further enquiry, and particular opportunities lie in exploring the effects of separated flow, particularly in developing new insights into the dynamics of collapsing ash fountains and ash flows, as well as other related geological phenomena.

Table 9.1. *List of Variables*

| | |
|---|---|
| $C$ | specific heat |
| $c$ | concentration of bubbles |
| $f$ | rotation frequency of the Earth |
| $F$ | heat flux |
| $g$ | gravitational acceleration |
| $h$ | magma layer depth/ thickness of intruding cloud |
| $H$ | depth of chamber |
| $L$ | latent heat |
| $dV$ | change in volume of a magma chamber |
| $M$ | momentum flux |
| $N$ | Brunt Vaiasala frequency |
| $n$ | gas mass fraction |
| $R$ | gas constant |
| $T$ | temperature |
| $t$ | time |
| $p$ | pressure |
| $Q$ | bubble production rate |
| $R$ | radius of the umbrella cloud |
| $Ra$ | Rayleigh number |
| $s$ | solubility coefficient in Henry's law |
| $y$ | solid fraction in crystal pile |
| $u$ | bubble rise speed |
| $V$ | volume of magma chamber |
| $v$ | crystal or ash particle fall speed |
| $w$ | width of the neutral cloud |
| $x$ | crystal content (Sections 9.2–9.4; defined beside equations) |
| $x$ | distance downwind (Section 9.5) |
| $\alpha$ | thermal expansion coefficient |
| $\beta$ | bulk modulus |
| $\rho$ | density |
| $\gamma$ | heat transfer coefficient |
| $\mu$ | viscosity |
| $\kappa$ | thermal diffusivity |
| $\Delta\rho$ | change in density |
| $\varphi$ | void fraction |
| *Subscript* | |
| $a$ | ambient |
| $l$ | lower |
| $m$ | magma |
| $o$ | initial value |
| $u$ | upper |

## References


Barmin, A., O. Melnick, and R. S. J. Sparks, 2002: Periodic behavior in lava dome eruptions. *Earth Planet. Sci. Lett.* 199, 175–184.

Blundy, J. and R. S. J. Sparks, 1992: Petrogenesis of mafic inclusions in Granitoids of the Adamello Massif, Italy, *J. Petrol.* **33**, 1039–1104.

Bower, S. and A. W. Woods, 1996: Control of magma volatile content and chamber depth on the mass erupted during explosive volcanic eruptions. *J. Geophys. Res.* **102** (B5), 10273–10290.

Bursik, M. and A. W. Woods, 1996: The dynamics and thermodynamics of ash flows. *Bull. Volcanol.* **58**, (2–3), 175–193.

Cardoso, S. S. S. and A. W. Woods, 1999: On convection in a volatile saturated magma. *Earth Planet. Sci. Lett.* **168**, (3–4), 301–310.

Conrad, W. and R. Kay, 1984: Ultramafic and mafic inclusions from Adak Island: Crystallization history, and implications for the nature of primary magmas and crustal evolution in the Aleutian Arc. *J. Petrol.* **25**, 88–125.

Cowan, A. and A. W. Woods, 2011a: Some dynamical constraints on hybrid magma formation in closed sills. *Earth Planet. Sci. Lett.* Sub-judice.

Cowan, A. and A. W. Woods, 2011b: An experiment to illustrate the impact of magma overturn during explosive eruptions. Sub-judice.

Gill, A. E. 1981: *Atmosphere Ocean Dynamics*. Wiley, London.

Hazen, A. 1904: On sedimentation. *Trans. ASCE* **53**, 45.

Helfrich, C. and T. M. Battisti, 1991: Experiments on baroclinic vortex shedding from hydrothermal plumes. *J. Geophys. Res.* **96**, 12511–12518.

Huppert, H. E. and R. S. J. Sparks, 1988: The generation of granitic magmas by intrusion of basalt into continental crust. *J. Petrol.* **29**, 588–624.

McBirney, A., 1995: *Igneous Petrology*. Academic Press, New York.

Mott, R. and A. W. Woods, 2010: A model of overturn of $CO_2$ laden lakes triggered by bottom mixing. *J. Volcanal. Geotherm. Res* **3–4**, 151–158.

Morton, G. I., G. I. Taylor, and J. S. Turner, 1956: Turbulent buoyant convection fro maintained and instantaneous sources. *Proc. Roy. Soc.* A **231**, 1–24.

Niemala, J., L. Skrbeck, K. Sreenivasan, and R.Donnelly, 2000: Turbulent convection at very high Rayleigh number. *Nature* **404**, 837–840.

Phillips, J. C. and A. W. Woods, 2002: Suppression of large-scale magma mixing by melt-volatile separation. *Earth Planet. Sci. Lett.* **204**, 47–60.

Sparks, R. S. J. The dynamics of bubble formation and growth in magmas: A review and analysis. *J. Volcanol. Geotherm. Res.* **3**, 1–37.

Sparks, R. S. J., M. Bursik, S. Carey, J. Gilbert, L. Glaze, H. Sigurdsson, and A. W. Woods, 1997: *Volcanic Plumes*, Wiley, London.

Tait, R. C. Jaupart, and S. Vergniolle, 1989: Pressure, gas content and eruption periodicity of a shallow crystallizing magma chamber. *Earth Planet. Sci. Lett.* **92**, 107–123.

Touloukian, M., S. Judd, and H. Roy, 1981: *Physical Properties of Rocks and Minerals,* McGraw Hill, New York.

Turner, J. S., 1979: *Buoyancy Effects in Fluids*, Cambridge University Press, London.

Turner, J. S. and I. Campbell, 1987: Temperature, density and buoyancy fluxes in "black smoker" plumes, and the criterion for buoyancy reversal. *Earth Planet. Sci. Lett.* 85–92.

Turner, J. S. and I. Campbell, 1997: Temperature, density and buoyancy fluxes in "black smoker" plumes, and the criterion for buoyancy reversal. *Earth Planet. Sci. Lett.* **86**, 85–92.

Woods, A.W., 1995: The dynamics of explosive volcanic eruptions. *Rev. Geophys.* **33**, 495–530.

Woods, A. W., 1988: The dynamics and thermodynamics of eruption columns. *Bull. Volcanal.* **50**, 169–193.

Woods, A. W., M. I. Bursik, and A. Kurbatov, 1998: The interaction of ash flows with ridges, *Bull. Volcanal.* **60**, 38–51.

Woods, A. W., 2010: Turbulent plumes in nature, *Ann. Rev. Fluid Mech.* **42**, 391–412.

Woods, A. W., and Bush, J. W. M. 1999: The dimensions and dynamics of megaplumes, *J. Geophys Res.*, **104**, 20495–20507.

Woods, A. W., and A. Cowan, 2009: Magma mixing triggered during volcanic eruptions. *Earth Planet. Sci. Lett.* **288**, (1–2), 132–137.
Woods, A. W., and J. Phillips, 1999: On bubble driven lake plumes. *J. Volcanal. Geotherm. Res.* **92**, 259–272.
Woods, A. W., and H. E, Huppert, 2003: On magma chamber evolution during low effusive eruptions. *J. Geophys. Res.* **108**, 2403–2407.
Woods, A. W., 2011: Dynamics of submarine eruptions. *Earth Planet. Sci. Lett. Sub-judice*.

# 10

# Gravity Flow on Steep Slope

CHRISTOPHE ANCEY

## 10.1 Introduction

Particle-laden, gravity-driven flows occur in a large variety of natural and industrial situations. Typical examples include turbidity currents, volcanic eruptions, and sandstorms (see Simpson 1997 for a review). On mountain slopes, debris flows and snow avalanches provide particular instances of vigorous dense flows, which have special features that make them different from usual gravity currents. Those special features include the following:

- They belong to the class of non-Boussinesq flows since the density difference between the ambient fluid and the flow is usually very large, whereas most gravity currents are generated by a density difference of a few percent.
- Whereas many gravity currents are driven by pressure gradient and buoyancy forces, the dynamics of flows on slope are controlled by the balance between the gravitational acceleration and dissipation forces. Understanding the rheological behavior of particle suspensions is often of paramount importance when studying gravity flows on steep slope.

This chapter reviews some of the essential features of snow avalanches and debris flows. Since these flows are a major threat to human activities in mountain areas, they have been studied since the late 19th century. In spite of the huge amount of work done in collecting field data and developing flow-dynamics models, there remain great challenges in understanding the dynamics of flows on steep slope and, ultimately, in predicting their occurrence and behavior. Indeed, these flows involve a number of complications such as abrupt surge fronts, varying free and basal[1] surfaces, and flow structure that changes with position and time.

Subaqueous landslides and debris avalanches have many similarities with subaerial debris flows and avalanches (Hampton et al. 1996). The correspondence, however, is not complete since subaqueous debris flows are prone to hydroplane and transform into density currents as a result of water entrainment (Elverhøi et al. 2005); the slope range over which they occur is also much wider than the slope range for subaerial

[1] The basal surface is the interface between the bottom of the flow and the ground/snowcover.

flows. Powder-snow avalanches are related to turbidity currents on the ocean floor (Parker et al. 1986) and pyroclastic flows from volcanoes (Huppert and Dade 1998; Bursik et al. 2005). Powder-snow avalanches sometimes experience a rapid deceleration of their dense cores, which eventually separate from their dilute clouds and form stepped thickness patterns in their deposits. This behavior is also seen with submarine flows and pyroclastic flows. In addition to being non-Boussinesq flows, powder-snow avalanches differ from submarine avalanches in that they are closer to fixed-volume, unsteady currents than to the steady density currents with constant supply.

## 10.2 A Physical Picture of Gravity Flows

### *10.2.1 Debris Flows*

Debris flows are mass movements of concentrated slurries of water, fine solids, rocks, and boulders (Iverson 1997; 2005). They are highly concentrated mixtures of sediments and water, flowing as a single-phase system on the bulk scale. Debris flows look like mudslides and landslides, but the velocities and the distances they travel are much larger. They differ from floods in sediment transport in that they are characterized by a very high solids fraction (mostly exceeding 80%).

There are many classifications of debris flows and related phenomena based on compositions, origins, and appearances. Many events categorized as "mudflows," "debris slides," lahars, and "hyperconcentrated flows" can be considered as particular forms of debris flows (Fannin and Rollerson, 1993; Iverson, 1997). Debris flows may result from the following:

- Mobilization from a landsliding mass of saturated unsorted materials, often after heavy and/or sustained rainfalls (Iverson et al. 1997)
- Transformation from a sediment-laden water flood into a hyperconcentrated flow, probably as a result of channel-bed failure (Tognacca 1997)
- Melting of ice and snow induced by pyroclastic or lava flows and accompanied by entrainment of large ash volumes (Voight 1990)
- Collapse of a moraine-dammed lake generating an outburst flood (Clague and Evans 2000).

The material volume mobilized by debris flows ranges from a few thousands cubic meters to a few millions, exceptionally a few billions. The velocity is typically a few meters per second, with peak velocities as high as 10 m/s (VanDine 1985; Major and Pierson 1992; Hürlimann et al. 2003). Debris flows usually need steep slopes (i.e., in excess of 20%) to be initiated and to flow, but occasionally they have been reported to travel long distances over shallow slopes (less than 10%).

Figure 10.1 shows two deposits of debris flows. In Figure 10.1a, a debris flow involving well-sorted materials embedded in a clayey matrix came to a halt on an

Figure 10.1. (a) Debris-flow deposit in the Ravin-des-Sables watershed (France); the bucket gives a scale of the deposit thickness. (b) Debris flow on the road to la-Chapelle-en-Valgaudemar (France). For a color version of this figure please see the color plate section.

alluvial fan[2]; note that there was no water seepage, which implies that the material was still water-saturated a few hours after stoppage. Figure 10.1b shows a car hit by a debris flow made up of coarse material; the conspicuous streaks of muddy water indicate that water and the finest grain fraction separated from the coarsest grain fraction as soon as the flow approached the arrested state.

### 10.2.2 Snow Avalanches

Avalanches are rapid, gravity-driven masses of snow moving down mountain slopes. Many, if not most, catastrophic avalanches follow the same basic principle: Fresh snow accumulates on the slope of a mountain until the gravitational force at the top of the slope exceeds the binding force holding the snow together. A solid slab of the surface layer of snow can then push its way across the underlying layer, resulting in an avalanche. The failure may also arise from a temperature increase, which reduces snow cohesion. Typically, most avalanches travel for a few hundred meters at a rather low velocity (a few meters per second), but some can move up to 15 km and achieve velocities as high as 100 m/s. They can also pack an incredible punch, up to several atmospheres of pressure. It is helpful to consider two limiting cases of avalanches depending on the flow features (de Quervain 1981):

- The *flowing avalanche*: A flowing avalanche is an avalanche with a high-density core at the bottom. Trajectory is dictated by the relief. The flow depth does not generally exceed a few meters (see Figure 10.2a). The typical mean velocity ranges from 5 to 25 m/s. On average, the density is fairly high, generally ranging from 150 to 500 kg/m$^3$.
- The *powder snow avalanche*: It is a very rapid flow of a snow cloud, in which most of the snow particles are suspended in the ambient air by turbulence (see Figure 10.2b). Relief has usually weak influence on this aerial flow. Typically, for the flow depth, mean velocity, and mean density, the order of magnitude is 10–100 m, 50–100 m/s, 5–50 kg/m$^3$, respectively.

[2] An alluvial fan is a fan-shaped deposit formed typically at the exit of a canyon, as a result of the sudden change in the bed gradient, which causes massive sediment deposition.

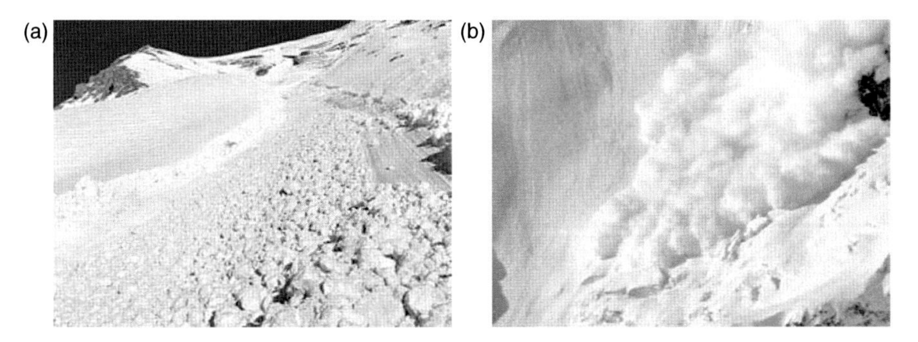


Figure 10.2. (a) Wet-snow avalanche deposit in the southern face of Grammont (Switzerland); the snowballs are approximately 10 cm in diameter. (b) Powder-snow avalanche in the northern face of Dolent (Switzerland); the typical flow depth is 20 m. For a color version of this figure please see the color plate section.

## 10.3 Anatomy of Gravity Currents on Slope

Knowing how a gravity-driven flow is organized is of paramount importance to understanding its properties. Contrary to most fluid-mechanics problems in which the fluid volume is bounded or infinite, a gravity current is characterized by moving boundaries:

- The free surface at the interface with the ambient air and
- The surface of contact with the ground (or snow cover), where much of the energy dissipation occurs.

These boundaries can be passive (i.e., they mark the boundaries of the volume occupied by the flowing material). On some occasions, they may be active (e.g., by promoting mass and momentum exchanges with the ambient fluid and/or the bed). Also, a gravity-driven flow often is split into three parts: the head at the leading edge, the body, and the tail. The structure of these regions depends on the material and flow properties. It is quite convenient to consider two end members of gravity flows to better understand their anatomy: debris flows are typical of dense granular flows, for which the ambient fluid has no significant dynamic role, whereas powder-snow avalanches are typical of flows whose dynamics are controlled to a large extent by the mass and momentum exchanges at the interfaces.

### *10.3.1 Anatomy of Debris Flows*

On the whole, debris flows are typically characterized by three regions, which can change with time (see Figure 10.3):

- At the leading edge, a granular front or snout contains the largest concentration of big rocks; boulders seem to be pushed and rolled by the body of the debris flow. The front is usually higher than the rest of the flow. In some cases, no front is observed because the body has

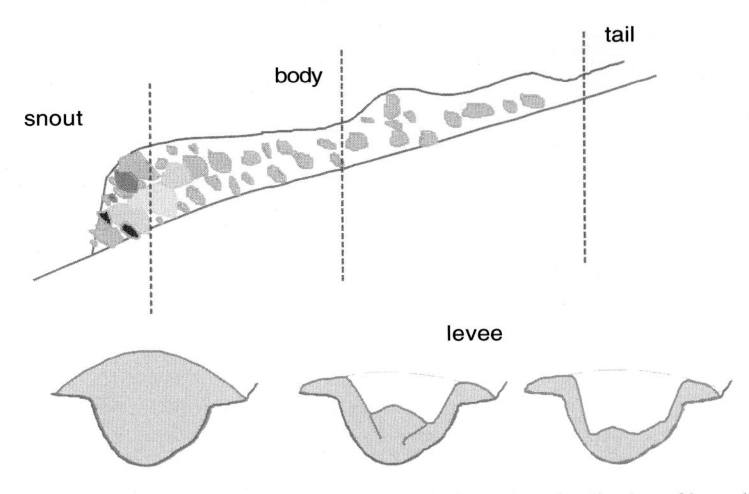


Figure 10.3. Idealized representations of a debris flow (longitudinal profile and cross-section). The different sections correspond to the dashed lines of the upper panel. Adapted from Johnson and Rodine (1984).

overtaken (a frequent occurrence when the debris flow spreads onto the alluvial fan) or because the materials are well sorted and no significant variation in the bulk composition can be detected.

- Behind the front, the body is the flow of a rock and mud mixture. Usually, the debris flow body is not in a steady state but presents unsteady surges (Zanuttigh and Lamberti 2007). It can transport blocks of any size. Many authors have reported that boulders of relatively small size seem to float at the free surface, while blocks of a few meters in size move merely by being overturned by the debris flow. The morphological characteristics of the debris flow are diverse depending on debris characteristics (size distribution, concentration, mineralogy) and channel geometry (slope, shape, sinuosity, width). Flowing debris can resemble wet concrete, dirty water, or granular material, but irrespective of the debris characteristics and appearance, viscosity is much higher than for water. Most of the time, debris flows move in a completely laminar fashion, but they can also display minor turbulence; on some occasions, part of the debris flow may be highly turbulent.
- In the tail, the solid concentration decreases significantly and the flow looks like a turbulent muddy water flow.

In recent years, many outdoor and laboratory experiments have shed light on the connections existing between particle-size distribution, water content, and flow features for fixed volumes of bulk material (Davies 1986; Iverson 1997; Parsons et al. 2001; Chambon et al. 2009). In particular, experiments performed by Parsons et al. (2001) and Iverson (1997) have shown that the flow of poorly sorted materials was characterized by the coexistence of two zones, each with a distinctive rheological behavior: the flow border was rich in coarse-grained materials, whereas the core was fine grained. This self-organization has a great influence on the flow behavior; notably, the flow core behaves more like a viscoplastic material, while the flow region close

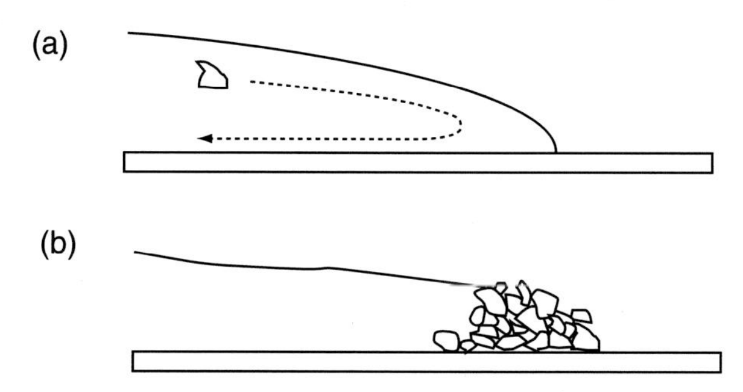


Figure 10.4. Schematic of the behavior contrast between fine-grained and coarse-grained flows. (a) Conveyer-belt-like flow at the front. (b) Formation of a frictional front. After Parsons et al. (2001).

to the levees (lateral deposits) are in a Coulomb frictional regime (sustained solid frictional contacts between grains). Moreover, the runout distance can be significantly enhanced as a result of levee formation limiting lateral spreading and energy dissipation.

Parsons et al. (2001) ran a series of experiments to investigate the effect of the composition (i.e., the importance of the finest- and coarsest-grain fractions). They used a semicircular inclined flume and measured the velocity profile at the free surface. Different slurries were prepared by altering the sand, clay, and silt fractions. They obtained muddy slurries when the matrix was rich in silt and clay, and poorly sorted mixtures when the silt and clay contents were reduced. Surprisingly enough, the change in the fine-particle content did not significantly modify the appearance of the body, whereas it markedly altered the composition of the front and its behavior. Reducing the fine fraction in the slurries induced a radical change of behavior for the front (see Figure 10.4):

- For muddy, fine-grained slurries, the front took the form of a blunt nose. Lack of slip along the flume bottom caused a conveyer-belt-like flow at the front.
- For coarse-grained slurries, the front took the form of a dry granular locked nose slipping along the bed as a result of the driving force exerted by the fluid accumulating behind the snout. Additional material was gradually incorporated into the snout, which grew in size until it was able to slow the body.

Interestingly enough, the changes in the rheological properties mainly affected the structure of the flow, especially within the tip region.

Iverson and his colleagues investigated slurries predominantly composed of a water-saturated mixture of sand and gravel, with a fine fraction of only a few percent (Iverson 1997, 2003a, 2005; Iverson et al. 2010). Experiments were run by releasing a volume of slurry (approximately 10 m$^3$) down a 31-degree, 95-m-long flume. At the base of the flume, the material spread out on a planar, nearly horizontal, unconfined runout zone. Flow-depth, basal normal stress, and basal interstitial-flow pressure were measured at different places along the flume. Iverson and his co-workers observed that at early times (just after the release), an abrupt front formed at the head of the flow, which was followed by a gradually tapering body and a thin, more watery tail. The front remained relatively dry (with pressure of interstitial water dropping to zero) and of constant thickness, while the body elongated gradually in the course of the flow. Over the longest part of the flume, the basal pore pressure (i.e., the pressure of the interstitial water phase) nearly matched the total normal stress, which means that shear strength was close to zero and the material was liquefied within the body (Iverson 1997). In their recent data compilation, Iverson et al. (2010) confirmed the earlier observations made by Parsons et al. (2001): Mud enhanced flow mobility by maintaining high pore pressures in flow bodies. They also observed that roughness reduced flow speeds, but not runout distances. The explanation for this apparently strange behavior lies in the particular role played by debris agitation and grain-size segregation. Indeed, if the bed is flat, particles slip along the bottom, and shear is localized within a thin layer close to the bed, with almost no deformation through the flow depth (i.e., uniform velocity profile). In contrast, if the bed is corrugated, particles undergo collisions and are more agitated, which promotes the development of a nonuniform velocity profile through the depth and causes the flow to slightly dilate and the particles to segregate (see Section 10.4.3). Velocity shear and dilatancy act together as a sieve that constantly and randomly opens gaps. The finest particles are more likely to drop down into the gaps under the action of gravity than the coarsest ones are, which eventually creates inversely graded layers of particles (coarse particles on top, fine particle near the base of the flow). Here, particle segregation has two effects: (i) it reduces bottom friction (small particles acting as rolling balls) as shown in granular avalanche experiments carried out by (Phillips et al. 2006; Linares-Guerrero et al. 2007); (ii) the largest particles concentrate in the fast-moving upper layers (next to the free surface) and are transported to the flow front, where they are shouldered to the side by the core (made up of more mobile fine particles) and create static coarse-grained lateral levees that channelize the flow.

Figure 10.5 shows a sequence of aerial photographs taken when the material spread out on the runout surface. Self-organization of the slurry flow into a coarse-grained boundary and a muddy core became conspicuous as the flow traveled the runout surface. Lateral levees were formed by the coarsest grains that reached the front, being continuously shouldered aside by the muddy core. These levees then confined the ensuing muddy body. Note the levee formation is probably not induced by particle segregation alone since it is also observed for dry granular flows involving spherical equal-size particles (Félix and Thomas, 2004). Figure 10.6(a) shows the lateral levees, which can be used to evaluate the crosssection of the flow, and Figure 10.6(b) shows a granular levee formed by a debris flow on the alluvial fan. Similar features are also observed for wet-snow avalanches (Jomelli and Bertran 2001) and pyroclastic flows (Iverson and Vallance 2001).

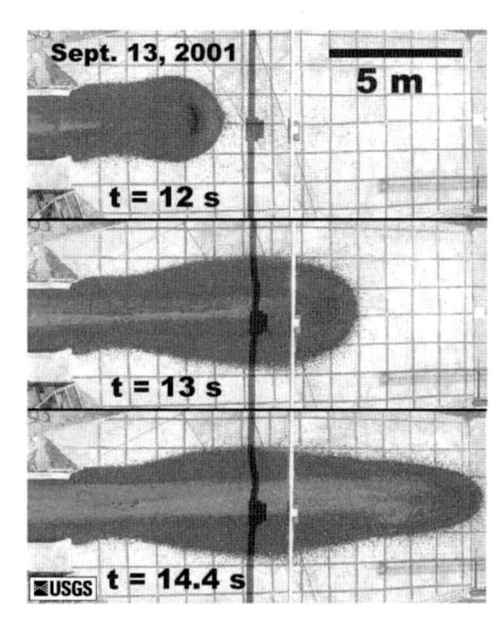


Figure 10.5. Snapshots showing slurry flow discharging from the U.S. Geological Survey Debris-Flow Flume and crossing the unconfined, nearly horizontal runout zone. The dark-toned material around the perimeter of the flow was predominantly gravel; the light-toned material in the center of the flow was liquified mud. Figure reproduced from Iverson (2003a); courtesy of Richard M. Iverson.

Figure 10.6. (a) Crosssection of the Malleval stream after a debris flow in August 1999 (Hautes-Alpes, France). (b) Levees left by a debris flow in the Dunant river in July 2006 (Valais, Switzerland); courtesy of Alain Delalune. For a color version of this figure please see the color plate section.

### *10.3.2 Anatomy of Powder-Snow Avalanches*

Although there is probably no unique typical outline, powder-snow avalanches are usually made up of two regions when they are in a flowing regime (see Figure 10.7).

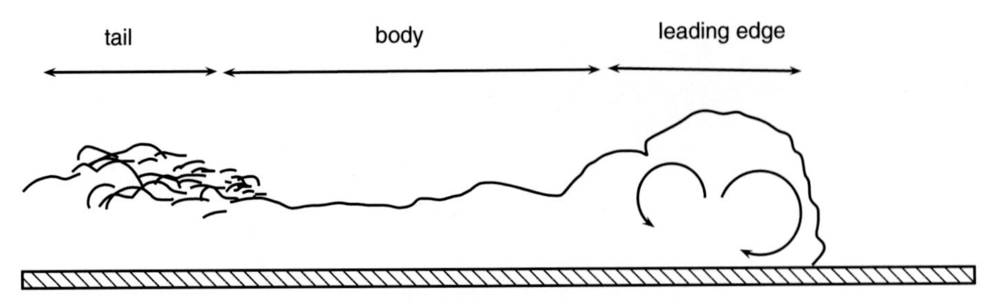


Figure 10.7. Anatomy of a powder-snow cloud.

- The leading edge is the frontal zone where intense mixing occurs. Motion is produced by the density contrast between the suspension and the surrounding fluid. Since the surrounding fluid is entrained into the current, the snow concentration decreases inside the current, leading, in turn, to a decrease in the buoyancy force unless the current is supplied by a sufficiently strong input of particles. Bed erosion and entrainment of the surrounding fluid into the head are therefore the two main processes that control the bulk dynamics. As long as this balance is maintained, the interface with the ambient fluid is a sharp surface that clearly delineates the avalanche and the ambient air. When air entrainment becomes the prevailing mechanism, the interface becomes a blurred layer. Turbulence is needed in the leading edge to counteract the particle settling; with sufficient turbulence, snow particles (ranging from snowflakes to snowballs) are maintained in suspension. The key condition for the formation and development of a powder cloud is that the vertical velocity fluctuations exceed the particle settling velocity, a condition that is reminiscent of the "ignition" of a gravity current on the ocean floor (Parker, 1982).
- The tail or "turbulent wake" is the volume of fluid behind the head, and it is often separated from the leading edge by a billow. The density contrast with the ambient fluid is usually much less marked than for the head. For some events, the powder-snow avalanche leaves behind a motionless cloud whose size may still be growing as a result of turbulent diffusion. It rapidly settles as turbulent energy falls off.

In the release and runout phase, the structure is usually very different. Indeed, in the release phase, the cloud is not formed (the avalanche looks like a flowing avalanche), whereas in the runout phase, the cloud collapses and settles to form a vast, thin deposit (thickness less than 1 m). For many events, it has been observed that the cloud separates from the dense core, which comes to a halt as soon as the slope gradient is too low (typically lower than 20–25%). This "decoupling" process is similar in many respects to the abrupt transition observed by Hallworth et al. (1998b) in their laboratory experiments on the instantaneous release of particle-driven gravity currents in a water-filled flume; it probably results from enhanced friction between particles, which implies higher dissipation rates in the core than in the dilute cloud. Figure 10.8a shows a powder-snow avalanche in a flowing regime. The trees on either side of the avalanche path give a scale of the depth of this avalanche. Its velocity was close to 60 m/s. Figure 10.8b also shows a powder-snow avalanche, but in its

Figure 10.8. (a) Powder-snow avalanche in a flowing regime. Photograph taken in the Vallée-de-la-Sionne field site (Switzerland) in January 2004; courtesy of François Dufour, SLF. (b) Runup of a cloud of a powder-snow avalanche in a runout phase. Photograph taken at le Roux-d'Abriès, France in January 2004; courtesy of Maurice Chave. (c) Deposit of the dense core for the same avalanche; courtesy of Hervé Wadier. For a color version of this figure please see the color plate section.

runout phase. Note that the depth is much higher than the trees. Although its velocity was quite high, this cloud did not cause any damage to the forest, which implies that the impact pressure, and thus the bulk density, were low. Figure 10.8c shows that for this avalanche, part of the avalanche mass was concentrated in a dense core, which stopped prior to reaching the valley bottom.

There are not many field observations of the internal structure of powder-snow avalanches (Issler 2003; Rammer et al. 2007) and much of our current knowledge stems from what we can infer from small-scale experiments in the laboratory, which were conducted with a partial similitude with real flows. Field observations and laboratory experiments reveal the following four important aspects:

- *Existence of eddies*: Field measurements (based on radar or pressure-sensor measurements) show that the internal velocity is higher than the front velocity and varies cyclically with time, which was interpreted as the hallmark of rotational flows. Experiments of gravity currents in tanks have shown that the leading edge is associated with a pair of vortices, one located at the leading edge and another one at the rear of the head (see Figure 10.9). In experiments conducted by Simpson (1972), the development of the flow patterns was made visible using a blend of dense fluid and fine aluminum particles. A stretching vortex occupying the tip region was clearly observed at the leading edge and produced an intense roll-up of fine aluminum particles, which makes it possible to visualize the streamlines and the two vortices; in the upper part of the head, a counterclockwise rotating vortex occurred. Experiments carried out by Ancey (2004) on finite-volume gravity currents moving down a slope also revealed that the particle cloud was composed of two evident eddies: when the surge involving a glass-bead suspension in water moved from left to right, he observed a small vortex ahead of the front, spinning clockwise, and a large counterclockwise eddy occupying most of the surge volume. Theoretically, this is in line with the paper of McElwaine (2005) who extended Benjamin's results by considering steady finite-volume currents flowing down a steep slope, which experience resistance from the surrounding fluid. Like Benjamin (1968), he found that the front makes a $\pi/3$ angle with the bottom line. More recently, Ancey et al. (2006; 2007) worked out analytical solutions to the depth-averaged equations and the Euler equations, which represent the flow of non-Boussinesq currents; it was also found that the flow must be rotational and that the head is wedge-shaped.

- *Vertical density stratification*: Turbulence is often not sufficient to mix the cloud efficiently and maintain a uniform density through the cloud depth. Instead, a dense layer forms at the bottom and the density decreases quickly upward (Issler 2003). For many events, it has also been observed that the dilute component of the avalanche flowed faster than the core and eventually detached from it, which leads us to think that there was a sharp transition from the dense basal layer to the dilute upper layer. For instance, from impact force measurement against static obstacles, it was inferred that the dense layer at the base of the flow was 1–3 m thick, with velocity and (instantaneous) impact pressure as high as 30 m/s and 400 kPa. The transition layer is typically 5 m thick, with kinetic pressure in the 50–100 kPa range. In the dilute upper layer, which can be very thick (as large as 100 m), the kinetic pressure drops to a few pascals, but the velocity is quite high, with typical values close to 60–80 m/s.
- *Snow entrainment*: It alters speed and runout distance. The primary mode of entrainment appeared to be frontal ploughing, although entrainment behind the avalanche front was also observed (Gauer and Issler 2003; Sovilla and Bartelt 2006). When there is snow entrainment, the front is wedge shaped. It can present lobes and clefts, more occasionally fingering patterns, which appear and quickly disappear. The total flow depth lies in the 10–50 m range and varies little with distance. Theoretical calculations predict a wedge angle of $\pi/3$, which seems consistent with field observations (McElwaine 2005). In the absence of entrainment, the front becomes vertical, with a typical nose shape. The surface is diffuse and smoother. The flow depth can be as large as 100 m and quickly varies with distance.
- *Air entrainment*: Changes in cloud volume result primarily from the entrainment of the surrounding air. Various mixing processes are responsible for the entrainment of an ambient, less-dense fluid into a denser current (or cloud). It has been shown for jets, plumes, and currents that (1) different shear instabilities (Kelvin-Helmoltz, Hölmböe, etc.) can occur at the interface between dense and less dense fluids, and (2) the rate of growth of these instabilities is controlled by a Richardson number (Turner 1973; Fernando 1991), defined here as

$$Ri = \frac{g' H \cos\theta}{U^2}, \tag{10.1}$$

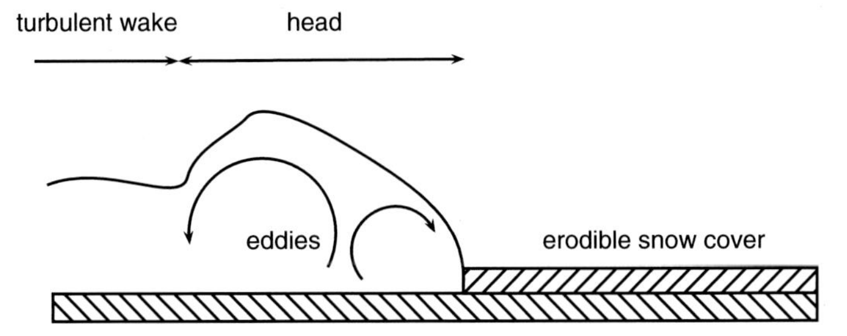


Figure 10.9. Typical structure of the head of a powder-snow avalanche as interpreted from field measurements and laboratory experiments.

where $g'$ denotes the reduced gravity $g' = g\Delta\bar{\varrho}/\varrho_a$ and $\Delta\bar{\varrho} = \bar{\varrho} - \varrho_a$ is the density mismatch between the cloud and the ambient fluid. Note that the Richardson number is the inverse square of the Froude number used in hydraulics. The Richardson number can be seen as the ratio of the potential energy ($g\Delta\bar{\varrho}H\cos\theta$) to the kinetic energy ($\varrho_a U^2$) of a parcel of fluid at the current interface. Usually a smaller $Ri$ value implies predominance of inertia effects over the restoring action of gravity, thus resulting in greater instability and therefore a higher entrainment rate; it is then expected that the entrainment rate is a decreasing function of the Richardson number. Mixing is observed to occur in gravity currents due to the formation of Kelvin-Helmoltz instabilities at the front, which grow in size, are advected upward, and finally collapse behind the head. The lobe-and-cleft instability is also an efficient mechanism of entrainment (Simpson 1997). Although the details of the mixing mechanisms are very complex, a striking result of recent research is that their overall effects can be described using simple relations with bulk variables (Turner 1973; Fernando 1991). For instance, as regards the volume balance equation, the most common assumption is to state that the volume variations result from the entrainment of the ambient fluid into the cloud and that the inflow rate is proportional to the exposed surface areas and a characteristic velocity $u_e$: $\dot{V} = E_v S u_e$ where $E_v$ is the bulk entrainment coefficient and $u_e = \sqrt{\bar{\varrho}/\varrho_a}U$ for a non-Boussinesq current.

## 10.4 Fluid-Mechanics Approach to Gravity Currents

Gravity-driven flows usually take the appearance of more or less viscous fluids flowing down a slope, and this observation has prompted the use of fluid-mechanics tools for describing their motion. However, the impediments to a full fluid-mechanics approach are many: a wide range of particle size (often in the $10^{-3}$–1-m range), composition that may change with time and/or position, poorly known boundary conditions (e.g., erodible basal surface) and initial conditions, time-dependent flows with abrupt changes (e.g., surge front, instabilities along the free surface), and so on. All these difficulties pose great challenges in any fluid-mechanics approach for modeling rapid mass movements and have given impetus to extensive research combining laboratory and field experiments, theory, field observation, and numerical simulations.

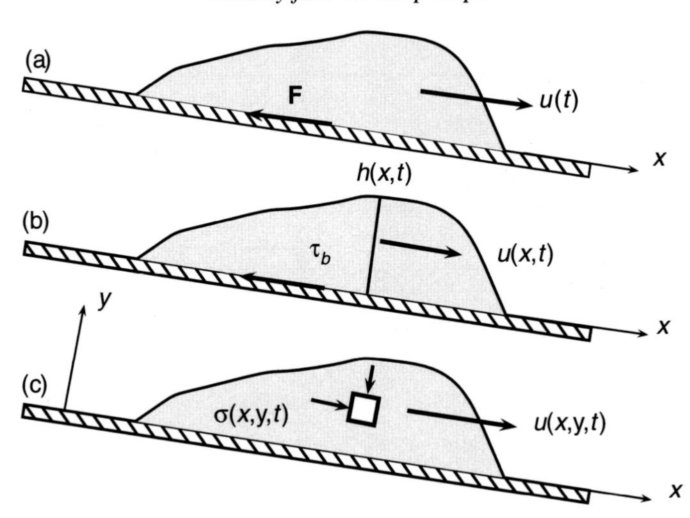


Figure 10.10. Different spatial scales used for describing avalanches (and related flows): (a) the avalanche as a rigid body moving at velocity $\mathbf{u}(t)$; (b) the avalanche as a stream of depth $h(x,t)$ and velocity $\mathbf{u}(x,t)$; (c) the avalanche as a continuum.

Avalanches and debris flows can be considered at different spatial scales (see Figure 10.10). The larger scale, corresponding to the entire flow, leads to the simplest models. The chief parameters include the location of the gravity center and its velocity. Mechanical behavior is mainly reflected by the friction force $F$ exerted by the bottom (ground or snowpack) on the avalanche. The smallest scale, close to the size of snow particles involved in the avalanches, leads to complicated rheological and numerical problems. The flow characteristics (velocity, stress) are calculated at any point of the occupied space. Intermediate models have also been developed. These models benefit from being less complex than three-dimensional numerical models and yet are more accurate than simple ones. Such intermediate models are generally obtained by integrating the equations of motion through the flow depth, as is done in hydraulics for shallow water equations.

We start our review of these three approaches with a discussion of the flow regimes (see Section 10.4.1). We then briefly describe the rheological behavior of natural materials involved in gravity flows in Section 10.4.2. Since most gravity flows are made up of different sized particles, the rheological properties usually on the solids concentration of each component. As a result of various processes such as kinetic sieving, particles can migrate and segregate. In Section 10.4.3, we tackle the difficult issue of particle segregation. The second part of this section is devoted to presenting three fluid-mechanics approaches. In Section 10.4.4, we outline the simplest approach: the sliding-block model, which can be used to give some crude estimates of the speed and dynamic features as well as scaling relations between flow variables and input parameters. A more involved approach consists of taking the depth average of the local governing equations (see Section 10.4.5), which enables us to derive a set of partial differential equations for the flow depth $h$ and mean velocity $\bar{u}$. In principle, the local governing equations could be integrated numerically, but the numerical cost is very high and the gain in accuracy is spoiled by the poor knowledge of the rheologic properties or the initial/boundary conditions. Here we confine attention to analytical treatments, which involves working out approximate solutions by using asymptotic expansions of the velocity field (see Section 10.4.6).

### 10.4.1 Scaling and Flow Regimes

Here we will examine how different flow regimes can occur depending on the relative strength of inertial, pressure, and viscous contributions in the governing equations. Dimensional analysis helps clarify the notions of *inertia-dominated* and *friction-dominated* regimes. In the analytical computations, we will use the shallowness of flows to derive approximate equations.

We consider a shallow layer of fluid flowing over a rigid impermeable plane inclined at an angle $\theta$ (see Figure 10.11). The fluid is incompressible; its density is denoted by $\varrho$ and its bulk viscosity by $\eta = \tau/\dot{\gamma}$, with $\dot{\gamma}$ the shear rate (i.e., in a simple shear flow, it is the velocity gradient in the $y$-direction). Note that this bulk viscosity may depend on $\dot{\gamma}$ (this is the general case when the behavior is non-Newtonian). The ratio $\epsilon = H_*/L_*$ between the typical vertical and horizontal length scales, $H_*$ and $L_*$, respectively, is assumed to be small. The streamwise and vertical coordinates are denoted by $x$ and $y$, respectively.

A two-dimensional flow regime is assumed (i.e., any cross-stream variation is neglected). The depth of the layer is given by $h(x, t)$. The horizontal and vertical velocity components of the velocity $\mathbf{u}$ are denoted by $u$ and $v$, respectively. The pressure is referred to as $p(x, y, t)$, where $t$ denotes time, whereas the extra stress tensor (or deviatoric stress tensor) is denoted by $\boldsymbol{\sigma}$. The surrounding fluid (assumed to be air) is assumed to be dynamically passive (i.e., inviscid and low density compared to the

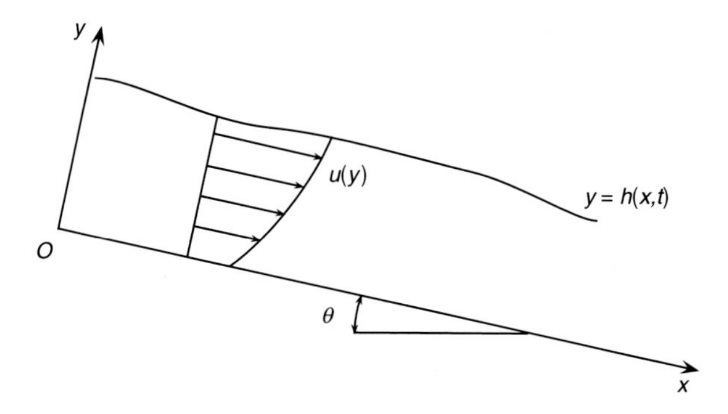


Figure 10.11. The configuration of the flow: $h(x,t)$ the flow depth, $u(y)$ the cross-stream velocity profile, and $\theta$ the bed inclination.

moving fluid) and surface tension is neglected, which implies that the stress state at the free surface is zero.

The governing equations are given by the mass and momentum balance equations

$$\nabla \cdot \mathbf{u} = 0, \tag{10.2}$$

$$\varrho \frac{\partial \mathbf{u}}{\partial t} + \varrho (\mathbf{u} \cdot \nabla)\mathbf{u} = \varrho \mathbf{g} - \nabla p + \nabla \cdot \boldsymbol{\sigma}, \tag{10.3}$$

supplemented by the following boundary conditions at the free and bottom surfaces

$$v(x,h,t) = \frac{dh}{dt} = \frac{\partial h}{\partial t} + u(x,h,t)\frac{\partial h}{\partial x}, \quad v(x,0,t) = 0\,. \tag{10.4}$$

There are many ways of transforming these governing equations into dimensionless expressions (Liu and Mei 1990a; Balmforth and Craster 1999; Keller 2003; Ancey and Cochard 2009). Here we depart slightly from the presentation given by Liu and Mei (1990a). The characteristic streamwise and vertical velocities, the timescale, the typical pressure, and the order of magnitude of bulk viscosity are referred to as $U_*$, $V_*$, $T_*$, $P_*$, and $\eta_*$, respectively. Moreover, in addition to the lengthscale ratio $\epsilon$, we introduce the following dimensionless numbers that characterize free-surface, gravity-driven flows: the flow Reynolds number and the Froude number

$$\mathrm{Re} = \frac{\varrho U_* H_*}{\eta_*} \text{ and } \mathrm{Fr} = \frac{U_*}{\sqrt{g H_* \cos\theta}}.$$

The following dimensionless variables will be used in this section:

$$\hat{u} = \frac{u}{U_*},\ \hat{v} = \frac{v}{V_*},\ \hat{x} = \frac{x}{L_*},\ \hat{y} = \frac{y}{H_*}, \text{ and } \hat{t} = \frac{t}{T_*}.$$

A natural choice for $T_*$ is $T_* = L_*/U_*$. The stresses are scaled as follows:

$$\hat{\sigma}_{xx} = \frac{\eta_* U_*}{L_*}\sigma_{xx},\ \hat{\sigma}_{xy} = \frac{\eta_* U_*}{H_*}\sigma_{xy},\ \hat{\sigma}_{yy} = \frac{\eta_* U_*}{L_*}\sigma_{yy}, \text{ and } \hat{p} = \frac{p}{P_*},$$

where $\sigma_{xx}$, $\sigma_{xy}$, and $\sigma_{yy}$ are the normal stress in the $x$-direction, the shear stress, and the normal stress in the $y$-direction, respectively. Here we are interested in free-surface flows. This leads us to set $P_* = \varrho g H_* \cos\theta$ since we expect that, to leading order, the pressure adopts a hydrostatic distribution (see later). If we define the vertical velocity scale as $V_* = \epsilon U_*$, the mass balance equation (10.2) takes the following dimensionless form

$$\frac{\partial \hat{u}}{\partial \hat{x}} + \frac{\partial \hat{v}}{\partial \hat{y}} = 0. \tag{10.5}$$

Substituting the dimensionless variables into the momentum balance equation (10.3) leads to

$$\epsilon\, \mathrm{Re}\, \frac{d\hat{u}}{d\hat{t}} = \frac{\epsilon\, \mathrm{Re}}{\mathrm{Fr}^2}\left(\frac{1}{\epsilon}\tan\theta - \frac{\partial \hat{p}}{\partial \hat{x}}\right) + \epsilon^2 \frac{\partial \hat{\sigma}_{xx}}{\partial \hat{x}} + \frac{\partial \hat{\sigma}_{xy}}{\partial \hat{y}}, \tag{10.6}$$

$$\epsilon^3 \,\mathrm{Re}\frac{d\hat{v}}{d\hat{t}} = \frac{\epsilon\,\mathrm{Re}}{\mathrm{Fr}^2}\left(-1 - \frac{\partial \hat{p}}{\partial \hat{y}}\right) + \epsilon^2 \frac{\partial \hat{\sigma}_{xy}}{\partial \hat{x}} + \epsilon^2 \frac{\partial \hat{\sigma}_{yy}}{\partial \hat{y}}. \tag{10.7}$$

The momentum balance equation expresses a balance between gravity acceleration, inertial terms, pressure gradient, and viscous dissipation, whose order of magnitude is $\varrho g \sin\theta$, $\varrho U_*^2/L_*$, $P_*/L_*$, and $\eta_* U_*/H_*^2$, respectively. Depending on the values considered for the characteristic scales, different types of flow regime occur. At least four regimes, where two contributions prevail, could be achieved in principle:

1. *Inertial regime*, where inertial and pressure-gradient terms are of the same magnitude. We obtain
$$U_* = \sqrt{g H_* \cos\theta}.$$
The order of magnitude of the shear stress is $\partial \sigma_{xy}/\partial y = \varrho g\, O(\epsilon^{-1}\,\mathrm{Re}^{-1})$. This regime occurs when $\epsilon\,\mathrm{Re} \gg 1$ and $\mathrm{Fr} = O(1)$.
2. *Diffusive regime*, where the pressure gradient is balanced by viscous stresses within the bulk. In that case, we have
$$U_* = \frac{\varrho g \cos\theta H_*^3}{\eta_* L_*}.$$
Inertial terms must be low compared to the pressure gradient and the slope must be shallow ($\tan\theta \ll \epsilon$). This imposes the following constraint: $\epsilon\,\mathrm{Re} \ll 1$. We deduced that $\mathrm{Fr}^2 = O(\epsilon\,\mathrm{Re}) \ll 1$.
3. *Visco-inertial regime*, where inertial and viscous contributions are nearly equal. In that case, we have
$$U_* = \frac{1}{\epsilon}\frac{\eta_*}{\varrho H_*}.$$
The pressure gradient must be low compared to the viscous stress, which entails the following condition: $\eta_* \gg \epsilon\varrho\sqrt{g H_*^3}$. We obtain $\epsilon\,\mathrm{Re} \sim 1$ and $\mathrm{Fr} = \eta_*/(\varrho\epsilon\sqrt{g H_*^3}) \gg 1$.
4. *Nearly steady uniform regime*, where the viscous contribution matches gravity acceleration. In that case, we have
$$U_* = \frac{\varrho g \sin\theta H_*^2}{\eta_*}.$$
Inertia must be negligible, which means $\epsilon \ll 1$ (stretched flows). We obtain $\mathrm{Re} = O(\mathrm{Fr}^2)$ and $\tan\theta \gg \epsilon$ (mild slopes).

In the *inertia-dominated regime*, the rheological effects are so low that they can be neglected and the final governing equations are the Euler equations; this approximation can be used to describe high-speed flows such as powder-snow avalanches in the flowing regime (Ancey et al. 2007). The *visco-inertial regime* is more spurious and has no specific interest in geophysics, notably because the flows are rapidly unstable. More interesting is the *diffusive regime* that may be achieved for very slow flows on gentle slopes ($\theta \ll 1$), typically when flows come to rest, or within the head (Liu and Mei 1990b; Balmforth et al. 2002; Ancey and Cochard 2009; Ancey et al. 2009). We will further describe this regime in Section 10.5.3. The *nearly-steady regime* will be exemplified in Section 10.5.2 within the framework of the kinematic-wave approximation.

Note that the partitioning into four regimes holds for viscous (Newtonian) fluids and non-Newtonian materials for which the bulk viscosity does not vary significantly with shear rate over a sufficiently wide range of shear rates. In the converse case, further dimensionless groups (e.g., the Bingham number) must be introduced, which makes this classification more complicated.

### *10.4.2 Rheology*

In geophysical fluid mechanics, there have been many attempts to describe the rheological behavior of natural materials (Ancey 2007). However, since rheometric experiments are not easy (see later), scientists use proxy procedures to characterize the rheological behavior of natural materials. Interpreting the traces of past events (e.g., shape of deposits), running small-scale experiments with materials mimicking the behavior of natural materials, and making analogies with idealized materials are common approaches to this issue. Because of a lack of experimental validation, there are many points of contention within the different communities working on geophysical flows. A typical example is provided by the debate surrounding the most appropriate constitutive equation for describing sediment mixtures mobilized by debris flows (Iverson 2003a). A certain part of the debris flow community uses soil-mechanics concepts (Coulomb behavior), whereas another part prefers viscoplastic models. A third part of the community merges the different concepts from soil and fluid mechanics to provide general constitutive equations.

Over the last 20 years, a large number of experiments have been carried out to test the rheological properties of natural materials. The crux of the difficulty lies in the design of specific rheometers compatible with the relatively large size of particles involved in geophysical flows. Coaxial-cylinder (Couette) rheometers and inclined flumes are the most popular geometries. Another source of trouble stems from additional effects such as particle migration and segregation, flow heterogeneities, fracture, layering, etc. These effects are often very pronounced with natural materials, which may explain the poor reproducibility of rheometric investigations (Major and Pierson 1992; Contreras and Davies 2000; Iverson 2003b). Poor reproducibility, complexity in the material response, and data scattering have at times been interpreted as the failure of the one-phase approximation for describing rheological properties (Iverson, 2003b). In fact, these experimental problems demonstrate above all that the bulk behavior of natural material is characterized by fluctuations that can be as wide as the mean values. As for turbulence and Brownian motion, we should describe not only the mean behavior, but also the fluctuating behavior to properly characterize the rheological properties. For concentrated colloidal or granular materials (Lootens et al. 2003; Tsai et al. 2001), experiments on well-controlled materials have provided evidence

Figure 10.12. Different types of snow observed in avalanche deposits. (a) Block of wet snow (size: 1 m). (b) Slurry of dry snow including weak snowballs formed during the course of the avalanche (the heap height was approximately 2 m). (c) Ice balls involved in a huge avalanche coming from the north face of the Mont Blanc (France); the typical diameter was 10 cm. (d) Sintered snow forming broken slabs (typical length: 40 cm, typical thickness 10 cm). For a color version of this figure please see the color plate section.

that to some extent; these fluctuations originate from jamming in the particle network (creation of force vaults sustaining normal stress and resisting against shear stress, both of which suddenly relax). Other processes such as ordering, aging (changes in the rheological behavior over time as a result of irreversible processes), and chemical alteration occur in natural slurries, which may explain their time-dependent properties (Marquez et al., 2006). Finally, there are perturbing effects (e.g., slipping along the smooth surfaces of a rheometer) that may bias measurement.

Snow is a very special material. To illustrate the diversity of materials involved in snow avalanches, Figure 10.12 reports different types of snow observed in avalanche deposits. Experiments have been done in the laboratory to characterize snow's rheological behavior. Authors such as Dent and Lang (1982) and Maeno (1993) have measured the velocity profile within snow flows and generally deduced that snow generates a non-Newtonian viscoplastic flow whose properties depend a great deal on density. Carrying these laboratory results over to real avalanches is not clearly reliable because of size-scale effects and similarity conditions. Furthermore, given the severe difficulties inherent in snow rheometry (sample fracture during shearing tests, variation in the snow microstructure resulting from thermodynamic transformations of crystals, etc.), properly identifying the constitutive equation of snow with modern rheometers is out of reach for the moment. More recently, Ancey and Meunier (2004) showed how avalanche-velocity records can be used to determine the bulk frictional force; a striking result is that the bulk behavior of most snow avalanches can be approximated using a Coulomb frictional model. Kern (2004; 2009) ran outdoor experiments to measure shear-rate profiles inside snow flows to infer rheological properties; this preliminary experiment is rather encouraging and clears the way for precise rheometrical investigations of real snow avalanches.

Since little sound field or laboratory data are available on the basic rheological processes involved in avalanche release and flow, all avalanche-dynamics models proposed so far rely on analogy with other physical phenomena: typical examples include analogies with granular flows (Savage and Hutter 1989; Savage 1989; Tai et al. 2001; Cui et al. 2007), Newtonian fluids (Hunt 1994), power-law fluids (Norem et al. 1986), and viscoplastic flows (Dent and Lang 1982; Ancey 2007). From a rheological point of view, these models rely on a purely speculative foundation. Indeed, most of the time, the rheological parameters used in these models have been estimated by matching the model predictions (such as the leading-edge velocity and the runout distance) with field data (Buser and Frutiger 1980; Dent and Lang 1980; Ancey et al. 2004). However, this procedure obviously does not provide evidence that the constitutive equation is appropriate.

For debris flows, natural suspensions are made up of a great diversity of grains and fluids. This observation motivates fundamental questions: How do we distinguish between the solid and fluid phases? What is the effect of colloidal particles in a suspension composed of coarse and fine particles? When the particle size distribution is bimodal (i.e, we can distinguish between fine and coarse particles), the fine fraction and the interstitial fluid form a viscoplastic fluid embedding the coarse particles, as suggested by Sengun and Probstein (1989); this leads to a wide range of viscoplastic constitutive equations, the most common being the Herschel-Bulkley model, described later. The bimodal-suspension approximation usually breaks for poorly sorted slurries. In that case, following Iverson and his co-workers (Iverson 1997; 2005), Coulomb plasticity can help understand the complex, time-dependent rheological behavior of slurries.

When the bulk is made up of fine colloidal particles, phenomenological laws are used to describe rheological behavior. One of the most popular is the Herschel-Bulkley model, which generalizes the Bingham law

$$\tau = \tau_c + K\dot{\gamma}^n, \tag{10.8}$$

with $\tau_c$ the yield stress and $K$ and $n$ two constitutive parameters; the linear case ($n = 1$) is referred to as the Bingham law. In practice, this phenomenological expression successfully describes the rheological behavior of many materials over a sufficiently wide range of shear rates, except at very low shear rates.

When the bulk is made up of coarse noncolloidal particles, Coulomb friction at the particle level imparts its key properties to the bulk, which explains (i) the linear relationship between the shear stress $\tau$ and the effective normal stress $\sigma' = \sigma - p$ (with $p$ the interstitial pore pressure, $\sigma$ the stress normal to the plane of shearing)

$$\tau = \sigma' \tan\varphi, \tag{10.9}$$

and (ii) the nondependence of the shear stress on the shear rate $\dot{\gamma}$. Some authors have suggested that in high-velocity flows, particles undergo collisions, which gives rise to a regime referred to as the frictional-collisional regime. The first proposition of bulk stress tensor seems to be attributable to Savage (1982), who split the shear stress into frictional and collisional contributions

$$\tau = \sigma \tan\varphi + \mu(T)\dot{\gamma}, \tag{10.10}$$

with $T$ the granular temperature (root mean square of grain velocity fluctuations). Elaborating on this model, Ancey and Evesque (2000) suggested that there is a coupling between frictional and collisional processes. Using heuristic arguments on energy balance, they concluded that the collisional viscosity should depend on the Coulomb number $\mathrm{Co} = \varrho_p a^2 \dot{\gamma}^2/\sigma$ (with $a$ the particle radius and $\varrho_p$ its density) to allow for this coupling in a simple way

$$\tau = \sigma \tan\varphi + \mu(\mathrm{Co})\dot{\gamma}. \tag{10.11}$$

Jop et al. (2005) proposed a slightly different version of this model, where both the bulk frictional and collisional contributions collapse into a single term, which is a function of the inertial number $I = \mathrm{Co}^{1/2}$ (i.e., a variant of Coulomb number)

$$\tau = \sigma\ \tan\varphi(I). \tag{10.12}$$

In contrast, Josserand et al. (2004) stated that the key variable in shear stress was the solid concentration $\phi$ rather than the Coulomb number

$$\tau = K(\phi)\sigma + \mu(\phi)\dot{\gamma}^2, \tag{10.13}$$

with $K$ a friction coefficient. Every model is successful in predicting experimental observations for some flow conditions, but to date, none is able to describe the frictional-collisional regime for a wide range of flow conditions and material properties.

### 10.4.3 Segregation and Particle Migration

Particle segregation refers to a sorting process that leads to separating a mixture containing free-flowing particles, the size distribution of which is sufficiently wide (at least a factor of 2 between the finest and coarsest grain sizes). It is an important feature of sheared granular flows, in which the coarsest particles rise to the top of the flow, while the finest percolate down to the bottom. By changing the local composition of the bulk, segregation has significant consequences on the behavior of granular avalanches made up of different-sized particles, for example, by increasing their runout distance (Legros 2002; Linares-Guerrero et al. 2007), forming bouldery fronts (Gray and Kokelaar 2010), and giving rise to segregation-mobility feedback effects (Gray and Ancey 2009). While the effects of particle segregation on bulk dynamics has been essentially investigated in the laboratory, there is substantial field evidence that segregation is a key mechanism in natural gravity flows such as snow avalanches (Bartelt and McArdell 2009) and debris flows (Iverson and Vallance 2001).

Among the numerous processes that cause segregation, kinetic sieving and squeeze expulsion are likely to be the most efficient in dense, dry, granular flows down sloping beds (Gray 2010): Velocity shear and dilatancy act together as a random fluctuating sieve that allows the finer particles to percolate to the bottom under the action of gravity, while squeezing larger particles upward. Figure 10.13 shows a typical experiment of particle segregation in a granular flow down a chute: Small beads were injected from above while large particles crept along the flume base. The small particles rapidly percolated to the bottom, whereas the large ones drifted to the top of the flow.

Segregation in dense granular flows has been investigated theoretically using different approaches, including information entropy theory, statistical mechanics, and binary-mixture theory (Gray 2010; Ottino and Khakhar 2000). For dense granular flows involving binary mixtures, the last theoretical approach is interesting in that it provides a relatively simple description of segregation-remixing in the form of a nonlinear advection diffusion equation for the concentration (Gray 2010):

$$\frac{\partial \phi}{\partial t} + \mathrm{div}(\phi \mathbf{u}) - \frac{\partial}{\partial z}\Big(q\phi(1-\phi)\Big) = \frac{\partial}{\partial z}\left(D\frac{\partial \phi}{\partial z}\right), \tag{10.14}$$

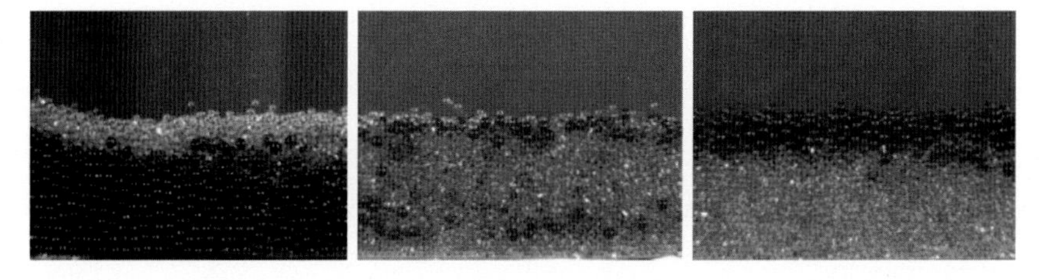

Figure 10.13. Snapshots showing particles segregating down a flume. Initially, when the particles enter the chute (image on the left), the mixture is normally graded, with all the small particles (1-mm-diameter glass beads, colored) on top of the coarse grains (2-mm-diameter glass beads in black). Segregation leads to a grading inversion, in which the smallest particles percolate to the bottom of the flow, while the largest rise toward the free surface (image in the middle). In the final state (in this experiment, approximately 1 m downward of the flume entrance), the particles separate out, with the large particles on top and small particles next to the bottom (image on the right).

where $\phi$ is the small particle concentration ($1-\phi$ is then the concentration in large particles), $\mathbf{u}$ is the bulk velocity field, $D$ is a coefficient of diffusion, and $q$ is the percolation rate (i.e., the rate at which small particles percolate to the base). The first term on the left-hand side of (10.14) is the time rate of change of the small particle concentration, and the second term is due to advection by the bulk velocity field $\mathbf{u}$. The third nonlinear term accounts for segregation, while the right-hand side introduces the diffusive effects of remixing. Mathematically, this equation is a second-order parabolic equation when $D > 0$. For $D = 0$, it reduces to a hyperbolic first-order partial differential equation and in this case, equation (10.14) may form shocks (i.e., waves across which the small particle concentration experiences a jump). When $D > 0$, diffusive remixing smears out the shock wave, replacing it by a smooth transition in the small-particle concentration, and equation (10.14) is then able to model experiments very realistically, but for practical purposes, neglecting the remixing (thus assuming $D = 0$) simplifies a great deal equation (10.14), which is then more amenable to analysis (Gray and Kokelaar 2010). This situation is reminiscent of water flows, in which sharp gradients in flow depth are replaced by discontinuities called *hydraulic jumps* (LeVeque 2002). Equation (10.14) does not depend on a particular form of governing equations for the bulk, and it is therefore compatible with most existing granular-flow models.

Particle migration refers to the diffusion of particles in sheared flows of particle suspensions, which produces inhomogeneous particle concentrations through the suspension (Stickel and Powell 2005). In a simple shear flow, the particles are driven toward the regions that are characterized by low shear rates, whereas regions dominated by high shear rates tend to become free of particles. For concentrated suspensions of particles in a viscous fluid, this diffusion process markedly affects the flow dynamics since the bulk viscosity $\mu$ depends on the particle concentration $\phi$, a dependence that is well captured by the empirical Krieger-Dougherty equation

$$\mu = \mu_0 \left(1 - \frac{\phi}{\phi_m}\right)^{-\frac{5}{2}\phi_m},$$

where $\phi_m$ is the maximum packing fraction and $\mu_0$ is the viscosity of the interstitial fluid (Chang and Powell 1994; Stickel and Powell 2005; Morris 2009). This change in the local rheological properties has profound impact on the bulk properties by giving rise to non-Newtonian properties such as normal-stress effects and apparent yield stress (Zarraga et al. 2000; Ovarlez et al. 2006). For suspension flows down sloping beds, particle migration causes the particles to rise to the free surface of the flow (Singh et al. 2006; Timberlake and Morrison 2005). The upper layers next to the free surface move faster and accumulate particles, which tend to be transported to the flow front, forming pasty flow fronts. This situation is reminiscent of the observations made by Parsons et al. (2001) (see Figure 10.4).

The time variations in the particle concentration are described by an advection diffusion equation

$$\frac{\partial \phi}{\partial t} + \langle \mathbf{u} \rangle \cdot \nabla \phi = -\nabla \cdot \mathbf{j}. \tag{10.15}$$

where $\mathbf{j} = \phi(\langle \mathbf{u} \rangle_p - \langle \mathbf{u} \rangle)$ denotes the particle flux (relative to the bulk flow), $\langle \mathbf{u} \rangle$ is the average bulk velocity field, and $\langle \mathbf{u} \rangle_p$ is the average particle velocity. As with other diffusion processes, particle migration has been modeled by relating the particle flux $\mathbf{j}$ to a driving force or potential. Since it was observed that particles migrate from regions of high to low shear rate, the first phenomenological law proposed by Leighton and Acrivos (1987) was to assume that $\mathbf{j} \propto -\dot{\gamma}$, with $\dot{\gamma}$ the shear rate. A more general formulation was then proposed by Morris and Boulay (1999), who showed using microstructural arguments that

$$\mathbf{j} \propto \nabla \cdot \langle \mathbf{\Sigma} \rangle_p,$$

where $\langle \mathbf{\Sigma} \rangle_p$ denotes the average particle stress tensor. This law performs fairly well for a number of flow configurations (Morris 2009).

### *10.4.4 Sliding-Block and Box Models*

The simplest model for computing the propagation speed of a gravity current proceeds by assuming that there is no downstream variation in flow properties (i.e., density, friction) within the flowing bulk. Several classes of models have been developed.

- *Sliding block model*: The flow is assumed to behave as a rigid block experiencing a frictional force. The early models date back to the beginning of the 20th century (Mougin 1922). Similar models have been developed for debris flows (Zimmermann et al. 1997). See Section 10.5.1.
- *Box model*: The model relaxes the rigidity assumption of the sliding block model by considering that the current behaves as a deformable rectangular box of length $\ell$ and height $h$ (Hogg et al. 2000; Ungarisch 2009). Mass conservation implies that the volume of this rectangle is known. For inertia-dominated flows, the Froude number at the leading edge is usually given by a boundary condition such as the von Kármán condition: $Fr = u_f/\sqrt{g'h} = constant$ (with $g'$ the reduced gravity acceleration, see (10.1), and $u_f$ the front velocity). Since box models have been developed for flows on horizontal surface, they are not well suited to studying flows on steep slope.
- Cloud model: The current is assumed to behave as a deformable body whose shape keeps the same aspect. The governing equations are given by the mass and momentum conservation equations for a mass-varying body (Kulikovskiy and Svehnikova, 1977). See Section 10.6.1.

For almost 80 years, simple models have been developed to provide crude estimations of avalanche features (velocity, pressure, runout distance). They are extensively used in engineering throughout the world. Despite their simplicity and approximate character, they can provide valuable results (Bozhinskiy and Losev 1998; Salm 2004; Ancey 2005).

### 10.4.5 Depth-Averaged Equations

The most common method for solving free-surface problems is to take the depth-average of the local equations of motion. In the literature, this method is referred to as the Saint-Venant approach since it was originally developed to compute floods in rivers.

We consider flows without entrainment of the surrounding fluid and without variation in density (see Section 10.6.2 for flow with entrainment). Accordingly, the bulk density may be merely replaced by its mean value. In this context, the equations of motion may be inferred in a way similar to the usual procedure used in hydraulics to derive the shallow water equations (or Saint–Venant equations). It involves integrating the momentum and mass balance equations over the depth. As such, a method has been extensively used in hydraulics for water flow (Chow 1959) as well for non-Newtonian fluids (Savage and Hutter 1991; Bouchut et al. 2003), we briefly recall the principle and then directly provide the resulting equations of motion. Let us consider the local mass balance: $\partial \varrho/\partial t + \nabla \cdot (\varrho \mathbf{u}) = 0$. Integrating this equation over the flow depth leads to

$$\int_0^{h(x,t)} \left(\frac{\partial u}{\partial x} + \frac{\partial v}{\partial y}\right) dy = \frac{\partial}{\partial x} \int_0^h u(x,y,t)dy - u(h)\frac{\partial h}{\partial x} + v(x,h,t) - v(x,0,t)\,, \tag{10.16}$$

where $u$ and $v$ denote the $x$- and $y$-component of the local velocity. At the free surface and at the bottom, the $y$-component of velocity satisfies the following boundary conditions:

$$v(x,h,t) = \frac{dh}{dt} = \frac{\partial h}{\partial t} + u(x,h,t)\frac{\partial h}{\partial x}\,, \tag{10.17}$$

$$v(x,0,t) = 0\,. \tag{10.18}$$

We easily deduce

$$\frac{\partial h}{\partial t} + \frac{\partial h\bar{u}}{\partial x} = 0\,, \tag{10.19}$$

where we have introduced depth-averaged values defined as

$$\bar{f}(x,t) = \frac{1}{h(x,t)} \int_0^{h(x,t)} f(x,y,t)dy\,. \tag{10.20}$$

The same procedure is applied to the momentum balance equation: $\varrho d\mathbf{u}/dt = \varrho\mathbf{g} + \nabla\cdot\boldsymbol{\sigma}$, where $\boldsymbol{\sigma}$ denotes the stress tensor. Without difficulty, we can deduce the averaged momentum equation from the $x$-component of the momentum equation:

$$\varrho\left(\frac{\partial h\bar{u}}{\partial t} + \frac{\partial h\overline{u^2}}{\partial x}\right) = \varrho g h \sin\theta + \frac{\partial h\bar{\sigma}_{xx}}{\partial x} - \tau_b\,, \tag{10.21}$$

where we have introduced the bottom shear stress: $\tau_b = \sigma_{xy}(x,0,t)$. In the present form, the motion equation system (10.19)–(10.21) is not closed since the number of variables exceeds the number of equations. A common approximation involves introducing a parameter (sometimes called the Boussinesq momentum coefficient), which links the mean velocity to the mean square velocity.

$$\overline{u^2} = \frac{1}{h}\int_0^h u^2(y)\,dy = \alpha\bar{u}^2\,. \tag{10.22}$$

Usually $\alpha$ is set to unity, but this may cause trouble when computing the head structure (Hogg and Pritchard 2004; Ancey et al. 2006; 2007). A point often neglected is that the shallow-flow approximation is in principle valid for flow regimes that are not too far away from a steady uniform regime. In flow parts where there are significant variations in the flow depth (e.g., at the leading edge and when the flow widens or narrows substantially), corrections should be made to the first-order approximation of stress. Recent studies, however, showed that errors made with the shallow-flow approximation for the leading edge are not significant (Ancey et al. 2007; Ancey and Cochard 2009; Ancey et al. 2009).

A considerable body of work has been published on this method for Newtonian and non-Newtonian fluids, including viscoplastic (Coussot 1997; Huang and Garcìa 1998; Siviglia and Cantelli 2005), power-law (Fernández-Nieto et al. 2010), and granular materials (Savage and Hutter 1989; Gray et al. 1998; Pouliquen and Forterre 2002; Iverson and Denlinger 2001; Bouchut et al. 2003; Chugunov et al. 2003; Pudasaini and Hutter 2003; Kerswell 2005). In this chapter, we will provide two applications for dense flows in Section 10.5.2: viscoplastic and friction Coulomb materials. Extension to dilute flows is outlined in Section 10.6.2. In Section 10.7.2, we will also show how the depth-averaged equations can be used to delineate the flow regimes and infer their main features.

### *10.4.6 Asymptotic Expansions*

On many occasions, flows are not in equilibrium, but deviate slightly from it. In this context, it is often convenient to use asymptotic expansions for the velocity field (Holmes 1995):

$$u(x,y,t) = u_0(x,y,t) + \epsilon u_1(x,y,t) + \epsilon^2 u_2(x,y,t) + \cdots,$$

where $\epsilon$ is a small number (e.g., the aspect ratio $\epsilon = H_*/L_*$ in equations (10.6)–(10.7)) and $u_i(x, y, t)$ are functions to be determined; usually, $u_0$ is the velocity field when the flow is at equilibrium and $u_i$ represents perturbations to this equilibrium state. Substituting $u$ by this expansion into the local governing equations such as equations (10.6)–(10.7) leads to a hierarchy of equations of increasing order. Most of the time, only the zero-order solution and the first-order correction are computed. Examples will be provided with the computation of an elongating viscoplastic flow in Section 10.5.3.

## 10.5 Dense Flows

We address the issue of dense flows, for which the effect of the surrounding air is neglected. We first illustrate the sliding block approach by outlining the Voellmy-Salm-Gubler model, which is one of the most popular models worldwide for computing the main features of extreme snow avalanches (Salm et al. 1990). In Section 10.5.2, we see two applications of the flow-depth averaged equations (frictional and viscoplastic fluids). We end this section with the use of asymptotic expansions to describe the motion of viscoplastic flows (see Section 10.5.3).

### *10.5.1 Simple Models*

The avalanche is assumed to behave as a rigid body that moves along an inclined plane. The position of the center of mass is given by its abscissa $x$ in the downward direction. The momentum equation is

$$\frac{du}{dt} = g \sin\theta - \frac{F}{m}, \tag{10.23}$$

with $m$ the avalanche mass, $u$ its velocity, $\theta$ the mean slope of the path, and $F$ the frictional force. In this model, the sliding block is subject to a frictional force combining a solid-friction component and a square-velocity component:

$$F = mg\frac{u^2}{\xi h} + \mu mg \cos\theta, \tag{10.24}$$

with $h$ the mean flow depth of the avalanche, $\mu$ a friction coefficient related to the snow fluidity, and $\xi$ a coefficient of dynamic friction related to path roughness. If these last two parameters cannot be measured directly, they can be adjusted from several series of past events. It is generally accepted that the friction coefficient $\mu$ depends only on the avalanche size and ranges from 0.4 (small avalanches) to 0.155 (very large avalanches) (Salm et al. 1990); in practice, lower values can be observed for large-volume avalanches (Ancey et al. 2004). Likewise, the dynamic parameter $\xi$ reflects the influence of the path on avalanche motion. When an avalanche runs down a wide-open rough slope, $\xi$ is close to 1,000 m s$^{-2}$. Conversely, for avalanches moving down confined straight gullies, $\xi$ can be taken as being equal to 400 m s$^{-2}$. In a steady state, the velocity is directly inferred from the momentum balance equation

$$u = \sqrt{\xi h \cos\theta\,(\tan\theta - \mu)}\,. \tag{10.25}$$

According to this equation, two flow regimes can occur depending on path inclination. For $\tan\theta > \mu$, (10.25) has a real solution and a steady regime can occur. For $\tan\theta < \mu$, there is no real solution; the frictional force (10.24) outweighs the downward component of the gravitational force. It is therefore considered that the flow slows. The point of the path for which $\tan\theta = \mu$ is called the characteristic point (point $P$). It plays an important role in avalanche dynamics since it separates flowing and runout phases. In the stopping area, we deduce from the momentum equation that the velocity decreases as follows:

$$\frac{1}{2}\frac{du^2}{dx} + u^2\frac{g}{\xi h} = g\cos\theta\,(\tan\theta - \mu)\,. \tag{10.26}$$

The runout distance is easily inferred from (10.26) by assuming that at a point $x = 0$, the avalanche velocity is $u_P$. In practice the origin point is point $P$ but attention must be paid to the fact that, according to (10.25), the velocity at point $P$ should be vanishing; a specific procedure has been developed to avoid this shortcoming (Salm et al. 1990). Neglecting the slope variations in the stopping zone, we find that the runout distance $x_s$ (point of farthest reach) counted from point $P$ is

$$x_s = \frac{\xi h}{2g}\ln\left(1 + \frac{u_P^2}{\xi h\cos\theta\,(\mu - \tan\theta)}\right). \tag{10.27}$$

This model enables us to easily compute the runout distance, the maximum velocities reached by the avalanche on various segments of the path, the flow depth (by assuming that the mass flow rate is constant and given by the initial flow rate just after the release), and the impact pressure.

### 10.5.2 Depth-Averaged Equations

The Saint-Venant equations consist of the following depth-averaged mass and momentum balance equations

$$\frac{\partial h}{\partial t} + \frac{\partial h\bar{u}}{\partial x} = 0, \tag{10.28}$$

$$\varrho\left(\frac{\partial h\bar{u}}{\partial t} + \frac{\partial h\overline{u^2}}{\partial x}\right) = \varrho g h\sin\theta - \frac{\partial h\bar{p}}{\partial x} + \frac{\partial h\bar{\sigma}_{xx}}{\partial x} - \tau_b, \tag{10.29}$$

where we have introduced the bottom shear stress $\tau_b = \sigma_{xy}(x,0,t)$ and we assume $\overline{u^2} = \bar{u}^2$; the flow-depth averaged pressure is found to be lithostatic

$$\bar{p} = \frac{1}{2}\varrho g h \cos\theta.$$

Within the framework of the long-wave approximation, we assume that longitudinal motion outweighs vertical motion; for any quantity $m$ related to motion, we have $\partial m/\partial y \gg \partial m/\partial x$. This allows us to consider that every vertical slice of flow can be treated as if it were locally uniform. In such conditions, it is possible to infer the bottom shear stress by extrapolating its steady-state value and expressing it as a function of $\bar{u}$ and $h$. For instance, for viscoplastic fluids, a common constitutive equation is the Herschel-Bulkley law (10.8). By relating the bottom shear rate to the flow-depth averaged velocity, Coussot (1997) showed that the bottom shear stress is a solution to the implicit equation

$$\frac{\bar{u}}{h} = \frac{n}{2n+1}\left(\frac{\tau_b}{K}\right)^{1/n}\left(1-\frac{\tau_c}{\tau_b}\right)^{1+1/n}\left(1+\frac{n}{n+1}\frac{\tau_c}{\tau_b}\right),$$

for Herschel-Bulkley fluids, with $n$, $K$, and $\tau_c$ the constitutive parameters introduced in equation (10.8). Note that this equation admits physical solutions provided that $\tau_b > \tau_c$. For $\tau_b \le \tau_c$, the material comes to a halt. For $n = 1/3$, Coussot (1997) provided the following approximation (accurate to within 5%),

$$\tau_b = \tau_c\left(1 + a\left(\frac{\tau_c}{K}\right)^{-9/10}\left(\frac{\bar{u}}{h}\right)^{3/10}\right),$$

with $a = 1.93$ for an infinitely wide plane and $a = 1.98$ for a semi-cylindrical flume.

For Coulomb materials, the same procedure can be repeated. The only modification concerns the momentum balance equation (10.29), which takes the form (Savage and Hutter, 1989; Iverson and Denlinger, 2001):

$$\varrho\left(\frac{\partial h\bar{u}}{\partial t} + \frac{\partial h\bar{u}^2}{\partial x}\right) = \varrho g h\left(\sin\theta - k\cos\theta\frac{\partial h}{\partial x}\right) - \tau_b, \tag{10.30}$$

with $k$ a proportionality coefficient between the normal stresses $\bar{\sigma}_{xx}$ and $\bar{\sigma}_{yy}$, which is computed by assuming a limited Coulomb equilibrium in compression ($\partial_x\bar{u} < 0$) or extension ($\partial_x\bar{u} > 0$); the coefficient is called the *active/passive pressure* coefficient. In equation (10.30), the bottom shear stress can be computed by using the Coulomb law $\tau_b = (\bar{\sigma}_{yy}|_{y=0} - p_b)\tan\varphi$, with $\bar{\sigma}_{yy}|_{y=0} = \bar{\varrho} g h\cos\theta$ and $p_b$ the pore pressure at the bed level.

Analytical solutions can be obtained for the Saint-Venant equations. Most of them were derived by seeking self-similarity solutions; see (Savage and Nohguchi 1988; Savage and Hutter 1989; Chugunov et al. 2003) for the Coulomb model and (Hogg and Pritchard, 2004) for viscoplastic and hydraulic models. Some solutions can also be obtained using the method of characteristics. We will present two applications based on these methods.

In the first application, we use the fact that the Saint-Venant equations for Coulomb materials are structurally similar to those used in hydraulics when the bottom drag can be neglected. The only difference lies in the nonhydrostatic pressure term and the source term (bottom shear stress). However, using a change in variable makes it possible to retrieve the usual shallow-water equations and seek similarity solutions to derive the Ritter solutions (Mangeney et al. 2000; Karelsky et al. 2000; Kerswell 2005). The Ritter solutions are the solutions to the so-called dam-break problem, where an infinite volume of material at rest is suddenly released and spreads over a dry bed (i.e., no material lying along the bed). Much attention has been paid to this problem, notably in geophysics, because it is used as a paradigm for studying rapid surge motion. We pose

$$x^* = x - \frac{\delta}{2}t^2,\ t^* = t,\ u^* = u - \delta t,\ \text{and } h^* = h,$$

where we introduced the parameter $\delta = g\cos\theta(\tan\theta - \mu)$. We deduce

$$\frac{\partial h^*}{\partial t^*} + \frac{\partial h^* u^*}{\partial x^*} = 0, \tag{10.31}$$

$$\frac{\partial u^*}{\partial t^*} + u^* \frac{\partial u^*}{\partial x^*} + gk\cos\theta \frac{\partial h^*}{\partial x^*} = 0. \tag{10.32}$$

For the dam-break problem, the initial and boundary conditions are

$$\begin{aligned} -\infty < x < \infty,\ u(x,0) &= 0, \\ x < 0,\ h(x,0) &= h_i, \\ x > 0,\ h(x,0) &= 0. \end{aligned} \tag{10.33}$$

The analytical solutions to equations (10.31)–(10.32) are the well-known Ritter solutions. We are looking for a similarity solution in the form (Gratton and Vigo 1994)

$$\bar{u}^* = t^{*\beta/\alpha} U(\zeta^*) \text{ and } h^* = t^{*\gamma/\alpha} H(\zeta^*),$$

with $\zeta^* = x^*/t^{*\alpha}$ the similarity variable, and $H$ and $U$ two unknown functions. Substituting $\bar{u}^*$ and $h^*$ with their similarity forms into (10.31)–(10.32), we find: $\beta + \alpha = 1$ and $\gamma + 2\alpha = 2$. For this solution to satisfy the initial and boundary conditions, we must pose $\beta = \gamma = 0$; hence, $\alpha = 1$. We then infer

$$\begin{pmatrix} H & U - \zeta^* \\ U - \zeta^* & kg\cos\theta \end{pmatrix} \cdot \begin{pmatrix} U' \\ H' \end{pmatrix} = 0,$$

where the prime denotes the $\zeta^*$-derivative. For this system to admit a nonconstant solution, its determinant must vanish, which leads to $kgH\cos\theta = (U - \zeta^*)^2$. On substituting this relation into the preceding system, we deduce $U'=2\zeta^*/3$, thus, $U = 2(\zeta^* + c)/3$, where $c$ is a constant of integration, $H = 4(c - \frac{1}{2}\zeta^*)^2/(9kg\cos\theta)$. The constant $c$ is found using the boundary conditions and by assuming that the undisturbed flow slides at constant velocity $\delta t$: $c = \sqrt{kgh_i \cos\theta}$. Returning to the original variables, we find

$$\bar{u}(x, t) = \bar{u}^* + \delta t = \frac{2}{3}\left(\frac{x}{t} + \delta t + c\right), \tag{10.34}$$

$$h(x, t) = \frac{1}{9kg\cos\theta}\left(-\frac{x}{t} + \frac{\delta}{2}t + 2c\right)^2. \tag{10.35}$$

The boundary conditions also imply that the solution is valid over the $\zeta$-range $[-c - \delta t,\ 2c + \delta t/2]$; the lower bound corresponds to the upstream condition $\bar{u} = 0$, while the upper bound is given by the downstream condition $h = 0$. It is worth noting that the front velocity $u_f = 2c + \delta t/2$ is constantly increasing or decreasing depending on the sign of $\delta$. When $\delta < 0$ (friction in excess of slope angle), the front velocity vanishes at $t = 4c/|\delta|$. Figure 10.14 shows that the shape of the tip region is parabolic at short times ($\delta t \ll c$), in agreement with experimental data (Balmforth and Kerswell 2005; Siavoshi and Kudrolli 2005). Solutions corresponding to finite released volumes were also obtained by Ancey et al. (2008), Hogg (2006), and Savage and Nohguchi (1988); and Savage and Hutter (1989).

In the second application, we use the method of characteristics to find a solution to the governing equations for Bingham flows that are stretched thin layers when they are nearly steady uniform. In a steady uniform regime, the velocity field can be obtained by using the Bingham law (10.8) and equating it to the shear stress distribution:

$$\tau = \varrho g(h - y)\sin\theta = \tau_c + K\frac{du}{dy}.$$

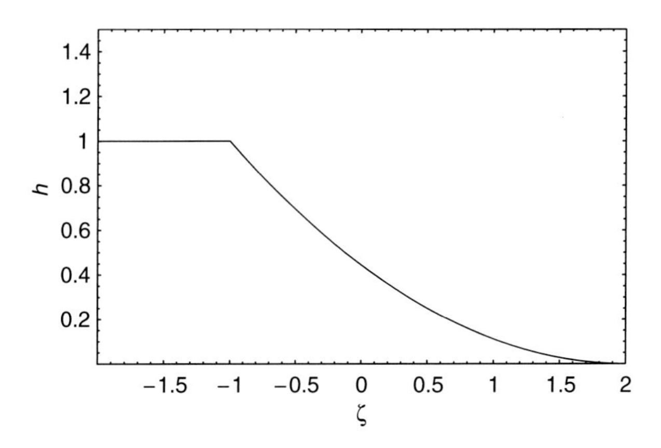


Figure 10.14. Flow-depth profile generated just after the wall retaining a granular material is removed. Computations made with $c = 1$ m/s. The similarity variable $\zeta$ is $\zeta = x/t$.

On integrating twice, we obtain the depth-averaged velocity

$$\bar{u}_s = u_p \left(1 - \frac{h_0}{3h}\right), \tag{10.36}$$

where $u_p$ is the plug velocity

$$u_p = \frac{\varrho g h_0^2 \sin\theta}{2K},$$

with $h$ the flow depth and $h_0 = h - \tau_c/(\varrho g \sin\theta)$ the yield-surface elevation; $h_0$ must be positive, or no steady flow occurs. For mild slopes, when the aspect ratio $\epsilon$ is very low, the inertial and pressure contributions can be neglected (see dimensional analysis in Section 10.4.1). This means that the depth-averaged velocity is very close to the mean velocity (10.36) reached for steady uniform flows. We then use the kinematic-wave approximation introduced by Lighthill and Whitham (1955) to study floods on long rivers; this approximation has been then extensively used in hydraulic applications (Hunt 1994; Huang and Garcìa 1997, 1998). It involves substituting the steady-state value $\bar{u}_s$ for the mean velocity into the mass balance equation (10.28)

$$\frac{\partial h}{\partial t} + \frac{\partial}{\partial x} u_p \left(h - \frac{h_0}{3}\right) = 0. \tag{10.37}$$

Introducing the plug thickness $h_p = h - h_0 = \tau_c/(\varrho g \sin\theta)$, we obtain an expression that is a function of $h$ and its time and space derivative

$$\frac{\partial h}{\partial t} + G\left(h^2 - h h_p\right)\frac{\partial h}{\partial x} = 0,$$

with $G = \varrho g \sin\theta / K$. The governing equation takes the form of a nonlinear advection equation, which can be solved using the method of characteristics (LeVeque 2002).

Using the chain rule for interpreting this partial differential equation (10.37), we can show that it is equivalent to the following ordinary equation

$$\frac{dh}{dt} = 0, \tag{10.38}$$

along the characteristic curve

$$\frac{dx}{dt} = \lambda(h), \tag{10.39}$$

in the $(x,\ t)$ plane, with $\lambda(h) = Gh\left(h - h_p\right)$. Equation (10.38) shows that the flow depth is constant along the characteristic curve; hence, the characteristic curves are straight lines, the slope of which are given by the right-hand side term $\lambda(h)$ in equation (10.39). These characteristic curves can be used to solve an initial value problem, where the initial value of $h$ is known over a given interval: $h = h_i(x_i)$ (at

$t = 0$). The value of $h$ along each characteristic curve is the value of $h$ at the initial point $x(0) = x_i$. We can thus write

$$h(x,\ t) = h_i(x_i) = h_i(x - \lambda(h_i(x_i))t).$$

It is worth noting that because of the nonlinearity of equation (10.37), a smooth initial condition can generate a discontinuous solution (shock) if the characteristic curves intersect, since at the point of intersection $h$ takes (at least) two values (LeVeque 2002).

### *10.5.3 Elongating Viscoplastic Flows*

Slow motion of a viscoplastic material has been investigated by Liu and Mei (1990a; 1990b), Mei et al. (2001), Coussot et al. (1996), Balmforth and Craster (1999); Balmforth et al. (2002), Matson and Hogg (2007), Ancey and Cochard (2009), and Hogg and Matson (2009).

Here we consider that the shear stress is given by (10.8) with $n = 1$. Taking the two dominant contributions in equations (10.6)–(10.7), integrating, and returning to the physical variables, we deduce

$$\tau = \sigma_{xy} = \varrho g \cos\theta (h - y)\left(\tan\theta - \frac{\partial h}{\partial x}\right), \tag{10.40}$$

$$p = \varrho g(h - y)\cos\theta. \tag{10.41}$$

The bottom shear stress is then found to be $\tau_b = \sigma_{xy}|_{y=0}$. For bottom shear stresses in excess of the yield stress $\tau_c$, flow is possible. When this condition is satisfied, there is a yield surface at depth $y = h_0$ within the bulk, along which the shear stress matches the yield stress

$$\tau|_{y=h_0} = \varrho g \cos\theta (h - h_0)\left(\tan\theta - \frac{\partial h}{\partial x}\right) = \tau_c. \tag{10.42}$$

The yield surface separates the flow into two layers (Liu and Mei 1990a; Balmforth and Craster 1999): the bottom layer, which is sheared, and the upper layer or plug layer, where the shear rate is nearly zero. Indeed, using an asymptotic analysis, Balmforth and Craster (1999) demonstrated that in the so-called plug layer, the shear rate is close to zero, but nonzero. This result may seem anecdotal, but it is in fact of great importance since it resolves a number of paradoxes raised about viscoplastic solutions (these paradoxes refer to the existence or nonexistence of true unyielded plug regions as described, for instance, by Piau (1996)).

On integrating the shear-stress distribution, we can derive a governing equation for the flow depth $h(x,\ t)$. For this purpose, we must specify the constitutive equation. For the sake of simplicity, we consider a Bingham fluid in one-dimensional flows as

Liu and Mei (1990a) did; the extension to Herschel-Bulkley and/or two-dimensional flows can be found in (Balmforth and Craster, 1999; Balmforth et al., 2002; Mei and Yuhi, 2001; Ancey and Cochard, 2009). In the sheared zone, the velocity profile is parabolic

$$u(y) = \frac{\varrho g \cos\theta}{K}\left(\tan\theta - \frac{\partial h}{\partial x}\right)\left(h_0 y - \frac{1}{2}y^2\right) \text{ for } y \leq h_0,$$

while the velocity is constant to leading order within the plug

$$u(y) = u_0 = \frac{\varrho g h_0^2 \cos\theta}{K}\left(\tan\theta - \frac{\partial h}{\partial x}\right) \text{ for } y \geq h_0,$$

The flow rate is then

$$q = \int_0^h u(y)dy = \frac{\varrho g h_0^2(3h - h_0)\cos\theta}{6K}\left(\tan\theta - \frac{\partial h}{\partial x}\right). \tag{10.43}$$

Integrating the mass balance equation over the flow depth provides

$$\frac{\partial h}{\partial t} + \frac{\partial q}{\partial x} = 0. \tag{10.44}$$

Substituting $q$ with its expression (10.43) and the yield surface elevation $h_0$ with equation (10.42) into equation (10.44), we obtain a governing equation for $h$, which takes the form of a nonlinear advection diffusion equation

$$\frac{\partial h}{\partial t} = \frac{\partial}{\partial x}\left[F(h,\, h_0)\left(\frac{\partial h}{\partial x} - \tan\theta\right)\right], \tag{10.45}$$

with $F = \varrho g h_0^2(3h - h_0)\cos\theta/(6K)$.

A typical application of this analysis is the derivation of the shape of a viscoplastic deposit. Contrary to a Newtonian fluid, the flow depth of a viscoplastic fluid cannot decrease indefinitely when the fluid spreads out along an infinite plane. Because of the finite yield stress, when it comes to rest, the fluid exhibits a nonuniform flow-depth profile, where the pressure gradient is exactly balanced by the yield stress. On an infinite horizontal plane, the bottom shear stress must equal the yield stress. Using equation (10.40) with $\theta = 0$ and $y = 0$, we eventually obtain (Liu and Mei 1990a)

$$\sigma_{xy}|_{y=0} = \tau_c = -\varrho g h \frac{\partial h}{\partial x}, \tag{10.46}$$

which, on integrating, provides

$$h(x) - h_i = \sqrt{\frac{2\tau_c}{\varrho g}(x_i - x)}, \tag{10.47}$$

where $h = h_i$ at $x = x_i$ is a boundary condition. This equation shows that the deposit-thickness profile depends on the square root of the distance. This is good agreement

Figure 10.15. Lobes of a debris-flow deposit near the Rif Paulin stream (Hautes-Alpes, France).

with field observations (Coussot et al. 1996); Figure 10.15 shows the lobe of a debris-flow deposit whose profile can be closely approximated by (10.47).

When the slope is nonzero, an implicit solution for $h(x)$ to equation (10.40) is found (Liu and Mei, 1990a)

$$\tan\theta(h(x)-h_i)+\frac{\tau_c}{\varrho g\cos\theta}\log\left[\frac{\tau_c-\varrho gh\sin\theta}{\tau_c-\varrho gh_i\sin\theta}\right]=\tan^2\theta(x-x_i). \qquad (10.48)$$

The shape of a static two-dimensional pile of viscoplastic fluid was investigated by Coussot et al. (1996), Mei and Yuhi (2001), Osmond and Griffiths (2001), and Balmforth et al. (2002). Balmforth et al. (2002) derived an exact solution, whereas Coussot et al. (1996) used numerical methods or ad hoc approximations to solve the two-dimensional equivalent to equation (10.40). Similarity solutions to equation (10.45) have also been provided by Balmforth et al. (2002) in the case of a viscoplastic flow down a gently inclined, unconfined surface with a time-varying source at the inlet. Ancey and Cochard (2009) used matched-asymptotic expansions to build approximate analytical solutions for the movement of a finite volume of Herschel-Bulkley fluid down a flume. Matson and Hogg (2007) and Hogg and Matson (2009) investigated the slumping motion of a fixed volume on a plane and down an inclined slope.

## 10.6 Dilute Inertia-Dominated Flows

### *10.6.1 Sliding Block Model*

The first-generation models of airborne avalanches used the analogy of density currents along inclined surfaces. Extending a model proposed by Ellison and Turner (1959) on the motion of an inclined plume, Hopfinger and Tochon-Danguy (1977) inferred the mean velocity of a steady current, assumed to represent the avalanche body behind the head. They found that the front velocity of the current was fairly independent of the bed slope. The second generation of models has considered the avalanche as a finite-volume, turbulent flow of a snow suspension. Kulikovskiy and Svehnikova (1977) set forth a fairly simple theoretical model (the KS model) in which the cloud was assimilated to a semi-elliptic body whose volume varied with time. The kinematics were entirely described by the mass center position and two geometric parameters of the cloud (the two semiaxes of the ellipsis). The cloud density can vary depending on air and snow entrainments. Kulikovskiy and Sveshnikova obtained a set of four equations describing the mass, volume, momentum, and Lagrangian kinetic energy balances. The idea was subsequently redeveloped by many authors, including Beghin et al. (1981), Beghin and Brugnot (1983), Fukushima and Parker (1990), Beghin and Olagne (1991), Fukushima et al. (2000), Ancey (2004), and Turnbull et al. (2007).

Here we outline the KSB model as presented and extended by Ancey (2004). We will consider the two-dimensional motion of a cloud along a plane inclined at an angle $\tan\theta$ with respect to the horizontal. Figure 10.16 depicts a typical cloud entraining particles from the bed. In the following, $H$ denotes the cloud height, $L$ its length, $m$ its mass, and $V$ its volume. The cloud velocity is $U = dx/dt$, but since the body is deformable, the velocity varies inside the body. The front position is given by the abscissa $x_f$ and its velocity is $U_f = dx_f/dt$. The volume solid concentration is $\phi$; it is assumed that the cloud is a homogeneous suspension of particles of density $\varrho_p$ (no density stratification) in the ambient fluid of density $\varrho_a$ and viscosity $\mu_a$. The bulk cloud density is then $\bar{\varrho} = \phi\varrho_p + (1-\phi)\varrho_a$. Ahead of the front, there is a particle bed made up of the same particles as the cloud and whose thickness is denoted by $h_n$. The apparent density of the layer is $\varrho_s = \phi_m\varrho_p + (1-\phi_m)\varrho_a$, where $\phi_m$ denotes the maximum random volume concentration of particles.

The surface area (per unit width) exposed to the surrounding fluid is denoted by $S$ and can be related to $H$ and $L$ as follows: $S = k_s\sqrt{HL}$, where $k_s$ is a shape factor. Here we assume that the cloud keeps a semielliptic form whose aspect ratio $k = H/L$ remains constant during the cloud run when the slope is constant. We then obtain

$$k_s = \mathrm{E}(1-4k^2)/\sqrt{k}, \tag{10.49}$$

where E denotes the elliptic integral function. Similarly, we can express the volume $V$ (per unit width) as $V = k_v HL$, where $k_v$ is another shape factor for a half ellipsis.

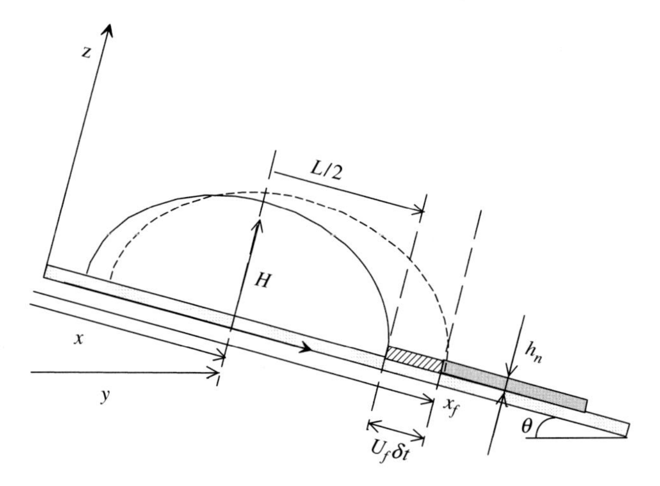


Figure 10.16. Sketch of the physical system studied here: The powder-snow avalanche is assumed to be a half ellipse whose volume grows with time. The major axis is the length $L(t)$, the semiminor axis is the flow depth $H(t)$. Since the body is deformable, the velocity varies with position: $U(t)$ refers to the velocity of the center of mass, while $U_f(t)$ is the velocity at the front $x = x_f(t)$. In the course of motion, the avalanche entrains snow from the snow cover; the thickness of the entrained snow layer is $h_n$.

Here we have

$$k_v = \pi/4. \tag{10.50}$$

In the following, we will also need to use the volume, height, and length growth rates

$$\alpha_v = \frac{1}{\sqrt{V}}\frac{dV}{dx},\ \alpha_h = \frac{dH}{dx},\ \alpha_l = \frac{dL}{dx}. \tag{10.51}$$

Experimentally, it is easier to measure the growth rates by deriving the quantity at hand by the front abscissa instead of by the mass center abscissa; we will refer to these rates as

$$\tilde{\alpha}_v = \frac{1}{\sqrt{V}}\frac{dV}{dx_f},\ \tilde{\alpha}_h = \frac{dH}{dx_f},\ \tilde{\alpha}_l = \frac{dL}{dx_f}. \tag{10.52}$$

Note that all these quantities are interrelated. For instance, using $x = x_f - L/2$, we find: $\tilde{\alpha}_h = (dH/dx)(dx/dx_f) = \alpha_h(1 - \tilde{\alpha}_l/2)$. Similarly, using the definition of $k$ and $k_v$, we obtain

$$\alpha_h = \frac{\alpha_v}{2}\sqrt{\frac{k}{k_v}} \text{ and } \alpha_l = \frac{\alpha_v}{2\sqrt{kk_v}}. \tag{10.53}$$

The KSB model outlined here includes three equations: volume, mass, and momentum balances. The volume variations mainly result from the entrainment of the ambient, less-dense fluid. To express the volume balance equation, the most common assumption is to state that the volume variations come from the entrainment of the ambient fluid into the cloud and that the inflow rate is proportional to the exposed surface area and a characteristic velocity $u_e$. This leads to the equation

$$\frac{dV}{dt} = E_v S u_e, \tag{10.54}$$

where $E_v$ is the bulk entrainment coefficient and is a function of the Richardson number (10.1). According to the flow conditions, different expressions of $E_v$ have been drawn from experiments. Interestingly enough, the value of $E_v$ has been expressed very differently depending on whether the current is steady or unsteady. There is, however, no clear physical reason that justifies this partitioning. Indeed, for most experiments, the currents were gradually accelerating, and mixing still occurred as a result of the development of Kelvin-Helmholtz billows, thus very similarly to the steady case. This observation prompted Ancey (2004) to propose a new expression of the entrainment coefficient for clouds, which holds for both steady and slightly unsteady conditions: Ancey (2004) related $E_v$ (or $\alpha_v$) as a function of $Ri$ (instead of $\theta$ as done by previous authors): for $Ri \leq 1$, $\alpha_v = e^{-1.6Ri^2}$, while for $Ri > 1$, $\alpha_v = 0.2/Ri$.

The cloud mass can vary as a result of the entrainment of the surrounding fluid and/or the entrainment of particles from the bed. The former process is easily accounted for. During a short time increment $\delta t$, the cloud volume $V$ is increased by a quantity $\delta V$ mainly as a result of the air entrainment, thus the corresponding increase in the cloud mass is $\varrho_a \delta V$. The latter process is less well known. Using an analogy with sediment erosion in rivers and turbidity currents, Fukushima and Parker (1990) assumed that particles are continuously entrained from the bed when the drag force exerted by the cloud on the bed exceeds a critical value. This implies that the particle entrainment rate is controlled by the surface of the bed in contact with the cloud and the mismatch between the drag force and the threshold of motion. Here, since in extreme conditions the upper layers of the snowcover made up of new snow of weak cohesion can be easily entrained, all the recent layer ahead of the cloud may be incorporated into the cloud. When the front has traveled a distance $U_f \delta t$, where $U_f$ is the front velocity, the top layer of depth $h_n$ and density $\varrho_s$ is entirely entrained into the cloud (see Figure 10.16). The resulting mass variation (per unit width) is written: $\varrho_s U_f h_n \delta t$. At the same time, particles settle with a velocity $v_s$. During the time step $\delta t$, all the particles contained in the volume $L v_s \delta t$ deposit. Finally, by taking the limit $\delta t \to 0$, we can express the mass balance equation as follows:

$$\frac{dm}{dt} = \varrho_a \frac{dV}{dt} + \varrho_s U_f h_n - \phi \varrho_s L v_s,$$

where $m = \bar{\varrho} V$ is the cloud mass. Usually the settling velocity $v_s$ is very low compared to the mean forward velocity of the front so that it is possible to ignore the third term on the right-hand side of the preceding equation. We then obtain the following simplified equation:

$$\frac{d\Delta\bar{\varrho}V}{dt} = \varrho_s U_f h_n. \tag{10.55}$$

The cloud undergoes the driving action of gravity and the resisting forces due to the ambient fluid and the bottom drag. The driving force per unit volume is $\bar{\varrho}g\sin\theta$. Most of the time, the bottom drag effect plays a minor role in the accelerating and steady-flow phases but becomes significant in the decelerating phase (Hogg and Woods 2001). Since we have set aside a number of additional effects (particle sedimentation, turbulent kinetic energy), it seems reasonable to also discard this frictional force. The action of the ambient fluid can be broken into two terms: a term analogous to a static pressure (Archimedes' theorem), equal to $\varrho_a Vg$, and a dynamic pressure. As a first approximation, the latter term can be evaluated by considering the ambient fluid as an inviscid fluid in an irrotational flow. On the basis of this approximation, it can be shown that the force exerted by the surrounding fluid on the half cylinder is $\varrho_a V\chi dU/dt$, where

$$\chi = k \tag{10.56}$$

is called the *added mass coefficient*. Since at the same time volume $V$ varies and the relative motion of the half cylinder is parallel to its axis of symmetry, we finally take $\varrho_a\chi d(UV)/dt$. Note that this parameter could be ignored for light interstitial fluids (e.g., air), whereas it has a significant influence for heavy interstitial fluids (basically, water). Thus, the momentum balance equation can be written as

$$\frac{d(\bar{\varrho}+\chi\varrho_a)VU}{dt} = \Delta\bar{\varrho}gV\sin\theta. \tag{10.57}$$

Analytical solutions can be found in the case of a Boussinesq flow ($\bar{\varrho}/\varrho_a \to 1$); for the other cases, numerical methods must be used. In the Boussinesq limit, since the final analytical solution is complicated, we provide only an asymptotic expression at early and late times. To simplify the analytical expressions without loss of generality, we take: $U_0 = 0$ and $x_0 = 0$ and assume that the erodible snowcover thickness $h_n$ and density $\varrho_s$ are constant. The other initial conditions are at $t = 0$ and $x = 0$, $H = H_0$, $L = L_0$, $V_0 = k_v H_0 L_0$, and $\bar{\varrho} = \bar{\varrho}_0$. At short times, the velocity is independent of the entrainment parameters and the initial conditions ($\bar{\varrho}_0$ and $V_0$):

$$U \propto \sqrt{2gx\sin\theta\frac{\Delta\varrho_0}{\Delta\varrho_0+(1+\chi)\varrho_a}} \approx \sqrt{2gx\sin\theta}, \tag{10.58}$$

where we used $\varrho_a \ll \Delta\bar{\varrho}_0$. This implies that the cloud accelerates vigorously in the first instants ($dU/dx\to\infty$ at $x=0$), and then its velocity grows more slowly. At long times for an infinite plane, the velocity reaches a constant asymptotic velocity that depends mainly on the entrainment conditions for flows in the air

$$U_\infty \propto \sqrt{\frac{2gh_n(1+\frac{\alpha_l}{2})\sin\theta\varrho_s}{\alpha_v^2(1+\chi)\varrho_a}}. \tag{10.59}$$

Because of the slow growth of the velocity, this asymptotic velocity is reached only at very long times. Without particle entrainment, the velocity reaches a maximum at approximately $x_m^2 = (2\varrho_0/3\varrho_a)\alpha_v^{-2}V_0/(1+\chi)$:

$$U_m^2 \approx \frac{4}{\sqrt{3}}\sqrt{\frac{\varrho_0}{\varrho_a}\frac{g\sqrt{V_0}\sin\theta}{\alpha_v\sqrt{1+\chi}}},$$

then it decreases asymptotically as

$$U \propto \sqrt{\frac{8\Delta\varrho_0}{3\varrho_a}\frac{gV_0\sin\theta}{x}\frac{1}{\alpha_v^2(1+\chi)}}. \tag{10.60}$$

In this case, the front position varies with time as

$$x_f \propto (g_0' V_0 \sin\theta)^{1/3} t^{2/3} \tag{10.61}$$

These simple calculations show the substantial influence of the particle entrainment on cloud dynamics. In the absence of particle entrainment from the bed, the fluid entrainment has a key role since it directly affects the value of the maximum velocity that a cloud can reach.

Here, we examine only the avalanche of 25 February 1999, for which the front velocity was recorded. In Figure 10.17, we have reported the variation in the mean front velocity $U_f$ as a function of the horizontal downstream distance $y_f$. The dots correspond to the measured data, and the curves represent the solution obtained by integrating equations (10.54)–(10.57) numerically and by assuming that the growth rate coefficient depends on the overall Richardson number (solid line). For the initial conditions, we assume that $u_0 = 0$, $h_0 = 2.1$ m $ł_0 = 20$ m, and $\varrho_0 = \varrho_s = 150$ kg/m$^3$. Because of the steep slope between the origin and the elevation $z = 1,800$ m ($y = 1,250$ m) we have considered that, on average, the released snow layer $h_n$ is 0.7 m thick and is entirely entrained into the avalanche. Using $\alpha_v \propto Ri^{-1}$ for $Ri \gg 1$, we apply the following relationship: for $Ri \leq 1$, $\alpha_v = e^{-1.6Ri^2}$, whereas for $Ri > 1$, we take $\alpha_v = 0.2/Ri$.

As shown in Figure 10.17, the avalanche accelerated vigorously after the release and reached velocities as high as 80 m/s. The velocity variation in the release phase is fairly well described by the KSB model. The model predicts a bell-shaped velocity variation, whereas field data provide a flatter velocity variation. The computed flow depth at $z = 1,640$ m is approximately 60 m, which is consistent with the value estimated from the videotapes. To evaluate the sensitivity of the simulation results, we examined different values of the erodible mass. In Figure 10.17, we have reported the comparison between

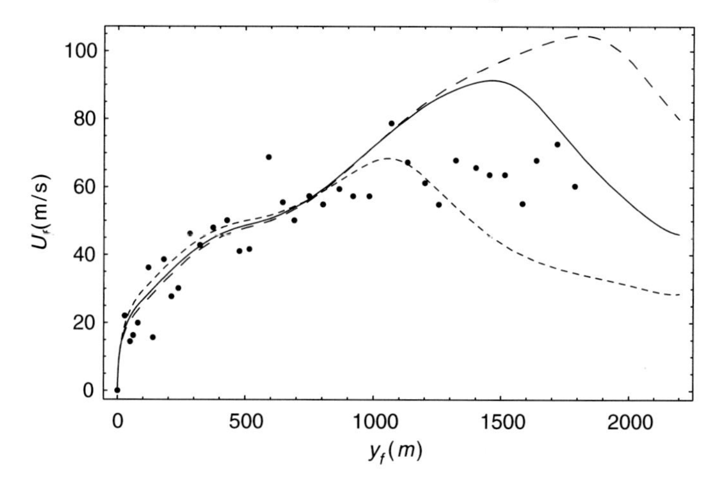


Figure 10.17. Dependence of the front velocity on the erodible mass. Solid line, $\varrho_s h_n = 105\,\text{kg/m}^2$; dashed line, $\varrho_s h_n = 50\,\text{kg/m}^2$; long-dashed line, $\varrho_s h_n = 150\,\text{kg/m}^2$. After Ancey (2004).

field data and computations made with three different assumptions: $\varrho_s h_n = 50$, 105, or $150\,\text{kg/m}^2$. It can be seen that there is no significant variation in the computed velocities in the accelerating phase, but both the maximum velocity and the position at which the maximum velocity is reached depend on the $\varrho_s h_n$ value. By increasing the erodible mass per unit surface from 50 to 150 $\text{kg/m}^2$, the maximum velocity is increased from 69 m/s to 105 m/s (i.e., by a factor of 1.5). Note that the dependence of the maximum velocity on the snowcover thickness is consistent with field measurements. For instance, the avalanche of 10 February 1999 was approximately half as large in terms of deposited volume as the avalanche of 25 February 1999, and its maximum velocity was 25% lower than the maximum velocity recorded on 25 February 1999. This result is of great importance in engineering applications since it means that the maximum velocity and therefore the destructive power of a powder-snow avalanche primarily result from its ability to entrain snow from the snowcover when it descends.

### *10.6.2 Depth-Averaged Equations*

An airborne avalanche is a very turbulent flow of a dilute ice–particle suspension in air. It can be considered as a one-phase flow as a first approximation. Indeed, the Stokes number, defined as the ratio of a characteristic time of the fluid to the relaxation time of the particles, is low, implying that particles adjust quickly to changes in the air motion. At the particle scale, fluid turbulence is high enough to strongly shake the mixture since the particle size is quite small. To take into account particle sedimentation, authors generally consider airborne avalanches as turbulent, stratified flows. Thus, contrary to flowing avalanches, bulk rheological behavior is well identified in the case of airborne avalanches. The main differences between the various models result from the different boundary conditions, use of the Boussinesq approximation, and the closure equations for turbulence. Parker et al. (1986) developed a complete depth-averaged model for turbidity currents. The equations of motion proposed by these authors are more complicated than the corresponding set for dense flows presented in Section 10.5.2, since they include additional equations arising from the mass balance for the dispersed phase, the mean and turbulent kinetic energy balances, and the boundary conditions related to the entrainment of sediment and surrounding fluid.

$$\frac{\partial h}{\partial t} + \frac{\partial hu}{\partial x} = E_a u \,, \tag{10.62}$$

$$\frac{\partial (Ch)}{\partial t} + \frac{\partial (hUC)}{\partial x} = v_s E_s - v_s c_b \,, \tag{10.63}$$

$$\frac{\partial hu}{\partial t} + \frac{\partial hu^2}{\partial x} = RCgh\sin\theta - \frac{1}{2}Rg\frac{\partial Ch^2}{\partial x} - u_*^2 \,, \tag{10.64}$$

$$\frac{\partial hK}{\partial t} + \frac{\partial huK}{\partial x} = \frac{1}{2}E_a u^3 + u_*^2 u - \varepsilon_0 h - \frac{1}{2}E_a uRCgh - \frac{1}{2}Rghv_s\left(2C + E_s - c_b\right), \tag{10.65}$$

where $u$ is the mean velocity, $h$ the flow depth, $K$ the mean turbulent kinetic energy, $C$ the mean volume concentration (ratio of particle volume to total volume), $E_a$ a coefficient of entrainment of surrounding fluid into the current, $v_s$ the settlement velocity, $E_s$ a coefficient of entrainment of particles from the bed into the current, $c_b$ the near-bed particle concentration, $R$ the specific submerged gravity of particles (ratio of buoyant density to ambient fluid density), $u_*^2$ the bed shear velocity, and $\varepsilon_0$ the depth-averaged mean rate of dissipation of turbulent energy due to viscosity. The main physical assumption in Parker *et al.*'s model is that the flow is considered as a one-phase flow in terms of momentum balance, but treated as a two-phase flow concerning the mass balance. Equation (10.62) states that the total volume variation results from entrainment of surrounding fluid. In (10.63), the variation in the mean solid concentration is due to the difference between the rate of particles entrained from the bed and the sedimentation rate. Equation (10.64) is the momentum balance equation The momentum variation results from the driving action of gravity and the resisting action of bottom shear stress; depending on the flow depth profile, the pressure gradient can contribute to either accelerating or decelerating the flow. Equation (10.65) takes into account the turbulence expenditure for the particles to stay in suspension. Turbulent energy is supplied by the boundary layers (at the flow interfaces with the surrounding fluid and the bottom). Turbulent energy is lost by viscous dissipation ($\varepsilon_0 h$ in (10.65)), mixing the flow (fourth and fifth terms in (10.65)), and maintaining the suspension against sedimentation flow mixing (last term on the right-hand side of (10.65)).

Although originally devoted to submarine turbidity currents, this model has been applied to airborne avalanches with only small modifications in the entrainment functions (Fukushima and Parker, 1990). A new generation of powder-snow avalanche models has recently appeared (Hutter, 1996). Some rely on the numerical resolution of local equations of motion, including a two-phase mixture approximation and closure equations, usually a $k - \epsilon$ model for turbulence (Hutter 1996). A number of researchers believe that a powder-snow avalanche is tightly related to a denser part that supplies the airborne part with snow; these researchers have thus tried to establish the relation existing between a dense core and an airborne avalanche (Eglit 1983; Nazarov 1991; Issler 1998). Though these recent developments are undoubtedly a promising approach to modeling powder-snow avalanches, their level of sophistication contrasts with the crudeness of their basic assumptions as regards the momentum exchanges between phases, turbulence modification due to the dispersed phase, and so on. At this level of our knowledge of physical and natural processes, it is still interesting to continue using simple models and fully exploring what they can describe and explain.

## 10.7 Comparison with Data

In this chapter, emphasis has been given to presenting the physical features of gravity flows on steep slope and outlining various fluid-mechanics approaches to computing their flow behavior. Since the ultimate goals are to predict how materials are mobilized on a steep slope, how the resulting gravity flow behaves, and how this flow eventually comes to a halt, it is of paramount importance to address the predictive capability of the mathematical models outlined from Section 10.4 to Section 10.6. The question of prediction, a central topic in physical and natural sciences has attracted considerable attention not only from scientists but also from philosophers and sociologists. In the field of geomorphic and geophysical models, contemporary modeling faces special challenges, some of them controversial, such as the relevance of model calibration, the problem of scale between laboratory models and natural events, the uncertainty on the initial and boundary conditions, the random nature versus deterministic behavior of processes, and the making of models (what science philosophers refer to as reductionism and constructionism) and their testing (Wilcock and Iverson 2003).

### *10.7.1 Comparison with Laboratory Data*

In any fluid-mechanics approach, a major impediment to natural flows is the limited availability of relevant data to test models. Ideally, since gravity flows on steep slope involve physical processes that are accessible to direct measurement and observation, we may think of monitoring natural sites and collecting field data to test the models. In some instances (e.g., for snow avalanches), it is even possible to trigger flows, which opens up the way to quantitative tests that are similar to those of classical physics. Yet, in practice, this road is paved with difficulties of many kinds, some resulting from taking measurements in natural conditions, others stemming from the very nature of the test, for which experimental control is barely possible. Iverson (2003b) best summarized this situation: "The traditional view in geosciences is that the best test of a model is provided by data collected in the field, where processes operate at full complexity, unfettered by artificial constraints. [...] If geomorphology is to make similarly rapid advances, a new paradigm may be required: mechanistic models of geomorphic processes should be tested principally with data collected during controlled, manipulative experiments, not with field data collected under uncontrolled conditions." In this respect, laboratory experiments have the overwhelming advantage of testing models in a well-controlled environment and for various flow conditions. The disadvantages are related to the scale and similarity issues (Iverson 1997). When working on small scales, it is difficult to guarantee that the experiments are in full similarity with the natural phenomena. (In similitude theory, this implies that all dimensionless numbers that characterize the composition of the materials and the flow dynamics take similar values.) Moreover, in laboratory experiments, natural materials are replaced by simple materials such as glass beads or sand, which may appear as an absurd simplification of natural materials.

As a typical example, we shall focus on the motion of an avalanche of "mud" on a sloping bed. In Section 10.4.2, we have seen that mudflows involve clay-rich materials whose rheological behavior can be considered viscoplastic as a first approximation. To reproduce these flows on the laboratory scale, we first have to find a good candidate that mimics the rheological behavior of clayey materials. For a long time, pure clays such as kaolin and bentonite were used as a substitute for natural clays, but their rheological behavior departs from an ideal viscoplastic fluid described by the Herschel-Bulkley model (10.8), in particular because they exhibit thixotropic behavior. Today polymeric gels such as Carbopol Ultrez 10 are routinely used as Herschel-Bulkley fluids. We conducted experiments with this material in which we released a fixed volume of material (initially at rest in a reservoir) down an inclined flume and tracked the flowing mass using cameras. An ingenious system combining a high-speed camera, and a pattern projector made it possible to reconstruct the free surface of the flow and determine the position of the front as a function of time (Cochard and Ancey 2008). Figure 10.18 shows an example of surface reconstruction in our flume.

Carbopol is a polymer that forms a viscoplastic gel when mixed with water. A volume concentration as low as 0.4% can produce viscoplastic fluids with a yield stress in excess of 100 Pa (i.e., a consistency close to that of cosmetic products like hair gels). To gain insight into the flow dynamics of viscoplastic fluids, we carried out the experiments with different geometries (inclined plane/flume), inclinations (from 0 deg to 24 deg), and initial volumes. Here, we provide only the time variations in the front position for two flume inclinations and two Carbopol concentrations

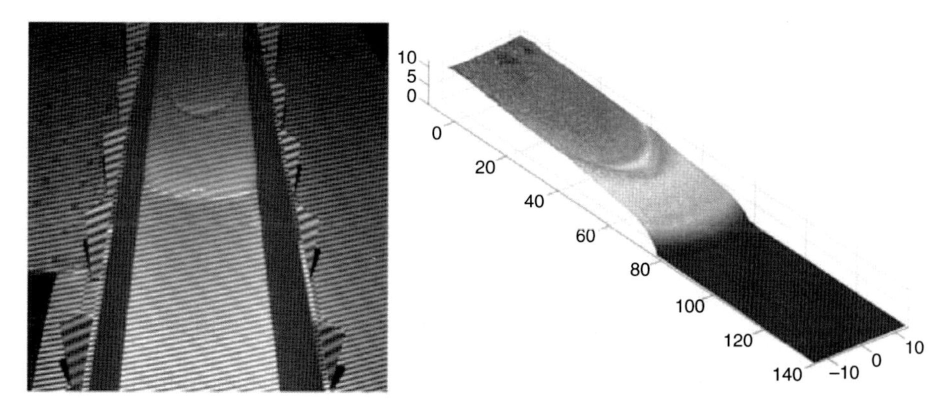


Figure 10.18. Reconstruction of the free surface using image processing. The photograph on the left shows the setup when patterns (here regularly spaced strips) are projected. The picture on the right shows the reconstructed free surface. Data from Cochard and Ancey (2008).

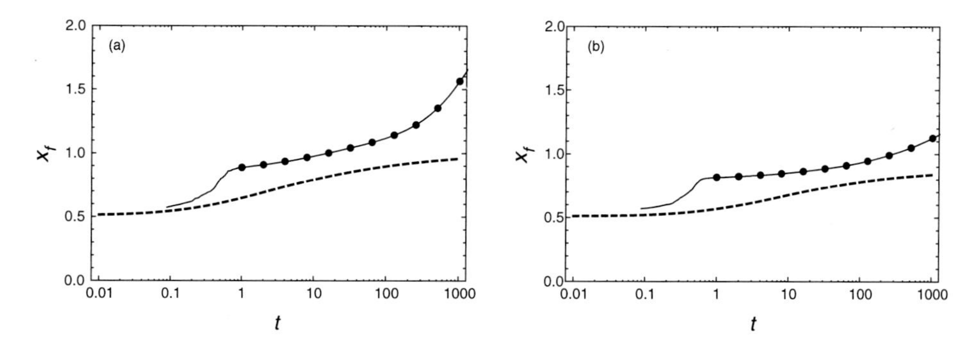


Figure 10.19. Position of the front $x_f(t)$ (in meters) as a function of time $t$ (in seconds) for a flume inclination $\theta = 6$ deg and for two Carbopol concentrations: (a) the concentration is 0.25%; (b) the concentration is 0.3%. The solid line with dots represents the experimental curve, and the dashed line is the analytical solution. Data from Ancey and Cochard (2009).

(0.25 and 0.3%): 6 deg (Figure 10.19) to 24 deg (Figure 10.20); see Ancey and Cochard (2009) and Cochard and Ancey (2009) for additonal data. The initial mass was 23 kg. The rheological properties were investigated using a coaxial-cylinder rheometer. A Herschel-Bulkley equation (10.8) was found to properly represent the rheological behavior, and the constitutive parameters were estimated from the rheometrical data. For a Carbopol concentration of 0.25%, $\tau_c = 78$ Pa, $n = 0.39$, and $K = 32.1$ Pa s$^{-n}$, while for a concentration of 0.3%, $\tau_c = 89$ Pa, $n = 0.42$, and $K = 47.7$ Pa s$^{-n}$.

Surprisingly enough, the numerical resolution to the full three-dimensional problem (10.2)–(10.3) is prohibitively complex and requires high-performance computing systems (Rentschler 2010). Time dependence, peculiarities due to rheological properties, along with the existence of a free surface and a contact line, are some of the

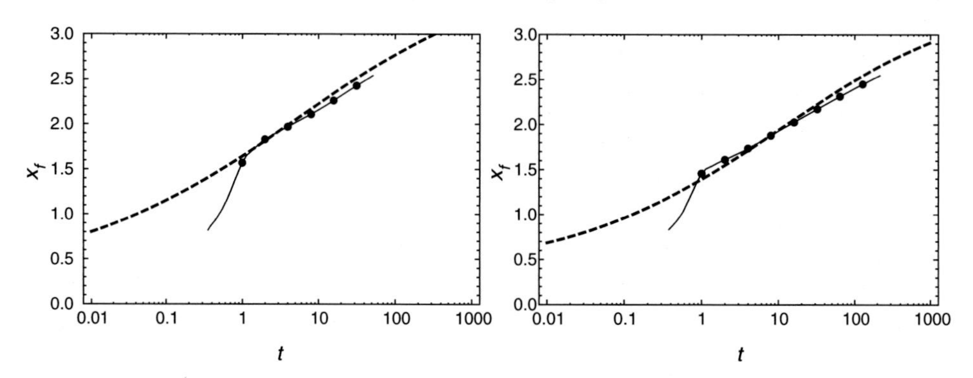


Figure 10.20. Position of the front $x_f(t)$ (in meters) as a function of time $t$ (in seconds) for a flume inclination $\theta = 24$ deg and for two Carbopol concentrations: (a) the concentration is 0.25%; (b) the concentration is 0.3%. The solid line with dots represents the experimental curve, and the dashed line is the analytical solution. Data from Ancey and Cochard (2009).

many complications that arise in the analysis. In spite of this complexity, it is possible to obtain approximate analytical solutions by simplifying the governing equations. In Section 10.5.3, we showed how the assumption of slow and shallow flows made it possible to simplify the governing equations and derive a nonlinear advection diffusion equation for the flow depth (10.45). In this form, this equation is not yet tractable, but using matched asymptotic expansions, we can obtain approximate solutions. For instance, at sufficiently steep slopes, the flow-depth gradient $\partial h/\partial x$ is small compared to the slope $\tan\theta$ in the body region (this approximation no longer holds within the tip region because $h$ drops to zero). As a first approximation (which will form the 0-order term in the asymptotic expansion), we can simplify (10.45) into a nonlinear advection equation:

$$\frac{\partial h}{\partial t} + \tan\theta \frac{\partial}{\partial x}\left[F(h,\, h_0)\frac{\partial h}{\partial x}\right] = 0, \tag{10.66}$$

with $F = \varrho g h_0^2(3h - h_0)\cos\theta/(6K)$. This equation can be solved analytically, for instance, using the methods of characteristics as shown in Ancey and Cochard (2009). Similar techniques can be used to derive approximations that hold for shallow slopes. In Figures 10.19 and 10.20, we report the theoretical curves together with the experimental data. Recall that the constitutive parameters have been obtained independently; thus, there is no curve adjustment in these plots. Whereas excellent agreement is found at steep slopes (see Figure 10.20), poor agreement is obtained at shallow slopes (see Figure 10.19). Experimental curves systematically exhibited convex shapes at sufficiently long times, whereas the theoretical curves were concave and tended toward an asymptotic value (corresponding to the arrested state). An explanation for this flow acceleration at shallow slope may lie in the formation of lateral levees. At the very beginning, after the material started flowing down the plane, the core of the flow was strongly sheared, whereas the fluid near the lateral rims was weakly sheared. Once the flow width reached a nearly constant value, the rims "froze" almost instantaneously and formed thick levees. At the same time, a pulse originating from the flow rear overtook the front and gave new impetus to the head. This produced the kink that can be seen in all $x_f(t)$ curves. Indeed, the flow rate remaining nearly constant over some period of time, flow narrowing caused by lateral levees led to swiftly increasing the mean velocity. If this scenario is correct, our two-dimensional analysis is too crude to capture the flow properties, notably the change in the front velocity induced by the levee formation. This explanation, however, remains speculative and calls for more work to elucidate this point.

These experiments illustrate the strengths and weaknesses of many theoretical approaches. Although theory can perform very well for some flow conditions, its predictive capability can be spoiled for other flow conditions unless clear reasons can be found. Many trials are usually needed to evaluate how well a theoretical model performs. Note also that seemingly simple problems such as the motion of a flowing mass of fluid may offer great resistance to analysis and that even today, with high-performance computers, these problems are difficult to solve numerically.

### *10.7.2 Comparison with Field Data*

When testing flow-dynamics models against field data, we face additional challenges. Most of the time, there is no way to measure the constitutive parameters independently (as we did earlier for the Carbopol avalanches), and these must be adjusted from field data, which may bias comparison. Moreover, many input parameters such as the initial volume, the composition of the material, and the volume of entrained/deposited materials are poorly known; some parameters (e.g., the snow density in avalanches) are known to vary a great deal in the course of the flow and setting them to a fixed value (as requested in most models) may make no sense. Finally, taking measurements in natural flows such as avalanches and debris flows, in particular measurements inside the flow, remains a difficult challenge and, as a consequence, we do not have many events fully documented.

Yet, in spite of these numerous impediments, it is possible to obtain relatively accurate descriptions of flow dynamics using simple models as long as one adjusts the model parameters to field data. We illustrate this with two examples. We use field measurements obtained by Gubler et al. (1986) on two snow avalanches in the Aulta and Fogas paths (near Davos, Switzerland). These avalanches exhibited two different behaviors, both of which deserve special mention. We first give the major features of these avalanches, then comment on their flow behavior.

The Aulta avalanche was a large, high-speed, dry-snow avalanche that involved 50,000 $m^3$ of snow. Part of the path is confined within a gully, and the initial flow thickness was in the range 0.7–1 m. The Fogas avalanche was a small, dry-snow

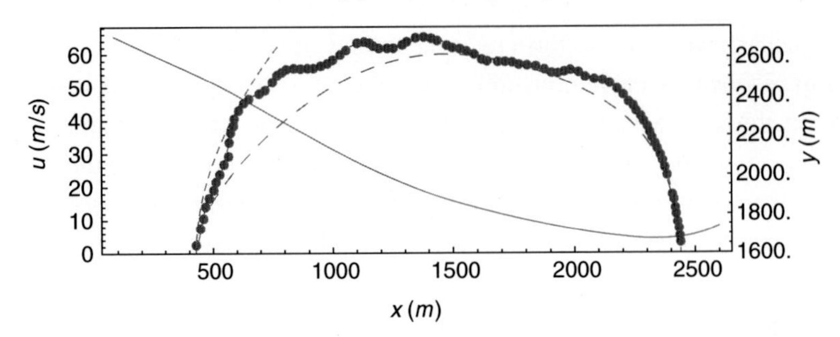


Figure 10.21. Path profile (solid line) of the Aulta site and front velocities (dots) for the 8 February 1984 avalanche. The long dashed curve on the left represents the velocity in a purely inertial regime, computed using equation (10.68), the dashed line illustrates the velocity variation when the Coulomb model is selected (with $\mu = 0.4$ adjusted on the runout distance). Data from Gubler et al. (1986) and Ancey and Meunier (2004).

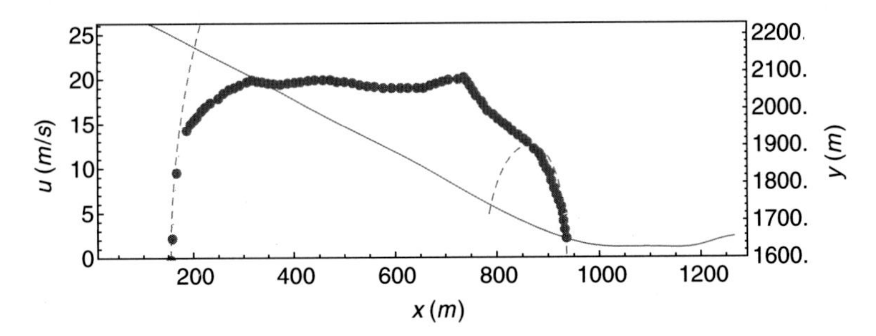


Figure 10.22. Path profile (solid line) of the Fogas site and front velocities (dots) for the 7 March 1985 avalanche. The long dashed curve on the left represents the velocity in a purely inertial regime, computed using equation (10.68), the dashed line on the right illustrates the velocity variation when the Coulomb model is selected (with $\mu = 0.8$ adjusted on the runout distance). Data from Gubler et al. (1986) and Ancey and Meunier (2004).

avalanche that involved 500 $m^3$ of snow. The path is an unconfined slope, with a fairly regular inclination close to 34 deg. The initial flow thickness was 30 cm. Gubler et al. (1986) measured the front velocity using a Doppler radar but did not provide the flow-depth variation. Figures 10.21 and 10.22 show the path profiles together with the variation in the front velocity for the Aulta and Fogas avalanches, respectively.

To model these avalanches, we use the flow-depth averaged equations (10.30) with a simplified Coulomb model as constitutive equation: $\tau_b = \mu \varrho g h \cos\theta$ and $k = 1$ (Ancey and Meunier 2004; Platzer et al. 2007). We then obtain

$$\frac{\partial \bar{u}}{\partial t} + \bar{u}\frac{\partial \bar{u}}{\partial x} = g\sin\theta - \frac{\tau_b}{\varrho h} + g\cos\theta\frac{\partial h}{\partial x}, \tag{10.67}$$

with $\theta$ the local path slope, $t$ time, and $x$ the curvilinear abscissa along the path. Note that for curvilinear paths, additional terms should have been added to the equation (Savage and Hutter 1989), but when the topography changes are slow, these terms are negligible. Just after the release, a large amount of material is suddenly entrained and accelerates vigorously. In the momentum balance equation (10.67), inertia and the pressure gradient must be of the same magnitude, whereas the bottom shear stress has negligible effects. In that case, equation (10.67) is formally identical to the momentum equation of the inviscid shallow-water equations and thus, the dam-break solution (for sloping beds) provides a fairly good approximation of the flow behavior at short times. We expect, most notably, a front-velocity variation in the form

$$u_f \sim gt\sin\theta + u_0 \approx u_0 + \sqrt{2g(x-x_0)\sin\theta}, \tag{10.68}$$

with $h_0$ the initial flow depth, $u_0 = 2\sqrt{gh_0\cos\theta}$ the initial velocity (dam-break approximation), $x_0$ the initial position of the front. This regime is usually referred to as the *inertial phase*. Indeed, it can be shown that once motion of the head begins, the boundary propagates downslope with an acceleration identical to that of a frictionless point mass moving along the slope (Ancey et al. 2008; Mangeney et al. 2010). This finding implies that the boundary speed is uninfluenced by the presence of adjacent fluid after motion commences.

At the end of its course, the flow experiences a *runout phase*, where all its energy is dissipated by friction. The flow behavior is then governed by a balance between the pressure gradient and shear stress (on shallow slopes), which leads to a significant drop in velocity over a fairly short distance. It is straightforward to show that the front velocity $u_f$ decelerates as

$$u_f \sim \sqrt{2g(x_s-x)\cos\theta(\tan\theta-\mu)}, \tag{10.69}$$

where $x_s$ denotes the runout distance (the point of farthest reach).

In Figures 10.21 and 10.22, we have reported the curves corresponding to the inertial and runout phases given by equations (10.68) and (10.69). For the Aulta avalanche, equation (10.68) provides a fairly good approximation of the inertial phase over the first 200 m of the avalanche course. For the Coulomb model, we adjusted the bulk friction coefficient $\mu$ for the computed runout distance to match the recorded value. As seen in Figure 10.21, there is relatively good agreement between computed and measured velocities for the runout phase and the agreement is still correct for the earlier flow phases. Note that the flow geometry has been significantly simplified in the computation: In particular, changes in the flow section between the starting area and the flow zone were not taken into account, which may explain why differences between the recorded and measured velocities for $x$ in the range 500–1500 m can be observed. Snow entrainment, which was likely to occur and affect the front velocity, was not considered here. Despite these substantial simplifications, a one-parameter model as simple as the Coulomb model is able to reproduce the front-velocity variations for the Aulta avalanche. The avalanche may have reached a nearly steady regime, but the general trend is an initial acceleration followed by a deceleration after reaching a maximum velocity.

For the Fogas avalanche, the transition from the inertial to nearly steady regimes is quite abrupt and the same holds for the transition between the steady and runout regimes. The inertial and Coulomb approximations hold for a very narrow range of $x$-values (i.e., in the starting and deposition areas). The important point is that between the inertial and runout phases, equilibrium seems to have been reached over a 400-m length (i.e., half the distance traveled by the avalanche) since velocity was nearly constant and path slope was quite regular.

The relatively good agreement achieved with the Coulomb model is not an isolated occurrence. In a recent survey in which 15 documented avalanches were analyzed (Ancey and Meunier 2004), it was found that for ten events, the Coulomb model was a suitable approximation of the bulk behavior, whereas for five events, there was clearly a velocity dependence on the bulk friction coefficient. For a few events, the bulk frictional force exhibits a dependence on the mean velocity, but no clear trend in the $\tau_b(u)$ dependence was found (Ancey and Meunier 2004). An interesting property of this simple Coulomb block model is that knowing the runout distance (point of furthest reach) of an avalanche makes it possible to infer the $\mu$ value. Because avalanche events have been recorded over a long time period at different sites in different alpine regions, we can deduce the statistical properties of the $f$ distribution at different places. If the bulk friction coefficient $\mu$ were a true physical parameter, its statistical properties should not vary with space. Ancey thus conducted a statistical analysis on $\mu$ values by selecting data from 173 avalanches collected from seven sites in France. These sites are known to produce large avalanches, and their activity has been followed since the beginning of the 20th century. Figure 10.23

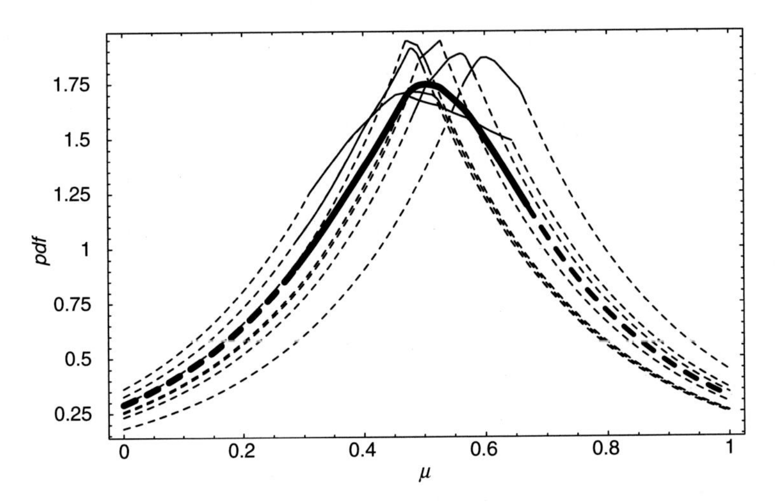


Figure 10.23. Empirical probability distribution functions (pdf) of the 173 $\mu$ values collected from seven paths. The thick line represents the distribution function of the total sample, whereas the thin lines are related to individual paths. Each curve has been split into three parts: the central part (solid line) corresponds to the range of computed $\mu$ values, and the end parts have been extrapolated. After Ancey (2005).

shows the probability distribution of $\mu$ for each site together with the entire sample. Although the curves are close and similar, they are not statistically identical. This means that the probability distribution function of $\mu$ is not uniquely determined and depends on other parameters such as snow properties and site configuration. Within this approach, the Coulomb model successfully captures the flow features, but its friction parameter is not a true physical parameter. This, however, should not negate interest of the Coulomb model because, given the number of approximations underpinning the sliding block model, the statistical deviance may originate from crude assumptions.

## 10.8 Concluding Remarks and Perspectives

Since the pioneering work of Coaz (1881), who initiated the first avalanche survey in the Swiss Alps, and Mougin (1922), who proposed the first avalanche-dynamics model, a huge amount of work has been done to collect field data, develop mathematical models to predict flow behavior, and conduct experiments on various scales (from laboratory to field scales). The subject has become sufficiently mature for conclusions and prospects to be drawn.

The phenomenological knowledge and the modeling of gravity flows such as snow avalanches and debris flows have essentially been motivated by land management issues and engineering applications. Indeed, predicting the runout distance, the impact forces, and the occurrence frequency of rare events are of paramount importance to risk mapping and protection against natural hazards. Even though other tools such as statistical techniques (data correlation, extreme value theory, Bayesian simulation) also have been used to predict extreme events (McClung and Lied 1987; Rickenmann 1999; Meunier and Ancey 2004; Keylock 2005; Eckert et al. 2008), the fluid-mechanics approach has emerged as the most fruitful way of computing the salient characteristics of gravity flows. Since the early 1920s, different generations of models, with increasing levels of sophistication, have been developed. For instance, the earliest dynamic models of avalanche considered snow avalanches as a rigid block (10.23) that experiences a frictional force that accounts for the resistance from the ambient air, the energy dissipation, and possibly momentum transfers (Mougin, 1922; Voellmy, 1955; Bozhinskiy and Losev, 1998). Although the first developments dated back to the 1920s, these very simple models were in use until the 1990s. The second generation of models used the analogy with water floods, which led to governing equations in the form of the flow-depth averaged equations referred to as the Saint-Venant equations (10.28)–(10.29). Although the earliest developments dated back to the 1960s with the work of Salm (1966) and Soviet researchers Grigorian and Eglit (see Bozhinskiy and Losev 1998 for a historical account), it was not until the 1980s that computers and numerical techniques were sufficiently powerful to solve hyperbolic differential equations such as the Saint-Venant equations (Brugnot and Pochat 1981; Vila 1986).

In the early 2000s, the first commercial products based on these equations were made available to engineers (Christen et al., 2002). Models for debris flows followed an evolution similar to that of avalanche-dynamics models, but with a lag time of 10–20 years.

Strikingly, while substantial progress has been achieved over the last 30 years in terms of physical modeling, the gain in accuracy for land management and engineering applications appears much more limited (Salm, 2004). Indeed, a number of problems (e.g., model calibration and values of input parameters) that already existed in the first generation of models have not been fixed and persist, often hidden by the level of complexity of current models, but sometimes exacerbated by the growing differences between variants of the same original model. In the sliding-block models, the frictional parameters could not be measured and were thus fit on field data (Buser and Frutiger 1980; Salm et al. 1990). There is clear evidence that these parameters are more conceptual than physical in that they do not represent a physical process but combine many different physical processes into a single, simple mathematical expression (Meunier et al. 2004). The comparison with field data in Section 10.7.2 provides an example. There is still an avid debate about the rheological law to be used in the depth-averaged equations. A number of models used a Coulomb or a Voellmy empirical law to model bed resistance and internal energy dissipation (Savage and Hutter 1989; Pudasaini and Hutter 2006), which amounts to positing that the rheological behavior can be described using a simple, single-valued expression of the bottom shear stress as a function of the depth-averaged velocity and flow depth. Iverson (2003a) provided evidence that for debris mixtures, the rheological properties cannot be captured using simple constitutive equations (e.g., Newtonian or Herschel-Bulkley laws) since they depend on additional parameters such as the pore pressure or the particle concentration, which may vary significantly within the bulk. The direct consequence is that the depth-averaged equations (10.28)–(10.29) must be supplemented by additional equations that describe the evolution of inner variables such as pore pressure or solids concentration (Iverson and Denlinger 2001; Iverson et al. 2010). Outdoor experiments and field surveys confirmed the substantial time variations in the basal pore pressure in debris flows– variations that reflect changes in the bulk dynamics (McArdell et al. 2006; Iverson et al. 2010). This search of more versatile and robust constitutive equations of natural materials has also entailed the development of models that introduced a large number of parameters. For coarsely parameterized models, some parameters have no physical meaning, and since they cannot be measured independently, their value must be adjusted for the model to match the observations. If this exercise is feasible when the number of parameters is low, it becomes increasingly difficult when this number exceeds three or four parameters. Parsimony and physical meaning of the model parameters are thus two stringent constraints that have made the development of reliable models difficult (Iverson 2003b).

In short, the first generation of models (e.g., the Voellmy model for flowing avalanches, the KS model for powder-snow avalanches) forms a simple but consistent framework that has been extensively used by scientists and engineers. The second generation of models (based on depth-averaged equations) opens many new directions for more realistic predictions, but there are still many points that deserve clarification and further work.

- The constitutive equation of saturated and unsaturated granular mixtures remains a difficult topic. Although there are empirical relations that provide fairly good descriptions of laboratory experiments (Delannay et al. 2007; Forterre and Pouliquen 2008), their generalization and applicability to natural flows are still open questions.
- In most cases, there is mass exchange between the flow and sloping bed. The processes involved in entrainment and deposition have been investigated, in particular through laboratory experiments and numerical simulations (Issler 1998; Princevac et al. 2005; Sovilla and Bartelt 2006; Mangeney et al. 2007, 2010), but many points (e.g., the role of pore pressure fluctuations) are still unclear.
- As highlighted earlier, additional equations that describe the evolution of the material composition are required. Making allowance for particle segregation, variation in the solids concentration, and evolution of the pore pressure are some of the challenges to be addressed.
- There is clear evidence that natural flows can develop complex internal structures (e.g., levees that channelize the flow, bouldery front that may retain the flow behind it, digitate lobate terminations, fingering instabilities, density stratification). (Iverson 1997; Félix and Thomas 2004; Deboeuf et al. 2006; Gray 2010). In the framework of depth-averaged equations, is it possible to account for these inner structures despite the averaging process, or does this mean that a next generation of models is in order?
- Combining stochastic tools and deterministic flow-dynamics models has been attempted to answer a number of questions such as those related to model calibration and the influence of the uncertainties on the input variables and model parameters of stochastic modeling. (Harbitz et al. 2001; Barbolini and Keylock 2002; Meunier and Ancey 2004; Ancey 2005; Dalbey et al. 2008; Gauer et al. 2009). There is crucial need for integrated models, in particular for land use planning and risk mapping, that are able to provide not only the flow features but also the uncertainties on these predictions.

## References

Ancey, C., 2004: Powder-snow avalanches: Approximation as non-Boussinesq clouds with a Richardson-number-dependent entrainment function. *J. Geophys. Res.* **109**, F01005.

Ancey, C., 2005: Monte Carlo calibration of avalanches described as Coulomb fluid flows. *Phil. Trans. Roy. Soc. London A* **363**, 1529–1550.

Ancey, C., 2007: Plasticity and geophysical flows: A review. *J. Non-Newtonian Fluid Mech.* **142**, 4–35.

Ancey, C., and S. Cochard, 2009: The dam-break problem for Herschel-Bulkley fluids down steep flumes. *J. Non-Newtonian Fluid Mech.* **158**, 18–35.

Ancey, C., S. Cochard, and N. Andreini, 2009: The dam-break problem for viscous fluids in the high-capillary-number limit. *J. Fluid Mech.* **624**, 1–22.

Ancey, C., S. Cochard, S. Wiederseiner, and M. Rentschler, 2006: Front dynamics of supercritical non-Boussinesq gravity currents. *Water Resour. Res.* **42**, W08424.

Ancey, C., S. Cochard, S. Wiederseiner, and M. Rentschler, 2007: Existence and features of similarity solutions for supercritical non-Boussinesq gravity currents. *Physica D* **226**, 32–54.

Ancey, C., and P. Evesque, 2000: Frictional-collisional regime for granular suspension flows down an inclined channel. *Phys. Rev. E* **62**, 8349–8360.

Ancey, C., C. Gervasoni, and M. Meunier, 2004: Computing extreme avalanches. *Cold Reg. Sci. Technol.* **39**, 161–180.

Ancey, C., R. M. Iverson, M. Rentschler, and R. P. Denlinger, 2008: An exact solution for ideal dam-break floods on steep slopes. *Water Resour. Res.* **44**, W01430.

Ancey, C., and M. Meunier, 2004: Estimating bulk rheological properties of flowing snow avalanches from field data. *J. Geophys. Res.* **109**, F01004.

Balmforth, N. J., and R. V. Craster, 1999: A consistent thin-layer theory for Bingham plastics. *J. Non-Newtonian Fluid Mech.* **84**, 65–81.

Balmforth, N. J., R. V. Craster, and R. Sassi, 2002: Shallow viscoplastic flow on an inclined plane. *J. Fluid Mech.* **470**, 1–29.

Balmforth, N. J., and R. R. Kerswell, 2005: Granular collapse in two dimensions. *J. Fluid Mech.* **538**, 399–428.

Barbolini, M., and C. J. Keylock, 2002: A new method for avalanche hazard mapping using a combination of statistical and deterministic models. *Nat. Hazard Earth. Sys. Sci.* **2**, 239–245.

Bartelt, P., and B. W. McArdell, 2009: Granulometric investigations of snow avalanches. *J. Glac.* **55**, 829–833.

Beghin, P., and G. Brugnot, 1983: Contribution of theoretical and experimental results to powder-snow avalanche dynamics. *Cold Regions Sci. Technol.* **8**, 63–73.

Beghin, P., E. J. Hopfinger, and R. E. Britter, 1981: Gravitational convection from instantaneous sources on inclined boundaries. *J. Fluid Mech.* **107**, 407–422.

Beghin, P., and X. Olagne, 1991: Experimental and theoretical study of the dynamics of powder snow avalanches. *Cold Reg. Sci. Technol.* **19**, 317–326.

Benjamin, T. B., 1968: Gravity currents and related phenomena. *J. Fluid Mech.* **31**, 209–248.

Bouchut, F., A. Mangeney-Castelnau, B., Perthame, and J.-P. Vilotte, 2003: A new model of Saint Venant and Savage-Hutter type for gravity driven shallow flows. *C. R. Acad. Sci. Paris Sér. I* **336**, 531–536.

Bozhinskiy, N., and K. S. Losev, 1998: The fundamentals of avalanche science. *Tech. Rep.* 55. EISFL.

Brugnot, G., and R. Pochat, 1981: Numerical simulation study of avalanches. *J. Glaciol.* **27**, 77–88.

Bursik, M., A. Patra, E. B. Pitman, C. C. Nichita, J. L. Macias, R. Saucedo, and O. Girina, 2005: Advances in studies of dense volcanic granular flows. *Rep. Prog. Phy.* **68**, 271–301.

Buser, O., and H. Frutiger, 1980: Observed maximum runout distance of snow avalanches and determination of the friction coefficients $\mu$ and $\xi$. *J. Glaciol.* **26**, 121–130.

Chambon, G., A. Ghemmour, and D. Laigle, 2009: Gravity-driven surges of a viscoplastic fluid: an experimental study. *J. Non-Newtonian Fluid Mech.* **158**, 54–62.

Chang, C., and R. L. Powell, 1994: Effect of particle size distributions on the rheology of concentrated bimodal suspensions. *J. Rheol.* **38**, 85–98.

Chow, V. T. (ed.), 1959: *Open-Channel Hydraulics*. Mc Graw Hill, New York.

Christen, M., P. Bartelt, U. and Gruber, 2002: *AVAL1D: Numerische Berechnung von Fliess- und Staublawinen*. Eidgenössisches Institut für Schnee- und Lawinenforschung, Davos.

Chugunov, V., J. M. N. T. Gray, and K. Hutter, 2003: Group theoretic methods and similarity solutions of the Savage-Hutter equations. In: K. Hutter and N. Kirchener (eds.), *Dynamics Response of Granular and Porous Materials under Large and Catastrophic Deformations*, pp. 251–261. Springer, Berlin.
Clague, J. J., and S. G. Evans, 2000: A review of catastrophic drainage of moraine-dammed lakes in British Columbia. *Quaternary Sci. Rev.* **19**, 1763–1783.
Coaz, J. W., 1881: *Die Lawinen der Schweizer Alpen*. Schmid-Franke, Bern.
Cochard, S., and C. Ancey, 2008: Tracking the free surface of time-dependent flows: Image processing for the dam-break problem. *Exper. Fluids* **44**, 59–71.
Cochard, S., and C. Ancey, 2009: Experimental investigation into the spreading of viscoplastic fluids on inclined planes. *J. Non-Newtonian Fluid Mech.* **158**, 73–84.
Contreras, S. M., and T. R. H. Davies, 2000: Coarse-grained debris-flows: Hysteresis and time-dependent rheology. *J. Hydraul. Eng.* **126**, 938–941.
Coussot, P., 1997: *Mudflow Rheology and Dynamics*. Balkema, Rotterdam.
Coussot, P., S. Proust, and C. Ancey, 1996: Rheological interpretation of deposits of yield stress fluids. *J. Non-Newtonian Fluid Mech.* **66**, 55–70.
Cui, X., J. M. N. T. Gray, and T. Jóhannesson, 2007: Deflecting dams and the formation of oblique shocks in snow avalanches at Flateyri, Iceland. *J. Geophys. Res.* **122**, F04012.
Dalbey, K., A. Patra, E. B. Pitman, M. I. Bursik, and M. Sheridan, 2008: Input uncertainty propagation methods and hazard mapping of geophysical mass flows. *J. Geophys. Res.* **113**, B05203.
Davies, T. R. H., 1986: Large debris flow: a macro-viscous phenomenon. *Acta Mech.* **63**, 161–178.
Deboeuf, S., E. Lajeunesse, O. Dauchot, and B. Andreotti, 2006: Flow rule, self-channelization, and levees in unconfined granular flows. *Phys. Rev. Lett.* **97**, 158303.
Delannay, R., M. Louge, P. Richard, N. Taberlet, and A. Valance, 2007: Towards a theoretical picture of dense granular flows down inclines. *Nature Mater.* **6**, 99–108.
Dent, J. D., and T. E. Lang, 1980: Modelling of snow flow. *J. Glaciol.* **26**, 131–140.
Dent, J. D., and T. E. Lang, 1982: Experiments on the mechanics of flowing snow. *Cold Reg. Sci. Technol.* **5**, 243–248.
Eckert, N., E. Parent, M. Naaim, and D. Richard, 2008: Bayesian stochastic modelling for avalanche predetermination: From a general system framework to return period computations. *Stoch. Env. Res. Risk Ass.* **22**, 185–206.
Eglit, E. M., 1983: Some mathematical models of snow avalanches. In: M. Shahinpoor (ed.) *Advances in the Mechanics and the Flow of Granular Materials*, pp. 577–588. Trans Tech Publications, Clausthal-Zellerfeld.
Ellison, T. H., and J. S. Turner, 1959: Turbulent entrainment in stratified flows. *J. Fluid Mech.* **6**, 423–448.
Elverhøi, A., D. Issler, F. V. De Blasio, T. Ilstad, C. B. Harbitz, and P. Gauer, 2005: Emerging insights into the dynamics of submarine debris flows. *Nat. Hazard Earth. Sys. Sci.* **5**, 633–648.
Fannin, R. J., and T. P. Rollerson, 1993: Debris flows: Some physical characteristics and behaviour. *Can. Geotech. J.* **30**, 71–81.
Félix, G., and N. Thomas, 2004: Relation between dry granular flow regimes and morphology of deposits: Formation of levées in pyroclastic deposits. *Earth Planet. Sci. Lett.* **221**, 197–213.
Fernández-Nieto, E. D., P. Noble, and J.-P. Vila, 2010: Shallow water equation for non-Newtonian fluids. *J. Non-Newtonian Fluid Mech.* **165**, 712–732.
Fernando, H. J. S., 1991: Turbulent mixing in stratified fluids. *Annu. Rev. Fluid Mech.* **23**, 455–493.

Forterre, Y., and O. Pouliquen, 2008: Flows of dense granular media. *Annu. Rev. Fluid Mech.* **40**, 1–24.
Fukushima, Y., T. Hagihara, and M. Sakamoto, 2000: Dynamics of inclined suspension thermals. *Fluid Dyn. Res.* **26**, 337–354.
Fukushima, Y., and G. Parker, 1990: Numerical simulation of powder-snow avalanches. *J. Glaciol.* **36**, 229–237.
Gauer, P., and D. Issler, 2003: Possible erosion mechanisms in snow avalanches. *Ann. Glaciol.* **38**, 384–392.
Gauer, P., Z. Medina-Cetina, K. Lied, and K. Kristensen, 2009: Optimization and probabilistic calibration of avalanche-block models. *Cold Reg. Sci. Technol.* **59**, 251–258.
Gratton, J., and C. Vigo, 1994: Self-similar gravity currents with variable inflow revisited: plane currents. *J. Fluid Mech.* **258**, 77–104.
Gray, J. M. N. T., 2010: Particle size segregation in granular avalanches: A brief review of recent progress. In: J. D. Goddard, J. T. Jenkins and P. Giovine (eds.), *IUTAM-ISIMM Symposium on Mathematical Modeling and Physical Instances of Granular Flows*, vol. 1227, pp. 343–362. AIP, Melville, NY.
Gray, J. M. N. T., and C. Ancey, 2009: Particle size-segregation, recirculation, and deposition at coarse particle rich flow fronts. *J. Fluid Mech.* **629**, 387–423.
Gray, J. M. N. T., and B. P. Kokelaar, 2010: Large particle segregation, transport and accumulation in granular free-surface flows. *J. Fluid Mech.* **652**, 105–137.
Gray, J. M. N. T., M. Wieland, and K. Hutter, 1998: Gravity-driven free surface flow of granular avalanches over complex basal topography. *Proc. R. Soc. London Ser. A* **455**, 1841–1874.
Gubler, H., M. Hiller, G., Klausegger, and U. Suter, 1986: Messungen an Fliesslawinen. *Tech. Rep.* 41. SLF.
Hallworth, M. A., A. Hogg, and H. E. Huppert, 1998b: Effects of external flow on compositional and particle gravity currents. *J. Fluid Mech.* **359**, 109–142.
Hampton, M. A., H. J. Lee, and J. Locat, 1996: Submarine landslides. *Rev. Geophys.* **34**, 33–59.
Harbitz, C., A. Harbitz, and F. Nadim, 2001: On probability analysis in snow avalanche hazard zoning. *Ann. Glaciol.* **32**, 290–298.
Hogg, A. J., 2006: Lock-release gravity currents and dam-break flows. *J. Fluid Mech.* **569**, 61–87.
Hogg, A. J., and G. P. Matson, 2009: Slumps of viscoplastic fluids on slopes. *J. Non-Newtonian Fluid Mech.* **158**, 101–112.
Hogg, A. J., and D. Pritchard, 2004: The effects of hydraulic resistance on dam-break and other shallow inertial flows. *J. Fluid Mech.* **501**, 179–212.
Hogg, A. J., M. Ungarish, and H. E. Huppert, 2000: Particle-driven gravity currents: asymptotic and box model solutions. *Eur. J. Mech. B. Fluids* **19**, 139–165.
Hogg, A. H., and A. W. Woods, 2001: The transition from inertia- to bottom-drag-dominated motion of turbulent gravity current. *J. Fluid Mech.* **449**, 201–224.
Holmes, M. H., 1995: *Introduction to Perturbation Methods*. Springer Verlag, New York.
Hopfinger, E. J., and J.-C. Tochon-Danguy, 1977: A model study of powder-snow avalanches. *J. Glaciol.* **81**, 343–356.
Huang, X., and M. H. Garcìa, 1997: A perturbation solution for Bingham-plastic mudflows. *J. Hydraul. Eng.* **123**, 986–994.
Huang, X., and M. H. Garcìa, 1998: A Herschel-Bulkley model for mud flow down a slope. *J. Fluid Mech.* **374**, 305–333.
Hunt, B., 1994: Newtonian fluid mechanics treatment of debris flows and avalanches. *J. Hydraul. Eng.* **120**, 1350–1363.

Huppert, H. E., W. B., and Dade, 1998: Natural disasters: explosive volcanic eruptions and gigantic landslides. *Theor. Comput. Fluid Dyn.* **10**, 202–212.

Hürlimann, M., D. Rickenmann, and C. Graf, 2003: Field and monitoring data of debris-flow events in the Swiss Alps. *Can. Geotech. J.* **40**, 161–175.

Hutter, K., 1996: Avalanche dynamics. In: V. P. Singh (ed.), *Hydrology of Disasters*, pp. 317–392. Kluwer Academic Publications, Dordrecht.

Issler, D., 1998: Modelling of snow entrainment and deposition in powder-snow avalanches. *Ann. Glaciol.* **26**, 253–258.

Issler, D., 2003: Experimental information on the dynamics of dry-snow avalanches. In: K. Hutter and N. Kirchner (eds.), *Dynamic Response of Granular and Porous Materials Under Large and Catastrophic Deformation* pp. 109–160. Springer, Berlin.

Iverson, R. M., 1997: The physics of debris flows. *Rev. Geophys.* **35**, 245–296.

Iverson, R. M., 2003a: The debris-flow rheology myth. In: C. L. Chen and D. Rickenmann (eds.), *Debris Flow Mechanics and Mitigation Conference*, pp. 303–314. Davos: Mills Press.

Iverson, R. M., 2003b: How should mathematical models of geomorphic processus be judged? In: P. R. Wilcock and R. M. Iverson (ed.), *Prediction in Geomorphology*, pp. 83–94. American Geophysical Union, Washington, DC.

Iverson, R. M., 2005: Debris-flow mechanics. In: M. Jakob and O. Hungr (eds.), *Debris-Flow Hazards and Related Phenomena*, pp. 105–134. Springer, Berlin.

Iverson, R. M., and R. P. Denlinger, 2001: Flow of variably fluidized granular masses across three-dimensional terrain. 1. Coulomb mixture theory. *J. Geophys. Res.* **106**, 537–552.

Iverson, R. M., M. Logan, and R. G. LaHusen, 2010: The perfect debris flow? Aggregated results from 28 large-scale experiments. *J. Geophys. Res.* **115**, F03005.

Iverson, R. M., M. E. Reid, and R. G. LaHusen, 1997: Debris-flow mobilization from landslides. *Annu. Rev. Earth. Planet. Sci.* **25**, 85–138.

Iverson, R. M., and J. Vallance, 2001: New views of granular mass flows. *Geology* **29**, 115–118.

Johnson, A. M., and J. R. Rodine, 1984: Debris flow. In: D. Brunsden and D. B. Prior (eds.), *Slope Instability*, pp. 257–362. John Wiley and Sons, Chichester.

Jomelli, V., and P. Bertran, 2001: Wet snow avalanche deposits in the French Alps: Structure and sedimentology. *Geografiska Annaler Series A—Physical Geography* **83**, 15–28.

Jop, P., Y. Forterre, and O. Pouliquen, 2005: Crucial role of side walls for granular surface flows: Consequences for the rheology. *J. Fluid Mech.* **541**, 167–192.

Josserand, C., P.-Y. Lagrée, and D. Lhuillier, 2004: Stationary shear flows of dense granular materials: A tentative continuum modelling. *Eur. Phys. J. E* **14**, 127–135.

Karelsky, K. V., V. V. Papkov, A. S. Petrosyan, and D. V. Tsygankov, 2000: Particular solutions of shallow-water equations over a non-flat surface. *Phys. Lett. A* **271**, 341–348.

Keller, J. B., 2003: Shallow-water theory for arbitrary slopes of the bottom. *J. Fluid Mech.* **489**, 345–348.

Kern, M. A., P. Bartelt, B. Sovilla, and O. Buser, 2009: Measured shear rates in large dry and wet snow avalanches. *J. Glaciol.* **55**, 327–338.

Kern, M. A., F. Tiefenbacher, and J. N. McElwaine, 2004: The rheology of snow in large chute flows. *Cold Reg. Sci. Technol.* **39**, 181–192.

Kerswell, R. R., 2005: Dam break with Coulomb friction: a model for granular slumping? *Phys. Fluids* **17**, 057101.

Keylock, C. J., 2005: An alternative form for the statistical distribution of extreme avalanche runout distances. *Cold Reg. Sci. Technol.* **42**, 185–193.

Kulikovskiy, A. G., and E. I. Svehnikova, 1977: Model dlja rascheta dvizhija pilevoi snezhnoi lavini (a model for computing powdered snow avalanche motion) [in Russian]. *Materiali Glatsiologicheskih Issledovanii [Data of Glaciological Studies]* **31**, 74–80.

Legros, F., 2002: The mobility of long-runout landslides. *Eng. Geol.* **63**, 301–331.
Leighton, D., and A. Acrivos, 1987: The shear-induced migration of particles in a concentrated suspensions. *J. Fluid Mech.* **181**, 415–439.
LeVeque, R. J., 2002: *Finite Volume Methods for Hyperbolic Problems*. Cambridge University Press, Cambridge.
Lighthill, M. J., and G. B. Whitham, 1955: On kinematic waves. I. Flood movement in long rivers. *Proc. R. Soc. London Ser. A* **229**, 281–316.
Linares-Guerrero, E., C. Goujon, and R. Zenit, 2007: Increased mobility of bidisperse granular avalanches. *J. Fluid Mech.* **593**, 475–504.
Liu, K. F. and C. C. Mei, 1990a: Approximate equations for the slow spreading of a thin sheet of Bingham plastic fluid. *Phys. Fluids* **A 2**, 30–36.
Liu, K. F., and C. C. Mei, 1990b: Slow spreading of a sheet of Bingham fluid on an inclined plane. *J. Fluid Mech.* **207**, 505–529.
Lootens, D., H. Van Damme, and P. Hébraud, 2003: Giant stress fluctuations at the jamming transition. *Phys. Rev. Lett.* **90**, 178301.
Maeno, O., 1993: Rheological characteristics of snow flows. In: *International Workshop on Gravitational Mass Movements* (ed. L. Buisson and G. Brugnot), pp. 209–220. Cemagref, Grenoble.
Major, J. J., and T. C. Pierson, 1992: Debris flow rheology: Experimental analysis of fine-grained slurries. *Water Resour. Res.* **28**, 841–857.
Mangeney, A., P. Heinrich, and R. Roche, 2000: Analytical solution for testing debris avalanche numerical models. *Pure Appl. Geophys.* **157**, 1081–1096.
Mangeney, A., O. Roche, O. Hungr, N. Mangold, G. Faccanoni, and A. Lucas, 2010: Erosion and mobility in granular collapse over sloping beds. *J. Geophys. Res.* **115**, F03040.
Mangeney, A., L. S., Tsimring, I. S., Volfson, and F. Bouchut, 2007: Avalanche mobility induced by the presence of an erodible bed and associated entrainment. *Geophys. Res. Lett.* **34**, L22401.
Marquez, M., A. Robben, B. P. Grady, and I. Robb, 2006: Viscosity and yield stress reduction in non-colloidal concentrated suspensions by surface modification with polymers and surfactants and/or nanoparticle addition. *J. Colloid. Interface Sci.* **295**, 374–387.
Matson, G. P., and A. J. Hogg, 2007: Two-dimensional dam break flows of Herschel-Bulkley fluids: The approach to the arrested state. *J. Non-Newtonian Fluid Mech.* **142**, 79–94.
McArdell, B. W., P. Bartelt, and J. Kowalski, 2006: Field observations of basal forces and fluid pore pressure in a debris flow. *Geophys. Res. Lett.* **34**, L07406.
McClung, D. M., and K. Lied, 1987: Statistical and geometrical definition of snow avalanche runout. *Cold Reg. Sci. Technol.* **13**, 107–119.
McElwaine, J. N., 2005: Rotational flow in gravity current heads. *Phil. Trans. Roy. Soc. London A* **363**, 1603–1623.
Mei, C. C., K. F. Liu, and M. Yuhi, 2001: Mud flows—Slow and fast. In: N. J. Balmforth and A. Provenzale (eds.), *Geomorphological Fluid Mechanics: Selected topics in geological and geomorphological fluid mechanics*, pp. 548–577. Springer, Berlin.
Mei, C. C., and M. Yuhi, 2001: Slow flow of a Bingham fluid in a shallow channel of finite width. *J. Fluid Mech.* **431**, 135–159.
Meunier, M., and C. Ancey, 2004: Towards a conceptual approach to predetermining long-return-period avalanche run-out distances. *J. Glaciol.* **50**, 268–278.
Meunier, M., C. Ancey, and J.-M. Taillandier, 2004: Fitting avalanche-dynamics models with documented events from the Col du Lautaret site (France) using the conceptual approach. *Cold Regions Sci. Technol.* **39**, 55–66.
Morris, J. F., 2009: A review of microstructure in concentrated suspensions and its implications for rheology and bulk flow. *Rheol. Acta* **48**, 909–923.

Morris, J. F., and F. Boulay, 1999: Curvilinear flows of noncolloidal suspensions: The role of normal stresses. *J. Rheol.* **43**, 1213–1238.

Mougin, P., 1922: *Les avalanches en Savoie*, vol. IV. Ministère de l'Agriculture, Direction Générale des Eaux et Forêts, Service des Grandes Forces Hydrauliques, Paris.

Nazarov, A. N., 1991: Mathematical modelling of a snow-powder avalanche in the framework of the equations of two-layer shallow water. *Fluid Dyn.* **12**, 70–75.

Norem, H. F., F. Irgens, and B. Schieldrop, 1986: A continuum model for calculating snow avalanche velocities. In: H. Gubler and B. Salm (eds.), *Avalanche Formation, Movement and Effects*, pp. 363–379. IAHS, Wallingford, Oxfordshire, UK, Davos.

Osmond, D. I., and R. W. Griffiths, 2001: The static shape of yield strength fluids slowly emplaced on slopes. *J. Geophys. Res.* **B 106**, 16241–16250.

Ottino, J. M., and D. V. Khakhar, 2000: Mixing and segregation of granular materials. *Annu. Rev. Fluid Mech.* **32**, 55–91.

Ovarlez, G., F. Bertrand, and S. Rodts, 2006: Local determination of the constitutive law of a dense suspension of noncolloidal particles through magnetic resonance imaging. *J. Rheol.* **50**, 259–292.

Parker, G., 1982: Conditions for the ignition of catastrophically erosive turbidity currents. *Mar. Geol.* **46**, 307–327.

Parker, G., Y. Fukushima, and H. M. Pantin, 1986: Self-accelerating turbidity currents. *J. Fluid Mech.* **171**, 145–181.

Parsons, J. D., K. X. Whipple, and A. Simoni, 2001: Experimental study of the grain-flow, fluid-mud transition in debris flows. *J. Geol.* **109**, 427–447.

Phillips, J. C., A. J. Hogg, R. R. Kerswell, and N. Thomas, 2006: Enhanced mobility of granular mixtures of fine and coarse paricles. *Earth Planet. Sci. Lett.* **246**, 466–480.

Piau, J.-M., 1996: Flow of a yield stress fluid in a long domain. Application to flow on an inclined plane. *J. Rheol.* **40**, 711–723.

Platzer, K., P. Bartelt, and M. A. Kern, 2007: Measurements of dense snow avalanche basal shear to normal stress ratios (S/N). *Geophys. Res. Lett.* **34**, L07501.

Pouliquen, O., and Y. Forterre, 2002: Friction law for dense granular flow: Application to the motion of a mass down a rough inclined plane. *J. Fluid Mech.* **453**, 133–151.

Princevac, M., H. J. S. Fernando, and C. D. Whiteman, 2005: Turbulent entrainment into natural gravity-driven flows. *J. Fluid Mech.* **533**, 259–268.

Pudasaini, S. P., and K. Hutter, 2003: Rapid shear flows of dry granular masses down curved and twisted channels. *J. Fluid Mech.* **495**, 193–208.

Pudasaini, S. P., and K. Hutter, 2006: *Avalanche Dynamics*. Berlin, Springer.

de Quervain, R., 1981: *Avalanche Atlas*. Unesco, Paris.

Rammer, L., M. A. Kern, U. Gruber, and F. Tiefenbacher, 2007: Comparison of avalanche-velocity measurements by means of pulsed Doppler radar, continuous wave radar and optical methods. *Cold Reg. Sci. Technol.* **54**, 35–54.

Rentschler, M., 2010: Simulating viscoplastic avalanches. PhD thesis, École Polytechnique Fédérale de Lausanne.

Rickenmann, D., 1999: Empirical relationships for debris flows. *Nat. Hazard* **19**, 47–77.

Salm, B., 1966: Contribution to avalanche dynamics. In: *Scientific Aspects of Snow and Ice Avalanche*, pp. 199–214. IAHS Press, Wallingford, Oxfordshire, UK, Davos.

Salm, B., 2004: A short and personal history of snow avalanche dynamics. *Cold Reg. Sci. Technol.* **39**, 83–92.

Salm, B., A. Burkard, and H. Gubler, 1990: Berechnung von Fliesslawinen, eine Anleitung für Praktiker mit Beispielen. *Tech. Rep.* No 47. Eidgenössisches Institut für Schnee- und Lawinenforschung (Davos).

Savage, S. B., 1982: Granular flows down rough inclined—Review and extension. In: J. T. Jenkins and M. Satake (eds.), *U.S./ Japan Seminar on New Models and*

*Constitutive Relations in the Mechanics of Granular Materials*, pp. 261–282. Elseviers Science Publishers, Amsterdam, Ithaca.

Savage, S. B., 1989: Flow of granular materials. In: P. Germain, J.-M. Piau, and D. Caillerie (eds.), *Theoretical and Applied Mechanics*, pp. 241–266. Elsevier, Amsterdam.

Savage, S. B., and K. Hutter, 1989: The motion of a finite mass of granular material down a rough incline. *J. Fluid Mech.* **199**, 177–215.

Savage, S. B., and K. Hutter, 1991: The dynamics of avalanches of granular materials from initiation to runout. Part I: Analysis. *Acta Mech.* **86**, 201–223.

Savage, S. B., and Y. Nohguchi, 1988: Similarity solutions for avalanches of granular materials down curved bed. *Acta Mech.* **75**, 153–174.

Sengun, M. Z., and R. F. Probstein, 1989: Bimodal model of slurry viscosity with applications to coal slurries. Part 1. Theory and experiment. *Rheol. Acta* **28**, 382–393.

Siavoshi, S., and A. Kudrolli, 2005: Failure of a granular step. *Phys. Rev. E* **71**, 051302.

Simpson, J. E., 1972: Effects of a lower boundary on the head of a gravity current. *J. Fluid Mech.* **53**, 759–768.

Simpson, J. E., 1997: *Gravity Currents in the Environment and the Laboratory*. Cambridge University Press, Cambridge.

Singh, A., A. Nir, and R. Semiat, 2006: Free-surface flow of concentrated suspensions. *Int. J. Multiphase Flow* **32**, 775–790.

Siviglia, A., and A. Cantelli, 2005: Effect of bottom curvature on mudflow dynamics: Theory and experiments. *Water Resour. Res.* **41**, W11423.

Sovilla, B., and P. Bartelt, 2006: Field experiments and numerical modeling of mass entrainment in snow avalanches. *J. Geophys. Res.* **111**, F03007.

Stickel, J. J., and R. L. Powell, 2005: Fluid mechanics and rheology of dense suspensions. *Annu. Rev. Fluid Mech.* **37**, 129–149.

Tai, Y.-C., K. Hutter, and J. M. N. T. Gray, 2001: Dense granular avalanches: mathematical description and experimental validation. In: N. J. Balmforth and A. Provenzale (eds.), *Geomorphological Fluid Mechanics*, pp. 339–366. Springer, Berlin.

Timberlake, B. D., and J. F. Morrison, 2005: Particle migration and free-surface topography in inclined plane flow of a suspension. *J. Fluid Mech.* **538**, 309–341.

Tognacca, C., 1997: Debris-flow initiation by channel-bed failure. In: C. L. Chen (ed.), First International Conference on Debris Flow Hazards Mitigation, pp. 44–53. ASCE, New-York.

Tsai, J.-C., W. Losert, and G. A. Voth, 2001: Two-dimensional granular Poiseuille flow on an incline: Multiple dynamical regimes. *Phys. Rev. E* **65**, 011306.

Turnbull, B., J. McElwaine, and C. Ancey, 2007: The Kulikovskiy–Sveshnikova–Beghin model of powder snow avalanches: Development and application. *J. Geophys. Res.* **112**, F01004.

Turner, J. S., 1973: *Buoyancy Effects in Fluids*. Cambridge University Press, Cambridge.

Ungarisch, M., 2009: *An Introduction to Gravity Currents and Intrusions*. CRC Publishers, Boca Raton.

VanDine, D. F., 1985: Debris flows and debris torrents in the Southern Canadian Cordillera. *Can. Geotech. J.* **22**, 44–68.

Vila, J. P., 1986: Sur la théorie et l'approximation numérique des problèmes hyperboliques non-linéaires, application aux équations de Saint-Venant et à la modélisation des avalanches denses. Ph. D. thesis, Paris VI.

Voellmy, A., 1955: Über die Zerstörungskraft von Lawinen. *Schweizerische Bauzeitung* **73**, 159–162, 212–217, 246–249, 280–285.

Voight, B., 1990: The 1985 Nevado del Ruiz volcano catastrophe: Anatomy and restropection. *J. Volcan. Geoth. Res.* **44**, 349–386.

Wilcock, P. R., and R. M. Iverson, eds., 2003: *Prediction in Geomorphology*. American Geophysical Union, Washington, DC.

Zanuttigh, B., and A. Lamberti, 2007: Instability and surge development in debris flows. *Rev. Geophys.* **45**, RG3006.

Zarraga, I. E., D. A. Hill, and D. T. Leighton, 2000: The characterization of the total stress of concentrated suspensions of noncolloidal spheres in Newtonian fluids. *J. Rheol.* **44**, 185–221.

Zimmermann, M., P. Mani, P. Gamma, P. Gsteiger, O. Heiniger, and G. Hunizker, 1997: *Murganggefahr und Klimaänderung – ein GIS-basierter Ansatz*. VDF, Zürich.

# Index

Printed in the United States
By Bookmasters